Numerical Flow Simulation I

Edited by
Ernst Heinrich Hirschel

Notes on Numerical Fluid Mechanics (NNFM) Volume 66

Volume 66: Numerical Flow Simulation I (E. H. Hirschel, Ed.)
Volume 65: Computation and Comparison of Efficient Turbulence Models for Aeronautics – ETMA project (A. Dervieux / M. Braza / J.-O. Dussauge, Eds.)
Volume 64: Computation and Visualization of Three-Dimensional Vortical and Turbulent Flows (R. Friedrich / P. Bontoux, Eds.)
Volume 63: Vortices and Heat Transfer (M. Fiebig / Nimai K. Mitra, Eds.)
Volume 62: Large-Scale Scientific Computations of Engineering and Environmental Problems (M. Griebel / O. P. Iliev / S. D. Margenov / P. S. Vassilevski, Eds.)
Volume 61 Optimum Aerodynamic Design & Parallel Navier-Stokes Computations, ECARP-Europear Computational Aerodynamics Research Project (J. Periaux / G. Bugeda / P. Chaviaropoulos K. Giannokoglou / S. Lanteri / B. Mantel, Eds.)
Volume 60 New Results in Numerical and Experimental Fluid Mechanics. Contributions to the 10th A(STAB/DGLR Symposium Braunschweig, Germany 1996 (H. Körner / R. Hilbig, Eds.)
Volume 59 Modeling and Computations in Environmental Sciences. Proceedings of the First GAMM-Seminar at ICA Stuttgart, October 12–13, 1995 (R. Helmig / W. Jäger / W. Kinzelbach / P. Knabner / G. Wittum, Eds.)
Volume 58 ECARP – European Computational Aerodynamics Research Project: Validation of CFD Codes and Assessment of Turbulence Models (W. Haase / E. Chaput / E. Elsholz / M. A. Leschziner / U. R. Müller, Eds.)
Volume 57 Euler and Navier-Stokes Solvers Using Multi-Dimensional Upwind Schemes and Multigrid Acceleration. Resu..s of the BRITE/EURAM Projects AERO-CT89-0003 and AER2-CT92-00040, 1989–1995 (H. Deconinck / B. Koren, Eds.)
Volume 56 EUROSHOCK-Drag Reduktion by Passive Shock Control. Results of the Project EUROSHOCK, AER2-CT92-0049. Supported by the European Union, 1993–1995 (E. Stanewsky / J. Délery / J. Fulker / W. Geißler, Eds.)
Volume 55 EUROPT – A European Initiative on Optimum Design Methods in Aerodynamics. Proceedings of the Brite/Euram Project Workshop „Optimum Design in Aerodynamics", Barcelona, 1992 (J. Periaux / G. Bugeda / P. K. Chaviaropoulos / T. Labrujere / B. Stoufflet, Eds.)
Volume 54 Boundary Elements: Implementation and Analysis of Advanced Algorithms. Proceedings o the Twelfth GAMM-Seminar, Kiel, January 19–21, 1996 (W. Hackbusch / G. Wittum, Eds.)
Volume 53 Computation of Three-Dimensional Complex Flows. Proceedings of the IMACS-COST Conference on Computational Fluid Dynamics, Lausanne, September 13–15, 1995 (M. Deville / S. Gavrilakis / I. L. Ryhming, Eds.)
Volume 52 Flow Simulation with High-Performance Computers II. DFG Priority Research Programm(Results 1993–1995 (E. H. Hirschel, Ed.)

Volumes 1 to 51 are out of print.
The addresses of the Editors are listed at the end of the book.

Numerical Flow Simulation I

CNRS-DFG Collaborative Research Programme,
Results 1996–1998

Edited by
Ernst Heinrich Hirschel

vieweg

Die Deutsche Bibliothek – CIP-Einheitsaufnahme

Numerical flow simulation:
CNRS DFG collaborative research programme;
results 1996–1998 / ed. by Ernst Heinrich Hirschel. –
Braunschweig; Wiesbaden: Vieweg

1 (1998)
(Notes on numerical fluid mechanics; Vol. 66)
ISBN 978-3-642-53590-1 ISBN 978-3-540-44437-4 (eBook)
DOI 10.1007/978-3-540-44437-4

Softcover reprint of the hardcover 1st edition 1998

Vieweg ist a subsidiary company of Bertelsmann Professional Information.

http://www.vieweg.de

Produced by Geronimo GmbH, Rosenheim
Printed on acid-free paper

ISSN 0179-9614
ISBN 978-3-642-53590-1

Foreword

This volume contains twenty contributions of work, conducted since 1996 in the French-German Research Programme "Numerical Flow Simulation" of the Centre National de la Recherche Scientifiqué (CNRS) and the Deutsche Forschungsgemeinschaft (DFG).

The main purpose of this publication is to give an overview over the work conducted in this programme, and to make the results obtained available to the public. The reports are grouped under the four headings "Development of Solution Techniques", "Crystal Growth and Melts", "Flows of Reacting Gases" and "Turbulent Flows". All contributions to this publication were reviewed by a board consisting of T. Alziary de Roquefort (Poitiers, France), P. Bontoux (Marseille, France), J.A. Desideri (Sophia-Antipolis, France), W. Kordulla (Göttingen, Germany), R. Peyret (Nice, France), R. Rannacher (Heidelberg, Germany), G. Warnecke (Magdeburg, Germany), and the editor. The responsibility for the contents of the reports nevertheless lies with the authors.

E.H. Hirschel
Editor

Preface

The Colloquium on "Numerical Simulation of Flows", Marseille, November 21 and 22, 1997, was the 6th Joint CNRS-DFG Colloquium organized in the frame of the French-German Research Collaboration on Computational Fluid Dynamics. This Collaborative Program was elaborated progressively since 1991, when the two major research groups were brought together: the Priority Program "Flow Simulation with Super Computers" from the DFG in Germany and the Groupement de Recherche de "Mécanique des Fluides Numérique" (GDR MFN) from the CNRS in France. Both groups were already well identified and strongly structured, which considerably helped in the constitution of a Research Program "Numerical Simulation of Flows" that gathers now (since 1996) about twenty linked projects, strongly coupling French and German research teams.

Representatives of the Scientific Direction of the Department "Sciences de l'Ingenieur" of CNRS and of the Reviewing Board of the DFG Priority Research Program "Flow Simulation with High-Performance Computers" took the initiative during their first meeting in January 1991 in Paris to promote a co-operation between the two programs.

Three Joint Colloquia followed the next two years; - the first was held in Sophia-Antipolis in March 1992 on "Parallel Computing in Fluid Mechanics"; - the second in May 1993 in the Pfalzakademie in Lambrecht on "Three-Dimensional Flow – Alternative Formulations and Solutions of the Conservation Equations"; - the third in December 1993 in Stuttgart on "Computational Fluid Dynamics on Parallel Systems" with published proceedings (S. Wagner, (ed.): "Computational Fluid Dynamics on Parallel Systems", NNFM 50, Vieweg, Braunschweig/Wiesbaden, 1995).

The second official meeting between representatives of CNRS and DFG was held during the second colloquium in Lambrecht and was presided by Prof. S. Wittig, the Vice President of the DFG at that time. At this stage the decision was made to provide travel support for the already co-operating projects and also to involve a French representative in the German Reviewing Board of the DFG Program and vice versa.

The proposal for a French-German Program on Numerical Flow Simulation was introduced anew in June 1994 at the Annual Colloquium of the DFG Priority Program in Bonn. In November 1994, twenty research proposals were presented and discussed at the fourth CNRS-DFG Colloquium in Sophia-Antipolis. On this basis, a joint French-German Committee formulated a general proposal, which was submitted simultaneously to the CNRS and to the DFG in March 1995. Soon later in April, the Directeur Général of the CNRS, Prof. G. Aubert, and the President of the DFG, Prof. W. Frühwald, welcomed the proposed initiative and agreed on providing the financial means necessary for carrying out the research work in the various projects. After the Senat of the DFG recommended a frame for the support, both the CNRS and the DFG announced the program.

The program was limited to projects that rely on the co-operation of a French and a German partner. The strengthening of the co-operation between engineers and applied mathematicians was a second important goal of the program. The algorithms to be developed should be geared to the fast changes in the architecture of high-performance computers. Relevant topics of fluid dynamics referred to turbulence, combustion, convection and interface problems.

The joint French-German reviewing group, with Prof. Alziary de Roquefort in the chair, met in Aachen in November 1995 and recommended 20 of the 26 proposals submitted for a two year period beginning April 1996.

First results of the program were discussed at the fifth CNRS-DFG Colloquium on the "Simulation of Three-Dimensional Turbulent and Vortical Flows" that was held in December 1996 in München, also with published proceedings (R. Friedrich, P. Bontoux (eds.): "Computation and Visualization of Three-Dimensional Vortical and Turbulent Flows", NNFM 64, Vieweg, Braunschweig/Wiesbaden, 1998). One year later in November 1997 all the results were presented at the sixth CNRS-DFG Colloquium in Marseille. Most of the papers presented are contained in this volume.

The French-German reviewing board met again, at this colloquium under the chairmanship of Dr. W. Kordulla, in order to evaluate the work carried out until then and to examine the new proposals. It was agreed on to hold the next review meeting of the Joint French-German Research Program on Numerical Flow Simulation in November 1999 in Dresden. Two CNRS-DFG Colloquia will be held in the second half of 1998: The first is being organized by Prof. W. Rodi in Karlsruhe. Its main topic is again the simulation of turbulent flows, and the second is organized by Prof. G. Wittum in Stuttgart. This colloquium concentrates on the development of algorithms.

The undersigned take this opportunity to thank the DFG, the CNRS and the French Ministry for Education, Research and Technology for supporting the program over the past two years. Grateful acknowledgment is also due to the reviewers of the program, whose invaluable expertise helped to shape the program in many ways. It was with great enthusiasm that they stimulated interdisciplinary discussion between engineers and mathematicians participating in the program. It is hoped, that their efforts are reflected in the contributions published in this volume. We also wish to thank Dr. W. Lachenmeier (DFG) and Prof. D. Vandromme (Charge de mission at CNRS) for administering the program since its initiation. We also thank Dr. R. Peyret from CNRS for his continuous involvement in the development of the program.

Finally we thank the Vieweg Verlag for publishing the results obtained so far in the Notes on Numerical Fluid Mechanics, and Prof. E. H. Hirschel for editing this volume.

Aachen and Marseille, July 1998

E. Krause P. Bontoux

CONTENTS

CONTENTS (continued)

I. DEVELOPMENT OF SOLUTION TECHNIQUES

A Parallel Hybrid Highly Accurate Elliptic Solver for Viscous Flow Problems

W. Borchers*, M.Y. Forestier**, S. Kräutle*, R. Pasquetti**, R. Peyret**, R. Rautmann***, N. Roß***, C. Sabbah**

*) Institut für Angewandte Mathematik I, Universität Erlangen-Nürnberg, Martensstr. 3, 91058 Erlangen, Germany

**) Lab. J.A. Dieudonné, UMR CNRS 6621, Université de Nice-Sophia Antipolis, Parc Valrose, 06108 Nice cedex 2, France

***) Fachbereich Mathematik/Informatik, Universität-GH Paderborn, Warburgerstr. 100, 33098 Paderborn, Germany

Summary

We present a new parallel hybrid method to solve numerically elliptic equations on a channel-like domain. The method combines the highly accurate Chebyshev - spectral method with a standard finite difference one, via the CGBI - domain decomposition procedure. By this approach the solution of linear elliptic boundary value problems is reduced to a minimization principle for the unknown Neumann boundary data distributed on the subdomain interfaces. The subdomain solvers are based on Chebyshev spectral / finite difference methods, but finite elements, instead of finite differences, could be used to deal with more complicated geometries.

1 Introduction

The aim of our present researches is to combine the high accuracy of (Chebyshev) spectral solvers in simply shaped regions with the high flexibility of finite element and operator splitting methods in complex geometries, to develop a parallel hybrid solver of the full Navier-Stokes equations. In the spirit of projection and splitting methods (see [2], [3], [4], [8], [9], [21], [22], [23], [25], [26], [27]) each time-step will be decomposed into:

- a transport step running implicitly with the method of characteristics, i.e. starting at any grid-point we follow numerically the characteristic of the velocity field crossing this grid point ([2], [3], [10], [16], [20], [24]),

- a diffusion step, working implicitly with respect to the velocity, for which we have to solve a resolvent (Helmholtz) equation,
- a pressure correction step, which requires the solution of a Poisson equation. The divergence of the velocity, which appears as a source term, has just been calculated in the previous step.

In such a splitting method the full Navier– Stokes initial/boundary value problem is decomposed into simpler problems for which efficient parallelization methods are currently constructed. One of them concerns the multi-domain multi-method solution of elliptic equations. For such kind of elliptic boundary value problems we have introduced a new domain decomposition technique (the Conjugate Gradient Boundary Iteration method, CGBI– method) which was shown to be very efficient compared to other known schemes. The hybrid CGBI solver will provide two kinds of local solvers, a Chebyshev solver of spectral accuracy and a flexible FEM one. But presently, for the sake of simplicity, the method is tested by using standard second order Finite Differences instead of Finite Elements. In this paper we present results pointing out the efficiency of the multi-domain CGBI approach for the Helmholtz/Poisson equations.
In section 2 we describe the spectral Chebyshev Helmholtz/Poisson solver. Section 3 is devoted to the theoretical background and to the implementation of the CGBI method. Finally, numerical results on the efficiency of this multi-domain multi-method approach are given in section 4.

2 The Spectral Chebyshev Poisson/Helmholtz Solver

Here we present the spectral solver used in the multi-domain multi-method CGBI approach described in section 3. First we recall how the basic discrete equations are derived using the Chebyshev collocation method. Then we focus on the solution of these algebraic equations for which we are using a diagonalization technique (efficient when multiple resolutions are needed). The ill-posed Poisson problem which may occur when using a projection method is considered then. Finally, we present some tests on the accuracy of the solver.

2.1 The Basic Discrete Equations

In a rectangular domain Ω of boundary $\partial\Omega = \cup_{k=1}^{4} \partial\Omega_k$, one considers the elliptic Helmholtz or Poisson equation with linear boundary conditions:

$$\begin{aligned}(\Delta - \sigma)u + f &= 0, \qquad \text{in } \Omega \\ (\alpha_k + \beta_k \frac{\partial}{\partial \nu}) u_{|\partial\Omega_k} &= g_k, \qquad 1 \leq k \leq 4\end{aligned} \tag{2.1}$$

with a nonnegative constant $\sigma \geq 0$ and $\partial/\partial\nu$ the outward normal derivative. The constants α_k, β_k are chosen such that (2.1) is well posed. Our aim is now to solve (2.1) by using a collocation Chebyshev method. To this end, it is assumed that Ω is mapped one to one onto the unit square $]-1,1[^2$ and the Gauss-Lobatto mesh is defined by

$$\overline{\Omega}_{I,J} = \left\{X_i = \cos\left(\frac{\pi i}{I}\right), Y_j = \cos\left(\frac{\pi j}{J}\right) : \ 0 \leq i \leq I,\ 0 \leq j \leq J\right\}. \tag{2.2}$$

We also use

$$\Omega_{I,J} = \{X_i, Y_j : \ 0 < i < I,\ 0 < j < J\} \tag{2.3}$$

and $\partial\Omega_{I,J} = \overline{\Omega}_{I,J} \backslash \Omega_{I,J}$.
Then the Chebyshev collocation method for (2.1) is defined by the following discrete problem: With $P_{I,J}$ denoting the vector space of the polynomials of degree at most I in X and J in Y, we find $u \in P_{I,J}$ such as

$$\begin{aligned} (\Delta - \sigma)u + f &= 0, \qquad \text{in } \Omega_{I,J} \\ (\alpha_k + \beta_k \frac{\partial}{\partial\nu})u &= g_k, \qquad 1 \leq k \leq 4 \qquad \text{on } \partial\Omega_{I,J}. \end{aligned} \tag{2.4}$$

The Chebyshev collocation method requires the use of differentiation matrices in the physical space. Thus, the calculation of the derivative with respect to X of the polynomial $u(X,Y)$ is given by

$$\partial_X u_{ij} = \sum_{l=0}^{I} d_{i,l}^{I} u_{lj} \tag{2.5}$$

where $u_{ij} = u(X_i,Y_j)$ and $\partial_X u_{ij} = \partial_X u(X_i,Y_j)$.
Different ways are possible for the computation of the coefficients $d_{i,l}^{I}$ (trigonometric or algebraic formulas). Our numerical tests on the influence of round-off errors have shown that the following expressions, proposed in [1], were suitable:

$$d_{i,l}^{I} = \frac{\overline{c}_i}{\overline{c}_l} \frac{(-1)^{i+l}}{X_i - X_l} \qquad 0 \leq i,l \leq I,\ i \neq l \tag{2.6}$$

$$d_{i,i}^{I} = -\sum_{l \neq i} d_{i,l}^{I} \tag{2.7}$$

with $\overline{c}_0 = \overline{c}_I = 2$ and $\overline{c}_l = 1$ for $0 < l < I$. Especially, relation (2.7) ensures that the derivative of any constant is exactly zero.
The coefficients $d_{i,l}^{I}$ are the elements of the differentiation matrix $D_X = [d_{i,l}^{I}]$, $0 \leq i, l \leq I$. Obviously, for the calculation of $\partial_Y u(X_i,Y_j)$ one needs to introduce a similar matrix D_Y.

The Gauss-Lobatto mesh is often very suitable in fluid mechanics, since the collocation points are accumulating near the boundary where boundary layers generally occur. Nevertheless, in some specific physical situations it is necessary to use coordinate transforms in order to map a given Cartesian mesh $\{x_i, y_j : 0 \le i \le I,\ 0 \le j \le J\}$, to the domain Ω. Denote by $\mathcal{T}_X$ and $\mathcal{T}_Y$ the corresponding coordinate transforms

$$x_i = \mathcal{T}_X(X_i), \quad 0 \le i \le I, \quad \text{and} \quad y_j = \mathcal{T}_Y(Y_j), \quad 0 \le j \le J.$$

For the first order derivative with respect to x one gets

$$\partial_x u_{ij} = \frac{1}{\mathcal{T}'_X(X_i)} \partial_X u_{ij} \tag{2.8}$$

with $\mathcal{T}'_X(X)$ for the derivative of $\mathcal{T}_X(X)$. This leads to the definition of the diagonal matrix

$$T_X = diag\left\{\frac{1}{\mathcal{T}'_X(X_i)},\ 0 \le i \le I\right\} \tag{2.9}$$

and the differentiation matrix with respect to x

$$D_x = T_X D_X. \tag{2.10}$$

In the same way one introduces the differentiation matrix with respect to y, D_y. For the calculation of the second-order derivatives the use of non-linear coordinate transforms may cause difficulties. When discretising the expression

$$\frac{\partial^2 u}{\partial x^2} = \left(\frac{\partial X}{\partial x}\right)^2 \frac{\partial^2 u}{\partial X^2} + \left(\frac{\partial^2 X}{\partial x^2}\right) \frac{\partial u}{\partial X} \tag{2.11}$$

the basic property

$$\frac{\partial^2 u}{\partial x^2} = \frac{\partial}{\partial x}\left(\frac{\partial u}{\partial x}\right) \tag{2.12}$$

generally gets lost. For this reason, the second-order differentiation matrices must be defined as the square of the first-order ones:

$$D_{xx} = D_x^2 \quad \text{and} \quad D_{yy} = D_y^2 \tag{2.13}$$

and, more generally, differentiation matrices of order p are defined as the power p of differentiation matrices of order 1.
Then the algorithm is classical [11]:

- Write the system (2.4) using the elements d_{il}^{xx}, $0 \le i, l \le I$, and d_{jm}^{yy}, $0 \le j, m \le J$, of the differentiation matrices D_{xx} and D_{yy} :

$$\sum_{l=0}^{I} d_{il}^{xx} u_{lj} + \sum_{m=0}^{J} d_{jm}^{yy} u_{im} - \sigma u_{ij} + f_{ij} = 0, \quad 0 < i < I,\ 0 < j < J. \tag{2.14}$$

- Use the boundary conditions in order to express u at the collocation points of $\partial\Omega$, as linear combinations of the values of u at the inner collocation point. For example on $\partial\Omega_1$ $(i = 0)$ we have

$$u_{0j} = \sum_{l=1}^{I-1} \gamma_l^1 u_{lj} + \gamma_{0j}^1, \tag{2.15}$$

where the γ_l^1 only depend on the pairs (α_1, β_1) $(i = 0)$ and (α_3, β_3) $(i = I)$.

- Eliminate these boundary values from system (2.14) in order to get a matrix equation only involving the inner collocation points

$$AU + UB - \sigma U = S, \tag{2.16}$$

where the matrices U and S are associated with the inner grid points ($[U]_{ij} = u_{ij}$, $[S]_{ij} = s_{ij}$, $0 < i < I$, $0 < j < J$) and where the matrices A and B have the dimensions $(I-1) \times (I-1)$ and $(J-1) \times (J-1)$, respectively.

2.2 Solution Technique

For evolution problems, the matrix equation (2.16) has to be solved at each time step. Therefore it is interesting to use the diagonalization technique. If the eigenvalues of A and B are real, with Λ_A and Λ_B the diagonal matrices of the eigenvalues and M_A and M_B the matrices of the eigenvectors, from $\Lambda_A = M_A^{-1} A M_A$ and $\Lambda_B = M_B^{-1} B M_B$ one gets

$$\Lambda_A U' + U' \Lambda_B - \sigma U' = S' \tag{2.17}$$

with $U' = M_A^{-1} U M_B$ and $S' = M_A^{-1} S M_B$.
This yields immediately

$$u'_{ij} = \frac{s'_{ij}}{\lambda_i^A + \lambda_j^B - \sigma}, \qquad 0 < i < I,\ 0 < j < J, \tag{2.18}$$

where λ_i^A and λ_j^B are the eigenvalues of the matrices A and B. With U' known, one computes $U = M_A U' M_B^{-1}$.
When the coordinate transforms are linear, it is well known that the eigenvalues are real [14]. But when using non-linear coordinate transforms, complex conjugate pairs may occur. Then in the diagonalization technique, the matrix equation (2.17) should be treated as complex rather than real.This is a time and memory space consuming drawback. The 'quasi-diagonalization technique' [19], described briefly now, permits to overcome this difficulty. As an example let us assume that $\lambda_{i=1}^A$ and $\lambda_{i=2}^A$ are conjugate complex and associated with the two eigenvectors V_1^A and $V_2^A = \overline{V_1^A}$ of the matrix M_A. If one introduces a partition of the matrices Λ_A and M_A in the following way:

$$\Lambda_A = \begin{bmatrix} \Lambda_A^{(1)} & 0 \\ 0 & \Lambda_A^{(2)} \end{bmatrix} \text{ with } \Lambda_{\text{A}}^{(1)} = \begin{bmatrix} \lambda_1^A & 0 \\ 0 & \overline{\lambda_1^A} \end{bmatrix} \tag{2.19}$$

and

$$M_A = \begin{bmatrix} M_A^{(1)} & M_A^{(2)} \end{bmatrix} \text{ with } \text{M}_{\text{A}}^{(1)} = \begin{bmatrix} V_1^A & \overline{V_1^A} \end{bmatrix}, \tag{2.20}$$

one can easily prove [28] that the real matrices J_A and P_A defined by

$$J_A = \begin{bmatrix} J_A^{(1)} & 0 \\ 0 & \Lambda_A^{(2)} \end{bmatrix}, \text{ with } \text{J}_{\text{A}}^{(1)} = \begin{bmatrix} Re(\lambda_1^A) & Im(\lambda_1^A) \\ -Im(\lambda_1^A) & Re(\lambda_1^A) \end{bmatrix} \tag{2.21}$$

and

$$P_A = \begin{bmatrix} P_A^{(1)} & M_A^{(2)} \end{bmatrix}, \text{ with } \text{P}_{\text{A}}^{(1)} = \begin{bmatrix} Re(V_1^A) & Im(V_1^A) \end{bmatrix} \tag{2.22}$$

have the property

$$AP_A = P_A J_A. \tag{2.23}$$

Such a procedure is easily extended to the case of a set of conjugate complex pairs of eigenvalues. Each pair has to be replaced by a real 2×2 matrix to constitute the matrix J_A and the corresponding eigenvectors by their real and imaginary parts to constitute the matrix P_A. In the same way one has to introduce the matrices J_B and P_B so that (2.17) implies:

$$J_A U'' + U'' J_B - \sigma U'' = S'' \tag{2.24}$$

with $U'' = P_A^{-1} U P_B$ and $S'' = P_A^{-1} S P_B$.
Different cases are possible:

- the eigenvalues λ_i^A et λ_j^B are real, then u''_{ij} is calculated as in (2.18);
- the eigenvalue pair $(\lambda_i^A,\lambda_{i+1}^A)$ or $(\lambda_j^B,\lambda_{j+1}^B)$ is conjugate complex, then $(u''_{ij},u''_{(i+1)j})$ or $(u''_{ij},u''_{i(j+1)})$ solves a 2×2 system;
- the eigenvalue pairs $(\lambda_i^A,\lambda_{i+1}^A)$ and $(\lambda_j^B,\lambda_{j+1}^B)$ are conjugate complex, then $(u''_{ij},u''_{(i+1)j},u''_{i(j+1)},u''_{(i+1)(j+1)})$ solves a 4×4 system.

2.3 The Ill-posed Poisson/Neumann Problem

When solving the Navier-Stokes equations with a projection method, the pressure correction step may require the resolution of a Poisson equation ($\sigma = 0$) with Neumann boundary conditions ($\alpha_k = 0, \beta_k = 1, k = 1,4$), e.g. in case of no-slip conditions for the velocity. As it is well known, such a problem is ill-posed: the solution is either only unique up to a constant or does not exist. But the non-existence case is not to be encountered, because the compatibility condition between the source-term and the boundary conditions is fulfilled, at least at the level of the continuous problem. Nevertheless, after discretization or when using an iterative procedure, the case of non-existence may also occur. Then it can be interesting to recover the situation of non-unicity by adding a (small) constant to the body-force term. As outlined now, our solution technique is well adapted to such an algorithm.

At the level of the discrete equations the matrix A (B) shows a null eigenvalue associated with a constant eigenvector, spanning the basis of its kernel. The corresponding element $u''_{i_0 j_0}$ solves

$$\left(\lambda^A_{i_0} + \lambda^B_{j_0}\right) u''_{i_0 j_0} = 0 \cdot u''_{i_0 j_0} = s''_{i_0 j_0}.$$

As expected, two cases must be considered:

(1) $s''_{i_0 j_0} = 0$, then $u''_{i_0 j_0}$ is arbitrary.

(2) $s''_{i_0 j_0} \neq 0$, then $u''_{i_0 j_0}$ does not exist.

In case (1) one can give an arbitrary value to $u''_{i_0 j_0}$ (e.g. zero, like in the computer code) to compute U up to a constant. Note that, consequently, if $\delta U''$ is the matrix with all its elements being equal to zero except one being equal to $u''_{i_0 j_0}$, then the matrix $\delta U = P_B^{-1} \delta U'' P_A$ is constant.

In case (2) the solution does not exist, but in order to recover the situation (1) one can substitute $s''_{i_0 j_0}$ by zero. This induces an additive constant to the body force term of the Poisson equation, in such a way that the compatibility condition is satisfied. Indeed, substituting a zero value to $s''_{i_0 j_0}$ means subtracting from S'' a matrix $\delta S''$ with all its elements being equal to zero except one being equal to $s''_{i_0 j_0}$. Since $\delta S''$ has a similar structure as $\delta U''$, then $\delta S = P_B^{-1} \delta S'' P_A$ is, as δU, a constant matrix. But once U is known, the polynomial $u(x,y)$ in (2.4) is completely defined and so δS can only result from the addition of a constant to the body force term f at the inner grid points.

2.4 Tests on Accuracy

In order to check the accuracy of the spectral solver and the interest of coordinate transforms, we have tried to recover the exact solution

$$u(x,y) = \tanh(10x)\tanh(8y) + 1 \qquad \text{in} \quad \bar{\Omega} = [-1,1]^2 \tag{2.25}$$

which is stiff along the coordinate axis. Calculations have been made

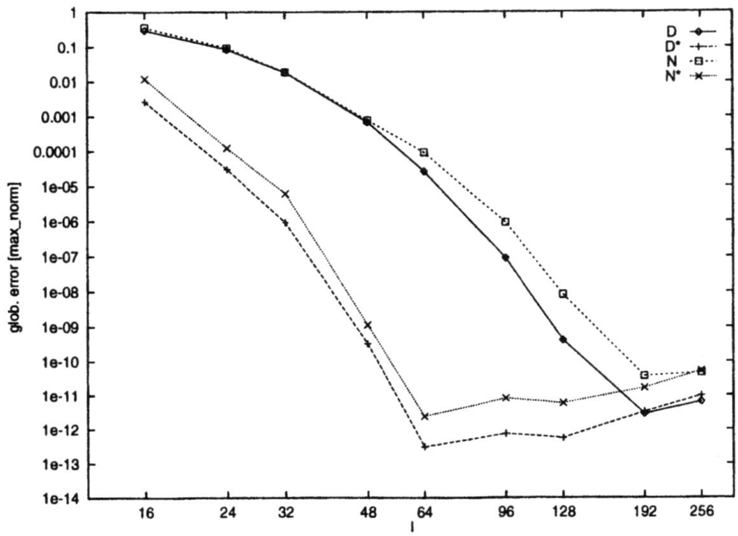

Figure 1 Error in the max norm vs the degree I (= J) of the polynomial approximation, for the Dirichlet (D) and the Neumann (N) problems without and with (*) coordinate transforms

(i) without coordinate transforms, and

(ii) with the following one in X (and similarly in Y):

$$x = \alpha X^3 + (1-\alpha)X \qquad \text{with} \quad \alpha = 0.6. \tag{2.26}$$

The results obtained for the elliptic Helmholtz equation, with $\sigma = 1$, for both the Dirichlet and Neumann problems are shown in fig. 1. Let us mention that complex eigenvalues have been detected for the Neumann problem, with $I = 24$.

For the Poisson equation, we have considered the exact solution:

$$u(x,y) = x(4-x)y(1-y)\sin(10x)\cos(10y) \qquad \text{in} \quad \bar{\Omega} = [0,4] \times [0,1] \tag{2.27}$$

and solved, using only linear coordinate transforms, the Dirichlet and ill-posed Neumann problems. The results are given in fig. 2. For the Neumann problems, situations 1) (non-unicity, curve N) and 2) (non-existence, curve N #, a constant equal to 1 has been added to the exact body-force term) have been illustrated.The additive constant has been simply adjusted by matching the exact and numerical solutions at a particular (central) grid-point. One observes the exponential decrease of the error and, as expected, the matching of the curves N and $N\#$ out of the round-off error range ($J \geq 24$).

The last example (fig. 2) is also investigated in section 4, when using the multidomain CGBI method with four subdomains.

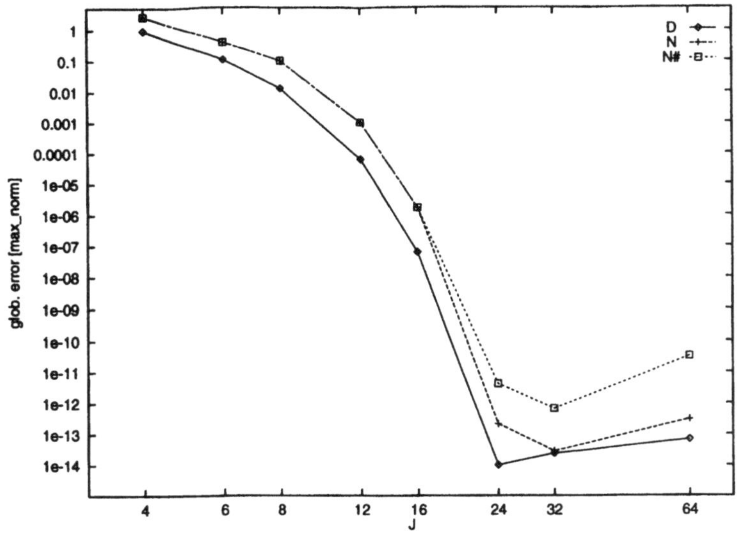

Figure 2 Error in the max norm vs J and with $I = 4J$, for the Dirichlet (D) and the Neumann (N, N#) problems

3 The Conjugate Gradient Boundary Iteration (CGBI)

3.1 The CGBI Method

The CGBI is a new and efficient method to solve a Helmholtz resolvent type equation

$$\begin{aligned} Lu := \sigma u - \Delta u &= f \text{ in } \Omega \quad (\sigma \geq 0) \\ u &= 0 \text{ on } \partial\Omega \end{aligned} \tag{3.28}$$

in a parallel mode (note that problem (3.28) is of type (2.1)). The name "boundary iteration" comes from the fact that the unknowns are distributed only on the artificial boundaries Γ_i (see figure below). This method reduces the above problem (3.28) to a number of mixed Dirichlet-Neumann problems on the subdomains which can be solved in parallel. Other kinds of domain decomposition methods can be found in [6], [7], [12], [13], [17]. Comparisons of numerical results with the CGBI method are given in [2]. The method given in [12], [13] also leads to mixed Dirichlet-Neumann problems. There the global solution is a saddle point of a suitable Lagrangian leading to projection type CG - algorithms.
Our approach is based on an unconstrained minimization problem (see (3.32)). The rectangular domain Ω in (3.28) is decomposed into subdomains Ω_i, $i = 1,...,n$ (see figure below). For simplicity we restrict ourselves to the case without interior crosspoints.

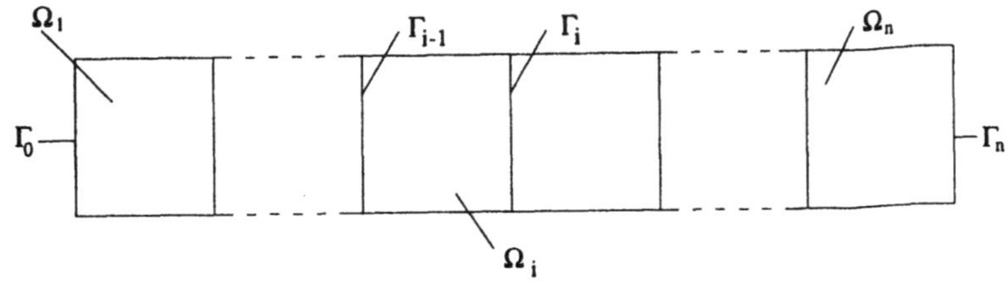

In sections 3.1 and 3.2 we consider homogeneous Dirichlet boundary conditions. In 3.4 we examine the modifications which are necessary in the case of Neumann boundary conditions.
The main idea is to represent the solution u as the sum

$$u = v_i + u_i, \quad i = 1,\dots,n, \tag{3.29}$$

on each Ω_i, where v_i is the solution of the following pre-step:

$$\begin{aligned} Lv_i &= f \text{ in } \Omega_i \\ v_i &= 0 \text{ on } \partial\Omega_i \setminus (\Gamma_{i-1} \cup \Gamma_i) \\ \frac{\partial v_i}{\partial \nu} &= 0 \text{ on } \Gamma_{i-1} \cup \Gamma_i \end{aligned} \tag{3.30}$$

ν is the outward normal direction on $\partial\Omega_i$.
Then, obviously u_i has to be fulfil $Lu_i = 0$ on Ω_i. Denoting by φ_i the unknown normal derivative of u_i and requiring its continuity we get the following problem for determining u_i:

$$\begin{aligned} Lu_i &= 0 \text{ in } \Omega_i \\ u_i &= 0 \text{ on } \partial\Omega_i \setminus (\Gamma_{i-1} \cup \Gamma_i) \\ \frac{\partial u_i}{\partial \nu} &= -\varphi_{i-1} \text{ on } \Gamma_{i-1} \\ \frac{\partial u_i}{\partial \nu} &= \varphi_i \text{ on } \Gamma_i, \end{aligned} \tag{3.31}$$

with obvious modifications of the boundary conditions on Ω_1 and Ω_n. The method is now to determine $\varphi = (\varphi_1,\dots,\varphi_{n-1})$ on the boundaries Γ_i. This is done by minimizing the quadratical functional

$$J(\varphi) = \sum_{i=1}^{n-1} \int_{\Gamma_i} \left[v_i - v_{i+1} + \frac{1}{2}\left(u_i(\varphi) - u_{i+1}(\varphi)\right)\right] \varphi_i \, do \longrightarrow \text{Min.}, \tag{3.32}$$

with u_i depending linearly on φ_{i-1} and φ_i. One can show that J can be extended to a continuous functional $J : H^{-\frac{1}{2}}(\Gamma) = (H^{\frac{1}{2}}_{0,0}(\Gamma))^* \to \mathbb{R}$, where Γ is the union of all Γ_i and $H^{-1/2}(\Gamma) := \bigtimes_{i=1}^{n-1} H^{-1/2}(\Gamma_i)$ (see [18] for the definition of $H^{1/2}_{0,0}$).
Since J (resp. J'') is coercive,

$$\gamma_0\|\varphi\|^2_{-\frac{1}{2}} \le J''(\varphi) \le \gamma_1\|\varphi\|^2_{-\frac{1}{2}} \quad \text{for all } \varphi\in H^{-\frac{1}{2}}(\Gamma) \quad (0<\gamma_0,\gamma_1), \tag{3.33}$$

the above minimization principle has a unique solution

$$\varphi^0 = (\varphi_1^0,...,\varphi_{n-1}^0) \in H^{-1/2}(\Gamma).$$

From

$$0 \;=\; J'(\varphi^0)\varphi \;=\; \sum_{i=1}^{n-1} \int_{\Gamma_i} [(v_i + u_i(\varphi^0)) - (v_{i+1} + u_{i+1}(\varphi^0))]\,\varphi_i\, do \quad \text{for all } \varphi\in H^{-\frac{1}{2}}(\Gamma) \tag{3.34}$$

it follows that $v_i + u_i$ is continuous across Γ_i at least in the sense of traces. Furthermore, the normal derivative is continuous as well by the above construction. Therefore the ansatz $u_{|\Omega_i} = v_i + u_i$ leads to the solution.

To apply the conjugate gradient algorithm to the above minimization problem, we identify $H^{-1/2}$ with $H^{1/2}_{0,0}$ by the Riesz representation $C : H^{1/2}_{0,0} \to H^{-1/2}$. Using this representation, the gradient of J is simply given by $C(u_i - u_{i+1})$ on Γ_i.

The condition number of this CG-iteration is determined by the constants γ_0, γ_1 in (3.33) which are independent of the mesh size. They do depend on the norms of the local trace and extension operators and thus the condition number is also independent of the number of subdomains. However, in the discretized version of this algorithm the constants depend additionally on the accuracy of the realization of the Riesz operator C which identifies $H^{-1/2}$ with $H^{1/2}_{0,0}$. In section 3.3 and 4.3 we tested various possibilities.

3.2 The CGBI Algorithm

The method described in section 3.1 leads to the following algorithm. Step 1 corresponds to the pre-step (3.30), in step 2 the boundary condition φ which fulfils the minimization principle (3.32) is calculated, and in step 3 the corresponding $u_i = u_i(\varphi)$ is calculated.

<u>Procedure CGBI(f):</u>

(Done in a parallel mode by processors $i = 1,...,n$.)

<u>1st step:</u> Solve (in a parallel mode)

$$\begin{aligned} &Lv = f \text{ on } \Omega_i,\\ &\tfrac{\partial v}{\partial \nu} = 0 \text{ on } \Gamma_i,\ \Gamma_{i-1}\\ &v = 0 \text{ on } \partial\Omega_i \setminus (\Gamma_i \cup \Gamma_{i-1})\\ &\text{with FDM or Chebyshev-method.} \end{aligned} \tag{3.35}$$

Let $g := [v]$ be the jump of v on the artificial boundaries; g is defined on $\bigcup_{i=1}^{n-1} \Gamma_i$.

2nd step: $\varphi := CG(g)$; *(see below)*

φ is defined on $\bigcup_{i=1}^{n-1} \Gamma_i$.

3rd step: Solve (in parallel)

$$\begin{aligned} &Lu = 0 \text{ on } \Omega_i,\\ &\tfrac{\partial u}{\partial \nu} = +\varphi \text{ on } \Gamma_i,\ \tfrac{\partial u}{\partial \nu} = -\varphi \text{ on } \Gamma_{i-1}\\ &u = 0 \text{ on } \partial\Omega_i \setminus (\Gamma_i \cup \Gamma_{i-1})\\ &\text{with FDM or Chebyshev-method.} \end{aligned} \tag{3.36}$$

$u := u + v$ is the solution of $Lu = f$ on Ω, $u = 0$ on $\partial\Omega$.

Procedure $CG(g_i)$:

(Done in parallel; each processor $i = 1...n$ calculates on one artificial boundary Γ_i and on one subdomain Ω_i.)

$\varphi_i := 0$

$d_i := -Cg_i$

$\delta_0 := -\sum_i \langle g_i, d_i \rangle$ ($\rightarrow$ *communication*)

get d_{i-1} from process $i-1$ ($\rightarrow$ *communication*)

repeat

{ solve

$$\begin{aligned} &Lu = 0 \text{ on } \Omega_i,\\ &\tfrac{\partial u}{\partial \nu} = +d_i \text{ on } \Gamma_i, \quad \tfrac{\partial u}{\partial \nu} = -d_{i-1} \text{ on } \Gamma_{i-1}\\ &u = 0 \text{ on } \partial\Omega_i \setminus (\Gamma_i \cup \Gamma_{i-1})\\ &\text{with FDM or Chebyshev-method} \end{aligned} \tag{3.37}$$

$\gamma_i := u_{i+1} - u_i$ on Γ_i ('jump') ($\rightarrow$ *communication*)

$\tau := \frac{\delta_0}{\sum_i \langle d_i, \gamma_i \rangle}$ ($\rightarrow$ *communication*)

$\varphi_i := \varphi_i + \tau d_i$

$g_i := g_i + \tau\gamma_i$, $h_i := Cg_i$

$\delta_1 := \sum_i \langle g_i, h_i \rangle$ ($\rightarrow$ *communication*), $\beta := \frac{\delta_1}{\delta_0}$

$d_i := -h_i + \beta d_i$

get d_{i-1} from process $i-1$ ($\rightarrow$ *communication*)

} until $\delta_1 < \epsilon$

return φ_i

The different local solvers use different grids. (See section 4.2 for an example for the grids.) Since we must compute the jumps between the subdomains along the artificial boundaries after (3.35) and (3.37) we have implemented different methods for 1d interpolation.

For details on the operator C in the procedure CG see the following section.

3.3 Preconditioning Techniques

In the CG procedure in section 3.2 a transmission operator C is used. We have tested different realizations of C:

- **Trivial Case**
  (condition number $\sim O(N^{1/2})$)
  Use the Identity Operator for transmission: $C = I$.

- **Finite Difference Realization of Transmission**
  (condition number $\sim O(N^{1/4})$)
  For $N+1$ equidistant grid points on each artificial boundary, use the $(N+1) \times (N+1)$ - matrix

$$C_f = \begin{pmatrix} 2 & -1 & & \\ -1 & \ddots & \ddots & \\ & \ddots & \ddots & -1 \\ & & -1 & 2 \end{pmatrix} \qquad (3.38)$$

  and set $C = C_f + \alpha I$, were an optimal $\alpha = \alpha(N)$ has to be determined.

- **Exact Transmission** (e.g. in case $\sigma = 0$ in (3.28))
  (condition number $\sim O(1)$)
  Having the function g on the artificial boundary [0,1] in the representation

$$g = \sum_{k=1}^{N-1} c_k \sin \pi k x,$$

  set

$$C_s g = \sum_{k=1}^{N-1} k\, c_k \sin \pi k x. \qquad (3.39)$$

  The calculation of the c_k and $C_s g$ is done by FFT (taking into account the redundancies in the input data of FFT to get a speed-up).

For numerical results see section 4.3.

3.4 The Local Neumann Problems

In the pressure correction step (mentioned in section 1) we have to solve (3.28) with $\sigma = 0$ and with boundary conditions of both Neumann and Dirichlet type, i.e.:

$$\begin{aligned} -\Delta u &= f \text{ in } \Omega \\ u &= 0 \text{ on } \Gamma_n \\ \frac{\partial u}{\partial \nu} &= 0 \text{ on } \partial\Omega \setminus \Gamma_n . \end{aligned} \tag{3.40}$$

Note that the boundary condition $u = 0$ on Γ_n stands for the idea of using a natural condition for the pressure at the outlet of the channel.
The global solution of (3.40) is unique because of mixed Dirichlet - Neumann boundary conditions. However, the local problems of type

$$\begin{aligned} -\Delta u &= f \text{ in } \Omega_i, \quad i = 1,...,n-1 \\ \frac{\partial u}{\partial \nu} &= g \text{ on } \partial\Omega_i \end{aligned} \tag{3.41}$$

corresponding to (3.35), (3.36), (3.37) are only solvable if the compatibility conditions

$$\int_{\Omega_i} f\, dx = -\int_{\partial\Omega_i} g\, do \quad \forall i = 1,...,n-1 \tag{3.42}$$

are fulfilled. To ensure (3.42) we solve

$$\begin{aligned} -\Delta v &= f \text{ in } \Omega_i \\ \frac{\partial v}{\partial \nu} &= 0 \text{ on } \partial\Omega \setminus (\Gamma_{i-1} \cup \Gamma_i) \\ \frac{\partial v}{\partial \nu} &= -c_{i-1} \text{ on } \Gamma_{i-1} \\ \frac{\partial v}{\partial \nu} &= c_i \text{ on } \Gamma_i \quad (u = 0 \text{ on } \Gamma_n) \end{aligned} \tag{3.43}$$

instead of (3.35) and

$$\begin{aligned} -\Delta u &= 0 \text{ in } \Omega_i \\ \frac{\partial u}{\partial \nu} &= 0 \text{ on } \partial\Omega \setminus (\Gamma_{i-1} \cup \Gamma_i) \\ \frac{\partial u}{\partial \nu} &= -\varphi_{i-1} \text{ on } \Gamma_{i-1} \\ \frac{\partial u}{\partial \nu} &= \varphi_i \text{ on } \Gamma_i \end{aligned} \tag{3.44}$$

instead of (3.36), (3.37), where $c_0 = 0$, and $c_1,...,c_{n-1}$ are determined by the local mean values of f on Ω_i calculated by numerical integration. The discretized version of problem (3.44) has automatically the compatibility property if we take C from (3.38) modified due to Neumann boundary conditions. Note that this choice of C implies $\int_{\Gamma_i} \varphi_i\, do = 0$ for $i = 1,...,n-1$.
To guarantee the uniqueness of the solutions of (3.43), (3.44) we additionally require the continuity of the solutions at prescribed special points P_i on Γ_i $(i = 1,...,n-1)$. To achieve this continuity we add suitable constants to the local solutions.

4 Numerical Results for the CGBI

4.1 The CGBI-Method for the Poisson Equation with Different Local Solvers

Test Configuration:

- 4 subdomains, $(N + 1) \times (N + 1)$ grid points on each subdomain with $N = 4, 8, 16, 32, 64, 128$.
- Solving the Poisson equation with the exact solution

$$u(x,y) = x\ (4 - x)\ y\ (1 - y)\ \sin(10x)\ \cos(10y)$$

  on $\bar{\Omega} = [0,4] \times [0,1]$ and Dirichlet boundary conditions.
- The CGBI is stopped when the residual δ_1 (see section 3.2) becomes stationary.
- Comparison FDM local solver on all subdomains vs. Chebyshev local solver on all subdomains.

For this test configuration the following two figures represent the global L^∞-error; the first one in dependence of the number of grid points and the second one in dependence of the calculation time.

Results:

- The CGBI with FDM - local solver has 2nd order convergence in N (see fig. 3).
- The CGBI with Chebyshev - local solver has spectral accuracy (see fig. 3).
- Although the Chebyshev method requires more computing time than the FDM for the same number of grid points it is more efficient if a high accuracy is required (see fig. 4).
  Extrapolating our results for FDM we see: In order to achieve an accuracy of 10^{-8} we would have to use $2^{14} \times 2^{14}$ grid points ($\simeq$ several Gigabytes of memory) instead of 32×32 and $10^4\times$ more computing time if we would use FDM/FEM instead of the Chebyshev local solver.
  Moreover, for *evolution problems* the local Chebychev solver gaines efficiency because the time consuming preliminary calculations (matrix diagonalizations and inversions, see section 2) only have to be done once.
- The comparison of the mono-domain and multi-domain results (fig. 2 and 3) shows that outside the range of numerical noise ($N \geq 16$) no loss of accuracy occurs when using the multi-domain approach (at least with Dirichlet boundary conditions).

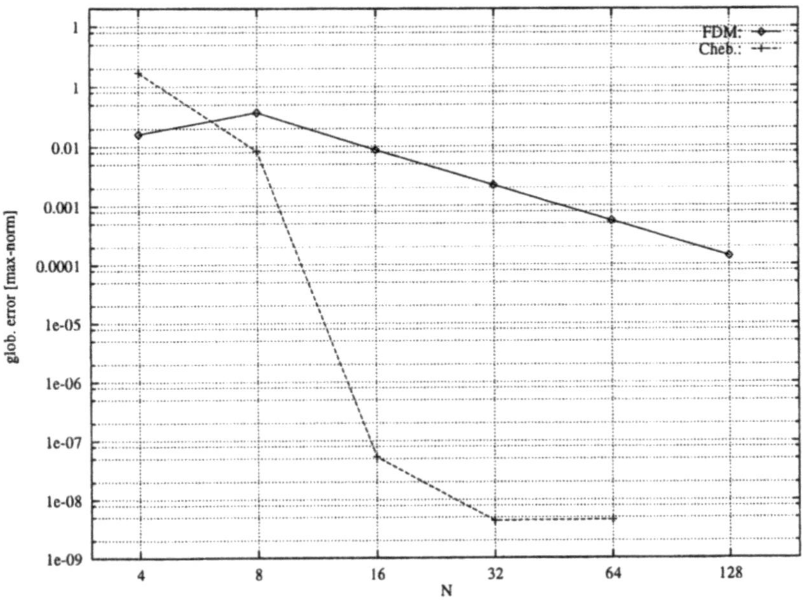

Figure 3 Error in the max norm vs N for the Finite Differences and Chebyshev approximations

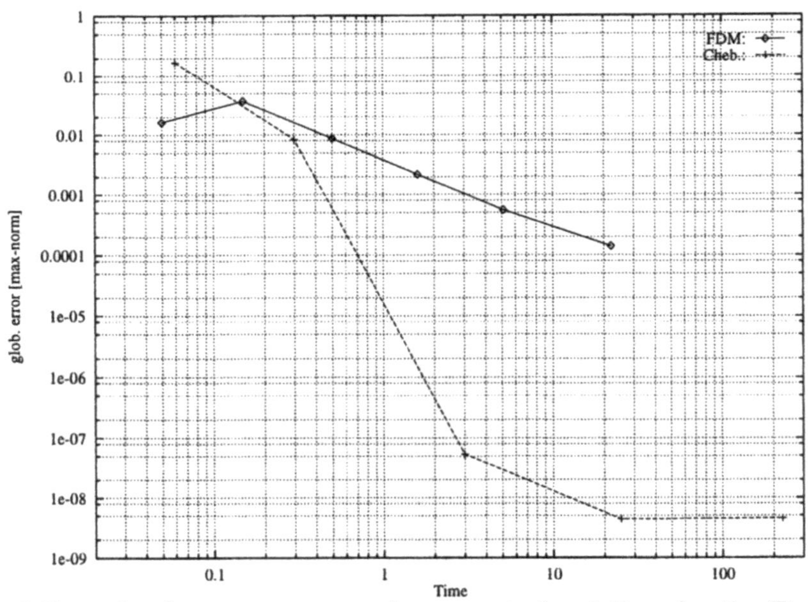

Figure 4 Error in the max norm vs the computational time for the Finite Differences and Chebyshev approximations

4.2 Error Decay Induced by a FDM-Subdomain along the Channel

Now we are *combining* the two different local solvers in order to investigate the influence of local solvers without spectral accuracy. In particular we compare the Poisson and the Helmholtz resolvent equation. We consider the following test configuration:

- 6 subdomains, 17 × 17 grid points each.
- Equation:
  - Poisson equation (fig. 5)
  - Helmholtz equation (fig. 6)

  with the given exact solution

  $$u(x,y) = \frac{1}{9}\ \sin(10x)\ \cos(10y)\ x\ (6-x)\ y\ (1-y)$$

  on $\bar{\Omega} = [0,6] \times [0,1]$ and Dirichlet boundary conditions.
- Local Solvers:
  a) FDM on 1st subdomain, Chebyshev on the others (decreasing line in fig. 5 and 6)
  b) FDM on all subdomains (lower line in fig. 5 and 6)
  c) Chebyshev method on all subdomains (upper line in fig. 5 and 6).

The following figure shows the mesh for case a) (but with less grid points):

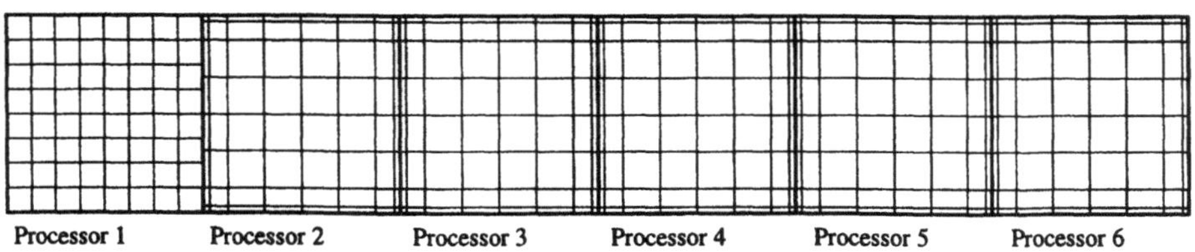

The following two figures show the local L^∞-error on each subdomain, especially the decay of the error along the channel in case a). In this case the FD-method on the 1st subdomain certainly causes a higher local error than the Chebyshev method. ***But this effect does not destroy the spectral accuracy on subdomains far away from the first!*** As predicted by theory, the error decreases exponentially along the channel (see fig. 5 and 6), for the Helmholtz equation (fig. 6) even faster than for the Poisson equation (fig. 5).

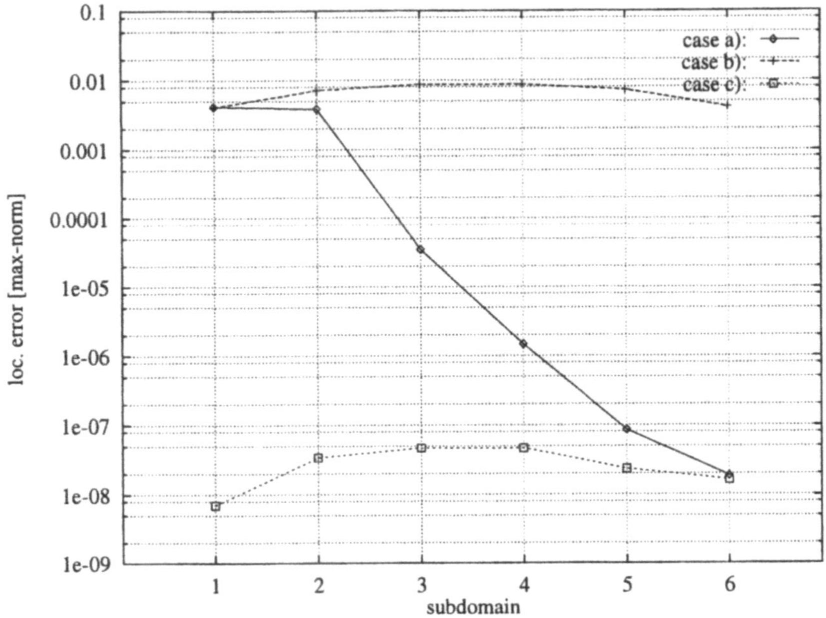

Figure 5 Error in the max-norm vs the subdomain index for the Poisson equation

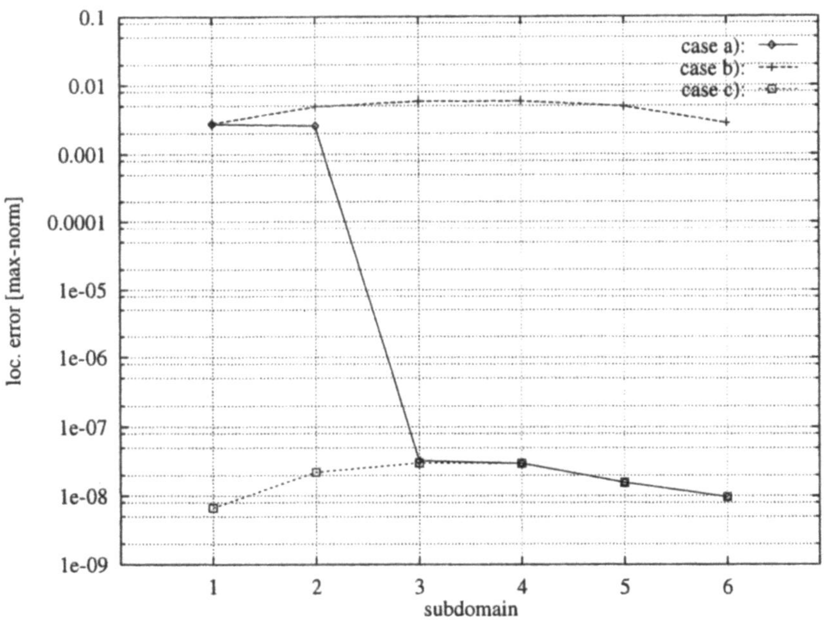

Figure 6 Error in the max-norm vs the subdomain index for the Helmholtz equation

4.3 Test of Various Transmission Operators C for the CGBI Method

We are solving the Poisson equation $-\Delta u = f$ on $\bar{\Omega} = [0,4] \times [0,1]$ with $u = 0$ on $\partial\Omega$ with the exact solution

$$u(x,y) = \frac{1}{4} \sin(10x) \cos(10y)\, x\, (4-x)\, y\, (1-y).$$

The code is running on 4 processors using the FDM local solver on all subdomains.
Each subdomain has $(N+1) \times (N+1)$ grid points with $N = 32$ (fig. 7), rsp. $N = 128$ (fig. 8), rsp. $N = 512$ (fig. 9).
The calculations were done with the three kinds of transmission conditions described in section 3.3:

upper line in fig. 7-9:	$\mathcal{C} = I$
middle line in fig. 7-9:	$\mathcal{C} = C_f + \alpha I$
lower line in fig. 7-9:	$\mathcal{C} = C_s$.

These cases are compared in fig. 7-9:
Using no preconditioning (uppermost lines in fig. 7-9) the number of iteration steps increases with refinement of discretization (fig. 7→fig. 8→fig. 9). According to theory (bounded condition number), the spectral transmission operator (lower lines in fig. 7-9) only requires a *constant* number of iteration steps. (Only 2 steps are sufficient!)

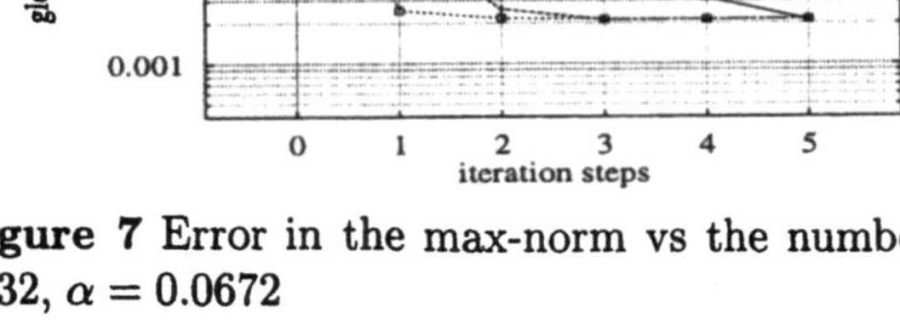

Figure 7 Error in the max-norm vs the number of iterations for a mesh size $1/32$, $\alpha = 0.0672$

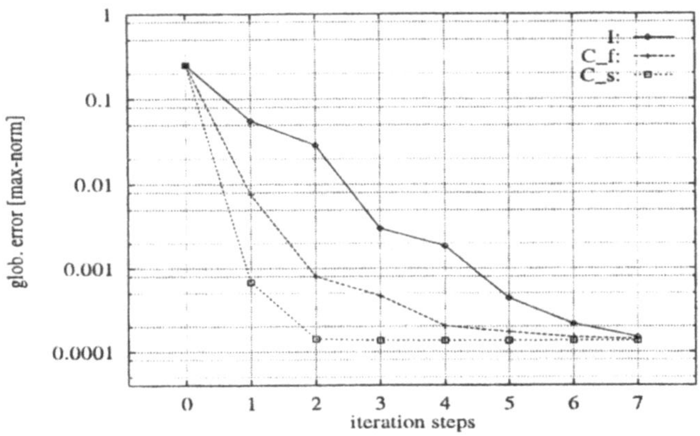

Figure 8 Error in the max-norm vs the number of iterations for a mesh size $1/128$, $\alpha = 0.0048$

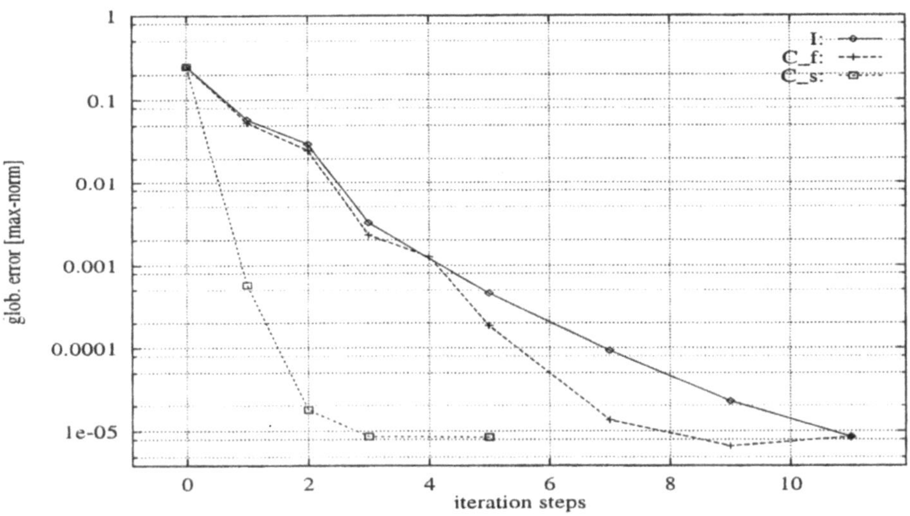

Figure 9 Error in the max-norm vs the number of iterations for a mesh size $1/512$, $\alpha = 0.0047$

5 Conclusion

The hybrid Spectral / Finite difference - CGBI method turns out to be a robust and efficient method to solve elliptic problems. The spectral accuracy of the local Chebyshev solver is preserved and, moreover, the error induced by lower order finite difference local solvers displays exponential decay. Thus the method is suitable to solve the elliptic problems which occur in the framework of projection methods for solving the unsteady Navier-Stokes equation. In the future we will improve the flexibility of the approach by using finite elements on some subdomains. Then it will be possible to deal with flows in more complex geometries.

References

[1] Bayliss, A., Class, A., Matkowsky, B.J.: Round off error in computing derivatives using the Chebyshev differentiation matrix, J. Comput. Phys. 116 (1994), 380-383.

[2] Blazy, S., Borchers, W., Dralle, U.: Parallelization methods for a characteristic's pressure correction scheme, in: Flow simulation with high-performance computers II (Hrs. E.H. Hirschel), Notes on Numerical Fluid Dynamics, Vol. 38, 305-321, Vieweg, Braunschweig, Vieweg 1996.

[3] Borchers, W.: On the characteristic's method for the incompressible Navier-Stokes-equations, In: Finite Approximations in Fluid Mechanics II; E. H. Hirschel (Ed.), Notes on Numerical Fluid Mechanics, Vol. 25, 43-50, Vieweg, Braunschweig 1989, 43-50.

[4] Borchers, W.: Zur Stabilität und Faktorisierungsmethode für die Navier-Stokes Gleichungen inkompressibler viskoser Flüssigkeiten. Habilitationsschrift für das Fach Mathematik im Fachbereich Mathematik-Informatik der Universität-GH Paderborn (1992).

[5] Borchers, W.: A new conjugate gradient boundary iteration method for parallel numerical computation of elliptic boundary value problems. In preparation.

[6] Bramble, J. H., Pasciac, J.E., Schatz, A.H.: The construction of preconditioners for elliptic problems by substructering I, Math. Comp. 47 (1986), 103-134.

[7] Bramble, J. H., Pasciac, J.E., Schatz, A.H.: The construction of preconditioners for elliptic problems by substructering II, Math. Comp. 49 (1987), 1-16.

[8] Chorin, A. J., Hughes, T.J.R., Mc Cracken, M.F., Marsden, J.E.: Product formulas and numerical algorithms, Comm. Pure Appl. Math. XXXI (1978), 205–256.

[9] Chorin, A. J.: Numerical solution of the Navier-Stokes equation. Math. Comp. 22,(1968), 745-762.

[10] Douglas, J., Russel, T.F.: Numerical methods for convection-dominated diffusion problems based on combining the method of characteristics with Finite Element or Finite Difference procedures. Siam J. Numer. Anal 19 (1982), 871-885.

[11] Ehrenstein, U., Peyret, R.: A Chebyshev collocation method for the Navier-Stokes equations with application to double diffusive convection, Int. Journal for Numerical Methods in Fluids 9 (1989), 427-452.

[12] Farhat, C., Roux, F.-X.: An unconventional domain decomposition method for an efficient parallel solution of large-scale finite element systems, SIAM J. Sci. Stat. Comput. 13 (1992), 379-396.

[13] Farhat, C., Po-Shu Chen, Roux, F.-X., The dual Schur complement method with well-posed local Neumann problems: Regularization with a perturbed Lagrangian formulation, SIAM J. Sci. Comput. 14 (1993), 752-759.

[14] Gottlieb, D., Lustman, L., The spectrum of the Chebyshev collocation operator for the heat equation, SIAM J. Numer. Anal. 20 (1983), 909-921.

[15] Hackbusch, W., Trottenberg, M.(eds.): Multigrid methods. Lecture Notes in Mathematics. Vol. 960. Springer-Verlag, Berlin-Heidelberg-New York, 1982.

[16] Hebeker, F.-K.: Analysis of a characteristics method for some incompressible and compressible Navier-Stokes problems. Preprint 1126 des Fb. Mathematik, TH Darmstadt (1988).

[17] Langer, U., Haase, G.: The approximate Dirichlet domain decomposition method. II. Application to 2nd-order elliptic BVPs. Comp. Arch. f. Inf. a. Numer. Comp. 47, (1991), 153-167.

[18] Lions, J.L., Magenes, E.: Non-homogeneous boundary value problems and applications I, Springer Verlag, Berlin-Heidelberg-New York, 1982

[19] R. Pasquetti, R. Bwemba: A spectral algorithm for the Stokes problem in vorticity-vector potential formulation and cylindrical geometry, Comput. Methods Appl. Mech. Engrg. 117 (1994), 71-90.

[20] Pironneau, O.: On the transport diffusion algorithm and its applications to the Navier-Stokes equations, Num. Math. 38 (1982), 309-332.

[21] Rannacher, R.: On Chorin's projection method for the incompressible Navier-Stokes equations, Lecture Notes in Math. Vol. 1530, 167-183, (Heywood, J.G., Masuda, K., Rautmann, R., Solonnikov, S.A. Eds.), The Navier-Stokes Equations II - Theory and Numerical Methods, Proceedings, Oberwolfach (1991), Springer-Verlag, Berlin - Heidelberg - New York.

[22] Rautmann, R., Masuda, K.: H^2-convergent approximation schemes to the Navier-Stokes equations. Comm. Math. Univ. Sancti Pauli 43 (1994), 55-108.

[23] Rautmann, R.: A regularizing property of Rothe's method to the Navier-Stokes equations, in: Navier-Stokes Equations and Related Nonlinear Problems, A. Sequeira (ed), Plenum Press New York (1995), 377-391.

[24] Rautmann, R.: Approximate Lagrangian coordinates in Navier- Stokes problems, Proceedings of the Conference "The Navier-Stokes equations, theory and numerical methods", Varenna June 1.-6., 1997, R. Salvi (ed.) to appear.

[25] Shen, J.: On error estimates of projection methods for the Navier-Stokes equations: First order schemes, SIAM J. Numer. Anal. 29 (1992), 57-77.

[26] Süli, E.: Convergence and nonlinear stability of the Lagrange-Galerkin method for the Navier-Stokes equations, Numer. Math. 53 (1988) 459–483.

[27] Temam, R.: Sur l'approximation de la solution des équations de Navier-Stokes par la méthode des pas fractionaires (II), Arch. Rational Mech. Anal. 32 (1969), 377-383.

[28] Wilkinson, M.A.: The algebraic eigenvalue problem, Clarendon Press, Oxford, 1965.

Adaptively Refined Cartesian Grid Generation and Euler Flow Solutions for Arbitrary Geometries

F. Deister[1], D. Rocher[2], E.H. Hirschel[1] and F. Monnoyer[2]
[1] Universität Stuttgart, I.A.G., Pfaffenwaldring 21, D-70550 Stuttgart, Germany
[2] Université de Valenciennes, L.M.E., Le Mont Houy - B.P. 311, F-59304 Valenciennes Cedex, France

Summary

An automatic Cartesian grid generator is presented together with Euler solutions of flows around complicated geometries. The computational grid is generated based on an octree-data structure. Solid bodies merely blank out areas of the background Cartesian grid. The part of the surface-intersecting cells is cut off, which lies inside the body. As a result arbitrarily shaped cut-cells arise around the geometry. The remaining cells inside the flow field benefit from their regular cubic shape. The strategy developed for the grid generation prevents numerical instabilities and treats geometric degeneracies in a generic way. Thus, cut-cells of too small size are merged into appropriate neighbour cells to ensure numerical stability of the flow solver. Two different flow solvers are applied for the Euler equations: the first solver represents a novel multi-grid upwind flow solver, which benefits directly from the octree-data structure. The second flow solver is a modified unstructured solver of Jameson type. The parameters of the adaptation criterion are calculated for every cell, thus cells are identified for adaptive refinement or coarsening.

1 Introduction

Grid generation is still one of the bottlenecks in application-oriented numerical fluid dynamics. The geometrical complexity of a realistic configuration makes lengthy and costly grid generation processes necessary, [1]. Cartesian grid methods appear to have the potential to overcome this. In this paper a completely self-organizing Cartesian grid generation system is presented, that reduces the CPU-time on a workstation (R10000 processor) to a minute for Euler grid generation around complicated three-dimensional geometries without any user intervention.

Conventional grid generation methods are typically grouped in one of the two categories: structured and unstructured. The recently developed Cartesian grid methods combine the advantages of both categories. On the one hand they are as flexible as unstructured methods concerning the complexity of the body geometry. Besides that, neither the body complexity nor the number of bodies and the position to each other are restricted, except that the surfaces must be water-tight. On the other hand background tree-data structures provide a global structure, which describes the connectivity between the Cartesian cells. Particularly the hierarchical nature of tree-data structures simplifies the adaptive refinement and coarsening of the cells. It also generically provides coarse grids for multi-grid calculations. But it must be noted that Cartesian methods cannot take over all advantageous characteristics of both categories. The regular cubic shape of the Cartesian cells brings in a decreased capability of anisotropic resolution of flow characteristics in comparison to classic structured and unstructured methods. Moreover, the data structure is not suitable for the actual flow solution calculation to the same degree as that of structured methods. Thus the data structure of the flow solver is comparable to that of unstructured methods. However, a factor of two speed-up of the flow solution can be obtained in comparison to unstructured tetrahedron grids. This is due to the ratio of cell-faces to cells, which amounts approximately 3.5 for Cartesian and 7 for tetrahedron grids.

In the approach presented here, the three-dimensional Cartesian grid generator is based on the octree-data structure, [2], [3] and [4]. It is used as a global background data structure. The cells of the octree represent the Cartesian cells. Following the approach of Quirk, [5] and [6], the cells are classified in three generic kinds: cells intersected by the body surface, cells located inside the flow field and cells located inside the body. The part inside the body of the intersected cells is cut off. Therefore only the Cartesian cells or rather parts of them inside the flow domain are taken into account for flow calculations. The presented novel Euler flow solver is based on the octree-structure directly, whereas the second solver is a classical unstructured Euler flow solver. After the solution is obtained on the initial grid, the grid is adaptively refined or coarsened. Several adaptation loops result in a nearly optimal grid. That means that a reasonable resolution of all flow characteristics is achieved with a minimum number of cells.

Other three-dimensional approaches apply a binary tree-data structure [7] or use a data base instead of a tree-data structure [8]. In the literature the two-dimensional Cartesian grid methods [9], [10] and [11] are all based on the quadtree-data structure.

In this project at the Universität Stuttgart the Cartesian grid generation and the solution-based adaptation is made. For testing purpose an unstructured flow solver from DASA has been modified and used. At the Université de Valenciennes, a novel flow solver benefiting from the octree-data structure has been developed. Details are provided in the intermediate progress reports of the authors, [12] and [13].

This report is structured in the following way. In Chapter 2 the generation of the three-dimensional Cartesian grid is described. Chapter 3 presents the two different Euler flow solvers, of which the first one directly benefits from the octree-data structure. The solution-adaptation procedure is explained in Chapter 4. In Chapter 5 Euler applications to two- and three-dimensional configurations prove the validity of the concept. The future work is sketched in Chapter 6 and a short conclusion closes the report.

2 Three-Dimensional Cartesian Grid Generation

The surface of the geometry is described in form of a surface triangulation, which is adapted to the local curvature radius. The triangulation requires the attribute of watertightness. Prior to the grid generation, the body is translated, rotated and scaled to achieve the requested conditions.

2.1 Spatial Decomposition of the Flow Domain

The flow domain is represented in a Cartesian space, therefore the physical and the computational space are identical. At the beginning of the domain decomposition, it consists of only one huge initial hexahedral cell. Now the flow domain is divided into smaller subdomains using the octree-structure, [14] and [15]. The cells of the octree are identical to the Cartesian cells.

The octree-structure is based upon the principle of recursive decomposition. The initial (universe) cell is hierarchically divided into 8 children cells of equal size, Figure 1(a). Each octant can be successively subdivided into sub-octants until a local refinement criterion is satisfied. The process is represented by a tree of degree eight. Figure 1(b) shows the octree, where the universe cell is drawn as filled oval, the tree cell as shaded circle and leaf cells as unfilled squares.

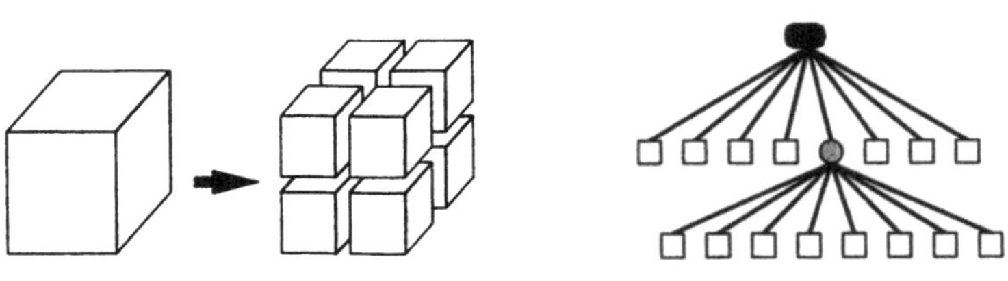

(a) The parent cell is split into 8 children cells.

(b) Schematic octree-data structure.

Figure 1 The octree-data structure.

The refinement criterion is taken over from Fujimura et al. [16]: a leaf cell is not allowed to contain more than one vertex of the surface triangulation. In practice all triangle vertices are taken successively and the leaf cell is identified enclosing the current vertex. If the leaf cell does not contain any other triangle vertex, the current vertex is inserted into this leaf cell. But if the leaf cell already contains another vertex, then the cell is refined until the former vertex and the current vertex are not located inside the same leaf cell any more.

For the following calculation of the surface-intersected cells, the sophisticated and efficient neighbour finding algorithm of Samet [17] is applied.

2.2 Calculation of the Surface-Intersected Cells (Cut-Cells)

The body surface splits the intersected Cartesian cells in parts inside and outside the flow field. Only the parts inside are considered and they are called cut-cells. The shape of the cut-cells is as arbitrary as the number of cut-cells, in which a Cartesian cell is divided. It must be noted that the triangulated surface of the body is cut out exactly. The cut-cell calculation consists of the following three steps.

As a first step, the cut-areas between the triangles and the single Cartesian cells are calculated, Figure 2(a). Therefore the Sutherland-Hodgeman polygon clipping algorithm is applied, [18] and [19]. The resulting cut-areas are called cut-faces and are stored as polygons with arbitrary numbers of vertices. The shape of the cut-faces is not restricted at all. As an example, Figure 2(b) shows the cut-faces of two Cartesian cells near the sharp trailing edge of a wing.

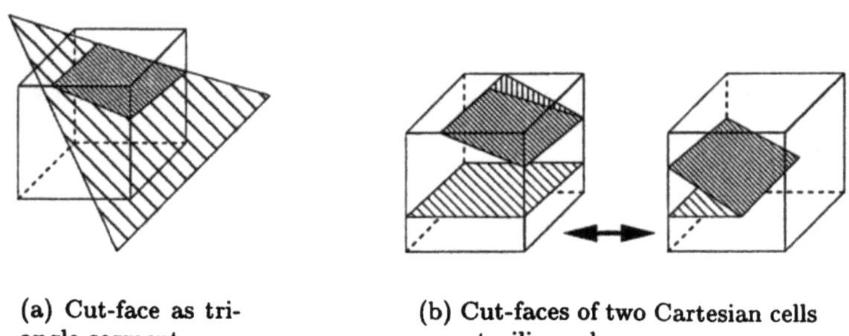

(a) Cut-face as triangle segment.

(b) Cut-faces of two Cartesian cells near trailing edge.

Figure 2 Cut-faces as cut-areas between triangles and Cartesian cells.

One major pre-requisite for numerical stability and efficient memory use is to calculate and store every intersection point between the triangles and the Cartesian cells only once. Therefore, the intersection points are indirectly addressed. It must be tested whether the triangle does intersect the current Cartesian clipping face indeed. This test is crucial for an appropriate treatment of geometric degeneracies and therewith for the overall stability and reliability. The test uses the Boolean concept of signed volumes, [7] and [12].

The second step consists of the calculation of the inter-faces. The surface triangulation intersects some faces of the Cartesian cells. These parts of the intersected Cartesian faces are called inter-faces, which are inside the flow field. Thus the parts of the Cartesian face inside the body are simply cut off. Again the inter-faces are represented as polygons of arbitrary shape and vertex number. Also a Cartesian face can be divided into zero, one or more inter-faces. The inter-faces are assembled from the adjacent cut-face edges and the segments of adjacent Cartesian cell edges.

As a final step the cut-cells are composed from the cut-faces, inter-faces and the cart-faces. A cart-face is a face of the Cartesian cell, which is not intersected by the surface triangulation. Figure 3(a) shows the inter-faces and cart-faces of the two Cartesian cells near the sharp trailing edge of Figure 2(b). The inter-faces are drawn shaded with dots and the cart-faces shaded with curly lines. As can be seen in Figure 3(b), only the cut-cells inside the flow domain are constructed. The left Cartesian cell contains two cut-cells, whereas the right Cartesian cell owns only one cut-cell. The sharp trailing edge is cut out of the Cartesian cells exactly.

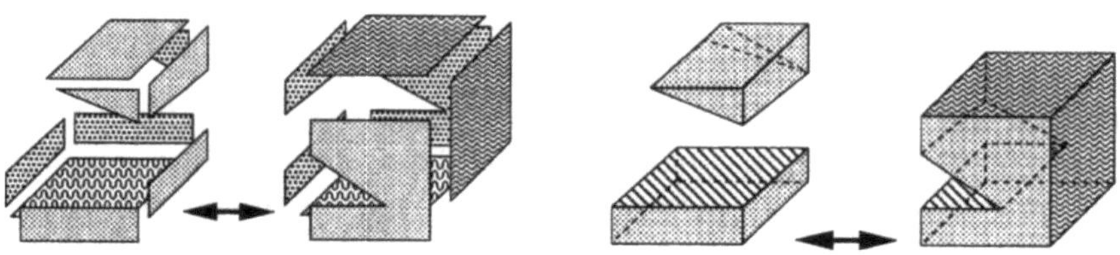

(a) Inter-faces and cart-faces of two Cartesian cells near trailing edge.

(b) Cut-cells of two Cartesian cells near trailing edge.

Figure 3 Cut-cells are composed from cut-faces, inter-faces and cart-faces.

2.3 Data Structure of the Finished Grid

The remaining Cartesian cells inside the flow domain, which are not intersected by the body triangulation, are identified through a frontal-wave algorithm: the neighbour cell of a cell inside the flow domain must be inside as well, if both cells are connected by an inter-face or a cart-face.

The data structure of the finished grid is edge-based: every inter-face and cart-face inside the flow domain is related to both adjacent cells. For boundary cells, the boundary face is connected to its boundary cell. The uncut Cartesian cells and the cut-cells represent directly the control volumes for flow calculations. In comparison to unstructured methods, [20], the Cartesian grid describes the dual grid of an imaginary grid. The imaginary grid consists of the edges connecting both cells adjacent to the corresponding inter-face or cart-face.

The size of the cut-cells may become arbitrarily small. In order to avoid numerical instabilities of the flow solver, these small cut-cells are merged in appropriate neighbour cells. The faces between constituent cells of a merged cell are simply deleted from the data structure.

All inter-faces and cart-faces are sorted into three main groups for the three coordinate directions perpendicular to the face. This results in simplifications of the flux calculation and saves memory. The faces of every main group are further sorted into groups in order to allow vectorization of the flow solver.

2.4 Extraction of Two-Dimensional Grid Data

Two-dimensional grid data is extracted from a three-dimensional grid. The approach is motivated by the need of obtaining fast results for testing purposes. It also offers the possibility of almost unlimited adaptive refinement. The idea consists in first generating a three-dimensional grid and then in taking only the grid data of the cells located at the symmetry plane. Therefore the two-dimensional object needs a three-dimensional representation for the grid generation. It is obtained by connecting two object profiles with a simple triangulation. The first profile has to be located at the symmetry plane, whereas the second profile is shifted parallel into the flow field and is triangulated.

The Cartesian grid generator remains unchanged except for adding a module for extracting the two-dimensional grid data and for replacing the module for the metric calculation. The unstructured flow solver remains completely unchanged because of the edge-based data structure. The two-dimensional grid just does not contain edges corresponding to faces perpendicular to the third coordinate direction. Also for two-dimensional calculations, the time for grid generation is negligible in comparison to the time needed by the flow solver.

3 Description of the Flow Solvers

3.1 New Cartesian Flow Solver

3.1.1 General Features

The compressible Euler equations in integral form are solved following the finite-volume method. A numerical scheme has been developed, based upon upwind discretization with cell-centered flow variables. The spatial second-order accuracy of the flux functions is achieved by means of the MUSCL approach [21]. This consists in linearly reconstructing the flow solution inside each cell.

For that purpose, the gradients of the solution are estimated. The calculation of the flux functions is detailed in Sub-Chapter 3.1.3. A fully explicit two-stage time-stepping scheme is used for integrating the solution to the steady state.

Local time stepping is performed in order to speed-up the convergence by computing the maximum allowed time step for each cell. The multi-grid process is used to further speed up the convergence [22]. Moreover, the stability is improved by limiting the relative change in both the density and the pressure for each cell at each time-step update of the solution vector [23]. Typically, the variation limit is taken equal to 10 per cent. After computing the flux balance of each cell, the potential solution update is determined, based on the maximum CFL-number. The CFL-number is then cut back to reach the 10 per cent limit. This procedure tends to damp the strong transient behaviour of the residual at the beginning of the solution process, which is due to the flow initialization with uniform conditions.

3.1.2 Cell-Based Approach

As seen in Sub-Chapter 2.3, all flow-cells are described in the Cartesian grid generator by means of edge-based informations. The cart-faces, inter-faces and cut-faces are defined by means of all their necessary geometric properties. Then, their attachment to flow-cells is given. The solver described here uses a cell-oriented description and benefits from the tree structure. Some *preprocessing* stage is required to translate the input grid information from surface-oriented to cell-oriented format. The algorithms of Samet [17] for neighbour finding allow a simplified definition of the multi-grid levels. The face-adjacent cells used for flux calculation as well as the neighbouring cells used for the gradient estimation and the coarser cells defining coarser grid levels can be easily identified. This latter feature is to be compared with complex methods used for purely unstructured grids, such as agglomeration techniques [24].

As far as the octree-structure is concerned, the grid is no more regular, as with multi-block grids. This is a major drawback for the solver and brings about several difficulties to circumvent. First, the gradients used for the linear reconstruction of the solution cannot be computed by means of classic finite-difference formulae. The gradient estimation is explained in Sub-Chapter 3.1.3. Secondly, the coarse grids cannot be defined by uniformly taking every two cells along the three space directions. If such a technique was used, the refinement level between two adjacent cells could differ from more than two, which is forbidden.

Different geometric features are determined and stored for the whole octree-structure, such as the cell volume, the cell center, the cart-face area and center. Similarly, variables used during the solution process are stored for the whole structure: solution, solution gradients, restricted solution and forcing function for the multi-grid process, residual and time step.

It is worth noticing that the flow-cells and their corresponding properties correspond to the leaf cells of the octree. Hence, informations for any cell in the octree can be obtained by simply following tree branches up to the flow-cells.

3.1.3 Flux Calculation

The finite-volume formulation of the discretized equations requires the estimation of the flux functions at the cell-face centers. For that purpose, the solution is needed on each side of the face. For first-order accurate fluxes, the values are taken identical to the value at the centers of both adjacent cells. For second-order accurate fluxes, the solution is linearly reconstructed using the gradients of the solution at the same cell centers as follows

$$U = U^C + \nabla U^C \cdot \begin{pmatrix} \Delta x \\ \Delta y \\ \Delta z \end{pmatrix} = U^C + \nabla U^C \cdot \Delta r,$$

where U denotes the set of conservative variables and superscript C designates the cell center.

As already mentioned in Sub-Chapter 3.1.2, classic finite-difference formulae can not be used for the estimation of the solution gradients. Two methods can be used instead, [9]: the *"path-integral" method* based upon the Green-Gauss theorem and the *least-square formulation*. Whereas the path-integral method would require less CPU-time and would be less complex for two-dimensional cases, it is more complex to implement for three-dimensional cases. Therefore, the least-square formulation has been implemented, because it is best suited for three-dimensional irregular grids. To estimate the gradient at a cell center, the solution is needed for all neighbouring cells. This means in all 26 neighbour directions: 6 faces, 12 edges and 8 corners of the cubic cell. It is sketched in Figure 4 for a two-dimensional quadtree where neighbours are needed in 8 neighbour directions (4 faces and 4 corners).

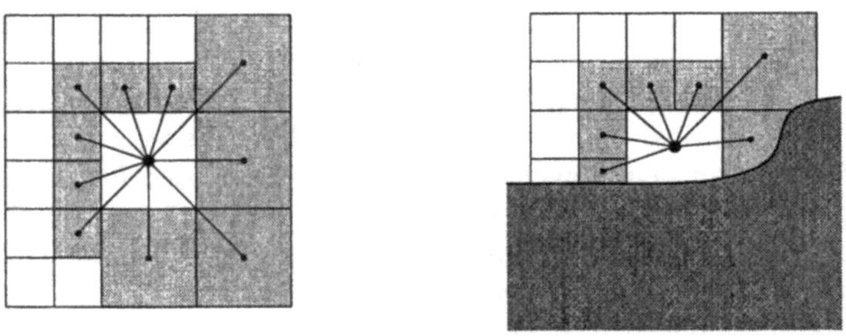

(a) Regular cell. (b) Cut-cell.

Figure 4 Neighbour cells used for the gradient calculation.

In the case of a cut-cell, neighbours are simply ignored which lie inside the body. The neighbour cut-cells are treated just as uncut cells. The method provides a good estimation of the gradient. Nonetheless, it is not guaranteed that it is suited for reconstructing the solution. Indeed, it can lead to under- or overestimations of the solution at cell faces. This means that the reconstructed piecewise linear solution will not be monotonic and will cause oscillations [21]. Prior to the reconstruction, the estimation of the gradients must be limited in order to fulfill the condition of monotonicity. For that purpose, the limiter function ϕ is introduced:

$$U = U^C + \phi \nabla U^C \cdot \Delta r.$$

The reconstructed solution must remain bounded by the values used for the least-square formulation. Whereas De Zeeuw [25],[9] and Coirier [23] test the values reconstructed at the cell corners, a more restrictive method has been developed which guarantees that neither under- nor overestimations will occur at cell faces. Instead, the solution is reconstructed and examined at the center of the cells used for the gradient estimation.

For cut-cells, a more restrictive expression is used: the gradients are limited in such a way that negative values of the reconstructed wall pressure and density are prevented, [23].

The flux functions are estimated by means of an approximate Riemann solver [26]. The formulation is based on the homogeneous acoustic wave theory and provides with an outstanding accuracy (vanishing entropy error). For best shock-capturing, Steger-Warming [27] splitted flux vectors are used and blended with the Riemann solution by means of the differentiable limiter defined in [26].

Because the grid is no more regular, the face of a cell can have more than one neighbouring cell on the other side (see Figure 4). Consequently, this feature has to be considered appropriately to ensure the flux conservation, and the consistency of the numerical scheme. The following procedure is used to treat such cell faces. The finer adjacent cells subdivide the coarser cell face into just as many fine faces. The solution is reconstructed onto each side of the fine face, and the corresponding flux functions are computed. As sketched in Figure 5, the resulting flux through the coarser cell face is obtained by summation of the individual flux functions over the impinging finer cells.

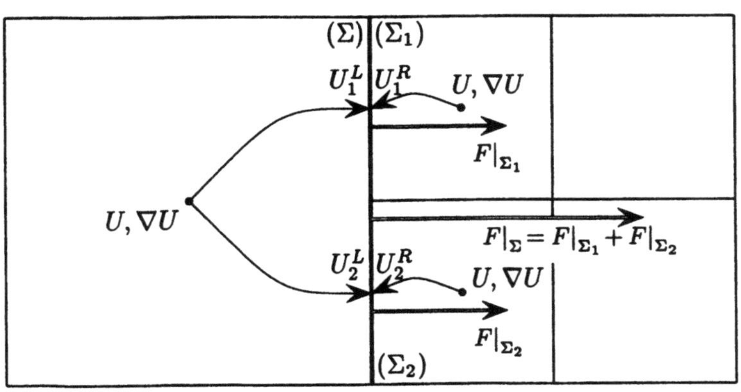

Figure 5 Flux conservation at refined cell faces.

3.1.4 Multi-Grid Strategy

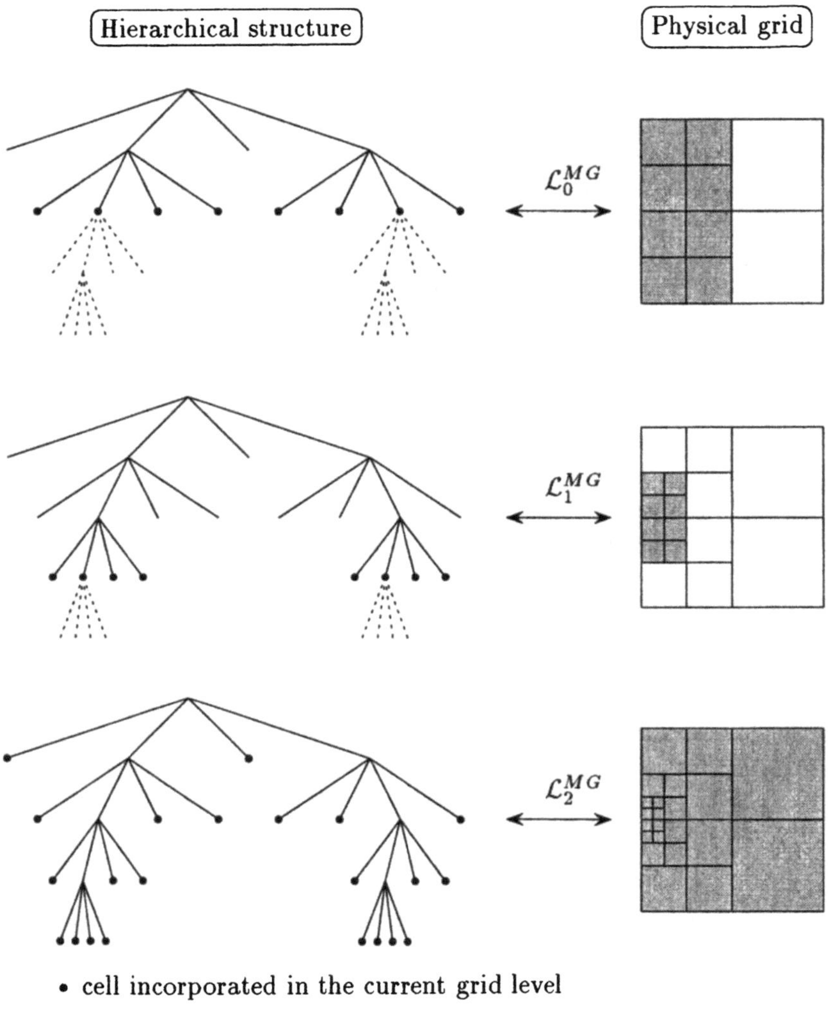

Figure 6 Definition of the multi-grid levels.

In addition to local time stepping, the multi-grid process is used to speed up the convergence process. The classic method consists in coarsening the cells in such a way that each coarse grid spreads over the whole computational domain. For structured grids, this consists in merging two adjacent cells in the three space-directions, i.e. eight cells. Because of the grid's irregular aspect, one cell can have one or four neighbouring cells at each cart- or inter-face. Therefore, a strategy taking advantage of this unfavourable feature has been considered. It is based upon the work of Mavriplis [28] and Connell [29] for unstructured meshes. Since nested coarse cells are already available in the hierarchical tree structure, they are used for the multi-grid process. Figure 6 shows an example. The level $\mathcal{L}_2^{MG}$ is the physical grid. It covers the whole computational domain. The coarser grids are defined by considering only the cells belonging to a given refinement level. Hence, the 8 finest cells of grid level $\mathcal{L}_2^{MG}$ are replaced by their parent cells. The coarser grid level $\mathcal{L}_1^{MG}$ is then completed by incorporating all the leaf cells of the new finest level. Similarly, the next coarser grid level $\mathcal{L}_0^{MG}$ is defined by replacing the cells of level $\mathcal{L}_1^{MG}$ with their parent cells and by incorporating the leaf cells belonging to the same refinement level.

The solution and the solution gradients in the cells at the boundary of these coarser grids are frozen. Thus, distinct problems are defined with particular boundary conditions. As a consequence, a *local* multi-grid strategy is defined. In the case of adaptively refined grids, strongly refined zones are zones where the transient and high frequencies of the error of the initial solution are dominant; poorly refined zones are zones where the transient and low frequencies of the solution error are weaker, and hence require less smoothing.

As can graphically be seen in Figure 6, by considering coarser and coarser grids according to the method, the subdomains expand more and more to occupy the whole computational domain. As a consequence, the frequency error to be damped by the smoothing time steps is lower and lower, and is smoothed over a wider and wider physical region. Since multi-grid is used to smooth the error of the solution for the possible wider frequency range, the method presented here fully respects, and even enhances the motivation for the multi-grid process.

Because of the tree structure, the solution transfers between grid levels only consist in following the tree branches, and are rather straightforward. Once the multi-grid levels have been defined for smoothing the solution, different cycles can be performed. The classic cycles have been implemented: V cycles, W cycles and saw-tooth cycles.

3.1.5 Code Organization

The multi-grid levels are defined as follows. The first (finest) level consists of all the flow-cells. It is the physical grid. The following coarser grids are successively defined by incorporating all cells located at a given refinement level (see Figure 6). For each grid level, informations necessary for the gradient and flux calculation are determined and stored in an object structure. These are:

- for each cell, the cells located at each one of the 6 cart-faces for the flux calculations,
- for each cell, the neighbouring cells in all 26 directions (face-, edge-, and vertex-adjacent) necessary for the gradient estimations,

- the list of the cells located at the domain boundary and requiring special treatment (inlet/outlet conditions, far field conditions, symmetry conditions, ...),
- for each cut-cell, the list of cut-faces describing the solid body intersection; it is built up by traveling through the tree-branch starting at the current cell.

This structure seems to require much memory and contain redundancies, but it improves efficiency. Indeed, changing the grid level during the solution process only requires to point at new appropriate arrays in the object structure, i.e. pointer or memory swapping. Moreover, this enables a fully vectorized flux computation. Because the cart-face normal directions are fixed and correspond to the three space directions, the inviscid fluxes are simplified and fully Cartesian formulated. Therefore, there are three generic treatments. The different loops can then be distributed on parallel processors to achieve multitasking and hence, to further reduce the execution CPU-time. Because of the irregular aspect of the grids, the number of neighbouring cells used for the least-square formulation is highly variable. It can vary between 3 and 56, considering the extreme case of a cut-cell with inter-faces in three directions and cart-faces in the remaining directions, where the neighbour cells are smaller in all 26 directions. Therefore, the gradient calculation can not be vectorized in an efficient way. Nonetheless, it can be parallel-tasked, giving a further speed-up.

3.2 Modified Unstructured Inviscid Flow Solver

The compressible Euler equations are solved with a cell-centered finite volume solver, which is of Jameson type, [30] and [31]. In order to achieve numerical stability, a blend of second and fourth differences in the flow variables is added to the convective fluxes. The integration to steady state is done with a fully explicit five-stage hybrid time stepping scheme of Runge-Kutta type. Convergence to steady state is accelerated by using the maximum permissible time step for every cell as determined by the local stability analysis (local time stepping scheme). The permissible time step is increased by applying implicit residual smoothing. Also enthalpy damping is used to accelerate the convergence.

The original flow solver is modified to take benefit of the simplified geometry of the Cartesian cells. It is worth mentioning, that the modified solver keeps the attribute of full vectorization. This is achieved by sorting all cell faces.

4 Solution-Based Grid Adaptation

The initial Cartesian grid is only of rough quality: the grid resolution decreases quickly with increasing distance to the body. Several solution adaptation loops are applied to achieve a nearly optimum grid, which provides a reasonable resolution of all flow characteristics with a minimum number of cells.

4.1 Adaptation Criterion

Fischer [32] and Hentschel et al. [33] studied several adaptation criteria for surface-oriented structured grids. The adaptation criterion applied in this paper is, however, similiar to those of [9] and [3], because it is better suited for Cartesian grids. The criterion consists of the three parameters: divergence and curl of velocity and the strength of the numerical entropy. The divergence of velocity is used to find shock waves and the curl of velocity to locate shear layers. High values of entropy tend to indicate grid areas that are simply under-resolved. For each cell, weighted forms of the divergence, τ_d, and curl, τ_c, of the velocity and the strength of the numeric entropy, τ_e, are computed:

$$\tau_{d_i} = |\nabla \cdot \vec{v_i}|\, L_i^{\frac{3}{2}}, \quad \tau_{c_i} = |\nabla \times \vec{v_i}|\, L_i^{\frac{3}{2}}, \quad \tau_{e_i} = \left|\nabla p_i - a_i{}^2 \nabla \rho_i\right| L_i^{\frac{3}{2}}.$$

L_i is the length-scale for each cell. The length-scale weight is used to find weaker features, which are in a coarser area of the flow. This allows the weaker features to be refined when the stronger features have been resolved, [9]. In order to set a threshold for refinement and coarsening, the standard deviation about zero is computed for each parameter of the criterion, with n the number of cells in the flow field:

$$\sigma_d = \sqrt{\frac{\sum_{i=1}^{n} \tau_{d_i}{}^2}{n}}, \quad \sigma_c = \sqrt{\frac{\sum_{i=1}^{n} \tau_{c_i}{}^2}{n}}, \quad \sigma_e = \sqrt{\frac{\sum_{i=1}^{n} \tau_{e_i}{}^2}{n}}.$$

A cell is flagged for refinement, if the condition holds:

$$\tau_{d_i} > \lambda_{d_{refine}} \cdot \sigma_d \quad or \quad \tau_{c_i} > \lambda_{c_{refine}} \cdot \sigma_c \quad or \quad \tau_{e_i} > \lambda_{e_{refine}} \cdot \sigma_e.$$

On the other hand, a cell is flagged for coarsening, if the following condition is satisfied:

$$\tau_{d_i} < \lambda_{d_{coarse}} \cdot \sigma_d \quad and \quad \tau_{c_i} < \lambda_{c_{coarse}} \cdot \sigma_c \quad and \quad \tau_{e_i} < \lambda_{e_{coarse}} \cdot \sigma_e.$$

The threshold values are usually set to $\lambda_{x_{refine}} = 1.0$ and $\lambda_{x_{coarse}} = 0.1$. But these values can be changed by the user to encourage or discourage adaptation around specific types of features, for example shock waves or shear layers.

5 Sample Calculations

5.1 Two-Dimensional Test Cases - Modified Unstructured Inviscid Flow Solver

The unstructured Euler flow solver is applied to two-dimensional test cases. They consider the flows over a NACA 0012 airfoil. The airfoil contour is discretized with 230 points. For all cases the downstream, upstream, upper and lower boundaries are set at 100 chords from the airfoil.

5.1.1 NACA 0012 Airfoil ($M_\infty = 0.95$, $\alpha = 0.0°$)

The first condition imposed on the NACA 0012 airfoil is Mach number $M_\infty = 0.95$ and angle of attack $\alpha = 0.0°$. The flow has two oblique shocks emanating from the trailing edge with a weak normal shock in the wake. The initial grid in Figure 7(a) contains 1277 cells, of which 261 cells are intersected by the surface. After 7 adaptation steps, the grid is refined strongly in the regions of the shocks and consists of 26207 cells, of which 281 cells are cut-cells, Figure 7(b).

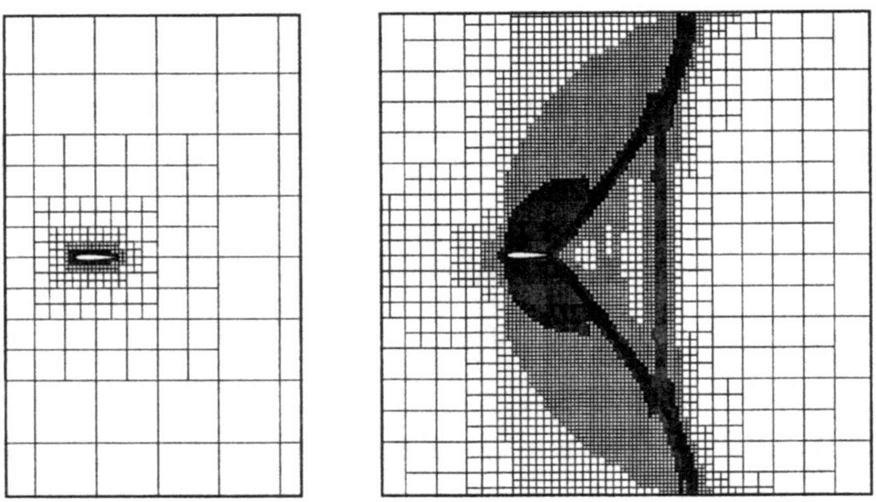

(a) Initial grid (inner part). (b) Adapted grid (inner part) after 7 adaptation steps.

Figure 7 Grids for NACA 0012 airfoil ($M_\infty = 0.95$, $\alpha = 0.0°$).

The Mach contours are presented in Figure 8(a). The two small subfigures show the grid and the Mach contours behind the trailing edge in a strongly magnified resolution. The pressure coefficient distribution is compared with computional data from [34], which are drawn as symbols in Figure 8(b). The produced drag coefficient is computed to $C_D = 0.11$. In comparison the drag coefficient in [34] is in the range between 0.1078 and 0.1086. This values are slightly smaller, because the methods are not adaptive and therefore especially the shock in the wake is not as well resolved as in this paper.

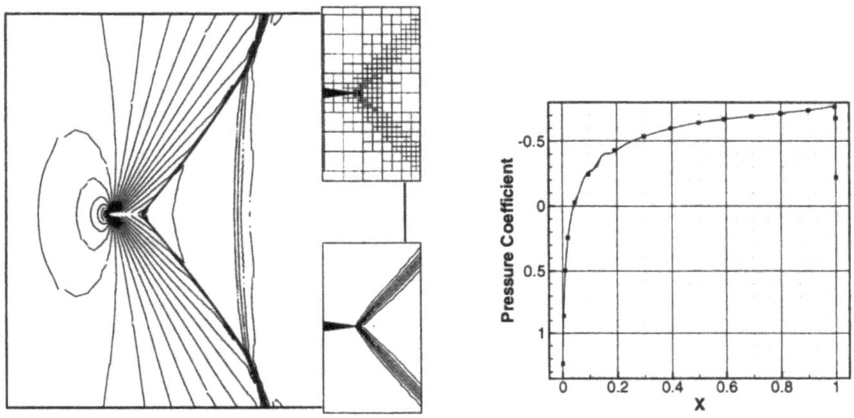

(a) Mach contours. (b) Pressure coefficient.

Figure 8 Inviscid flow results for NACA 0012 airfoil ($M_\infty = 0.95$, $\alpha = 0.0°$).

5.1.2 NACA 0012 Airfoil ($M_\infty = 1.2$, $\alpha = 7.0°$)

Next the flow solution is obtained for a NACA 0012 airfoil at Mach number $M_\infty = 1.2$ and angle of attack $\alpha = 7.0°$. At this supersonic condition a strong bow shock in front of the airfoil, an oblique shock and a shear layer characterize the flow, of which the latter two are emanating from the trailing edge. The initial grid has 1309 cells (285 cut-cells). Figure 9 shows the grid after 5 adaptation steps, which then contains 31885 cells, of which 351 cells are intersected by the body. The cell sizes in the region in front of the bow shock increase quickly: three chord lenghts in front of the airfoil, the cells are already much larger than the entire airfoil. The shear layer is refined strongly like the shocks.

In Figure 10(a) the Mach contours are drawn, and the pressure coefficient distribution is given in Figure 10(b) together with computional data from [34] for comparison drawn as symbols. The calculated aerodynamic coefficients are: $C_L = 0.524$, $C_D = 0.155$ and $C_M = -0.108$. They are in good aggrement with the computational results in [34]. For example the structured Euler calculation on the O-type mesh with 320×64 cells yields: $C_L = 0.5138$, $C_D = 0.1538$ and $C_M = -0.1087$.

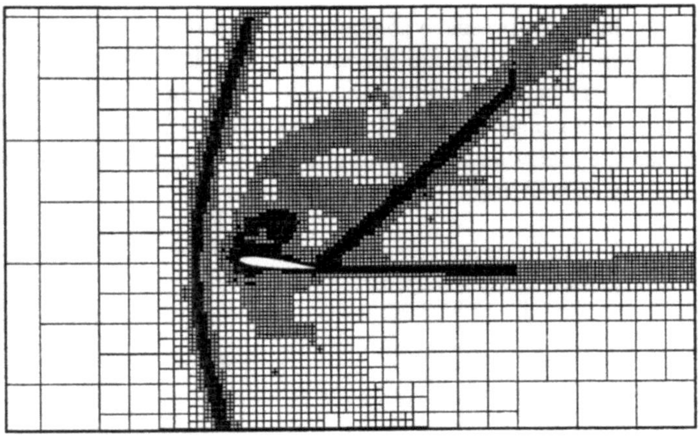

Figure 9 Grid (inner part) for NACA 0012 airfoil after 5 adaptation steps ($M_\infty = 1.2$, $\alpha = 7.0°$).

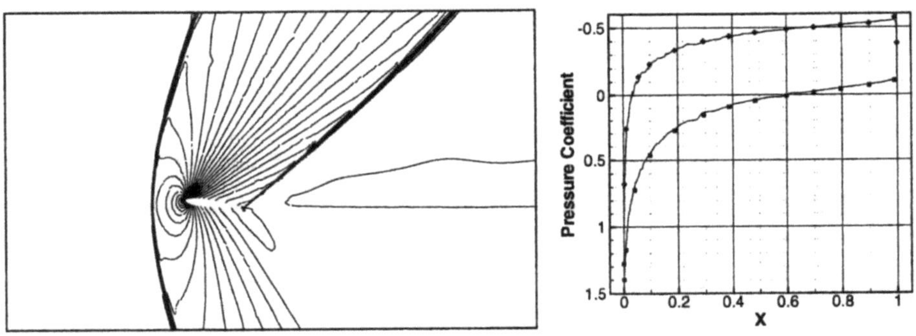

(a) Mach contours. (b) Pressure coefficient.

Figure 10 Inviscid flow results for NACA 0012 airfoil ($M_\infty = 1.2$, $\alpha = 7.0°$).

5.2 Three-Dimensional Test Cases - New Flow Solver

The test case has been computed using the supercomputer CRAY-J90 with 12 PVP processors of the University Valenciennes. The subsonic flow past a sphere at Mach number $M_\infty = 0.3$ is studied. An analytical solution is available for incompressible flow by inviscid incompressible flow theory. The pressure distribution is symmetric with respect to both the main and the transverse flow directions. For the computation, the surface discretization consists of 6080 triangles. All boundaries are set at 30 diameters from the sphere. The initial grid contains 20745 flow-cells with 7483 cut-cells. The solution process performs 50 multi-grid V-cycles with 5 coarser levels and 5 smoothing time steps on each level. Before multi-grid cycles, 5 time steps are performed on the fine grid, also. After the initial flow calculation, the Cartesian grid is adapted twice. Figures 11(a) and 11(b) show respectively the initial grid and the grid obtained after the second adaptation step.

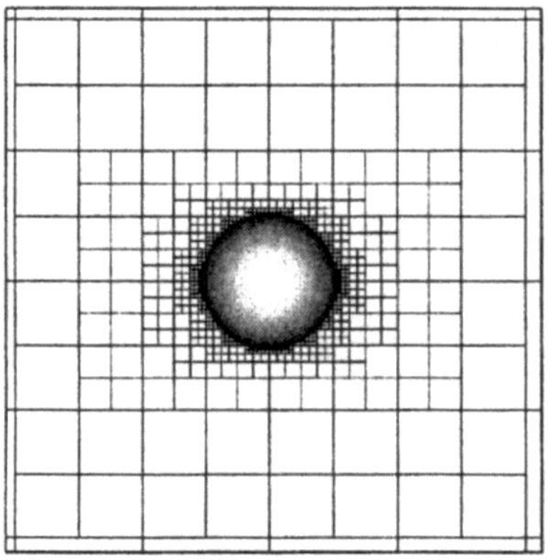

(a) Initial grid (inner part).

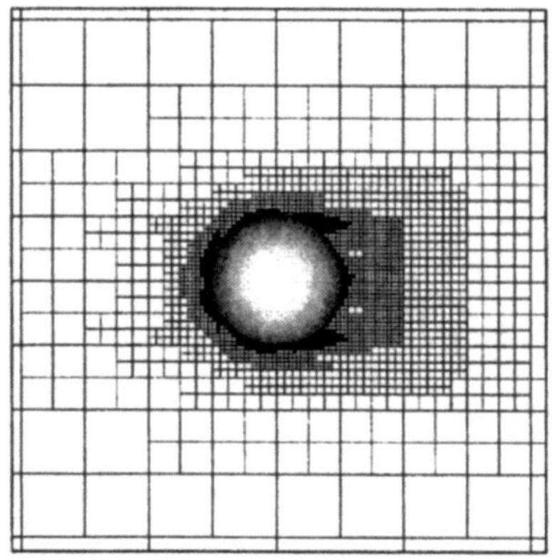

(b) Grid (inner part) after 2 adaptation steps.

Figure 11 Grids for the sphere ($M_\infty = 0.3$).

The number of flow-cells and cut-cells for each multi-grid level are summarized in Table 1 for each step in the adaptation process.

Table 1 Number of flow-cells and cut-cells for each multi-grid level.

Multigrid level	Number of flow-cells / Number of cut-cells		
	Initial grid	Adapted grid #1	Adapted grid #2
$\mathcal{L}_5$	20745 / 7483	41234 / 11517	112994 / 22444
$\mathcal{L}_4$	928 / 384	1308 / 684	828 / 508
$\mathcal{L}_3$	3332 / 2320	12001 / 7686	6885 / 4689
$\mathcal{L}_2$	10026 / 5122	17852 / 5312	56237 / 18359
$\mathcal{L}_1$	4592 / 1328	8888 / 1328	40220 / 5312
$\mathcal{L}_0$	1760 / 344	3632 / 344	15896 / 1328

The resulting pressure contours are plotted in Figures 12(a) and 12(b) respectively for the initial coarse grid and the grid obtained after 2 adaptation steps.

As far as the initial calculation is concerned, the 50 multi-grid cycles have reduced the residual by 3 orders of magnitude. The overpressure at the upstream stagnation point and the underpressure zone are well resolved, whereas the overpressure at the downstream stagnation point is still poorly resolved. The effect of adaptation is obvious. For the calculation after the second adaptation step, the residual has only been reduced by 2 orders of magnitude. The overpressure at the downstream stagnation point is much better resolved and the pressure in the other zones is improved. The solution is not symmetric with respect to the transverse direction, but adaptation has helped improving it. Nonetheless, the solution depicted here has not reached the steady state. The strategy of calculation has to be improved, in order to take the change in the number of flow-cells into account. Since adaptation tends to increase the number of flow-cells, the number of multi-grid cycles needed for achieving a comparable reduction of the residual increases, too.

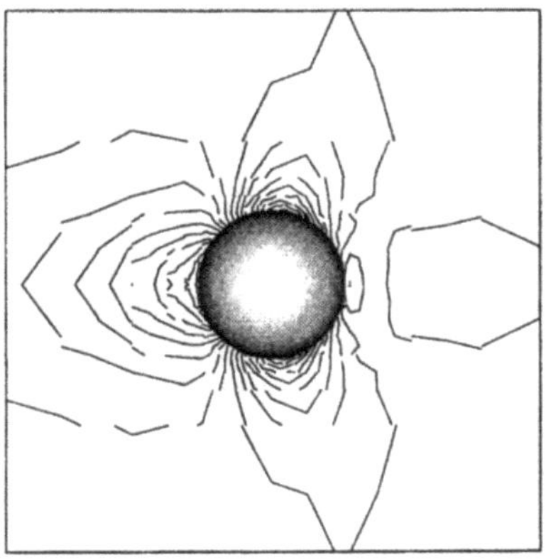

(a) Initial grid.

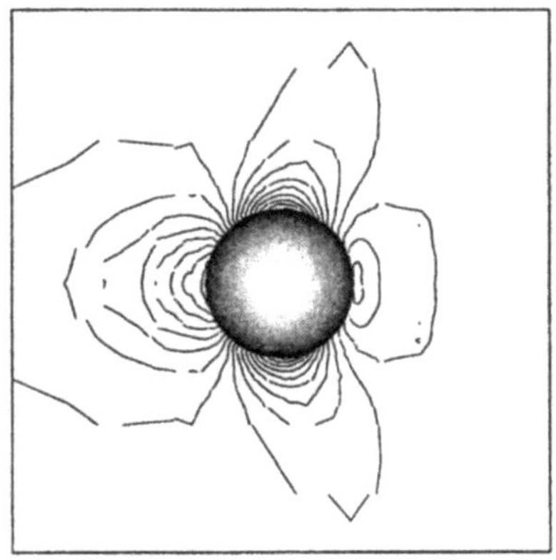

(b) Grid after 2 adaptation steps.

Figure 12 Pressure contours for the sphere ($M_\infty = 0.3$).

The solution obtained with the modified unstructured flow solver presents a better symmetry, see [12]. The grid used for that computation was generated with a much finer surface triangulation, and contained 150000 flow-cells with 74300 cut-cells, whereas the present adapted grid contains 110000 flow-cells with only 22000 cut-cells. The number of cut-cells obviously is much smaller, giving significantly coarser cells in the vicinity of the surface. Hence, the flow is potentially not well enough resolved. Consequently, the adaptation process should be performed with more steps to obtain a comparable flow resolution.

5.3 Three-Dimensional Test Cases - Modified Unstructured Inviscid Flow Solver

The two presented three-dimensional test cases are performed with the unstructured flow solver on the supercomputer NEC-SX4/32. They finally demonstrate the ability of the adaptive Cartesian grid system to handle complicated three-dimensional geometries with strong refinement. For both test cases all boundaries are set at 30 chords from the wing. A symmetry plane is used for the calculations.

5.3.1 ONERA-M6 Wing ($M_\infty = 0.84$, $\alpha = 3.06°$)

The flow field around the ONERA-M6 wing is computed at the standard test condition of Mach number $M_\infty = 0.84$ and angle of attack $\alpha = 3.06°$. It causes two shocks on the upper side of the wing. The first shock is parallel to the leading edge and builds together with the second shock the characteristic lambda configuration. At the side-tip of the wing, a smaller lambda configuration of shocks arises. The surface is discretized very finely with 103279 triangles. The initial grid contains 200859 cells inside the flow domain, of which 90075 cells are intersected by the wing surface. The cpu-time for the entire grid generation is only 95 seconds on a workstation with R10000 processor, including the data input, geometry preparation and the sorting procedures in order to allow vectorization. After three adaptation steps the grid consists of 1517356 cells (148119 cut-cells). Figure 13 shows the adapted grid, which is refined strongly in the regions of the shocks and the shear layer behind the wing.

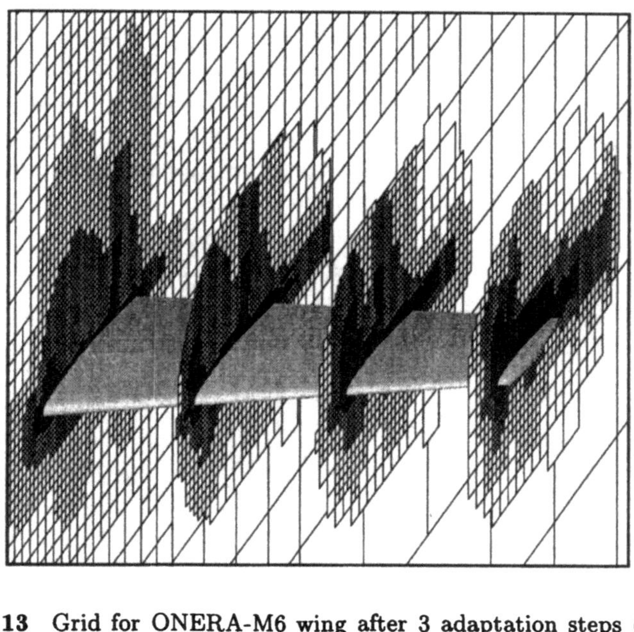

Figure 13 Grid for ONERA-M6 wing after 3 adaptation steps ($M_\infty = 0.84$, $\alpha = 3.06°$).

In Figure 14 the Mach contours are drawn. Especially the large two shocks forming the characteristic lambda configuration are well resolved. Besides, even the smaller lambda configuration of shocks is recognizable near the side-tip.

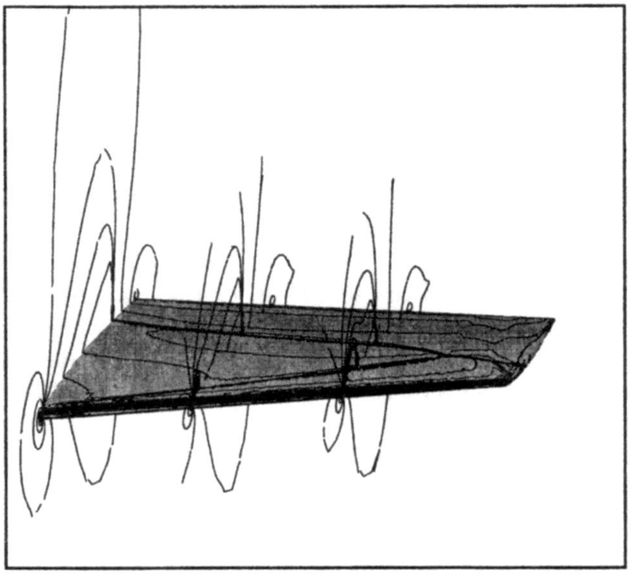

Figure 14 Mach contours for ONERA-M6 wing ($M_\infty = 0.84$, $\alpha = 3.06°$).

The distribution of the pressure coefficient is compared with experimental data [35], drawn as symbols in Figure 15. The differences between the computational results and the experimental data are mainly caused by the neglection of viscous effects in the solutions: the upper-side boundary layer effectively decambers the wing and shifts the shock positions in upstream direction. Also the pressure dips behind the shocks are typical for inviscid results.

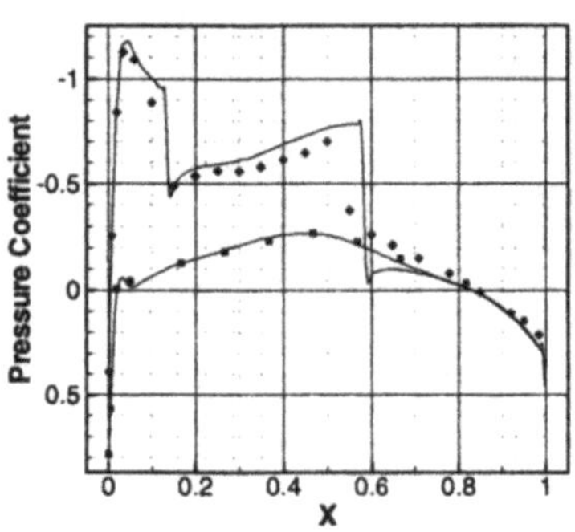

(a) $z = 44\%$ spanwise direction.

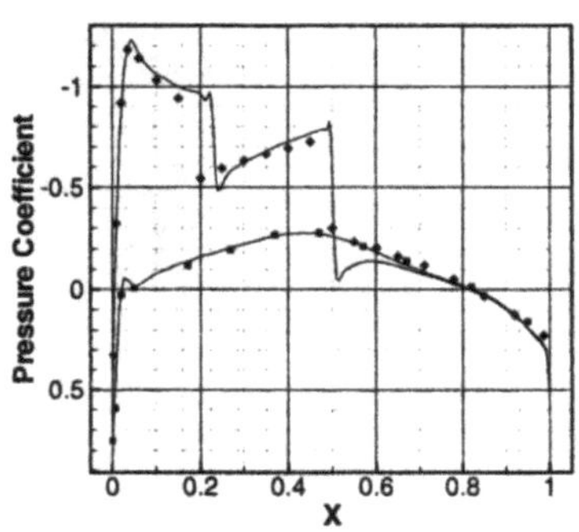

(b) $z = 65\%$ spanwise direction.

Figure 15 Pressure coefficients for ONERA-M6 wing ($M_\infty = 0.84$, $\alpha = 3.06°$).

The calculated aerodynamic coefficients are: $C_L = 0.290$, $C_D = 0.0117$ and $C_M = -0.08527$. They are within the range of previously computed results listed in [12].

5.3.2 Wing-Pylon-Nacelle Configuration ($M_\infty = 0.85$, $\alpha = 3.0°$)

The flow around a wing-pylon-nacelle configuration (DASA in-house test case) is computed as the second test case. The surface of the configuration consists of 80339 triangles. The condition imposed is Mach number $M_\infty = 0.85$ and angle of attack $\alpha = 3.0°$. The initial grid contains 204725 cells inside the flow domain, of which 86739 cells are intersected by the surface. Related to a R10000 processor the entire generation of the initial grid takes 90 seconds, including the data input, geometry preparation and the sorting procedures in order to allow vectorization. After four adaptation steps the grid consists of 1758089 cells, of which 138761 cells are cut-cells. The maximum memory used by the program is 1002 MByte. The grid is refined strongly in regions of the shocks and of the shear layer behind the wing, Figure 16(a). Figure 16(b) shows the nacelle, which is open for the flow. Inside the symmetric nacelle the flow is smooth, thus the cells remain coarse.

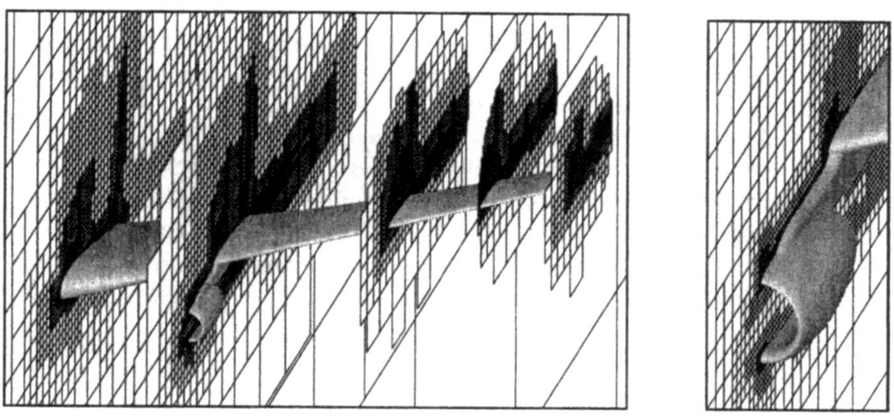

(a) Wing-pylon-nacelle configuration. (b) Nacelle.

Figure 16 Grids for wing-pylon-nacelle configuration after 4 adaptation steps ($M_\infty = 0.85$, $\alpha = 3.0°$).

The Mach contours are presented in Figure 17. Especially the strong shock near the trailing edge is well resolved.

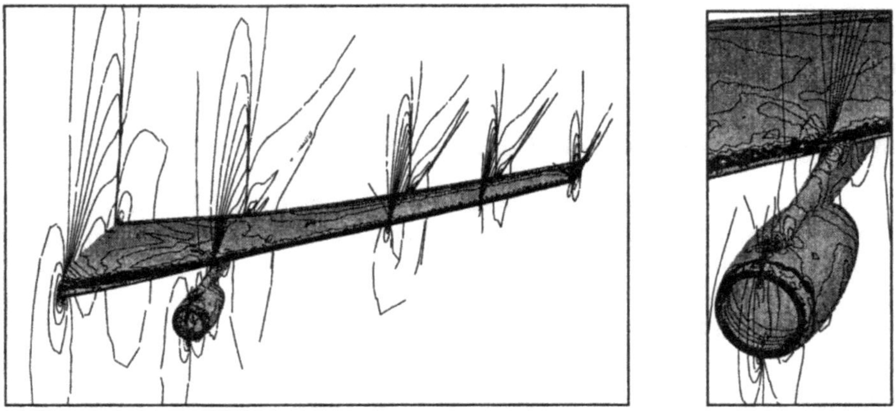

(a) Wing-pylon-nacelle configuration. (b) Nacelle.

Figure 17 Mach contours for wing-pylon-nacelle configuration ($M_\infty = 0.85$, $\alpha = 3.0°$).

6 Future Work

Work is underway to implement anisotropic grid refinement in order to reduce further the number of cells. In a post-processing step of the grid generation appropriate cells are identified and are merged together, [36]. The error estimators of Dompierre et al. [37] are considered.

Next the extension to viscous flows is planned with a hybrid approach. The boundary layer is discretized with a prismatic grid, which is cut out of the background Cartesian grid, [2] and [4]. Also the full Navier-Stokes equations are to be implemented into the flow solvers.

Finally, the entire Cartesian grid system will be extended to unsteady flows, including problems of moving and deforming bodies: these parts of the body are simply cut out of the background Cartesian grid again.

7 Conclusion

A Cartesian grid generator has been developed, which is based on the octree-data structure. Grids are created automatically even for complicated geometries. Two different Euler flow solvers are applied. The first flow solver is developed recently and benefits directly from the octree-data structure. Whereas the second solver represents a classic unstructured solver. The grid generation and the flow solver are embedded in an adaptation cycle.

The presented Cartesian grid system runs automatically without user intervention. In this sense a self-organizing Cartesian grid-generation system is on its way.

Acknowledgements

The authors would like to thank the CNRS and the DFG for grants under which the project "Self-Organizing Cartesian Grid-Generation System" is supported, embedded in the French- German project "Numerische Strömungssimulation - Simulation Numérique d'Ecoulements".

References

[1] E.H. Hirschel and W. Schwarz. "Mesh Generation for Aerospace CFD Applications". In Surveys on Mathematics for Industry, Volume 4, 1995, pp. 249-265.

[2] S.L. Jr. Karman. "Unstructured Cartesian/Prismatic Grid Generation for Complex Geometries". In Surface Modeling, Grid Generation and Related Issues in Computational Fluid Dynamics (CFD) Solutions. NASA CP-3291, 1995, pp. 251-270.

[3] E.F. Charlton. "An Octree Solution to Conservation-Laws over Arbitrary Regions (OSCAR) with Application to Aircraft Aerodynamics". PhD thesis, University of Michigan, 1997.

[4] Z.J. Wang, G.S. Hufford, and A.J. Przekwas. "Adaptive Cartesian/Adaptive Prism (ACAP) Grid Generation for Complex Geometries". AIAA-97-0860, 1997.

[5] J.J. Quirk. "An Alternative to Unstructured Grids for Computing Gas Dynamic Flows around Arbitrarily Complex Two-Dimensional Bodies". In NASA Contractor Report 189612. ICASE Report No. 92-7, 1992.

[6] J.J. Quirk. "A Cartesian Grid Approach with Hierarchical Refinement for Compressible Flows". In Computational Fluid Dynamics '94. John Wiley & Sons Ltd., 1994, pp. 200-209.

[7] M.J. Aftosmis. "Solution Adaptive Cartesian Grid Methods for Aerodynamic Flows with Complex Geometries". In Lecture Series 1997-02 (Computational Fluid Dynamics). Von Kármán Institute for Fluid Dynamics, 1997.

[8] J.E. Melton. "Automated Three-Dimensional Cartesian Grid Generation and Euler Flow Solutions for Arbitrary Geometries". PhD thesis, University of California, 1996.

[9] D.L. De Zeeuw. "A Quadtree-Based Adaptively-Refined Cartesian-Grid Algorithm for Solution of the Euler Equations". PhD thesis, University of Michigan, 1993.

[10] W.J. Coirier. "An Adaptively-Refined, Cartesian, Cell-Based Scheme for the Euler and Navier-Stokes Equations". PhD thesis, University of Michigan, 1994.

[11] Z.J. Wang. "A fast nested Multi-Grid Viscous Flow Solver for Adaptive Cartesian/Quad Grids". AIAA-96-2091, 1996.

[12] E.H. Hirschel and F. Deister. "Extension of Quirk's Approach to Three Dimensions". Internal Report IB 97-4, Universität Stuttgart, I.A.G., 1997.

[13] D. Rocher. "Technical Memorandum: Project Self-Organizing Cartesian Grid-Generation System". Intermediate Progress Report, Université de Valenciennes, L.M.E., 1997.

[14] H. Samet. "Hierarchical Data Structures and Algorithms for Computer Graphics - Part1: Fundamentals". In IEEE Computer Graphics and Applications, Volume 8(3), 1988, pp. 48-68.

[15] H. Samet. "Hierarchical Data Structures and Algorithms for Computer Graphics - Part2: Applications". In IEEE Computer Graphics and Applications, Volume 8(4), 1988, pp. 59-75.

[16] K. Fujimura and T.L. Kunii. "A Hierarchical Space Indexing Method". In T.L. Kunii, editor, Computer Graphics: Visual Technology and Art, Proceedings of Computer Graphics Tokyo '85 Conference. Springer Verlag, 1985, pp. 21-34.

[17] H. Samet. "Neighbor Finding in Images Represented by Octrees". In Computer Vision, Graphics and Image Processing 46, 1989, pp. 367-386.

[18] I.E. Sutherland and G.W. Hodgeman. "Reentrant Polygon Clipping". In Communications of the ACM, Volume 17(1), 1974, pp.32-42.

[19] J.D Foley, A. van Dam, S.K. Feiner, and J.F Hughes. "Computer Graphics - Principles and Practice". Addison Wesley, 2nd edition, 1996.

[20] D.J. Mavriplis. "Unstructured Mesh Generation and Adaptivity". In NASA Contractor Report 195069. ICASE Report No. 95-26, 1995.

[21] C. Hirsch. "Numerical Computation of Internal and External Flows, Volume 2: Computational Methods for Inviscid and Viscous Flows". John Wiley & Sons Ltd., 1990.

[22] P.W. Hemker and B. Koren. "Standard Multigrid Techniques for CFD". In Lecture Series 1997-02 (Computational Fluid Dynamics). Von Kármán Institute for Fluid Dynamics, 1997.

[23] W.J. Coirier. "An Adaptively-Refined, Cartesian, Cell-Based Scheme for the Euler and Navier-Stokes Equations". NASA TM-106754, 1994.

[24] F. Deister. "Analyse und Implementierung einer algebraisch orientierten Multigrid-Technik zur Konvergenzbeschleunigung eines unstrukturierten Eulerverfahrens". Diplomarbeit, Technische Hochschule Darmstadt (Germany), 1996.

[25] D.L. De Zeeuw and K.G. Powell. "An Adaptively Refined Cartesian Mesh Solver for the Euler Equations". Journal of Computational Physics, Volume 104, 1993, pp. 56-68.

[26] A. Eberle, M. Schmatz, and N. Bissinger. "Generalized Flux Vectors for Hypersonic Shock-Capturing". AIAA-90-0390, 1990.

[27] J.L. Steger and R.F. Warming. "Flux Vector Splitting of the Inviscid Gasdynamic Equations with Application to Finite-Difference Methods". Journal of Computational Physics, Volume 40, 1981, pp. 263-293.

[28] D.J. Mavriplis. "Multigrid Solution Strategies for Adaptive Meshing Problems". In NASA Contractor Report 195049. ICASE Report No. 95-14, 1995.

[29] S.D. Connell and D.G. Holmes. "Three-Dimensional Unstructured Adaptive Multigrid Scheme for the Euler Equations". AIAA Journal, Volume 32(8), 1994, pp. 1626-1632.

[30] A. Jameson, T.J. Baker, and N.P. Weatherill. "Calculation of Inviscid Transonic Flow over a Complete Aircraft". AIAA-86-0103, 1986.

[31] A. Jameson and T.J. Baker. "Improvements to the Aircraft Euler Methods". AIAA-87-0452, 1987.

[32] J. Fischer. "Sensors for Self-Adapting Grid Generation in Viscous Flow Computations". Volume 35, Notes on Numerical Fluid Mechanics, Vieweg, Braunschweig/Wiesbaden, 1992, pp. 365-375.

[33] R. Hentschel and E.H. Hirschel. "Self Adapted Flow Computations on Structured Grids". In Computational Fluid Dynamics '94. John Wiley & Sons Ltd., 1994, pp. 242-249.

[34] D.J. Jones. "Reference Test Cases and Contributors", Chapter 5. AGARD-AR-211, 1985.

[35] V. Schmitt and F. Charpin. "Pressure Distributions on the ONERA-M6 Wing at Transonic Mach Numbers", Appendix B1. AGARD-AR-138, 1979.

[36] F. Deister, D. Rocher, E.H. Hirschel, and F. Monnoyer. "Three-Dimensional Adaptively Refined Cartesian Grid Generation and Euler Flow Solutions for Arbitrary Geometries". In Computational Fluid Dynamics '98. John Wiley & Sons Ltd., to appear in 1998.

[37] J. Dompierre, M.-G. Vallet, Y. Fortin, M. Bourgault, and W.G. Habashi. "Anisotropic Mesh Adaptation: Towards a Solver and User Independent CFD". AIAA-97-0861, 1997.

COMBINED FINITE VOLUME AND SMOOTHED PARTICLE METHOD

A. Egelja, D. Kröner, R. Schwörer, N. Lanson*, M. Mancip*, J.P. Vila*

Institute for Applied Mathematics
Hermann-Herder-Str. 10
Freiburg University
D 79104 Freiburg

* Math. for Industry and Physics, UMR 5640
INSA, Domaine de Rangueil
F 31077 Toulouse Cedex

Summary

We want to develop an algorithm for the simulation of the cooling compressible flow in a given geometry of a real apparatus and for the free jet. The aim of this apparatus is to get a uniform jet of metal ions outside of this tube. Unfortunately one part of these ions will remain in the tube, which is bad for getting a uniform jet for the continuous coating process. Obviously the amount of these particles will depend on the geometry of the whole configuration. For the simulation of the flow inside the tube we will use a compressible Navier-Stokes solver and for the free jet the SPH method (Smoothed Particle Hydrodynamics). For the design of the physical system the algorithm should be used to study the influence of different geometries to the transport of the metal ions.

1 Introduction and the description of the physical problem

Let us start with the description of the physical experiment. We consider a cylindrical tube as shown in Figure 1.

By an electric light arc ions are resolved from the surface of a heated plate of metal. A cooling compressible flow of Argon coming from the left will transport these ions through the tube to a small hole at the right end of the tube. The ions which will leave the tube will built a free jet which is used to put a thin layer of metal to some surfaces. The atmosphere around the free jet consists of Helium. The realistic relations of the geometry are shown in Figure 7.

The paper is organized as follows. After the Introduction Section 2 is devoted to a detailed presentation of the discretization of the interior flow (mathematical

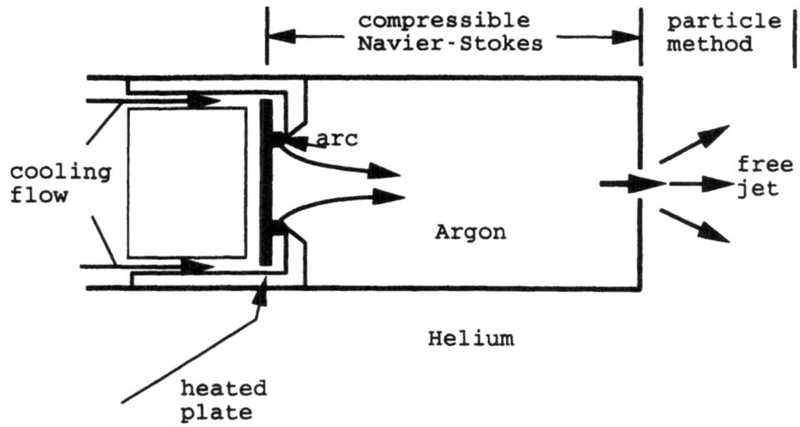

Figure 1 Testing plant

model, discretization of convective and viscous terms, comparison of different Riemann solvers on different grids, numerical results). In Section 3 the SPH method will be presented (discrete weak formulation, renormalization, Godunov type particle approximations). In the last part, Section 4, we will give some first numerical results for the combination of the Navier-Stokes solver in the interior of the tube and the particle method in the free jet.

2 Discretization of the interior flow

In this section we shall describe the mathematical model for the interior flow and the discretization of the system of the compressible Navier-Stokes equations, in particular the discretization of the viscous and convective terms. We will study a comparison between different Riemann solvers on different grids for a flow over a flat plate. The results of the comparison imply that the AUSMDV Riemann solver seems to be the most adapted one for our applications. Numerical results for thus will be given in the last part of this Section.

2.1 The mathematical model for the interior flow

As the mathematical model for this flow we take the system of the compressible Navier Stokes equations:

$$\partial_t \begin{pmatrix} \rho \\ \rho u \\ \rho v \\ e \end{pmatrix} + \partial_x \begin{pmatrix} \rho u \\ \rho u^2 + p \\ \rho uv \\ u(e+p) \end{pmatrix} + \partial_y \begin{pmatrix} \rho v \\ \rho uv \\ \rho v^2 + p \\ v(e+p) \end{pmatrix} -$$

$$\partial_x \begin{pmatrix} 0 \\ \tau_{11} \\ \tau_{12} \\ \tau_{11} u + \tau_{12} v + k\partial_x \Theta \end{pmatrix} - \partial_x \begin{pmatrix} 0 \\ \tau_{21} \\ \tau_{22} \\ \tau_{21} u + \tau_{22} v + k\partial_y \Theta \end{pmatrix} = 0 \quad (2.1)$$

where u and v are the components of the velocity, ρ the density, p the pressure, e total energy density, τ_{ij} the components of the viscous part of the stress tensor, Θ the absolute temperature and k is the heat conductivity. We will consider this system in the computational domain, shown in Figure 6.

On the inflow part of the boundary at the left end of the geometry we have prescribed a parabolic velocity profile and fixed pressure. On the walls of the container no-slip boundary condition have been assumed for the velocities and the temperature is prescribed (see Figure 7). On the outflow part of the boundary, which consists only of the hole at the right end of the tube we have prescribed the pressure. From the given data we expect a Reynolds-number of about 20000 and a Mach-number of 0.5.

2.2 The discretization

We discretize the system (2.1) of the compressible Navier – Stokes equation by an explicit first order finite volume scheme with a suitable Riemann solver for the convective terms.
The finite volume scheme for (2.1) can be written as

$$U_j^{n+1} := U_j^n + \frac{\Delta t}{|T_j|} \sum_{l=1}^{3} G_{jl}(U^n) - \frac{\Delta t}{|T_j|} \sum_{l=1}^{3} g_{jl}(U_j^n, U_l^n) \quad (2.2)$$

where U_j^n is assumed to become an approximation of the exact solution $(\rho, \rho u, \rho v, e)$ on the cell T_j at time t^n and $U^n := (U_j^n)_{j\in N}$. For the numerical flux $g_{jl}(U,V)$ we have chosen the AUSMDV Riemann solver, which will be defined in the following section. The discretization of the viscous terms G_{jl} will be described in Section 2.4.

2.3 The discretization of the convective terms by the AUSMDV Riemann solver

A flux splitting scheme (AUSMDV) is recognized as an improved advection upstream splitting method (AUSM) by Liou and Steffen [16]. The AUSMDV is a mixture of AUSMD and AUSMV, which will be defined in the following.

AUSMD scheme: The convective numerical flux is defined by:

$$g_{jl}(U_1,U_2) := \frac{1}{2}\left[(\rho u)_{1/2}(\psi_1+\psi_2) - |(\rho u)|_{1/2}(\psi_2-\psi_1)\right] + p_{1/2} \tag{2.3}$$

where $U_i = (\rho_i, \rho_i u_i, \rho_i v_i, e_i)$, $i = 1{,}2$;$(\rho u)_{1/2} := u_1^+ \rho_1 + u_2^- \rho_2$ is the mass flux.

$$u_1^+ := \begin{cases} \alpha_1 \left\{\frac{(u_1+c_m)^2}{4c_m}\right\} + (1-\alpha_1)\frac{u_1+|u_1|}{2} & \text{if} \quad |u_1| \le c_m \\ \frac{u_1+|u_1|}{2} & \text{otherwise ,} \end{cases}$$

and

$$u_2^- := \begin{cases} \alpha_2 \left\{-\frac{(u_2-c_m)^2}{4c_m}\right\} + (1-\alpha_2)\frac{u_2-|u_2|}{2} & \text{if} \quad |u_1| \le c_m \\ \frac{u_2-|u_2|}{2} & \text{otherwise ,} \end{cases}$$

and

$$\psi_1 := (1, u_1, v_1, H_1) \quad \text{and} \quad \psi_2 := (1, u_2, v_2, H_2) .$$

where H is the total enthalpy ,

$$\alpha_1 = \frac{2(p/\rho)_1}{(p/\rho)_1 + (p/\rho)_2} \quad \text{and} \quad \alpha_2 = \frac{2(p/\rho)_2}{(p/\rho)_1 + (p/\rho)_2} .$$

This choice of u_1^+, u_2^- implies an improved resolution of stationary and moving contact discontinuities. The pressure flux is:

$$p_{1/2} := (0, p_1^+ + p_2^-, 0, 0) ,$$

and

$$p_1^\pm := \begin{cases} \frac{p_1}{4}\left(\frac{u_1}{c_m} \pm 1\right)^2\left(2 \mp \frac{u_1}{c_m}\right) & \text{if} \quad \frac{u_1}{c_m} \le 1 \\ \frac{p_1}{2u_1}(u_1 \pm |u_1|) & \text{if} \quad \frac{u_1}{c_m} > 1 \end{cases} ,$$

and

$$c_m := max\{c_1, c_2\} .$$

AUSMV scheme: Discretization of the AUSMV scheme is similar as the discretization of AUSMD, with the following modification. Instead of:

$$\frac{1}{2}\left[(\rho u)_{1/2}(u_1+u_2) - (\rho u)_{1/2}(u_2-u_1)\right]$$

in (2.3) we use:

$$\left[u_1^+(\rho u)_1 + u_2^-(\rho u)_2\right] .$$

AUSMDV scheme: AUSMDV is a mixture of AUSMV and AUSMD, in which

$$\frac{1}{2}\left[(\rho u)_{1/2}(u_1+u_2) - (\rho u)_{1/2}(u_2-u_1)\right]$$

will be modified as follows:

$$\frac{1}{2}(1+s)\left[u_1^+(\rho u)_1 + u_2^-(\rho u)_2\right]_{AUSMV} +$$
$$+\frac{1}{2}(1-s)\left[(\rho u)_{1/2}(\psi_1+\psi_2) - (\rho u)_{1/2}(\psi_1-\psi_2)\right]_{AUSMD},$$

where s is the switching function of the pressure gradient. For the definition of s see Wada and Liou [30].

2.4 The discretization of the viscous terms

For smooth function u and $x \in T_j$ we have (see Figure 2):

$$\Delta u(x) \sim \frac{1}{|T_j|}\int_{T_j}\Delta u = \frac{1}{|T_j|}\int_{\partial T_j}\partial_n u = \frac{1}{|T_j|}\sum_l \int_{S_{jl}} n_{jl}\nabla u\,. \qquad (2.4)$$

This means for a consistent discretization of $\Delta u(x)$ on $|T_j|$ we need an approximation of ∇u on S_{jl}. The main problem is the definition of an approximation of ∇u such

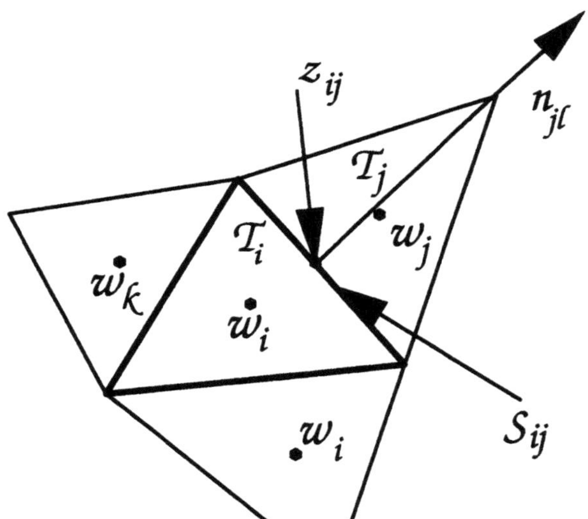

Figure 2 Control volume and used notation

that we get still good results on triangles with a large aspect ratios. Let w_i denote the center of gravity of triangle T_i (see Figure 2). Then the function

$$L(w) := u(w_i) + D_{ijk}(w - w_i) \qquad (2.5)$$

with

$$D_{ijk} \quad := \quad \frac{1}{detK_{ijk}} \begin{pmatrix} -(y_j - y_i) & y_k - y_i \\ x_j - x_i & -(x_k - x_i) \end{pmatrix} \begin{pmatrix} u(w_k) - u(w_i) \\ u(w_j) - u(w_i) \end{pmatrix} \tag{2.6}$$

and

$$K_{ijk} \quad := \quad \begin{pmatrix} x_j - x_i & y_j - y_i \\ x_k - x_i & y_k - y_i \end{pmatrix} \tag{2.7}$$

is the linear interpolant of $(w_s, u(w_s))$ for $s = i, j, k$ with $w_s = \binom{x_s}{y_s}$. Choose r such that:

$$|detK_{ijr}| = max\{|detK_{ijs}||s \in \{k,l\}\} , \tag{2.8}$$

define $D_{ij} = D_{ijr}$ and

$$L_{ij}(w) = u(w_i) + D_{ij}(w - w_i) . \tag{2.9}$$

The same can be done for the triangle T_j and we obtain:

$$L_{ji}(w) = u(w_j) + D_{ji}(w - w_j) . \tag{2.10}$$

Now as an approximation of $\nabla u(z_{ij})$ on S_{jl} we take:

$$\nabla u(z_{ij}) \sim \frac{1}{2}(D_{ij} + D_{ji}) . \tag{2.11}$$

Then the discretization of Δu on T_i is given by:

Definition 1 (version eII):

$$\Delta u \sim \frac{1}{2|T_i|} \sum_l n_i |S_i| (D_{il} + D_{li}) . \tag{2.12}$$

For the following modification of Definition 1 a discrete maximum principle holds [31].

Definition 2 (version eIII) Define (see Figure 2)

$$\omega_{ij} \quad := \quad \begin{pmatrix} y_j - y_i \\ x_i - x_j \end{pmatrix} \tag{2.13}$$

and

$$\begin{aligned} r &= k \quad \text{if} \quad \omega_{\mathrm{ij}} \cdot \mathrm{n_{ij}} \geq 0 \\ r &= l \quad \text{if} \quad \omega_{\mathrm{ij}} \cdot \mathrm{n_{ij}} < 0 , \end{aligned} \tag{2.14}$$

D_{ijr} as in (2.6), $D_{ij} = D_{ijr}$ and

$$L_{ij}(w) = u(w_i) + D_{ij}(w - w_i) . \qquad (2.15)$$

Then in this case we use

$$\Delta u \sim \frac{1}{2|T_i|} \sum_l n_i |S_i| (D_{il} + D_{li})$$

as a discretization for Δu. The comparison of version eII with eIII shows that eIII is much faster than eII on rectangles as well as on triangles (see Table 1 – 2).

2.5 Comparison of different Riemann solvers on different grids

Since we expect a Reynolds-number of order 20000 there will be boundary layers with large gradients and the discretization of the convection terms becomes most important. In [30] several Riemann-solvers have been studied in particular with respect for using them for the discretization in regions with strong shear flows. It was shown that the best results have been obtained with the Riemann solver AUSMDV as defined in Section 2.3. Other Riemann solvers which are claimed to be good for strong shear flows are presented in [8], [30], [16], [6], [22].

Table 1 Comparison of different Riemann solver on the grid in Figure 5

FLAT plate, CPU TIME, error $L^1 < 0.001$						
Grid	Unknowns	Viscous terms	Steger and Warming	Dick	Vijayasundaram	AUSMDV
triangles	60 × 40 × 2	eII	-	2534 s	2396 s	2125 s
rectangles	60 × 40	eII	1890 s	625 s	588 s	530s
triangles	60 × 40 × 2	eIII	-	1922 s	1831 s	1637 s
rectangles	60 × 40	eIII	1605 s	543 s	491 s	449 s

Table 2 Comparison of different Riemann solver on the triangular grid in Figure 5

FLAT plate, CPU TIME, error $L^1 < 0.0001$						
Grid	Unknowns	Viscous terms	Steger and Warming	Dick	Vijayasundaram	AUSMDV
triangles	80 × 60 × 2	eII	-	51265 s	47141 s	42187 s
triangles	80 × 60 × 2	eIII	-	47372 s	43168 s	38095 s

In order to check the numerical costs for this Riemann solver on different grids we have compared it with the fluxes of Steger and Warming [24], Dick [8], and Vijayasundaram [25]. As a test problem we have used the incompressible flow over a flat plate for which the exact solution [23] is known. We have used the grids as shown in Figure 3, 4, 5. Numerical experiments [23] show the enormous advantages of using stretched triangles in the boundary layer. The results of the calculation of the flow over the flat plate on Grid 1 (see Figure 3) are much better than using grid 2 (see Figure 4). Remark that Grid 2 consists of 170 times more triangles than Grid 1. In Table 3 the results for the viscous flow over a flat plate on grids 1 and 2 are shown. For the

Table 3 Comparison of different grid for Dick Riemann solver and eIII method

grid	number of cells	$\| u - u_h \|_{L^1}$	$\| v - v_h \|_{L^1}$	$\| T - T_h \|_{L^1}$	CPU-time
Grid 1	72	0.005286	0.000819	0.004453	19 s
Grid 2	12288	0.007927	0.002107	0.007753	7240 s

Table 4 Comparison of different grid for AUSMDV Riemann solver and eIII method

grid	number of cells	$\| u - u_h \|_{L^1}$	$\| v - v_h \|_{L^1}$	$\| T - T_h \|_{L^1}$	CPU-time
Grid 1	72	0.002686	0.000137	0.001686	7 s
Grid 2	12288	0.003900	0.000719	0.002669	2065 s

viscous terms we have used method eIII and for the convective terms the Dick Riemann solver. The results on both grids refer to final real time $t = 10s$. The Reynolds number for this test case was 1000. In Table 4 the corresponding results for similar problem with same stopping criterion with Reynolds number 10 000 and AUSMDV Riemann solver are shown. On the grid in Figure 5 we have obtained the best re-

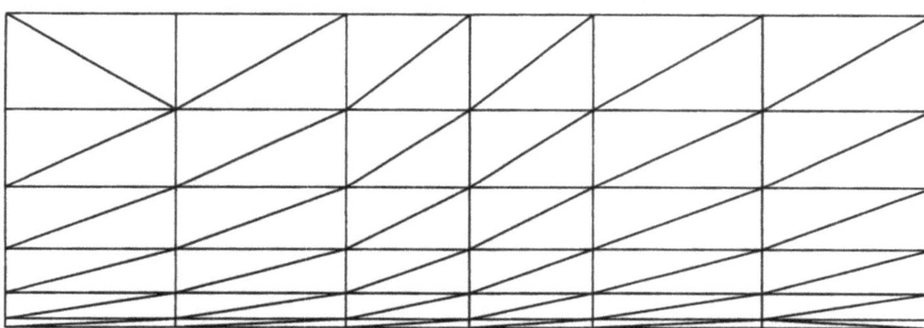

Figure 3 Grid 1

sults and in Table 1 - 2 we have documented the CPU times for the Riemann solvers mentioned above on the grid shown in Figure 5 as a triangular mesh and without the diagonal in each triangle as a rectangular grid. In all cases the iteration was stopped when a prescribed accuracy was obtained. The CPU-time for getting an L^1-error of u and v and T less than 0.001 is shown in Table 1, and for an L^1-error of u and v and T less than 0.0001 in Table 2.

Figure 4 Grid 2

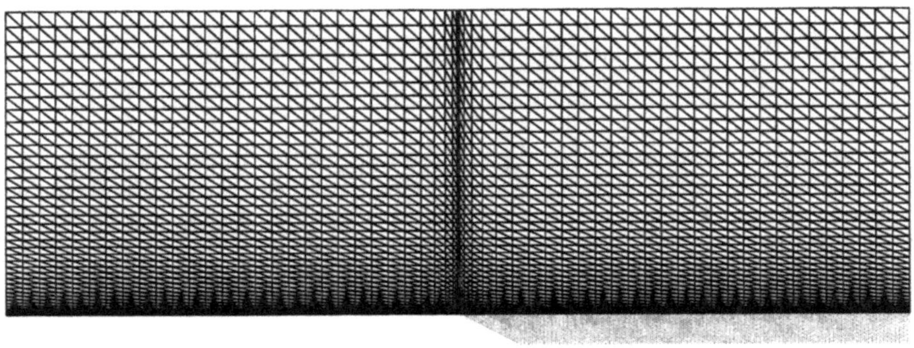

Figure 5 Grid 4

Figure 6 Calculated geometry

2.6 Numerical results for the interior flow

Because of the results mentioned in Section 2.5 we have used the AUSMDV Riemann solver for the convective terms and the discretization described in Section 2.4 for the viscous terms on a mesh consisting of triangles. In order to get well adapted initial

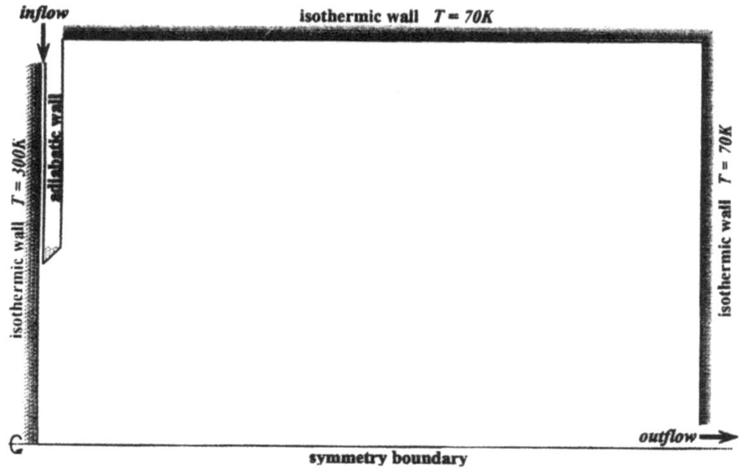

Figure 7 Realistic geometry

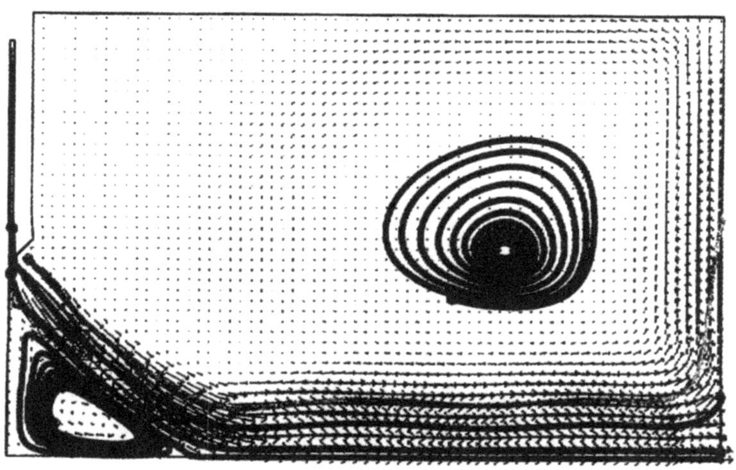

Figure 8 Vector field and particle traces

data we have first calculated the temperature field only. The boundary condition for the temperature are the same as in Section 2.1. Initially the temperature was assumed to be zero. This equation has been solved nearly up to the stationary state. Then these values for the temperature have been used as initial data for the computation

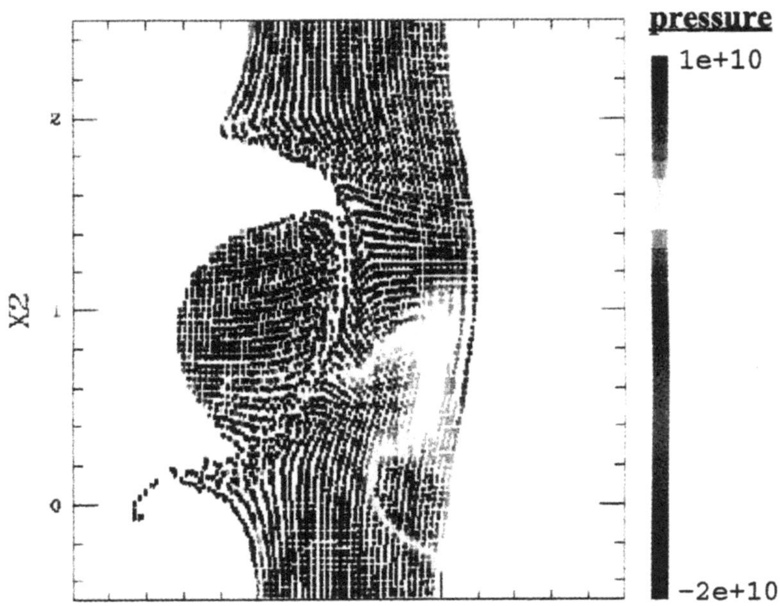

Figure 9 Standard method

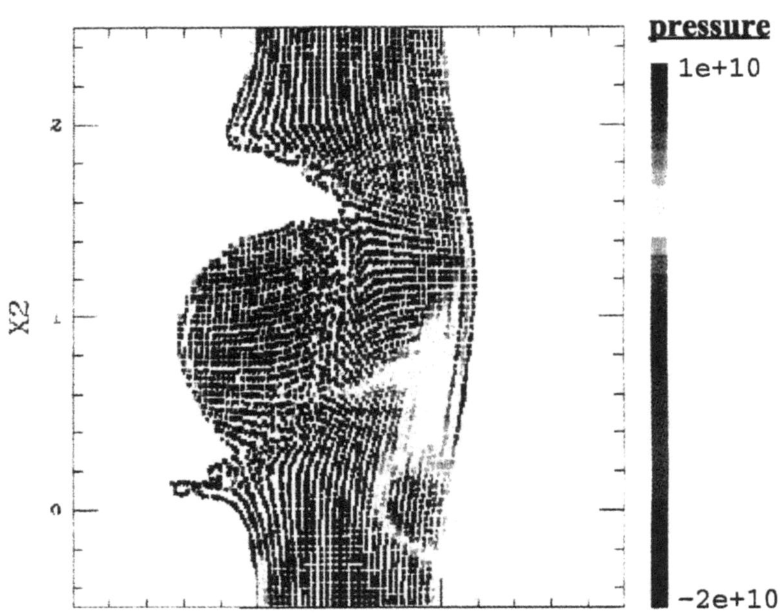

Figure 10 Renormalization

of (2.2). The numerical experiments indicate that the solution for the interior flow becomes stationary. In Figure 8 velocity arrows show the velocity after 2000 time-steps. It turns out that two big vortices appear. The particle traces in Figure 8 show that particle which reaches the regions of these vortices will stay there for some time and will not transported to the hole at the right end of the tube immediately. This confirms the behavior of the realistic particle in physical experiments.

3 Introduction to Particle Weighted methods

The particle method known as Smooth Particle Hydrodynamics was introduced in the end of seventies by Lucy [17] and Gingold and Monaghan [11]. SPH was first used in the area of astronomy and astrophysics. Recent applications in the field of high velocity impact have considerably enlarged the popularity of the method.

We give here a short presentation of this particle weighted numerical method based upon weak formulation. We refer to the recent paper of Vila ([29]) for a more detailed review.

3.1 Discrete Weak formulation

Let v a regular vector field in $\mathbb{R}^d$. We consider the following model PDE in conservation form :

$$L_v(\Phi) + divF(x,t,\Phi) = S \tag{3.16}$$

where F is the flux vector ($\in \mathbb{R}^d$) of the conservation law and L_v is the transport operator given by :

$$\Phi \longrightarrow L_v(\Phi) = \frac{\partial \Phi}{\partial t} + \sum_{l=1,d} \frac{\partial}{\partial x^l}(v^l \Phi).$$

To get a particle approximation of the equation (3.16), let us take a set of moving particles $(x_i(t),w_i(t))_{i\in P}$, indexed by $i \in P$, where $x_i(t)$ is the position of the particle and $w_i(t)$ its weight. We classically move the particles along the characteristic curves of the field $\vec{v}$ and also modify the weights in order to take account of deformations due to the field $\vec{v}$:

$$(i) \quad \frac{d}{dt}x_i = v(x_i,t) \qquad (ii) \quad \frac{d}{dt}w_i = div(\vec{v}(x_i,t))\, w_i \tag{3.17}$$

The accuracy of the approximation is connected with the quadrature formula over $\mathbb{R}^d$, given by the particles $(x_i(t),w_i(t))_{i\in P}$:

$$\int_{\mathbb{R}^d} g(x)dx \approx \sum_{j\in P} w_j(t)g(x_j(t)).$$

This formula is accurate for any $t > 0$ as soon as it is accurate initially and the particles and weights move according to (3.17) (see [21]). We provide the space with the discrete scalar product

$$(\varphi,\Psi)_h := \sum_{i\in P} w_i \varphi_i.\Psi_i = \int_{\mathbb{R}^d} \overline{\varphi}^\Delta.\overline{\Psi}^\Delta dx$$

which is clearly an approximation of the scalar product in $L^2(\mathbb{R}^d)^m$.

We also introduce the linear operator $D_{h,S}$ which is supposed to approximate strongly the derivative, i.e. for any φ regular enough

$$\sup_{i\in P} \|D_{h,S}\varphi_i - D\varphi_i\| \to 0 as h \to 0$$

and let us define $-D^*_{h,S}$ as the adjoint operator of $D_{h,S}$. We thus have

$$(D_{h,S}\varphi,\Psi)_h = -\left(\varphi,D^*_{h,S}\Psi\right)_h. \tag{3.18}$$

The weak formulation of (3.16) is defined by :

$$\begin{aligned}&\forall\varphi \in C_0^2(\mathbb{R}^d \times \mathbb{R}^{+,*})\\ &\int_{\mathbb{R}^d\times\mathbb{R}+} \left(\Phi.L_v^*(\varphi) + F(x,t,\Phi).\vec{\nabla}(\varphi) + S.\varphi\right) dxdt = 0\end{aligned} \tag{3.19}$$

where $-L_v^*$ is the adjoint operator of L_v defined as :

$$L_v^*(\varphi) = \frac{\partial\varphi}{\partial t} + \sum_{l=1,d} v^l \frac{\partial\varphi}{\partial x^l}.$$

A discrete version of (3.16) is provided by just replacing the integration over $\mathbb{R}^d$ by the discrete scalar product $(.,.)_h$ and the derivative $\vec{\nabla}(\varphi)$ by its approximation $D_{h,S}\varphi$:

$$\begin{aligned}&\forall\varphi \in \left[C_0^2(\mathbb{R}^d \times \mathbb{R}^{+,*})\right]^m\\ &\int_{\mathbb{R}+} \left[\left(\overline{\Phi}^\Delta,L_v^*(\varphi)\right)_h + \sum_{\alpha=1,\dots,d} \left(F^\alpha(\overline{\Phi}^\Delta),D^\alpha_{h,S}\varphi\right)_h + \left(S + R_h(\overline{\Phi}^\Delta),\varphi\right)_h\right] dt = 0.\end{aligned} \tag{3.20}$$

$R_h(\overline{\Phi}^\Delta)$ is an additional term which represents for example the artificial viscosity. Let us remark that, for any φ sufficiently regular

$$\frac{d}{dt}(\varphi_i) = \frac{d}{dt}(\varphi(x_i(t),t) = L_v^*(\varphi)_i \qquad \frac{d}{dt}(w_i\varphi_i) = w_i L_v(\varphi)_i.$$

Making an integration by part with respect to t, we get easily that (3.20) is true if and only if:

$$\begin{cases} (i) & \dfrac{d}{dt}x_i = v(x_i,t), \qquad\qquad x_i(0) = \xi_i \\ (ii) & \dfrac{d}{dt}w_i = div(v(x_i,t))\, w_i \qquad w_i(0) = \omega_i \\ (iii) & \dfrac{d}{dt}(w_i\Phi_i) + w_i \displaystyle\sum_{\alpha=1,\dots,d} D^{\alpha,*}_{h,S}(F^\alpha)_i = w_i\,(S_i + R_h(\Phi)_i) \\ (iv) & \Phi_j(0) = \Phi^0(\xi_j) \end{cases} \tag{3.21}$$

where F_i stands for $F(x_i,t,\Phi_i)$.

Remark 1 *Let us suppose that* $\left(R_h(\overline{\Phi}^\Delta),1\right)_h = 0$, *and that* $\varphi = 1$ *is in the kernel of* $D_{h,S}$, *then*

$$\frac{d}{dt}\left(\sum_{i\in P} w_i\Phi_i\right) = \sum_{i\in P} w_i S_i \tag{3.22}$$

which means that the discrete solution satisfies the same global conservation recipe as the exact one.

Like Lax and Wendroff for finite difference schemes ([15]) we are able to give sufficient conditions to get a good limit weak solution :

Theorem 2 *Let* $\overline{\Phi}^\Delta = \sum_{j\in P} \Phi_j(t)\chi_{B_j}(x)$ *the function associated with the sequence* $(\Phi_j(t))_{j\in P}$ *of regular functions of t defined by the scheme (3.21). We suppose that :*

(i) the function $\overline{\Phi}^\Delta$ *converges boundedly almost everywhere to* Φ *when h and* Δx *go to zero*

(ii) $\forall\varphi \in \left[C_0^2(\mathbb{R}^d \times \mathbb{R}^{+,*})\right]^m \sup_{i\in P} \|D_{h,S}\varphi_i - D\varphi_i\| \to 0$ *as h and* $\Delta x \to 0$

(iii) $\displaystyle\lim_{h \text{ and } \Delta x \to 0} \int_{\mathbb{R}^+} \left(R_h(\overline{\Phi}^\Delta),\varphi\right)_h dt = 0\,.$

Then Φ *is a weak solution of the model PDE (3.16) in the sense of Definition (3.19).*

Strictly speaking, Theorem 2 is not sufficient to get convergence of the numerical scheme since we are not able usually to get a priori estimates, sufficiently strong to get converging subsequences in L^1 by compactness arguments. By using L^∞ estimates and Diperna's uniqueness results ([9]) concerning measured valued solutions of scalar non linear conservation laws to get convergence of approximate solutions. In the section 3.3 we detail the construction of a large class of scheme for which we are able to prove that hypothesis (ii) is true, and also that the approximate solutions are bounded in L^∞. We refer to [3] for further details.

In order to define our numerical scheme we need to precise the definition of the operator $D_{h,S}$. Let us introduce a regularizing kernel $W(x,h)$ where the parameter h, the so-called *"smoothing length"* characterize the regularizing scale, and $W(x,h)$ converges towards the Dirac measure $\delta(x)$ (in a suitable sense) when $h \longrightarrow 0$. In practice we get a general kernel W from a given auxiliary function θ of a scalar variable :
$W(x,h) = \frac{1}{h^d}\theta(\frac{\|x\|}{h})$. We take usually θ as a positive function with compact support in [0,2], for example :

$$\theta(y) = C \times \begin{cases} 1 - \frac{3}{2}y^2 + \frac{3}{4}y^3 & \text{if } 0 \le y \le 1, \\ \frac{1}{4}(2-y)^3 & \text{if } 1 \le y \le 2 \end{cases}$$

where the coefficient C is $\frac{2}{3}, \frac{10}{7\pi}, \frac{1}{\pi}$ according to the space dimension (1,2 or 3). We shall use the following notations :

$$W_{ij} \equiv W(x_i(t) - x_j(t),h) \qquad \nabla W_{ij} \equiv grad_x[W(x_i(t) - x_j(t),h)].$$

Particle approximation of derivatives is taking usually at the point x_i :

$$\nabla \Pi^h(f)_i = \sum_{j \in P} w_j f(x_j) \nabla W_{ij} .$$

Note that (see e.g. [21]), in order to have $\nabla(\Pi^h \varphi) \longrightarrow \nabla\varphi$ (as $h \longrightarrow 0$), for a regular function φ, we need that the ratio $\frac{(\Delta x)}{h} \longrightarrow 0$. This is specific to classical particle methods. We define

$$D_{h,S}\varphi_i := \nabla \Pi^h(\varphi)_i - \varphi_i \nabla \Pi^h(1)_i = \sum_{j \in P} w_j \left(\varphi_j - \varphi_i\right) \nabla W_{ij}$$

in order to satisfy $D_{h,S}1 = 0$ and global conservation of the scheme.

Remark 3 *Here we use the standard notations in SPH literature. The smoothing length is h, the kernel is W and we have chosen Δx as the characteristic size of the mesh. In most of the mathematical papers related to particle weighted methods, the smoothing length is denoted ε, the kernel or cut-off function is ζ^ε and h is the characteristic size of the mesh. The reader has to take account of that in some of the references quoted in this paper.*

The statements (i) and (ii) in Theorem 2 differ from the standard ones in the classical Lax-Wendroff consistency theorem :

- the hypothesis (i) requires that the ratio $\frac{(\Delta x)}{h} \longrightarrow 0$. This hypothesis is often omitted in practical computation, it can be proved that the numerical solution may converge towards a wrong solution (see e.g. numerical results in [4]). This hypothesis can be relaxed, see section 3.2 below.

- the hypothesis (ii) means that the artificial viscosity is small in some weak sense.

3.2 Renormalization

Renormalization is a technic recently appeared in SPH literature (see [13] and [20]), it is supposed to improve accuracy of the method. We prove here that, with the help of the general setup of section 3.1 it is also conservative in the sense of (3.22). All the approximation and convergence results can be extended by using renormalized particle weighted approximation, this is precisely studied in [29]. In particular we are able to relax the assumption that the ratio $\frac{\Delta x}{h}$ goes to zero, and we just need that $\frac{\Delta x}{h} = O(1)$, see also the numerical tests in [28] (due to N. Lanson) which prove that classical SPH scheme converge to a wrong solution unless we take well chosen value of the ratio $\frac{\Delta x}{h}$ in order to almost satisfied condition (ii) of Theorem 2. We briefly resume the results.

Formally, Renormalization is a tool which provides new formulae for $D_{h,S}f(x)$ with the help of a weight matrix (the Renormalization matrix) in the following way:

$$D_{h,S}f(x) = B(x).\nabla\Pi^h(f)(x) - f(x)B(x)\nabla\Pi^h(1)(x).$$

We aim to increase the accuracy, thus instead of $P0 \subset \ker(D_{h,S})$ we ask that $D_{h,S}f = Df$ for any $f \in P1$. It can be easily proved that

Proposition 4 *We have $D_{h,S}f = \nabla f$ for any polynomial f of in $P1$, if and only if $B(x) = E(x)^{-1}$ with*

$$E(x)^{\alpha\beta} = \sum_{j\in P} w_j(x_j^\beta - x^\beta)\, \partial^\alpha W(x,x_j).$$

We thus have $E(x)^{\alpha\beta} = \partial^{\alpha,h}\Pi^h(x^\beta) - x^\beta\partial^{\alpha,h}\Pi^h(1)$, classical approximation results easily prove that $E(x)^{\alpha\beta} \simeq \delta^{\alpha\beta}$ and consequently that $B(x)$ makes sense, if $\frac{(\Delta x)}{h} \longrightarrow 0$. More precisely it can be proved (see [26]) that:

Proposition 5 *Let us suppose that B is uniformly bounded (with respect to h_0 and Δx) then we have*

$$||D_{h,S}\varphi(x) - D\varphi(x)|| \leq Ch_0||B(x)||||D^2\varphi||_\infty$$

where h_0 is the characteristic scale of the smoothing length (i.e. $\eta_- h_0 \leq h \leq \eta_+ h_0$ with η_- and η_+ two constants > 0)

The consistency of the method is thus satisfied at the only condition that the smoothing length goes to zero. Moreover, it can be proved that, if $\frac{\Delta x}{h}$ is bounded and if the initial distribution of particles is regular enough the matrix $B(x)$ is uniformly bounded and that :

$$\left|(\varphi, D_{h,S}(\varphi))_h\right| = \left|(\varphi, D^*_{h,S}(\varphi))_h\right| \leq C\|\varphi\|_h^2.$$

This also insure stability and convergence of the method (see [4] and [26]). The discrete operators are defined according to :

$$\begin{aligned} D_{h,S} f_i &= \sum_{j \in P} w_j (f_j - f_i) B_i.\nabla W_{ij} \\ D^*_{h,S} f_i &= \sum_{j \in P} w_j (f_i B_i.\nabla W_{ij} - f_j B_j.\nabla W_{ji}) \end{aligned} \tag{3.23}$$

and the scheme is :

$$\frac{d}{dt}(w_i \Phi_i) + w_i \sum_{j \in P} w_j (f_i B_i.\nabla W_{ij} - f_j B_j.\nabla W_{ji}) = w_i S_i .$$

We refer to [26] for more details (see [28] also for some numerical results).

The resulting method turns out to be more robust than standard methods, and also less expensive since we can use small values of the ratio $\frac{(\Delta x)}{h}$. This makes decrease the number of neighbors (a factor 2 or 3 is possible) and the cost of the method also decrease with a similar ratio. This is particularly true in situation with complex physics since the additional cost due to the computation of Renormalization matrices is no more than the computation of an additional physical unknown.

To show the efficiency of the renormalized S.P.H., we have computed the impact of a copper ball against an aluminum wall initially at rest. The diameter of the ball is 0.5 cm., the width of the wall is 1 cm. and the initially distance between the ball and the wall is 1 cm. The initial velocity of the ball is $2.10^5 \times \begin{pmatrix} 1 \\ -1 \end{pmatrix} cm.s^{-1}$. Results in Figure 9 are computed with standard S.P.H. method without Renormalization, results in Figure 10 are those of the modified method with Renormalization. On one hand, the renormalized computation gives smoother shape of the free surface, note also that it also removes the little gap at the interface between copper and aluminum. On the other hand it gives also a lower cost due to the decrease of the number of neighbors and provides efficient results.

3.3 Godunov type Particle Approximations

We develop here an alternative to artificial viscosity. It should be an opportunity to use this formalism in order to perform an efficient coupling of our two methods (i.e. Finite Volume and Particle Weighted methods). Use of Riemann solvers as been successful in the field of Finite Difference schemes and Finite Volume scheme to increase the robustness of numerical methods. We give a new formalism of particle weighted methods which includes such Riemann solvers, an improved stability is expected. We refer to [19] and [27] for numerical tests.

Note first that we can use a symmetric variant of the renormalization method developed in section 3.2.: we use $B_{ij} := \frac{1}{2}(B_i + B_j)$, and modify (3.23) according to

$$D_{h,S} f_i = \sum_{j \in P} w_j (f_j - f_i) B_{ij}.\nabla W_{ij}$$
$$D^*_{h,S} f_i = \sum_{j \in P} w_j \left(f_j + f_i\right) B_{ij}.\nabla W_{ij}\,.$$

It can be proved (see [4]) that this modification of the method retains the same consistency advantages and also improves the accuracy. Let us then introduce

$$A_{ij} := B_{ij}.\nabla W_{ij} \qquad n_{ij} := \frac{A_{ij}}{\|A_{ij}\|}\,.$$

We also suppose for simplicity, that $S \equiv 0$. Thus equation (3.21)(iii) reads as :

$$\frac{d}{dt}(w_i \Phi_i) + w_i \sum_{j \in P} w_j \left(F_i + F_j\right).n_{ij}\,\|A_{ij}\| = 0\,.$$

We thus compute the evolution of $w_i \Phi_i$ by summing up together the interactions of the particle located at x_i with its neighboring particles x_j. These interactions are computed along the direction n_{ij} connecting x_i with x_j. All these features introduces naturally at $x_{ij} = \dfrac{x_i + x_j}{2}$, the conservation law related to the direction n_{ij} :

$$\frac{\partial}{\partial t}(\Phi) + \frac{\partial}{\partial x}(F(x_{ij},t,\Phi).n_{ij}) = 0\,. \tag{3.24}$$

Therefore it is natural to introduce a 1-dimensional finite difference scheme in conservation form associated to (3.24), which brings a sufficient numerical viscosity. Such a scheme consists in replacing the centered approximation $(F(\Phi_i) + F(\Phi_j)).n_{ij}$ by the numerical flux of a Finite Difference scheme $2g(n_{ij},\Phi_i,\Phi_j)$, which is required to satisfy :

$$\begin{array}{ll} (i) & g(n,u,u) = F(u).n \\ (ii) & g(n,u,v) = -g(-n,v,u)\,. \end{array}$$

The numerical viscosity $Q(n,u,v)$ is classically defined in the scalar case (i.e. $\Phi \in \mathbb{R}$) as :

$$Q(n,u,v) = \frac{F(u).n - 2g(n,u,v) + F(v).n}{v - u}\,.$$

In the case of an explicit particle scheme the connection with finite 1D difference schemes is rather precise. We can establish (see [3] and [4]) in a way similar to the case of finite volume schemes (see for instance [1]) that the particle scheme is a convex combination of 1D finite difference schemes at which we add a controlled error term.

There is a lot of numerical flux well suited for such upwinding, among them we can quote the Lax Friedrichs and the Godunov schemes. By introducing the numerical viscosity our numerical scheme, which consists in finding functions $t \in I\!R^+ \longrightarrow u_i(t) \in I\!R, i \in P$ solutions of the differential system :

$$\frac{d}{dt}(w_i\Phi_i) + w_i \sum_{j\in P} w_l 2g(n_{ij},\Phi_i,\Phi_j)\, \|A_{ij}\| = 0 \qquad \Phi_i(0) = \Phi^0(\xi_i) \tag{3.25}$$

also reads as :

$$\begin{cases} \dfrac{d}{dt}(w_i\Phi_i) + w_i \sum\limits_{j\in P} w_j \left[(F(x_{ij},t,\Phi_i) + F(x_{ij},t,\Phi_j)).n_{ij} + \Pi_{ij}\right] \|A_{ij}\| = 0 \\ \Phi_i(0) = \Phi^0(\xi_i) \end{cases}$$

where $\Pi_{ij} = Q(n_{ij},\Phi_i,\Phi_j)(\Phi_i - \Phi_j)$. This form is very closed to the one classically used in SPH literature (for example, artificial viscosity in the momentum equation).

We refer to ([29]) for details on higher order extensions of the method based upon MUSCL technics classically used in the field of Finite Volume schemes.

Our approach is connected with the notion of Arbitrary Lagrange Euler approximation (A.L.E.), it is also in some sense a generalization of ideas develop by Harten and Hymann ([12]) in their work on self adjusting grid methods for conservation laws.

Let us give a regular transport field $v^0(x,t)$. We then consider the following conservative form of Euler equations in 2D (for simplicity)

$$L_{v^0}(\Phi) + \sum_{i=1,d} \frac{\partial}{\partial x^l}(F_E^l(\Phi) - v^{0,l}\Phi) = 0 \tag{3.26}$$

where the fluxes F_E^l are the convective flux given in (2.1). In order to compute the interaction between particle i and j, we consider the following 1D conservation law in the direction n_{ij} :

$$\begin{cases} \dfrac{\partial}{\partial t}(\Phi) + \dfrac{\partial}{\partial x}((F_E(\Phi).n_{ij} - v^0(x_{ij},t).n_{ij}\Phi)) = 0 \\ \Phi(x,0) = \begin{cases} \Phi_i \text{ if } x < 0 \\ \Phi_j \text{ if } x > 0\,. \end{cases} \end{cases}$$

We shall use Finite Volume fluxes based upon Riemann solvers such as those detailed in section 2, to compute the numerical flux.

Numerical experiments with these schemes, including 1D an 2D comparisons with standard SPH codes can be found in [27] and [19]. For details concerning the way to handle with boundary conditions we refer to [29] - see also the thesis by Ben Moussa B. [2] -.

4 Finite Volume methods and Particle Methods Coupling

4.1 Introduction

The aim of this part is to study the various coupling methods between particle weighted codes (noted PW) and finite volumes codes (noted FV). We limit our discussion to compressible Euler equations in two dimensions given by

$$L_v(\Phi) + \sum_{l=1,d} \frac{\partial}{\partial x^l}(F_E^l(\Phi)) = 0 \tag{4.27}$$

where the flux F_E^l are the convective fluxes given by (2.1).

Finite Volume Method

Let T_h be a triangulation of the computing domain $\Omega \subset I\!R^d$ ($\Omega = \bigcup_{T_i \in T_h} T_i$), $|T_i|$ be the measure in $I\!R^d$ of the cell T_i, $S_{ij} (j = 1,2,\ldots,n_i)$ the edges of the triangulation, n_{ij} the outward normal to the edge S_{ij} and $|S_{ij}|$ the measure in $I\!R^{d-1}$ of the edge S_{ij}. In order to simplify the presentation we only consider first order finite volume scheme :

$$\begin{cases} \Phi_h(x,t) = \Phi_i^n \qquad \forall\ (x,t)\ \in\ T_i \times [t^n,t^{n+1}[\\ \Phi_i^{n+1} = G_{FV}(\Phi_{FV}^n)_i = \Phi_i^n - \dfrac{\Delta t}{|T_i|} \displaystyle\sum_j g(n_{ij},\Phi_i^n,\Phi_j^n)\,|S_{ij}| \\ \Phi_h(x,0) = \dfrac{1}{|T_i|} \displaystyle\int_{T_i} \Phi(x,0)dx \equiv u_i^0 \qquad \forall\, x\ \in\ T_i\ . \end{cases} \tag{4.28}$$

Particle Method

We give for Euler equations the standard form of our Particle Renormalized scheme which leads for the momentum and energy equations to :

$$\begin{cases} \dfrac{d}{dt}(\vec{v_i}) = -\displaystyle\sum_{j\in P} m_j\Big(\frac{p_i + p_j}{\rho_i \rho_j} + \Pi_{ij}\Big) B_{ij}.\nabla W_{ij} \\ \dfrac{d}{dt}(u_i) = -p_i \displaystyle\sum_{j\in P} \frac{m_j}{\rho_i\rho_j}(\vec{v_j} - \vec{v_i}).B_{ij}.\nabla W_{ij} - \frac{1}{2}\sum_{j\in P} m_j \Pi_{ij}(\vec{v_j} - \vec{v_i}).B_{ij}.\nabla W_{ij} \end{cases} \tag{4.29}$$

where m_j is the mass (constant) of any particle moving with the velocity $\vec{v_j}$. The numerical dissipation Π_{ij} has to be taken from the numerical flux of a Riemann solver according to technics of Section 3.3.

For simplicity we limit ourselves to a first order time discretization :

$$(\Phi_{WP}^{n+1})_i = G_{WP}(\Phi_{WP}^n)_i = \Phi_i^n - \Delta t \sum_{j\in P} w_j^n (F_i^n + F_j^n + \Pi_{ij}^n).B_{ij}^n.\nabla W_{ij}^n .$$

4.2 Coupling modes

We propose 2 coupling modes :

- boundary condition coupling : The both codes compute the flow in two separate domains. The information that transits from one to the other is localized on a surface where suitable boundary conditions are implemented. This formulation is the simpler, it has been chosen in our first computational results presented below.
- domain overlapping : the domain of computation is in three disjoint domains :

  Ω_{WP} : domain of computation uniquely PW

  Ω_{FV} : domain of computation uniquely FV

  Ω_C : Both codes do the computation in this domain and exchange informations according to a specific procedure.
  This procedure brings certainly more robustness to the coupling, it will be carefully tested and analyzed in future works.

Coupling conditions

Physical conditions : We have to respect the global conservation :

$$\frac{d}{dt}\left(\int_\Omega \Phi \; dx\right) = \int_{\partial\Omega} F.n \; d\sigma + \int_{\partial\Omega} v.n\Phi \; d\sigma .$$

The both numerical methods described above respect exactly this condition at the discrete level.

Numerical Conditions A particle-grid and grid-particle projection system will be elaborated to control those balance sheet in the coupling zone Ω_C. We will note T_{GP}, the projection of the particles on the grid and T_{PG}, the projection of the grid on the particle positions.

These two projections need to satisfy exactly the conservation requirements. We can use tools previously developed in the field of PIC methods.

In the overlapping area we introduce a regular function $\theta(x)$ which characterize locally the amount of coupling

$$\begin{aligned}(\Phi_{FV}^{n+1})_i &= \theta_G T_{GP}(\Phi_{WP}^n)_i + (1-\theta_G) G_{FV}(\Phi_{FV}^n)_i \\ (\Phi_{WP}^{n+1})_j &= \theta_P T_{PG}(\Phi_{FV}^n)_j + (1-\theta_P) G_{SF}(\Phi_{WP}^n)_j .\end{aligned}$$

This procedure will be extensively studied in our future works. We only present here a preliminary test problem where we only investigate a crude coupling.

4.3 Numerical experiments for the combined problem

While we have described the Navier-Stokes solver for the interior flow in Section 2 and the particle method (SPH) in Section 3 we will put them together in this section.

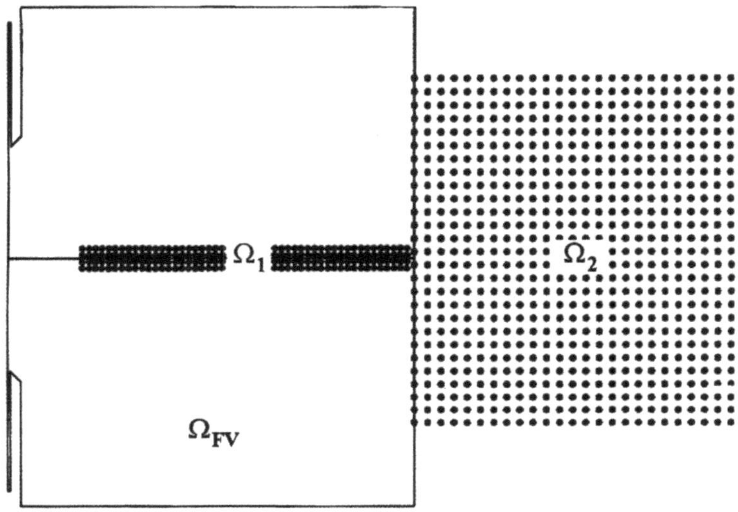

Figure 11 Computational domain

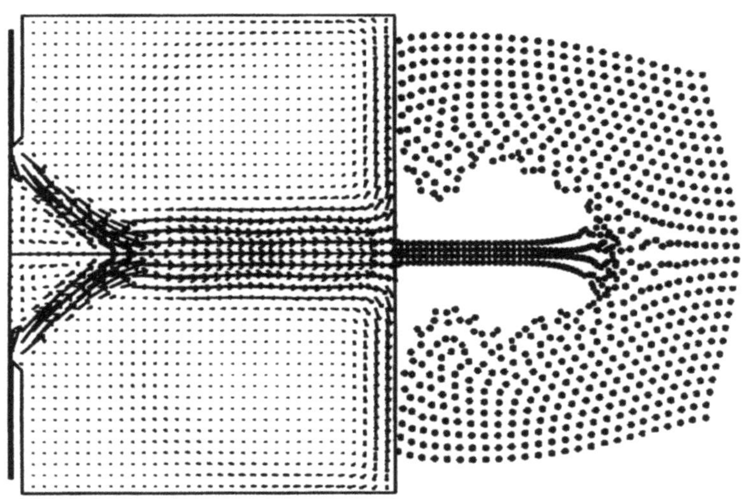

Figure 12 Calculated results

In the first step we have computed a stationary solution of the Navier-Stokes equation for the interior flow as described in Section 2. Then in the second step we consider the domain $\Omega_{WP} = \Omega_1 \cup \Omega_2$ as shown in Figure 11.

In Ω the following initial conditions for the SPH method have been assumed. In Ω_1 we take the data obtained from the stationary interior flow of Argon as initial data. In Ω_2 we assume that the Helium initially is at rest at some given pressure. Then the particle method describes the movement of the particles which are in Ω_1 at time $t = 0$, to the domain Ω_2 (Helium) during the time evolution. The results are visualized in Figure 12.

4.4 Remark

This coupling of course doesn't allow a time dependent interior flow and neglects the influences of the flow in Ω_2 to the interior flow in Ω_1. Later these additional feature will be taken into account. For these numerical experiments we have only used 875 particles. In this paper the heat and momentum exchange between Argon and the metal ions has been neglected. Furthermore we consider here only the 2-D case and a simplified coupling of the SPH method with the compressible Navier-Stokes solver. In a forthcoming paper we will extend these computations on the 3D geometry. In order to get also in this case reasonable results and effective computation we will use higher order discretization of MUSCL type [10], local adaption of the grid controlled by suitable error estimators and grid indicators [14], local alignment of the grid in the boundary layer, implicite discretization and parallelization.

The authors would like to thank Prof. Dr. Haberland, Faculty of Physics, University of Freiburg, for pointing out to us this problem.

References

[1] Benharbit S., Chalabi A., Vila J.P.: *Numerical viscosity, and convergence of finite volume Methods for conservation laws with Boundary conditions*. SIAM J. Numer. Anal., **32**, 3, (1995), pp. 775-796.

[2] Ben Moussa B.: PHD Thesis INSAT, (1998).

[3] Ben Moussa B., Vila J.P.: *Convergence of SPH method for scalar nonlinear conservation laws*. S.I.A.M Journal of Num. Analysis, to appear.

[4] Ben Moussa B., Lanson N., Vila J.P.: *Convergence of Meshless methods for conservation laws. Applications to Euler Equations*. VII Conf. Hyp. Pb., Zurich (1998).

[5] Bicknell G.V.: *The equations of motion of particles in smoothed particle hydrodynamics*. SIAM J. Sci. Stat. Comput., **12**, 5, (1991), pp.1198-1206.

[6] Coquel, F.; Liou, M.-S.: *Hybrid Upwind Splitting (HUS) by a Field-By-Field Decomposition.*

[7] Cottet G.H.: *Artificial viscosity for vortex and particle methods*. J. Comp. Physics, **127**, (1996), pp. 299-308.

[8] Dick, E.: *Multigrid Methods for Steady Euler- and Navier- Stokes Equations Based on Polynomial Flux-Difference Splitting.* International Series of Numer. Mathematics, Vol. 98, 1991.

[9] Diperna R.J.: *Measure-valued solution to conservation laws.* Arch. Rat. Mech. Anal., **88**, (1985), pp.223-270.

[10] Geiben, M.: *Convergence of MUSCL – Type Upwind Finite Volume Schemes on Unstructured Grids.* Preprint 318, SFB 256, Bonn, 1993.

[11] Gingold R.A., Monahghan J.J.: *Smoothed Particle Hydrodynamics : Theory and application to non spherical stars.* MNRAS, **181**, (1977), pp.181-375.

[12] Harten A., Hyman J.M.: *Self adjusting grid methods for one-dimensional hyperbolic conservation laws.* J. Comp. Physics, (1983).

[13] Johnson G.R., Beissel S.R.: *Normalized Smoothing functions for impact computations.* Int. Jour. Num. Methods Eng., **39**, (1996) , pp. 2725-2741.

[14] Kröner, D., Ohlberger, M.:*A posteriori error estimates for upwind finite volume schemes for nonlinear conservation laws in multi dimensions.* Preprint Nr. 2/1998, Freiburg, 1998.

[15] Lax P.D., Wendroff B.: *Systems of Conservation Laws.* CPAM, **13**, (1960), pp. 217-237.

[16] Liou, M.–S., Steffen, C.J.: *A New Flux Splitting Scheme.* J. Comput. Physics, 107, 23-39, (1993).

[17] Lucy .L.: *A numerical approach to the testing of the fission hypothesis*, Astrono. J.. **82**, 1013, (1977).

[18] Monaghan J.J.: *Smooth Particle Hydrodynamics.* Annu. Rev. Astron. Astrop., **30**, (1992), pp. 543-74.

[19] Paris L.: *Rapport de Fin d'étude.* INSA GMM, (1996).

[20] Randles R.W., Libertsky L.D.: *Smoothed Particle Hydrodynamics, Some recent improvements and Applications.* Comput. Methods Appl. Mech. Eng., **139**, (1996), pp. 375-408.

[21] Raviart P.A.: *An analysis of particle methods.* Num. method in fluid dynamics, F. Brezzi ed.. Lecture Notes in Math., 1127, Berlin, (Springer, 1985).

[22] Schwane, R. Hänel, D.: *An Implicit Flux–Vector Splitting Scheme for the Computation of Viscous Hypersonic Flow.* AIAA–paper No. 89–04, (1989).

[23] Schwörer, R.: *Entwicklung und Parallelisierung von Finite-Volumen-Verfahren zur Lösung der kompressiblen Navier-Stokes-Gleichungen in 2D.* Diplomthesis, (1997).

[24] Steger, J. L.; Warming R. F.: *Flux Vector Splitting of the Inviscid Gasdynamic Equations with Application to Finite–Difference Methods.* Journal of Computational Physics 40 (1981), 263-293.

[25] Vijayasundaram, G.: *Transonic Flow Simulations Using an Upstream Centered Scheme of Godunov in Finite Elements.* Journal of Computational Physics 63 (1986), 416-433.

[26] Vila JP.: *Meshless Methods and Particle Weighted methods for Conservation Laws.* submitted.

[27] Vila JP.: *Weighted Particle-Finite Volume Hybrid schemes.* Finite Volumes for Complex Applications ed. F. Benkhaldoun, R. Vilsmeier, (HERMES, Paris, 1996).

[28] Vila JP.: *Développements récents et Applications des Méthodes particulaires régularisées.* ESAIM Proceeding CANUM 97, (1997).

[29] Vila JP.: *Weighted Particle methods and Smooth Particle Hydrodynamic*. M3AS, (to appear).

[30] Wada, Y., Liou, M.-S.:*A Flux Splitting Scheme with High- Resolution and Robustness for Discontinuities*. AIAA-94-0083.

[31] Wierse (Geiben), M.:*Higher Order Upwind Schemes on Unstructured Grid for the Compressible Euler Equations in Timedependant Geometries in 3-D*. PhD Thesis, Freiburg, 1994.

The appropriate numbering for the multigrid solution of convection dominated problems

Wolfgang Hackbusch, Sabine Gutsch, Jean-François Maitre, François Musy

Christian-Albrechts-Universität zu Kiel, Mathematisches Seminar II, Lehrstuhl für Praktische Mathematik, Olshausenstr. 40, D-24098 Kiel

UMR CNRS 5585, Ecole Centrale de Lyon, Dept. Mathematiques Informatique, B.P. 163, F-69131 Ecully Cedex

Summary

The well-known smoothing iterations for scalar elliptic problems can usually not be applied to the Stokes- or Navier-Stokes equations. The reason is that the matrices arising from FE- or FD-discretizations of saddle-point problems are no longer positive definite. One remedy is to use the squared system. This leads to convergence in the symmetric case, but does not work if there is dominant convection. In this paper we will describe efficient smoothing methods that depend on the ordering of the unknowns. We provide numerical results illustrating the improved convergence properties of the multigrid algorithm.

1 Introduction

The incompressible Navier-Stokes equations are an important model in Computational Fluid Dynamics that describes instationary, incompressible, viscous flows. They are given by the following nonlinear system of partial differential equations:

$$\mathbf{u}_t - \epsilon\Delta\mathbf{u} + (\mathbf{u}\cdot\nabla)\mathbf{u} + \nabla p = \mathbf{f} \quad \text{in } \Omega \times (0,T), \tag{1.1}$$

$$\nabla\cdot\mathbf{u} = 0 \quad \text{in } \Omega \times (0,T), \tag{1.2}$$

complemented with appropriate boundary and initial conditions. Here Ω is a bounded domain in $\mathbf{R}^d$ where $d = 2$ or 3, and $(0,T)$ is a given time intervall.
The standard situation is that the convection is much stronger than the diffusion term ($\epsilon \ll 1$), i.e. the problem is singularly perturbed.
This leads to problems in the numerical solution of the Navier-Stokes equations such as finding a stable (upwind) discretisation as well as designing robust multigrid methods for the arising linear subproblems.

In this paper we propose smoothing iterations for linear systems of the form

$$L_l z_l = b_l \tag{1.3}$$

with

$$L_l = \begin{pmatrix} A_l & B_l^T \\ B_l & 0 \end{pmatrix}, \quad z_l = \begin{pmatrix} u_l \\ p_l \end{pmatrix} \quad \text{and} \quad b_l = \begin{pmatrix} f_l \\ g_l \end{pmatrix} \tag{1.4}$$

that arise in the solution of the Stokes or Navier-Stokes equations. The index l denotes the level in the multigrid iteration.

For the space discretization we use the stable modified Taylor-Hood element: The pressure is taken to be continuous piecewise linear, and for the velocity we use continuous piecewise linear functions on the triangulation obtained by dividing each triangle into four congruent subtriangles. For the convective part we employ an upwind technique that uses upstream differences as described in [11]. This leads to the block A_l being a block diagonal matrix with its diagonal blocks corresponding to the (nonsymmetric) stiffness matrix of a (possibly singularly disturbed) convection-diffusion problem $-\epsilon\Delta u + < b, \nabla u >= f$.
To demonstrate that the choice of discretization does not restrict the applicability of the smoothing iteration and ordering techniques that we will describe in this paper, we will also introduce a second discretization scheme (streamline upwind Petrov Galerkin) and apply the proposed ordering techniques to the resulting linear systems of equations.

For the solution of the linear subproblems we use a global multigrid method. The multigrid method is described in detail in [5]. Multigrid methods perform very well for elliptic problems or for problems where the convection is of the same order as the diffusion. When the convection becomes dominant, the convergence rate may deteriorate. This is due to a weakened smoothing property as well as to problems in the coarse grid correction.
Smoothing iterations developed for scalar problems like Gauß-Seidel or Jacobi are not applicable since the diagonal contains zeros in the lower block.
In order to obtain a multigrid method that is robust in the presence of dominant convection, we construct an efficient smoother based on a (block) LDU decomposition of the stiffness matrix combined with an appropriate ordering of the unknowns. A similar decomposition was also used to construct efficient smoothers for the Stokes problem in [1], [3]. Similar ordering techniques in relation with the multigrid algorithm have also been employed in [2], [10] and [12].

The remainder is structured as follows: Section 2 contains the definition of a matrix graph and the ordering techniques we employ. The smoothing method is described in section 3, and section 4 provides numerical results for the proposed ordering techniques and smoothing methods applied to the (scalar) convection-diffusion equation as well as for the Stokes equations with a convective term.

2 Ordering techniques for non-symmetric matrices

2.1 Definition of a matrix graph

We describe ordering algorithms in terms of graph theory: Let $G = G(A) = (V, E)$ be the matrix digraph (directed graph) of A with a vertex set V corresponding to the indices of A (which for our choice of discretization correspond to the set of vertices in the triangulation) and an edge set E with edges corresponding to the dominant entries of A (due to dominant convection). One possibility is to define the graph $G = (V, E)$ by edges

$$E = \{(p, q) \in V \times V : |a_{pq}| > \chi |a_{pp}|\} \tag{2.5}$$

with a mesh dependent parameter χ. Whereas in this definition we only compare the off-diagonal entries with the corresponding diagonal entry, definitions that take into account further entries per row or column or definitions that are based on a splitting of the matrix into its convective and diffusive part are also possible. Some choices for the set of edges have been discussed in [4].

In the special case that the resulting graph is acyclic, the numbering algorithm described in [6] provides an ordering following the strong connections so that the (re-ordered) matrix corresponding to the graph is a lower triangular matrix. Since the graph represents the strong matrix entries of A, one step of a Gauß-Seidel iteration as a smoother leads to good multigrid convergence.
However, in practical computations we often have to deal with cyclic graphs. In the following we describe ordering algorithms for this type of graphs.

2.2 Ordering techniques for cyclic graphs

We impose the following *one-flow-direction condition* on the planar graph G: The inward and outward edges E_i, $i = 1, \cdots, k$ associated with a vertex v can be ordered periodically with respect to the angle such that for some $m \in \{0, \cdots, k\}$ the edges E_i, $1 \leq i \leq m$, are outward and the edges E_i, $m < i \leq k$, are inward. Fig. 1 illustrates the one-flow-direction condition.
For graphs resulting from the discretisation of partial differential equations the one-flow-direction condition is typically fulfilled.

If the graph G has cycles, we propose three different ordering strategies:

- cyclic ordering
- parallel ordering
- feedback vertex set ordering

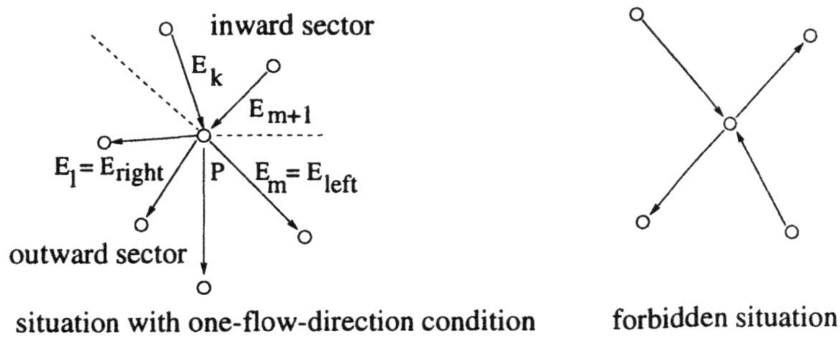

Figure 1 One-flow-direction condition

The *cyclic ordering* has been described in detail in [6] as well as in [7]. The idea is to decompose the graph into a set of disjoint (concentric) cycles and to perform a block Gauß-Seidel method with these cycles as blocks. In this case we need a block solver for the periodically bidiagonal blocks. Another possibility is to simply take a (backward) Gauß-Seidel step.

The *parallel ordering* combines two strategies: Following the flow while keeping the bandwidth of the resulting matrix low. The idea is to melt together neighbouring nodes if this preserves the one-flow-direction condition. This leads to a graph where each node represents one or more nodes of the original graph which are to be ordered as blocks. The blocks can be viewed as a collection of nodes neighboured in normal direction to the original cycle direction. The blocks of this reduced graph can be ordered by the cyclic ordering described above or the feedback vertex set ordering described below.

In the *feedback vertex set ordering* we determine a feedback vertex set, that is a set of vertices such that when we remove these vertices from the graph, the remaining graph becomes acyclic (and can be ordered by the algorithm described in [6]). The problem of finding a minimal feedback vertex set is known to be NP-complete. In the case of a planar graph, a quasi-optimal algorithm with linear amount of work is proposed in [6]. A heuristic feedback vertex set algorithm that is neither optimal nor linear in time but works well for many cases and is easy to implement is given in the following. It is similar to a feedback vertex set algorithm described in [9] that is optimal for two way reducible graphs. The idea is to successively perform one of the following transformations where $G \setminus \{v\}$ denotes the graph G without the vertex v and its adjacent edges:

- t_1 - Let v be a node with a loop. Set $G = G \setminus \{v\}$.
- t_2 - Let v be a node with no successor. Set $G = G \setminus \{v\}$.
- t_3 - Let v be a node with no predecessor. Set $G = G \setminus \{v\}$.
- t_4 - Let v be a node with exactly one successor. Set $G = G \setminus \{v\}$.
- t_5 - Let v be a node with exactly one predecessor. Set $G = G \setminus \{v\}$.

In the transformations t_4, t_5 we also add edges between the predecessors (successors) and the one successor (predecessor) of the eliminated vertex. Vertices that are eliminated in t_1 are added to the feedback vertex set. If there are still vertices left in the graph but no transformation is applicable, we eliminate an arbitrary node (or a node with a high number of adjacent nodes) and continue with the transformations locally. The graph algorithm is described formally by the following six Pascal-like procedures. It does not require any geometric information like vertex coordinates or the one-flow-direction condition as in the (quasi-optimal) algorithm described in [6].

For a directed graph $G = G(V, E)$, let $attr(i), i = 1, \cdots, |V|$ be an array which holds attributes D, A or L for the vertices. D stands for *deleted*, A for *active* and L for *listed*.

```
procedure findFVS( G(V,E) );
begin
FVS = ∅;
for P ∈ V do attr(P) = L;
for P ∈ V do begin
    if 'attr(P) = L' then transformZeroPred(P);
    if 'attr(P) = L' then transformZeroSucc(P)
end;
while (not all vertices are deleted) do begin
    while (∃ a node P with attr(P) = L) do
    begin
        transformOneSucc(P);
        if 'attr(P) = L' then transformOnePred(P);
        if 'attr(P) = L' then attr(P) := A
    end;
    if 'a node P with attr(P) = A exists' then
        addFVS(P)
end end;
```

```
procedure addFVS( P );
begin
M := {successors of P} \ P;
N := {predecessors of P} \ P;
FVS := FVS ∪ P;
G := G \ P;
attr(P) := D;
for Q ∈ M do
    if 'attr(Q) ≠ D' then begin
        attr(Q) := L;
        transformZeroPred(Q)
    end;
for Q ∈ N do
    if 'attr(Q) ≠ D' then begin
        attr(Q) := L;
        transformZeroSucc(Q)
end end;
```

```
procedure transformZeroPred( P );
begin
if 'P has no predecessor' then begin
  M := {successors of P};
  G := G \ P;
  attr(P) := D;
  for Q ∈ M do
      if 'attr(Q) ≠ D' then begin
        transformZeroPred(Q);
        if 'attr(Q) ≠ D' then
          attr(Q) := L
end end end;
```

```
procedure transformZeroSucc( P );
begin
if 'P has no successor' then begin
  M := {predecessors of P};
  G := G \ P;
  attr(P) := D;
  for Q ∈ M do
      if 'attr(Q) ≠ D' then begin
        transformZeroSucc(Q);
        if 'attr(Q) ≠ D' then
          attr(Q) := L
end end end;
```

```
procedure transformOnePred( P );
begin
if 'P has exactly one predecessor R' then
begin
  for Q ∈ {successor of P} do
      E := E ∪ (R,Q);
  G := G \ P;
  attr(P) := D
end;
if 'R has a self loop (R,R)' then
  addFVS(R)
end;
```

```
procedure transformOneSucc( P );
begin
if 'P has exactly one successor R' then
begin
  for Q ∈ {predecessor of P} do
      E := E ∪ (Q,R);
  G := G \ P;
  attr(P) := D
end;
if (R has a self loop (R,R))' then
  addFVS(R)
end;
```

The beginning of the procedure *findFVS* is just the numbering algorithm for acyclic graphs as described in [6]: All vertices without successors or predecessors are successively eliminated. The remaining part finds a feedback vertex set for the graph. Instead of choosing an arbitrary active node in the last but one line of the **procedure findFVS**, we suggest to choose an active node with high number of adjacent nodes.

We use the resulting ordering in a pointwise (backward) Gauß-Seidel method.

2.3 Some examples

In the following we give examples to illustrate the ordering techniques and resulting sparsity patterns of the reordered matrices for the convection-diffusion equation

$$-\epsilon\Delta u + < b, \nabla u >= f$$

on the unit square. We consider cyclic convection b with one cycle as well as with four cycles as depicted in Figure 2.

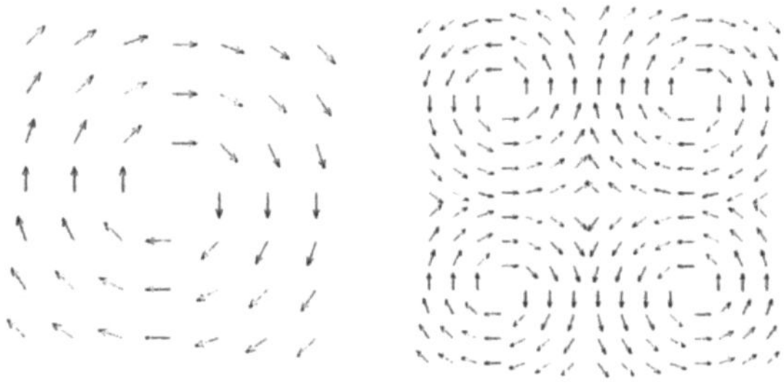

Figure 2 Sample cyclic convection directions

The graphs that are created with criterion (2.5) with $\chi = \frac{1}{3}$ are shown in Fig. 3.

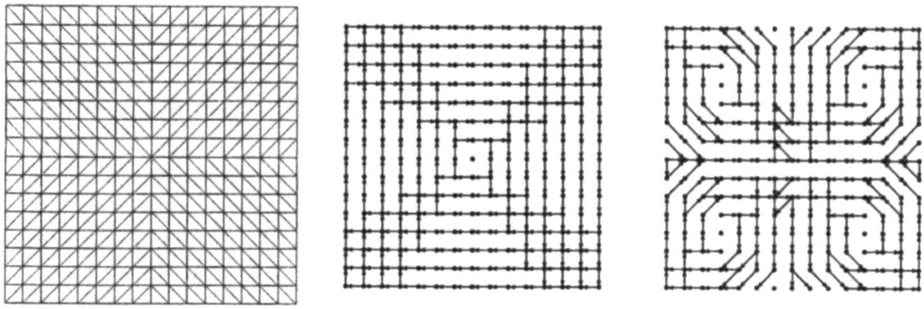

Figure 3 Triangulation and matrix graphs of strong convection

We apply the cyclic ordering and the feedback vertex set where we order the feedback vertex set vertices first and the remaining nodes in a downwind order following the flow. We also illustrate the parallel ordering where we first order the resulting blocks in a cyclic way and then with the feedback vertex set ordering.

The resulting sparsity patterns of the dominant entries are given in Fig. 4 for the one-cycle example and in 5 for the four-cycle example.

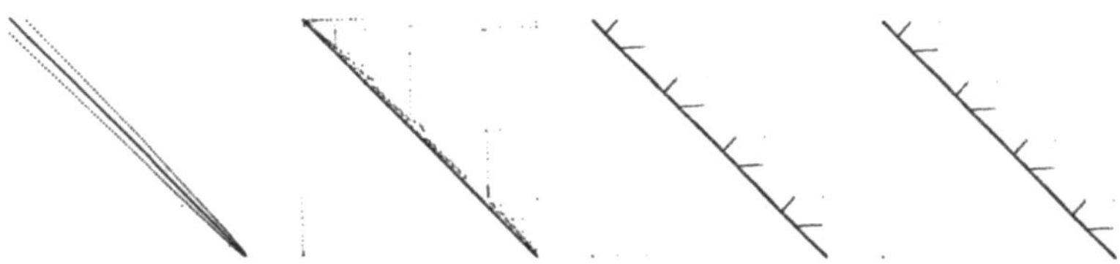

Figure 4 Resulting sparsity pattern for one-cycle example: cyclic, feedback vertex set, parallel (cyclic) and parallel (feedback vertex set) ordering

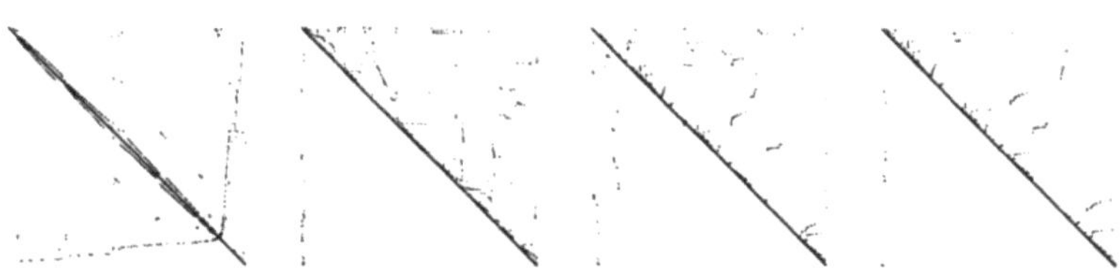

Figure 5 Resulting sparsity pattern for four-cycle example: cyclic, feedback vertex set, parallel (cyclic) and parallel (feedback vertex set) ordering

3 Smoothing iteration for a general system

The performance of the following smoother clearly depends on the ordering of the unknowns.
The starting point to this smoother is a block LDU-decomposition

$$K = \begin{pmatrix} A & B \\ B^T & 0 \end{pmatrix} = \begin{pmatrix} I & 0 \\ B^T A^{-1} & I \end{pmatrix} \begin{pmatrix} A & 0 \\ 0 & -S \end{pmatrix} \begin{pmatrix} I & A^{-1}B \\ 0 & I \end{pmatrix} \tag{3.6}$$

with Schur complement $S := BA^{-1}B^T$. Let W_S be an (easy to invert) approximation of S, and let W_L, W_A and W_U be (easy to invert) approximations of A. We obtain the smoothing iteration

$$x_{i+1} = x_i - W_K^{-1}(Kx_i - b) \tag{3.7}$$

where

$$W_K = \begin{pmatrix} I & 0 \\ B^T W_L^{-1} & I \end{pmatrix} \begin{pmatrix} W_A & 0 \\ 0 & -W_S \end{pmatrix} \begin{pmatrix} I & W_U^{-1}B \\ 0 & I \end{pmatrix}. \tag{3.8}$$

The performance of this smoother depends on the choice of approximations W_L, W_A, W_U and W_S. For W_L, W_A and W_U we simply take matrices that correspond to one step of a backward or symmetric (block) Gauß-Seidel method. If the unknowns are ordered such that all dominant entries lie in the upper (or lower) triangular part of the block A, then this becomes a very good and easy to invert approximation of A.

W_S will correspond to the application of several iterations of some iterative method for solving $B^T W_A^{-1} By = c$. Here we use a bicgstab solver with diagonal preconditioning, and W_A corresponds to one backward or symmetric Gauß-Seidel step. The number of inner bicgstab iterations improves the convergence of the overall method while increasing the work necessary per outer iteration step. Instead of taking a fixed number of inner iterations, we iterate until the residual is reduced by a certain factor, i.e. by 10^{-1}.

Assuming $W_A = W_L = W_U$, a straightforward calculation shows that we have for the iteration (3.7)

$$M_k = I - W_K^{-1}K = W_K^{-1} \begin{pmatrix} W_A - A & 0 \\ 0 & B^T W_A^{-1} B - W_S \end{pmatrix}.$$

The interesting observation is that, for a sufficiently accurate approximation $W_S \approx B^T W_A^{-1} B$, the error of the new approximate solution is (nearly) independent of the p-component of the approximation from the previous step. Following [3], this iterative method is called u-dominant. Typically, u-dominant iterations possess good smoothing effects.
If, on the other hand, $W_A \approx A$ then the error of the new approximate solution is (nearly) independent of the u-component and the method is called p-dominant.

We thus obtain a fast performable smoothing method. However, the effectiveness of this smoother depends crucially on the ordering of the unknowns since the order dependent Gauß-Seidel method is employed in several parts of the smoother.

4 Numerical results

4.1 The convection-diffusion equation[1]

We consider the two dimensional convection-diffusion problem

$$-\epsilon\Delta u + b\cdot\nabla u \;=\; f \quad \text{in } \Omega$$

with boundary conditions

$$u \;\equiv\; 0 \quad \text{on } \partial\Omega$$

where Ω is a polygonal domain. A triangulation $\mathcal{T}_h$ of Ω being given, the application of the streamline upwind Petrov Galerkin (SUPG) method [8] with piecewise linear functions leads to the formulation: Find $u_h \in V_h$ such that

$$\epsilon\int_\Omega \nabla u_u\cdot\nabla v_h + a_c(u_h, v_h) \;=\; \int_\Omega f v_h + L_c(v_h) \quad \forall v_h \in V_h$$

where

$$\begin{aligned} a_c(u_h, v_h) &= \int_\Omega (b\cdot\nabla u_h)v_h + \sum_{k\in\mathcal{T}_h}\delta_k\int_k (b\cdot\nabla u_h)(b\cdot\nabla v_h) \quad \text{and} \\ L_c(v_h) &= \sum_{k\in\mathcal{T}_h}\delta_k\int_k f(b\cdot\nabla v_h). \end{aligned}$$

δ_k is a parameter defined on each element k of $\mathcal{T}_h$ by $\delta_k = \frac{\delta\sqrt{mes(k)}}{|b|}$ if $\epsilon \leq \sqrt{mes(k)}|b|$ and $\delta_k = 0$ if $\epsilon > \sqrt{mes(k)}|b|$. $|b|$ is the Euclidean norm of b, and $\delta > 0$ is a sufficiently small real.
In the sequel we present numerical results which concern the influence of δ on the convergence of the multigrid method.

Let Ω be the unit square. We construct a family of nested triangulations by dividing each triangle into four congruent subtriangles. The initial triangulation is given in Fig. 6.

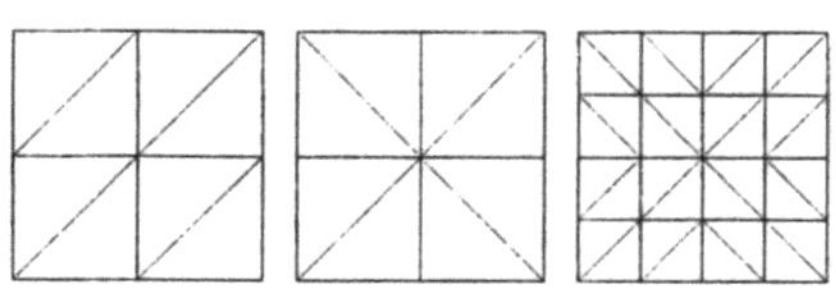

Figure 6 Coarse grid for convection-diffusion example (left); coarse grid for pressure (middle) and velocity (right) for Stokes example

We restrict our study to four examples of convection vector b:

[1] computed by the French project partners

- example 1: $b(x,y) = (1,0)^T$;
- example 2: $b(x,y) = (1,1)^T$;
- example 3: $b(x,y) = (-1,1)^T$;
- example 4: $b(x,y) = (1-y, x-1)^T$.

The first two examples correspond to the case where the direction of b at each node of the mesh coincides with the direction of an edge. The graph of the matrix A constructed from the bilinear form $a_c(\cdot,\cdot)$ is defined by

$$E = \left\{ (p,q) \in V \times V \;:\; |a_{pq}| > \frac{1}{2} \max_r |a_{pr}| \right\}.$$

For $\delta \leq 1.1$, we obtain acyclic graphs. The numbering algorithm described in [7] is applied. For the multigrid method, we use a V-cycle with six levels and two steps of Gauß-Seidel iteration as a smoother.

The following table gives the values of $\rho = \{||d_i||/||d_0||\}^{1/i}$ for $i = 5$ where d_i is the defect after i iterations and $||\cdot||$ corresponds to the L^2 norm in V_h.

Table 1 Convergence rates for different values of δ with $\epsilon = 0.0001$.

δ	0.40	0.45	0.50	0.55	0.60	0.70	0.80	0.90	1.00	1.10
ex 1	div	0.010	0.016	0.069	0.124	0.175	0.227	0.240	0.237	0.233
ex 2				div	0.150	0.005	0.039	0.088	0.127	0.156
ex 3			div	0.613	0.437	0.299	0.267	0.266	0.272	0.282
ex 4			div	0.632	0.311	0.246	0.230	0.237	0.249	0.278

Only for the first two examples we observe very small convergence rates:

$$\rho = 0.010 \quad \text{for} \quad \delta = 0.45 \text{ in example 1},$$
$$\rho = 0.005 \quad \text{for} \quad \delta = 0.70 \text{ in example 2}.$$

However, for values of δ in $[0.7, 1.1]$ we obtain good convergence rates ($\rho < 0.3$) in the four examples.

We refer to [4] where various further numerical results are provided for different initial triangulations, graph definitions, convection directions and ordering techniques.

4.2 The Stokes problem with a convective term[2]

Here numerical results are presented for the 2D Stokes problem with a convective term

$$\begin{aligned} -\epsilon\Delta\mathbf{u} + (\mathbf{b}\cdot\nabla)\mathbf{u} + \nabla p &= \mathbf{f} \quad \text{in } \Omega := (0,1)\times(0,1),\\ \nabla\cdot\mathbf{u} &= 0 \quad \text{in } \Omega,\\ u &= u_0 \quad \text{on } \Gamma := \partial\Omega. \end{aligned}$$

We have investigated the following test problem:

$$\begin{aligned} u(x,y) &= (\sin x \sin y, \cos x \cos y)^T\\ p(x,y) &= 2\cos x \cos y + C\\ f(x,y) &= (2(\epsilon-1)\sin x \sin y + b_0(x,y)\cos x \sin y + b_1(x,y)\sin x \cos y,\\ &\quad\; 2(1+\epsilon)\cos x \cos y - b_1(x,y)\cos x \sin y - b_0(x,y)\sin x \cos y)^T\\ u_0(x,y) &= u_\Gamma(x,y). \end{aligned}$$

We performed tests on the unit square with an initial triangulation as shown in Fig. 6. The finest level is obtained by five regular refinements resulting in 16129 unknowns for the velocity and 3969 unknowns for the pressure, yielding 36483 degrees of freedom on the finest level (in the two dimensional case). On the coarsest grid we have only 27 unknowns.

Throughout this section all errors are measured by the euclidean norm of the residuals. We report the number of iterations necessary to reduce the residual by a factor of 10^{-4} and the average convergence rates $\gamma = (\|r_i\|_2/\|r_0\|_2)^{1/i}$ where r_i is the residual after i steps. We carried out at most 40 steps. For inner iterations to solve the Schur complement problem we used a bicgstab iteration until the residual was reduced by 10^{-1}, but at most we performed 20 steps. If after 20 steps the desired reduction has not been achieved yet we take the iterate that corresponds to the so far smallest residual. All computations start with $p^0 = 100.0$ and $u_i^0 = 100.0$. The tests were performed on a Sparc Ultra 2.

In Table 4.2 we display numerical results for the Stokes problem (i.e. $b = 0$) where we used a multigrid iteration on five levels. We tested for several values for ϵ, several numbers of pre- and postsmoothing steps and for the V- as well as the W-cycle. The approximations W_L, W_A, W_U and W_S in (3.8) are taken as one step of a symmetric Gauß-Seidel method, and the unknowns are ordered by the reverse Cuthill-McKee algorithm.

From the results we conclude that the method is robust also for very small ϵ.

Table 4.2 shows convergence rates for the test problem where the convection is dominant. For all numerical examples we chose $\epsilon = 0.0001$. The convection directions CURVE, CIRCLE and 4CIRCLE are shown in Fig. 7.

The approximations W_L, W_A, W_U and W_S in (3.8) are taken as one step of a backward Gauß-Seidel method. The abbreviations stand for the following orderings:

[2] computed by the German project partners

Table 2 Number of iterations and convergence rates γ for Stokes problem.

ϵ	pre/post	V-cycle		W-cycle	
	smoothing	steps	rate	steps	rate
1.0	1/1	4	0.10	4	0.08
	2/2	3	0.03	3	0.02
10^{-2}	1/1	4	0.09	4	0.07
	2/2	3	0.03	3	0.02
10^{-4}	1/1	4	0.07	3	0.04
	2/2	2	0.01	2	0.01
10^{-6}	1/1	6	0.16	5	0.14
	2/2	3	0.05	3	0.04
10^{-8}	1/1	8	0.28	8	0.25
	2/2	5	0.13	5	0.11

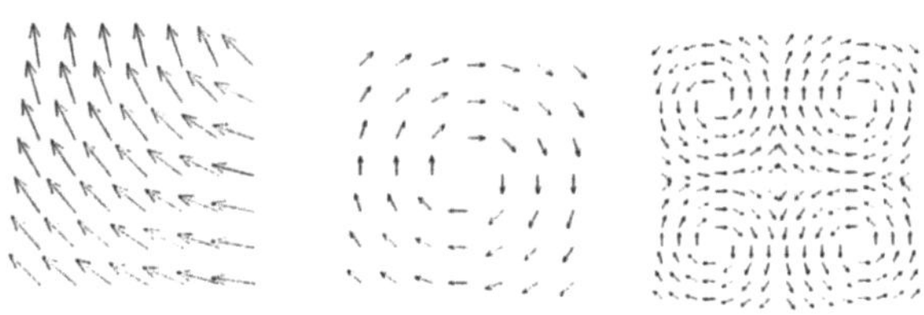

Figure 7 Convection directions CURVE, CIRCLE and 4CIRCLE, resp.

- *none*: no reordering, i.e. we keep the order that results from the grid refinement
- *cyclic*: cyclic ordering as described in section 2 and [6]; for acyclic graphs an upper triangular form is obtained
- *cyclic-parallel*: parallel ordering as described in section 2; the resulting blocks are ordered cyclic;
- *p-fvs-parallel*: parallel ordering as described in section 2; the resulting blocks are ordered with the quasi-optimal feedback vertex set algorithm
- *p-fvs*: ordering with a quasi-optimal feedback vertex set algorithm for planar graphs as described in [6]
- *h-fvs*: ordering with a heuristic feedback vertex set algorithm for general graphs as described in section 2

We observe that for nearly all test problems the proposed cyclic or feedback vertex set orderings improve the convergence behaviour of the method, in particular if only one smoothing step is applied. In the four cycle example the method even diverges without appropriate ordering.

Table 3 Number of iterations and convergence rates γ for Stokes equations with convective term.

convection	pre/post smoothing	ordering	V-cycle steps	V-cycle rate	W-cycle steps	W-cycle rate
CURVE	1/1	none	17	0.57	25	0.68
		cyclic	5	0.16	5	0.14
		cyclic-parallel	5	0.11	4	0.10
	2/2	none	10	0.37	8	0.30
		cyclic	3	0.03	3	0.03
		cyclic-parallel	3	0.02	3	0.03
CIRCLE	1/1	none	27	0.71	30	0.73
		cyclic	13	0.48	14	0.47
		cyclic-parallel	7	0.25	5	0.12
		p-fvs-parallel	7	0.25	6	0.20
		p-fvs	7	0.25	6	0.20
		h-fvs	6	0.21	6	0.19
	2/2	none	9	0.36	7	0.25
		cyclic	7	0.25	7	0.23
		cyclic-parallel	3	0.04	3	0.03
		p-fvs-parallel	4	0.08	4	0.07
		p-fvs	4	0.08	4	0.07
		h-fvs	4	0.09	4	0.07
4CIRCLE	1/1	none		div	24	0.67
		cyclic	10	0.39	8	0.29
		cyclic-parallel	13	0.48	11	0.41
		p-fvs-parallel	10	0.38	6	0.19
		p-fvs	10	0.38	6	0.19
		h-fvs	9	0.35	6	0.18
	2/2	none	9	0.36	7	0.24
		cyclic	6	0.20	5	0.14
		cyclic-parallel	7	0.25	4	0.10
		p-fvs-parallel	4	0.10	4	0.08
		p-fvs	4	0.10	4	0.08
		h-fvs	6	0.20	4	0.10

Comparing the results for the quasi-optimal and the heuristic feedback vertex set ordering, we observe that their performance is very similar.
As mentioned before, the convergence rates depend on the number of inner iterations that are performed. It is an area of current research to relate the convergence rate of the inner iteration with the outer convergence rate.

References

[1] R. E. Bank, B. D. Welfert, and H. Yserentant. A class of iterative methods for solving saddle point problems. *Numerische Mathematik*, 56(7):645–666, 1990.

[2] J. Bey and G. Wittum. Downwind numbering: Robust multigrid for convection diffusion problems. *Applied Numerical Mathematics*, 23(1):177–192, 1997.

[3] D. Braess and R. Sarazin. An efficient smoother for the Stokes problem. *Applied Numerical Mathematics*, 23:3–19, 1997.

[4] Sabine Gutsch and Thomas Probst. Cyclic and feedback vertex set ordering for the 2d convection-diffusion equation. Technical Report 97-22, Christian-Albrechts-Universität Kiel, 1997.

[5] W. Hackbusch. *Multi-grid methods and applications.* Springer, Berlin, 1988.

[6] W. Hackbusch. On the feedback vertex set problem for a planar graph. *Computing*, 58(2):129–155, 1997.

[7] W. Hackbusch and T. Probst. Downwind Gauß-Seidel Smoothing for Convection Dominated Problems. *Numerical Linear Algebra with Applications*, 4(2):85–102, 1997.

[8] C. Johnson. *Numerical solution of partial differential equations by the finite element method.* Cambridge University Press, 1987.

[9] Errol L. Lloyd, Mary Lou Soffa, and Ching-Chy Wang. On locating minimum feedback vertex sets. *Journal of Computer and System Sciences*, 37:292–311, 1988.

[10] H. Rentz-Reichert and G. Wittum. A comparison of smoothers and numbering strategies for laminar flow around a cylinder. In E.H. Hirschel, editor, *Flow Simulation with High-Performance Computers II*, volume 52 of *Notes on Numerical Fluid Mechanics*, pages 134–149. Vieweg, 1996.

[11] H.-G. Roos, M. Stynes, and L. Tobiska. *Numerical Methods for Singularly Perturbed Differential Equations.* Springer, 1996.

[12] S. Turek. On ordering strategies in a multigrid algorithm. In *Notes on Numerical Fluid Mechanics*, volume 41. Vieweg, 1997. Proc. 8th GAMM-Seminar, Kiel.

Development of Navier-Stokes Solvers on Hybrid Grids

D. Hänel[1], A. Dervieux[2], R. Vilsmeier[3], O. Gloth[4], C. Viozat[5], L.Fournier[6]
[1,3,4]: Uni-Duisburg, FB7/IVG, 47048 Duisburg, Germany
email: {hj454ha,hj000vi,oliver}@vug.uni-duisburg.de
[2,5,6]: INRIA, 2004, route des lucioles - 06902 Sophia-Antipolis, France
email: {Dervieux,Cecile.Viozat,Luc.Fournier}@Sophia.Inria.FR

Summary

The present paper is concerned with the ongoing joint work of two research groups to exploit the advantages of hybrid grids for the numerical simulation in fluid dynamics. Topics are related to the general outfit of the object oriented programming environment, the data-structures as well as the numerical methods used. Finally, mesh generation and modification techniques for hybrid grids are presented.

1 Introduction

Although many differing definitions might be usual, hybrid grids are defined here as grids of arbitrary element types chosen according to local requirements of the solution. These grids combine the advantages of geometrical flexibility and adaptivity of triangulated or tetrahedral grids with the higher accuracy of quadrilateral or hexahedral (structured-like) grids in regions of anisotropic flow behaviour. Anisotropic flow behaviour appears in essential in viscous shear layers but also near captured fronts. In these regions sufficient resolution is usually achieved by using highly stretched elements. However stretched simplex elements (triangles, tetrahedra) cause large numerical errors, while bilinear elements, i.e. quadrilaterals in 2-D and regular hexahedra as well as prisms in 3-D, remove essentially these difficulties. Depending on the local flow situation the use of these elements offer a higher accuracy with less computational efforts. Thus hybrid grid concepts, which combine both types, enable accurate and efficient computations of flows at high Reynolds numbers without impairing the flexibility of unstructured grids.
Since the meshing techniques enable to work on geometrically complex problems, it is useful to provide a comparable flexibility for the solution methods as well. Flows may be steady-state, cyclic or generally transient. According to the temporal behaviour and the flow-parameters, different codes are required, that however should rely on the same basic concepts.
Many questions arise in the realization of the proposed aim. These include the formulation of efficient numerical methods on generalized grids as well as the generation and adaptation of the meshes itself. Beside efficient numerical methods, computational efficiency is important with special consideration of modern computer architectures, as distributed memory or virtual shared memory parallel.
Along with the flexibility concerning the geometries and the numerical methods, a very flexible programming environment is of large interest. Although object oriented programming is not yet very common in use for numerical applications, it is a very well suited tool for modular software projects thus showing a strongly growing tendency, [1].
Within this paper a general overview, concerning the hybrid approach is presented. Starting on the equations for a fluid, the finite volume approach and the data structures employed are shown in more detail, followed by an overview concerning different time-integration methods and flux formulations with special interest on low Mach number flows. The approach for object oriented programming in this context is discussed. Finally, an approach for the automatic generation of grids consisting of arbitrary elements is presented.

2 Equations and Finite Volume Approach

The software concept presented here is not at all restricted to the developments concerning compressible flows. In fact, the efforts concerning grid generation, general pre- and postprocessing, time integration methods are so high, that this restriction would be inefficient, see also chapter 6. Therefore a very large part of the code is open for arbitrary purposes. However, within the frame of this paper, we may concentrate on compressible flows.

2.1 General Form of Governing Equations

Consider the general system of time dependent equations in integral form:

$$\int_V \vec{Q} \mathrm{d}V \quad + \quad \oint_{\delta V} \vec{H}\vec{n} \mathrm{d}A \quad = \quad \vec{S}\,. \tag{2.1}$$

$\vec{Q}$ is the vector of variables, $\vec{H}$ is the generalized flux vector for the transport terms and $\vec{S}$ is a possible source term. V is the control volume and δV is its boundary, whose local normal vector is $\vec{n}$.
In fact most time dependent physical problems may be written in this form, where $\vec{H} = \vec{H}(\vec{Q})$ and $\vec{S} = \vec{S}(\vec{Q})$. As an exception, we may at this point exclude integral terms over the whole domain, as for example required to describe the transport of energy due to radiation. These integral contributions could be included in the flux integral of the above equation, however the flux $\vec{H}$ would no longer be dependent only on the local state.

2.2 Finite Volume Approach

The integral form of the conservation equations 2.1 is valid for any control volumes V in the computational domain. Since computations have to be performed on a finite number of control volumes, a discrete formulation of the integral conservation equation is required. For any discrete Volume Vd it reads:

$$\left.\frac{\Delta\vec{Q}}{\Delta t}\right|_{Vd} \quad + \quad \vec{Res}_{\Delta,Vd} \quad = \quad 0\,. \tag{2.2}$$

Since unstructured grids do not offer a natural ordering of nodes and elements, an artificial ordering, a data structure, is required. Furthermore an element of the data structure will be called a molecule. To perform a time step for the equation given above, the discrete residual $\vec{Res}_{\Delta,Vd}$ has to be constructed upon such molecules:

$$\vec{Res}_{\Delta,Vd} \quad = \quad \frac{1}{V_{Vd}} \sum_{i=1}^{nr(Vd)} \vec{H}_{\bar{i}(i)} \vec{n}_{\bar{i}(i)} \Delta A_{\bar{i}(i)} \quad - \quad \vec{S}\,. \tag{2.3}$$

$nr(Vd)$ is the number of molecules contributing to the residual, V_{Vd} is the corresponding volume. $\vec{H}_{\bar{i}(i)}$ is the flux vector, $\vec{n}_{\bar{i}(i)}$ the normal vector and $\Delta A_{\bar{i}(i)}$ the area of the control interface supported by a molecule $\bar{i}(i)$.
The very general description of the finite volume approach opens up a large variety of possibilities how to define volumes and corresponding data structures. Since the data structures form the basis of all unstructured algorithms, they will be discussed in more detail within the next chapter.
Common to all variants is the problem, that the variables $\vec{Q}$ are not directly available at the positions, where the fluxes $\vec{H}$ have to be computed. Therefore a spatial reconstruction is required. The order of this reconstruction determines the spacial accuracy of the solver. In the case of a twin-sided upwind flux-computation, a linear projection from the nodes to the cell interfaces is used for second order accuracy, while the gradients at the nodes have previously been computed by a Gaussian integration on the control volumes themselves:

$$\nabla\vec{Q}_P = \frac{1}{V_{Vd}} \sum_{i=1}^{nr(Vd)} \vec{Q}_{\bar{i}(i)} \vec{n}_{\bar{i}(i)} \Delta A_{\bar{i}(i)}\,. \tag{2.4}$$

3 Data structures for logically unstructured grids

3.1 Some general aspects about data structures and terminology

The terminology "unstructured grids" is a misleading expression for the grid-type meant. In fact, these grids do not offer natural ordering, thus their description requires artificial ordering, so called data structures. These data structures provide a local orientation within the grid. The access to data, located farther away, requires usually a recursive use of the structures.
From a very general point of view, meshes consist of nodes, elements, faces and edges, that will furthermore be called components of a mesh. A structure may be referred to as an X-Y structure, meaning that the data Y are stored for the component X. Typical data structures are:

a) **Node-Data, Element-Data and Edge-Data structures:** These provide computational data at the nodes, elements or edges respectively. This is a non-logical data structure, in fact known form all grid types.

b) **Element-Node structure:** For all elements E the set of nodes $EP = \{P_1(E), P_2(E), \ldots, P_{np(E)}(E)\}$ is stored.

c) **Element-Element structure:** For all elements E the set of neighbouring elements is stored.

d) **Edge-Node structure:** For all edges K the two nodes $KP = \{P_1(K), P_2(K)\}$ are stored.

e) **Node-Node structure:** For all nodes P the set of neighbouring nodes $PP = \{P_1(P), P_2(P), \ldots, P_{np(P)}(P)\}$ are stored.

f) **Node-Edge structure:** For all nodes P the set of appending edges $PK = \{K_1(P), K_2(P), \ldots, K_{np(P)}(P)\}$ are stored.

g) **Node-Element structure:** For all nodes P the set of elements $PE = \{E_1(P), E_2(P), \ldots, E_{ne(P)}(P)\}$ sharing the node P as common vertex are stored.

Due to the large computations intended, memory consumption for the structures is crucial. The table below shows the memory needs per node in bytes, supposing that all floating data are 8 byte words, all integers are 4 byte.

Table 1 Memory consumption for different data structures used per node of a mesh. Assumes infinite meshes and 6 tetrahedra, respectively 7 edges per node for tetrahedral meshes.

	Memory per node in Bytes			
X-Y structure	Simplex (2-D)	Quad (2-D)	Simplex (3-D)	Hex (3-D)
Node-Data	8	8	8	8
Element-Data	16	8	48	8
Edge-Data	24	16	56	24
Element-Node	2x3x4 = 24	1x4x4 = 16	6x4x4 = 96	1x8x4 = 32
Element-Element	2x3x4 = 24	1x4x4 = 16	6x4x4 = 96	1x6x4 = 24
Edge-Node	3x2x4 = 24	2x2x4 = 16	7x2x4 = 56	3x2x4 = 24
Node-Node	6x1x4 = 24	4x1x4 = 16	14x1x4 = 56	6x1x4 = 24
Node-Edge	"	"	"	"
Node-Element	6x1x4 = 24	4x1x4 = 16	24x1x4 = 96	8x1x4 = 32

Resident, temporary and hidden data structures

The above data structures are, if in use, available at every time on every corresponding component of the mesh and thus called resident. Besides, temporary data structures may be used. For example, consider an algorithm, providing the Node-Node structure, but requiring an element reconstruction around a node for a specific task. It is possible to reconstruct the element neighbourship around the node upon the information given, thus producing a temporary data structure. Temporary data structures are most useful, if they do not require full storage across an entire mesh.

Temporary data structures do not have to be visible as a corresponding data construction within a program code. For example, the assumption that an element is a simplex might be required to perform a specific task within an unstructured solver. Without ever computing the nodes of an element, this corresponds to the existence of an element based data structure. Since it is not visible, we'll call it hidden.

Uniform and non-uniform data structures
A data structure is considered to be uniform, if the appearance of the storage record is independent from the grid. Among the mentioned list of data structures, all but the element-based data structures are uniform.
If, further on, the storage record is identical for all components of a given type, the structure will be called strictly uniform. Strictly uniform structures are very useful for the construction of efficient computational codes, specially when using vector machines. The above X-Data structures and the Edge-Node structure are strictly uniform.
Due to the independence concerning the grid type, unstructured solution algorithms employing uniform data structures (resident, temporary or hidden), may be used on hybrid grids with arbitrary element types without further changes. By way of contrast this is not the case for algorithms employing non uniform data structures, typical for most unstructured methods originally designed for simplex grids.

3.2 Operations to be performed

Before discussing the use of the above data structures more precisely, we may clarify which type of operations will have to be performed.

- Direct data operations: For example the division of the residual by the volume, for which it is used. These operations are very simple and are required in comparable form for all algorithms on every mesh.
- Indirect data operations: Data operations to be performed on an indirectly addressed subset of nodes or elements. For example, a specific consideration of a boundary condition for a subset of the boundary nodes is performed in this way.
- Cumulative operations: For example, the sum of edge contributions for a residual, equation 2.3. Operations of this type require indirect addressing in the frame of unstructured grids. These operations are thus essentially different, if dealing with structured or unstructured grids. Operations of this type are responsible for far most data access conflicts, as cache-misses and trashing and bear potential recursions for vector and parallel computers.
- Max/Min and search-operations: Consider the minimization of a time step for a non stationary calculation. Unless in the context of parallel or, depending on the compiler, vector computing, these operations are not of remarkable difficulty.
- Data structure altering operations: Operations to be performed within the frame of mesh-generation or adaptation. These operations bear very strong recursions.

Within the frame of the present paper we may concentrate on the cumulative operations, since these are the most time consuming in most flow-solvers. For the explicit codes treated here, the discrete equation 2.3 is very representative.

3.3 Edge-Node structure for nodal solution algorithms

For a nodal finite volume approach, the strictly uniform Edge-Node structure is the natural choice. Since a definition of the cell interfaces is required, it is very useful to store the normal vector $\vec{n}\Delta A$ of the cell interface area along with the edge. Figure 1 illustrates this.
The evaluation of a residual, equation 2.3, is consequently carried out in a loop for all edges, cumulating contributions for the indirectly addressed nodes of these edges. This is a very well known method and corresponding codes are usually called "nodal and edge-based".

```
for(K in all_edges) {
  compute edge contributions
  contribute to P1(K)
  contribute to P2(K)
}
```

The drawback of the algorithm is its low flexibility. In fact it is very inefficient, if the residual is to be computed only for a subset of nodes, typical for parallel computing or enhanced time-integration methods.

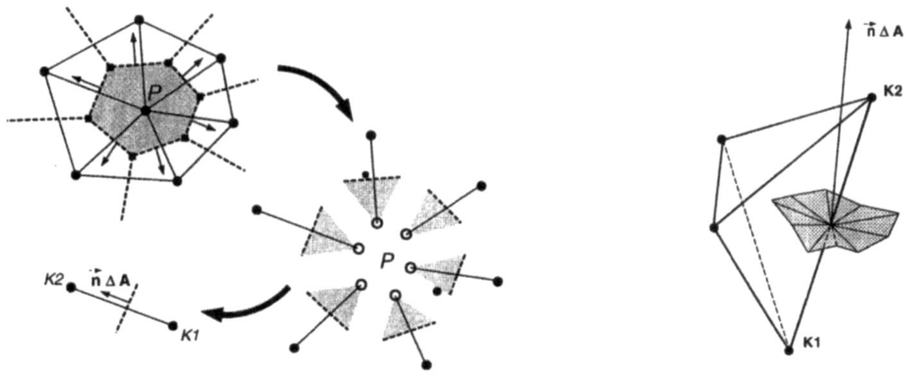

Figure 1: Decomposition of a 2-D control volume and edge-based molecule (left). 3-D counterpart (right).

3.4 A Node-Node structure for the solver

As an alternate for the Edge-Node structure above, a slightly modified version of the Node-Node structure can be used. Regarding the original Node-Node structure as previously introduced, we note that each edge corresponds to two point-neighbourhood relations. The set of neighbouring nodes PP for a node P is thus split in two parts, PP^+ and PP^-, where PP^+ contains all node addresses greater than the address of the node P and PP^- the lower addresses respectively. Obviously, the sets satisfy $PP = PP^+ \cup PP^-$. To avoid double storage, the area-normal vectors $\vec{n}\Delta A$ are stored along with PP^+ only, however allowing a direct memory access as with the Edge-Node structure. Figure 2 illustrates the storage concept.

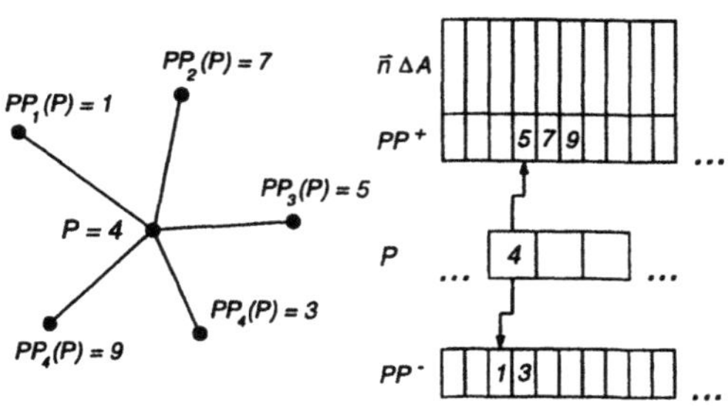

Figure 2: Sketch of the Node-Node structure with cell-interface data. Sample molecule for the node with address 4.

A full evaluation of the residual

Employing the above structure, a full residual evaluation on all nodes is possible, by simply spooling all edges represented by the subsets PP^+ of all nodes:

```
for(P in all_nodes) {
  for(P2 in PP+(P)) {
```

```
      compute edge contributions
      contribute to P1(K)
      contribute to P2(K)
    }
  }
```

Note, however, that this structure is not strictly uniform, since the number of neighbouring nodes may vary. Therefore, the above double loop for the evaluation of residuals or other cumulating operations is required. The use of this data structure is thus not useful on vector computers.

A partial evaluation of the residual

Suppose, a partial evaluation of the residual takes place. That is, only for a subset S of nodes a residual has to be computed. The data structure allows doing so, if the computed contribution to node P2 is made dependent on a condition. A problem arises, if an edge-wise computation would have to be performed for an edge (a neighbourship relation) represented in the set $PP^-(P)$. This means, the corresponding interface normal vector $\vec{n}\Delta A$ is stored along with the node P2 at the opposite side of the connecting edge. The reverse relation can be found employing a search among the entries in $PP^+(P2)$, an affordable operation since the number of neighbouring nodes is usually quite small.

```
for(P in S) {
  for(P2 in PP+(P)) {
    compute edge contributions
    contribute to P
    if (P2 in S) {
      contribute to P2
    }
  }
  for(P2 in PP-(P)) {
    if(P2 not in S) {
      for(PH in P+(P2) {   \\ this is the search loop
        if(PH == P) {
          get cell interface from reverse relation
          compute edge contributions
          contribute to P
          break search loop
        }
      }
    }
  }
}
```

Such a partial evaluation is extremely useful, when using parallel computers, where S is a subset for a processor. Another application would be the Multi-sequence Runge-Kutta scheme, where partial applications enable remarkable savings in CPU-time. Also, on a shared-memory machine the combination of both is possible.

Access to remote data

Remember the discussion about the local orientation, provided by the data structures. The recursive use of the Node-Node structure allows access to more distant data. This property is very useful in many cases, for example for the logical setup for the formulation of an agglomeration multigrid-scheme.

3.5 A combined Edge-Node and Node-Edge structure for the solver

Both above structures, the Edge-Node as well as the Node-Node structure enable complete information about the computational grid. Obviously, any additional data is in principal obsolete. To carry further logical informations can only to be justified by a simpler and thus faster access to data required.

Considering again a partial evaluation of the residual for a subset of nodes S, the additional Node-Edge structure can be used to indicate which edges to process, while the edge contributions can be computed in a simple loop as sketched in section 3.3 where only the contribution to the residual is made dependent on a condition.

```
\\
\\.. Build set of edges to be treated
  for(P in S) {
    for(K in KK(P)) {
      Put K in set of edges SK
    }
  }
\\
\\.. Compute residual for set of nodes in S
  for(K in SK) {
    compute edge contributions
    if(P1(K) in S) {contribute to P1(K)}
    if(P2(K) in S) {contribute to P2(K)}
  }
```

Although memory consumption is higher, the use of this combined structure showed to be a very flexible and efficient compromise.

4 Time Integration Methods

Integration in time of the resulting discrete conservation laws is based presently on the existing, different solution concepts, previously developed at the sites of the authors. Among these are implicit [2] and some explicit schemes.
The advantage of implicit schemes is, first of all, its better stability, resulting usually in very efficient codes. In contrast, explicit algorithms are very simple and require usually far less memory. One of the obvious trials is thus to combine the advantages of both methods. The present paper is thus restricted to the basically explicit versions with some further enhancements.

4.1 Fully Explicit Time Integration

The fully explicit time integration is the simplest version. The current implementation is based on the Runge-Kutta scheme. The method is specially useful for transient flow problems, if the characteristic time scales and the time scales required for reasons of accuracy are in the same order of magnitude. Such conditions are typical for flows at moderate or high Mach numbers, inviscid or at moderate Reynolds numbers of about 10 to 1000. Low Mach numbers, chemically reacting flows or strongly adaptive grid result in stiff systems, that may not be treated efficiently by a fully explicit scheme.

4.2 Multi-sequence Runge-Kutta Scheme

This method, developed in the context of a previous work [3] is specially suitable, if the time scales for stability or accuracy reasons vary essentially throughout the computational domain, typical for highly adaptive grid computations. The method is not yet implemented in the hybrid grid solver. However, for its implementation the Node-Node structure as proposed in chapter 3.4 is especially useful, since partial evaluation is required.

4.3 Dual time stepping method

The idea of the scheme is to solve the discretized equations for an implicit time step, solving a modified steady state problem. This method has originally been developed for incompressible flows, [4], employing a term of artificial compressibility, that vanishes as the iteration converges. However, the method

generally valuable, if a non time-consistent solution can be obtained with far higher efficiency, while a time accurate, transient solution is required. In the frame of compressible flows, this is especially true for low Mach numbers and systems with stiff source terms.
Consider the implicit discretized time step arising from the equation 2.2:

$$\left.\frac{\vec{Q}^{n+1}-\vec{Q}^{n}}{\Delta t}\right|_{Vd} + \vec{Res}^{\,n+1}_{\Delta,Vd} = 0. \tag{4.5}$$

Since the residual is computed upon the variables $\vec{Q}$, one may write:

$$\left.\frac{\vec{Q}^{n+1}-\vec{Q}^{n}}{\Delta t}\right|_{Vd} + \vec{Res}_{\Delta,Vd}(\vec{Q}^{n+1}) = 0. \tag{4.6}$$

Replacing the time level $n+1$ by an iteration index α and introducing a discrete difference operator for a pseudo time step $\Delta\tau$, one obtains the following modified equation:

$$\left.\frac{\vec{Q}^{\alpha+1}-\vec{Q}^{\alpha}}{\Delta\tau}\right|_{Vd} + \left.\frac{\vec{Q}^{\alpha}-\vec{Q}^{n}}{\Delta t}\right|_{Vd} + \vec{Res}_{\Delta,Vd}(\vec{Q}^{\alpha}) = 0. \tag{4.7}$$

Supposing, that a steady-state solution can be obtained in the artificial time τ, one obtains

$$\vec{Q}^{\alpha+1}-\vec{Q}^{\alpha}\rightarrow 0 \quad \text{and} \quad \vec{Q}^{\alpha}\rightarrow\vec{Q}^{n+1}$$

and thus the implicit relation is obtained again. Note that no stability restriction applies for the physical time step. For $\Delta t\rightarrow\infty$ the physical steady state solution is obtained, if existent.
For the iteration in the pseudo-time τ any method may be used. At present, an explicit Runge-Kutta with preconditioning is employed. Since for the pseudo-time iteration no time-accuracy is required, any acceleration technique is applicable. Therefore the use of an explicit multigrid scheme, will probably be a good choice.
The authors estimate, that the method with multigrid relaxation will be competitive with classical implicit codes for time accurate calculations. At present, the implementation is of first order in time. A second order extension is possible using a multi-stage scheme for the physical time.

4.4 Explicit Multigrid Methods

Multigrid methods have been developed by the authors in the frame of previous works for structured grids, [5] as well as for fully unstructured grids, [6]. Subject of an ongoing investigation is thus the implementation of an agglomeration multigrid scheme for hybrid grids. Corresponding experiences on this grid type are not yet available. The agglomeration type is chosen as a first attempt and the development is based on the combined Edge-Node/Node-Edge structure, as sketched in chapter 3.5.

5 Navier Stokes Equations and Flux Formulation

Starting at equation 2.1, chapter 2, the time-dependent equations for the conservation of mass, momentum and energy in integral form are obtained with:

$$\vec{Q}=\begin{bmatrix}\rho\\ \rho\vec{v}\\ \rho E\end{bmatrix} \qquad \vec{H}=\begin{bmatrix}\rho\vec{v}\\ \rho\vec{v}\vec{v}+\sigma\\ \rho\vec{v}E+\sigma\vec{v}+\vec{q}\end{bmatrix} \qquad \vec{S}=0\,.$$

ρ, $\vec{v}$ and E are the density, speed vector and specific total Energy. $\vec{q}$ is the heat-flux vector and σ the stress tensor. Note that $\vec{S}$ is not zero in the cases of external forces or energy sources. In the case of reactive media, the continuity equation is split up and corresponding source terms describing production of species and energy apply. This paper is however restricted to the conservation laws without sources. The flux vector $\vec{H}$ can be subdivided in an inviscid and a viscous part, $\vec{H}=\vec{H}_{inv}+\vec{H}_{visc}$, with:

$$\vec{H}_{inv}=\begin{bmatrix}\rho\vec{v}\\ \rho\vec{v}\vec{v}+pI^d\\ \rho\vec{v}(E+p)\end{bmatrix} \qquad \vec{H}_{visc}=\begin{bmatrix}0\\ \sigma-pI^d\\ (\sigma-pI^d)\vec{v}+\vec{q}\end{bmatrix}$$

where I^d is the unit matrix in the dimension d of the problem. More detailed explanations may be found elsewhere.

5.1 Flux Formulation

Since the code is intended to be most flexible possible, the software is open for arbitrary flux formulations. Both research groups have long term experiences in development, handling and application of different numerical flux approximation. Among these, special interest is in the flux difference formulations of Roe's type, [7], which was extended to the low Mach number limit using the Roe-Turkel scheme [8]. Besides, the Advective Upwind Splitting method (AUSM) by Liou and Steffen [9] and former approaches [10] showed to be interesting alternates. Since these formulations are well known, they're not repeated here. Aim of this part of the studies is the investigation and development of accurate and robust Riemann solvers for applications in a wide range of Mach numbers. Of particular interest is the behaviour of the flux approaches at small Mach numbers.

5.2 Sample results, Flow around a circular cylinder

Of particular interest is the behaviour of the flux approaches at small Mach numbers. The test case is a rather simple one as far as the geometry is concerned but it is already a rather discriminating one for the flow under investigation. We consider a steady flow around a circular cylinder. The flow is inviscid and the Mach number at infinity will be taken between 0.1 and 0.01. The mesh chosen is an unstructured triangulation of Voronoi type using 3302 nodes and 6492 triangles without stretched elements. Isovalues of the velocity and of the pressure in these conditions are shown in Figure 3.

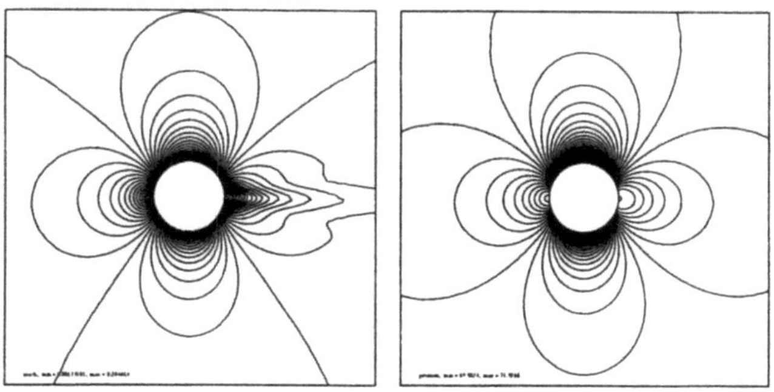

Figure 3 Mach (left) and pressure (right) contour plot at second-order accuracy for $M_\infty = 0.1$ with the Roe-Turkel scheme on a 3302 nodes mesh.

A tool to analyze the behaviour at low Mach number of upwind schemes has been proposed in the article of [11] in which it is shown that the Roe scheme [7] is unusable to compute low Mach number flow but that the Roe-Turkel scheme has a good behaviour in this range.
The application of this analysis to the AUSM scheme enabled us to state the following points :

- The AUSM scheme remains consistent as the Mach number goes to zero. As shown in Figure 4 the fluctuation of the pressure remains on the order of the Mach number square.
- However, below Mach one tenth, oscillations appear even in the first order version (see in Figure 5).

Comparison of the convergence of the AUSM and Roe-Turkel scheme for an explicit method

Concerning the study of the convergence to steady state, the convergence of all schemes degrades strongly as the Mach number is smaller and smaller. Indeed, the CFL (Courant-Friedrichs-Lewy) condition is based on the speed of the fastest wave which at low Mach number flow is the acoustic wave. The lower the Mach number, the smaller the length travelled through by the material wave compared to the length travelled through by the acoustic wave and therefore the more time steps are necessary to see the evolution of the material wave.

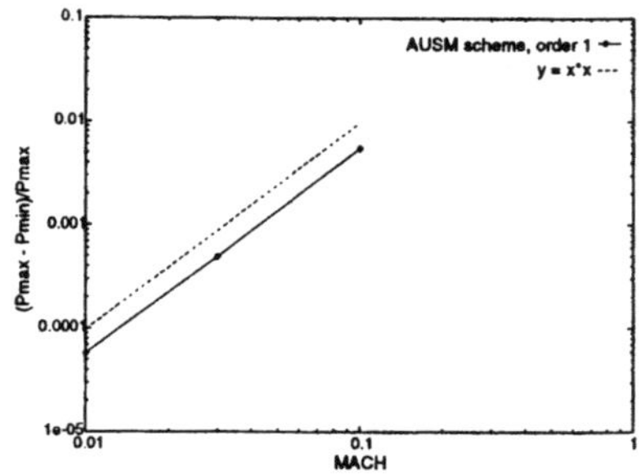

Figure 4 Pressure fluctuations versus inflow Mach number for the AUSM scheme. For comparison, the curve y=x*x is represented by a dotted line

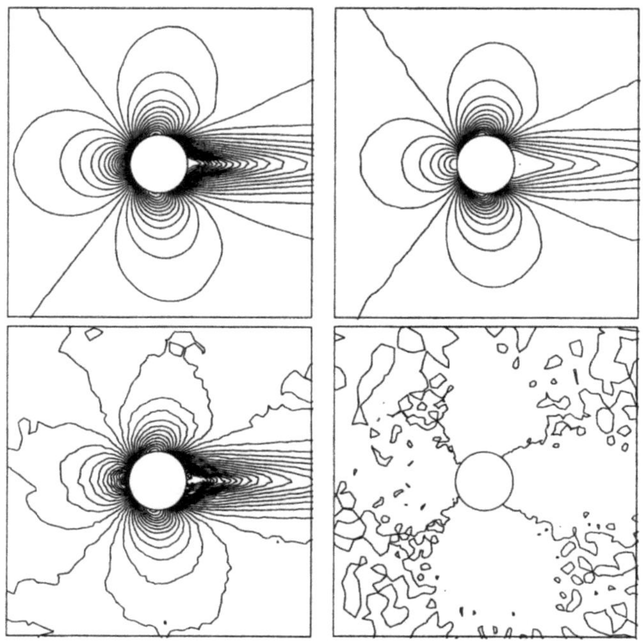

Figure 5 Mach (left) and pressure (right) contour plot at first-order accuracy for $M_\infty = 0.1$ (top) and $M_\infty = 0.01$ (bottom) with the AUSM scheme on a 3302 nodes mesh. Interval between pressure isovalues: 0.00025. Min/Max: 0.99, 1.003. Interval between Mach isovalues: 0.005. Min/Max: 0, 0.1, for $M_\infty = 0.1$.

However, it is possible to add a preconditioning matrix, as proposed in the article of [8], which is a 4 × 4 block diagonal matrix in 2D and makes the wave speeds uniform.

The effect of the preconditioning matrix on the Roe-Turkel scheme is (see in Figure 6 (left)) beneficial but does not enable us to use large time steps because of the loss of stability of the Roe-Turkel scheme as the Mach number goes to zero.
On the other hand, the iterative preconditioner has a totally satisfactory effect on the AUSM scheme since the convergence is independent of the Mach number (see in Figure 6 (right)).

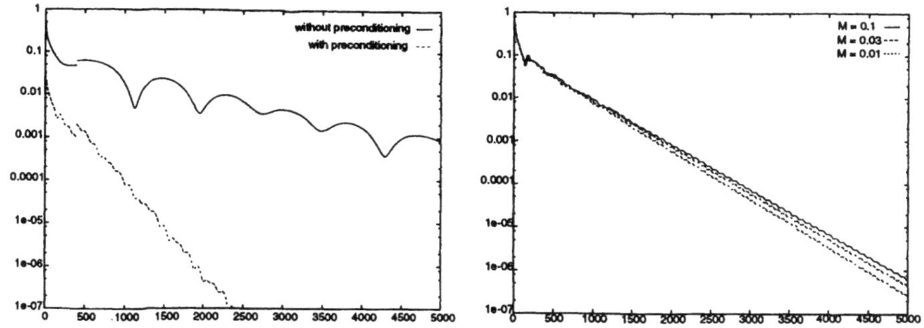

Figure 6 Convergence at first-order accuracy, momentum residual, according to the preconditioning of the equations, $M_\infty = 0.1$, Roe-Turkel scheme. (left). Convergence of the preconditioned AUSM scheme for different Mach numbers (right)

General conclusion

In conclusion, only the Roe-Turkel scheme enables to compute accurately very low Mach number flows. The preconditioner brings a noticeable improvement for the convergence, but this is not enough to enable a convergence to the steady state independent of the Mach number. As for the AUSM scheme, the convergence to the steady state is independent of the Mach number, but the robustness limitation (appearance of oscillations at low Mach number) leads to advise this scheme only for Mach number greater than 0.1.

6 Object oriented programming

The vast majority of numerical methods for the simulation of complex physical problems relies on procedural programming languages, mainly Fortran. The reason for that is, first, of historic nature. Fortran codes are being developed since several decades and new versions are usually based on older ones. A further reason for the continuing use of procedural programming languages is the availability of very good compilers for Fortran and also for C.

We may conclude, that procedural languages are very useful for the solution of a specific problem. However, the corresponding algorithms are less flexible. In practice this results in large numbers of program versions when solving different physical problems. Worse than this, the choice of different methods, for instance explicit or implicit, adaptive or not, often results in a multiplier for the number of versions. Maintenance of such software-systems becomes difficult, major changes are impossible.

For the present paper, only a short overview concerning the object oriented approach is given. A more detailed description of the code development can be found in [12].

6.1 Advantages and Problems of Object Oriented Programming

Considering the chapters above, only a very restricted choice of possible methods are presented, to be seen as a preliminary state. Specially to enable efficient developments in future, object oriented programming can be very helpful. Modules are written once and can be employed throughout the whole software. The development work can easily be split in different subtopics, since interfaces are simple. Data handling, crucial for large scale computations, is fully dynamic. Already in C, the flexible use of the memory is a remarkable advantage compared to Fortran. The above advantages are the essential reason, why the software described in this article is written in C++.

As a disadvantage at present, we may note some drawbacks in computational performance of C++, specially when compared to Fortran. The performance problem appears already in straight forward routines, written in Fortran-style. Compared to Fortran, the identical C-program shows around 90%,

a C++ program around 70% performance. Since these results are obtained without employing object oriented features of C++, we may conclude that, at present, optimizations performed by Fortran and C compilers is superior. From one side, these results are disappointing, on the other hand we may expect better optimizing compilers soon. A more detailed study, related to the use and performances of C++ and F90 for scientific computing can be found in [13].
Another aspect is the support for vectorization and parallelism, which is remarkable for the classical languages. At present the authors do not know any vectorizing C++ compiler, although a simple, non recurrent loop would be recognizable in the same way as in Fortran. Similar, but not so restrictive, is the support for parallel programming on most platforms. Again, since the restrictions do not relate to any judicious reason, we may expect improved support in future versions.

6.2 Algorithmic Structure

A major goal of this project is to decouple the logical parts, such as grid generation, grid refinement or parallel algorithms from the physical parts e.g. which problem is in fact to be solved and which formulations are to be employed. The whole software is therefore split in three major programming levels:

- In a first level, **basic storage classes**, including dynamic data handling have been created. These classes are independent of the kind of problems treated and thus do not even relate to the solution of differential equations.
- In a second level, a **basic functionality** is provided. These include the solution concepts, mesh-generators and adaptive concepts as well as some utility routines. For example, this level covers time-integration modules, which are not yet related to on a physical problem, but provide only the corresponding methods.
- The third level is the **physical level**. Herein a user may allocate data and create the processing-methods for his specific application.

All calls to procedures created in the physical level are virtual. That is, a user may choose methods of the basic functionality, suitable for its application, without knowing much about these.

6.3 The Solution of a Physical Problem

As an example, consider an explicit single time step scheme to be performed on equation 2.2, chapter 2.2. For this very simple task, the basic functionality provides the corresponding logic, consisting of the computation of Res_Δ and the advance in time. Some of the operations are automatic, that is, regardless of the variables in use, the corresponding method is always available. For the projection of variables from the nodes to the cell interfaces, a choice of automatic methods is available. The user of the system may thus concentrate on the computation of the time step length and the flux-formulation for his application, figure 7.

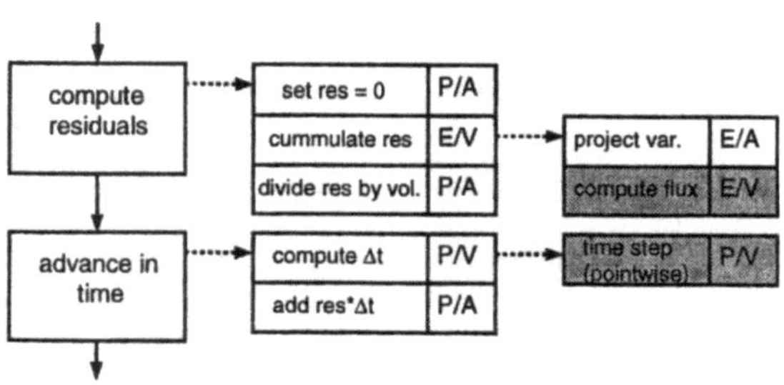

Figure 7 Sketch of a simple explicit time step. P: pointwise, E: edgewise, A: automatic, V: virtual. Grey fields: dependent on a specific application.

The dotted arrows indicate virtual function calls. These require an additional amount of computational time. To minimize this overhead it is useful to provide a sufficient amount of computational work within the virtually called methods. Therefore these methods contain loops for sets of points or edges. Since data is passed between subsequent virtual calls, it is not useful to perform the inner loops over all points or edges, but only for moderate sized groups, thus saving memory.
The one step time integration is the simplest version. Other time integration methods are constructed in a comparable way, while the physically relevant modules can be re-used without any further changes. The next subchapter illustrates the users interaction to alter the time integration sequence.

6.4 Building up a specific computation

To enable maximum flexibility, a macro type language for a master control file was developed. This language allows a user or to build up his individual application program by indicating the computational sequence in detail, or to rely on a predefined sequence.
To clarify this point, consider a transient compressible flow problem with a low Mach-number to be computed. The user decides to use an AUSM-flux. For a very first attempt, a one-step forward Euler time-integration is used. A corresponding control-syntax could look like this:

```
....
       OneStepEuler {#MACRO-PAR[all_ausm_expl]()}
....
```

In this context, the used parameter macro `all_ausm` refers to a previous definition of all actions performed within a single explicit sub-step. The empty parameter range () is supposed to allow further actions to be performed.
Since the chosen method shows to be very inefficient, the user decides to try out a Runge-Kutta integration. This can be obtained in the following way:

```
....
       RungeKutta {#MACRO-PAR[rk](3)
         OneStepEuler {#MACRO-PAR[all_ausm_expl]()}
       }
....
```

The macro-parameter `rk` contains, as written, all definitions concerning the looping within the integration cycle and carries a further parameter range, to allow describing the number of sub-steps chosen 3 here.
Still not satisfied, a Turkel type preconditioner may be employed. Since time accuracy would be lost, the dual time stepping method is used and the inner iteration is still done with the Runge Kutta algorithm. This could look as follows:

```
....
       DualTimeStep {#MACRO-PAR[dual]()
         RungeKutta {#MACRO-PAR[rk](3)
           OneStepEuler {#MACRO-PAR[all_ausm_expl](#MACRO-PAR[turkel])}
         }
       }
....
```

Again to improve the performance, a multigrid scheme, currently in development for the present software, can be applied. Since it is useful for the inner iteration process of the Dual time stepping, this is a shell between `DualTimeStep` and `RungeKutta`:

```
....
       DualTimeStep {#MACRO-PAR[dual]()
         MultiGrid {#MACRO-PAR[multigrid]()
           RungeKutta {#MACRO-PAR[rk](3)
             OneStepEuler {#MACRO-PAR[all_ausm_expl](#MACRO-PAR[turkel])}
           }
         }
       }
....
```

And terms correspond in analogy to the above said.
Consider this combination showed to be efficient and robust to solve a problem of the given type, thus the method might also be useful for another type of physical problem. It is thus worth while constructing a macro upon the whole sequence:

```
....
#DEF[good_method](%1,%2,%3,%4) {
  DualTimeStep {#MACRO-PAR[dual](%1)
    MultiGrid {#MACRO-PAR[multigrid](%2)
      RungeKutta {#MACRO-PAR[rk](%3)
        OneStepEuler {%4}
      }
    }
  }
}
```

Further on the method may be used in this macro version:

```
....
      #MACRO[good_method](,,3,#MACRO-PAR[all_ausm_expl](#MACRO-PAR[turkel]))
....
```

Note, that all these variations can be performed, without even recompiling the code. All modules are thus available all time and used according to specific needs.

7 Grid generation

Hybrid grid generation forms a major part of the whole project. It is obvious, that the development of solution methods on hybrid grids is only meaningful, if corresponding meshing methods are available. At the present stage, the developments are carried out by group G. To allow comparisons, file translators enable group F to access sample meshes.
Regarding the difficulties, associated with the generation of hybrid grids, we may justify the corresponding developments. A simple 2-D proof shows, that, employing a classical nodal approach, the positiveness of the diffusive terms on a 2-D simplex mesh is fulfilled, if the triangulations satisfy the Delaunay conditions. Reversing this proof, we find, that this property is lost for non-Delaunay triangulations.
This restriction is a crucial disadvantage of unstructured grids in CFD, since corresponding solutions show highly anisotropic scaling, e.g. boundary layers, shear layers or even shocks. For an efficient computation of such anisotropic features, also an anisotropic resolution of the mesh is required. To justify this claim, consider the following example:
Assume an isotropic, non adaptive mesh to solve a problem in the dimension d. Assume furthermore, that an anisotropic feature of the solution requires to reduce the scale length of the mesh by a factor of r in a distinct direction, and that this is required for a number Ne_1 of elements of the non adaptive mesh. The number of nodes (or cells) associated with this adaption is then:

$$\begin{aligned} Ne_{1,adapt} &\sim Ne_1 \cdot r && \text{for an anisotropic refinement} \\ Ne_{1,adapt} &\sim Ne_1 \cdot r^d && \text{for an isotropic refinement}. \end{aligned}$$

The factor r may easily reach values above 100, for example in boundary layers. We may thus conclude, that, specially in 3-D, isotropic meshes are not useful for most problems.
However, conventional meshing approaches on unstructured grids tend to produce flat triangles with obtuse angles in anisotropic regions, thus violating the Delaunay-criterion. To overcome the problem, the following strategies may be followed:

a) Improve the behaviour of the discretization of terms containing second derivatives on the discretization level.

b) Alignment of simplex elements in order to reduce the viscous contributions in critical directions.

c) The development of skew meshing methods, satisfying the Delaunay-criterion in anisotropic regions.

The technique of alignment b) showed to be useful for 2-D computations. However, it is not possible to satisfy the conditions for all viscous effects at once, if several diffusive mechanisms have to be modelled. The development of skewed simplex meshing methods is possible, however, the difficulties are comparable to the ones for the generation of multi-element meshes.

Demands for the generation methods

Before continuing, the following demands on the generation system are imposed:

- **Quality:** The estimation of the quality is a difficult task. We may however, in a first approach claim, that simplex neighbour relations must satisfy the Delaunay criterion, while bilinear elements are intended to approach orthogonality.
- **Flexibility:** Grid generation is intended, not to be restricted by geometrical means and also not to be limited in its capability for adaptation.
- **Automatism:** The user interaction must be minimized.
- **Ability for modification:** Grid generators are intended to be able to transform an existing mesh into a new one, avoiding complete regeneration, when adapting a mesh.

7.1 Overview of the basic unstructured generator for simplex grids

A review about unstructured grid techniques and generation has been published recently by Mavripilis [14]. Own developments in the same direction, previously published for simplex mesh generation, [15], play an essential role for the further development of the hybrid grid generator. Therefore the essentials of the existing grid generator is described in short form at next.

(a) Inputs

As inputs for the generation system a set of closed boundary lines (triangulated boundary faces) are required. The surfaces for the 3-D version can be obtained by a modified version of the 2D unstructured mesh generator. For more complex problems, an interface for a CAD-system is available.

(b) First triangulation of the computational domain

Starting at given boundary segments (triangulated surfaces), simplex elements are build using a front Delaunay, rising bubble type mesh generation algorithm. At this stage of the generation only the given nodes at the surfaces are used. The 3-D version however, requires a repair tool for the insertion of additional nodes due to geometrical reasons.

(c) Mesh optimization

After the initial triangulation the mesh is optimized employing a set of local tools. These are employed in a recurrent structure up to medium convergence of the mixed discrete analogue optimization problem. While iterating, the generator has to keep traces on the rated value of the local mesh density, called G_k. This local quantity provides the information required for a smooth change of the element sizes between the boundaries of the domain and stored at the nodes of the developing mesh. Since the size is given at the boundaries, the quantity is computed solving a boundary value problem during the development of the mesh.

The mesh modification tools themselves are the following:

a) Additional nodes are inserted in the center of elements or boundary segments (triangles) according to the local edge length compared to the mesh density function G_k. The reverse tool for the deletion of nodes is formulated by shrinking edges to single points.

b) Mesh reconnection by a swapping algorithm. The connection of local groups of simplex elements are analyzed. In 2-D, these are two neighbouring triangles, whose common edge may be swapped. In 3-D groups of tetrahedra with at all five nodes are analyzed. These are two neighbouring tetrahedra, three ones around a common edge or four ones around a common node. Possible swaps are able to transform the two tetrahedra versions in the one with three tetrahedra and vice-versa. Decisions are made upon the Delaunay criterion.

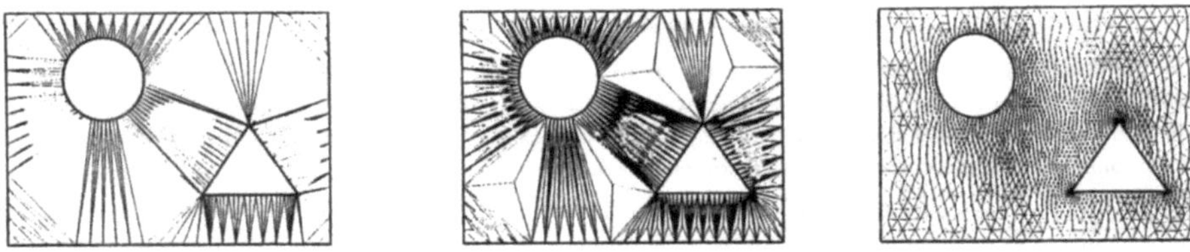

Figure 8 Simple example for mesh generation: First closed triangulation (left), mesh after the first insertion of additional nodes (center) and final, optimized mesh (right)

c) Smoothing by moving nodes to optimized positions. This tool makes up the analogue part of the optimization process. The corresponding smoothing criterion is formulated as the minimization of a sensitive quantity. For meshes, consisting only of simplex elements, criteria based on circumcircle areas (circumsphere volumes) and element measures showed to be very useful.

A very small 2-D example is shown in figure 8.

(d) Mesh adaptation

Adaptivity is introduced via virtual stretching, that is, according to some criteria, the physical space is transformed in a wider virtual space. The transformations are performed employing local symmetric 2x2 (3x3) matrices:

$$A = \begin{pmatrix} a11 & a12 & [a13] \\ a12 & a22 & [a23] \\ [a13] & [a23] & [a33] \end{pmatrix}. \tag{7.8}$$

Mesh optimization proceeds at the transformed virtual space employing the previously introduced generation algorithm, and in physical space the adaptive triangulation is obtained. Since the transformation matrices are very flexible, isotropic or anisotropic, adaptive regions can be produced. Since the stretching properties play an essential role in the latter hybridization tools, their stretching properties are explained in more detail. For this consider the transformation of a unit circle.

Three types of transformations are possible. An isotropic transformation, a unidirectional stretching and a generally anisotropic behaviour. Figure 9 shows the corresponding transformations. In the isotropic case, the scale lengths in physical space are reduced by a factor of R, in the other cases a reduction of the scale lengths is dependent on the direction. The extrema of the stretching function appear in the direction of the eigenvectors of A, with R_1, R_2 (and R_3) being the correspondent eigenvalues.

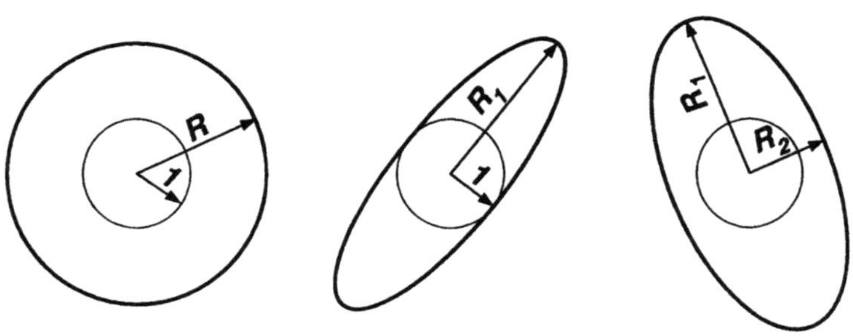

Figure 9 Transformation of a unit circle: isotropic (left), unidirectional (center) and generally anisotropic (right).

Since the quantity will play an important role later, the stretch relation is introduced, being the quotient of the maximum and minimum eigenvalue:

$$QS = \frac{R_{max}}{R_{min}}.$$

Since the generator produces isotropic elements in the transformed space, QS represents the aspect ratio of the elements in the physical space.

7.2 Properties of the element types

At present, the following elements are considered:

- 2-D: triangles and quadrilaterals
- 3-D: tetrahedra, pyramids, prisms and hexahedra

Properties in stretched regions

Suppose that using isotropic elements, it is possible to construct a discretization method which is able to treat a given problem sufficiently well. This assumption corresponds to the experiences, at least for triangles and quadrilaterals in 2-D as well as tetrahedra, prisms and regular hexahedra in 3-D. Knowing that the interior angles of the elements are crucial for the accuracy of the discretization, allows to choose the best suited element types according to the properties of the stretching matrix. The question is, for which element types is it possible to apply a corresponding transformation, keeping the interior angles invariant.

To answer the question, the internal orientation of the elements must be analyzed. The question is, how many perpendicular directions for a stretching of an element can be found, without impairing the interior angles. It appears immediately, that the simplex elements show zero directions with this property. Upon the above assumption, their use is only save in isotropic regions. Quadrilateral elements show two such directions, they may thus be stretched as long as their main axis correspond to the eigenvectors of the stretching matrix.

The same analysis for the 3-D elements, reveals, that a prism shows one distinct direction of angle-invariant stretching, it is thus well suited at locations of unidirectional stretching, if its normal direction is aligned to the stretching direction (eigenvector). This means, the prism is a suitable element for shear layers, boundary layers of stocky geometries and an adaptive treatment of shocks.

Hexahedra show three directions of angle-invariant stretching. These elements are thus suitable for locations with generally anisotropic scale lengths, however requiring an alignment of their directions with the corresponding eigenvectors of the stretching matrices.

Unfortunately, the pyramid shows no distinct internal orientation.

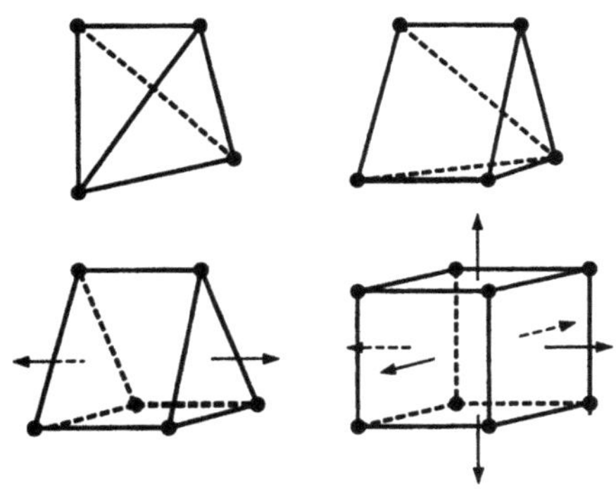

Figure 10: Internal orientation of 3-D elements.

Aspects of computational efficiency

Current algorithms of both groups use nodal arrangements of control volumes with edge based calculations of flux contributions. The CPU-time and memory consumption of the edge-based data structures are thus proportional to the number of edges in the mesh. Therefore, the relation between the number of edges and the number of nodes is an important measure for the efficiency of a solution method. The tables below show the relations and confirm, that meshes with quadrilateral or hexahedral elements yield lower computational cost.

2-D element-type	relation of edges / node
triangular	3
quadrilateral	2

3-D element-type	relation of edges / node
tetrahedral	7
prismatic	4
hexahedral	3

7.3 General approach for the hybrid grid generation

At present, two principal strategies are analyzed, a relatively simple zonal approach and a fully unstructured, elliptic approach. Both methods are restricted to sets of elements. In 2-D only triangles and quadrilaterals are used. Upon the above stated properties of the elements, prisms and hexahedra are

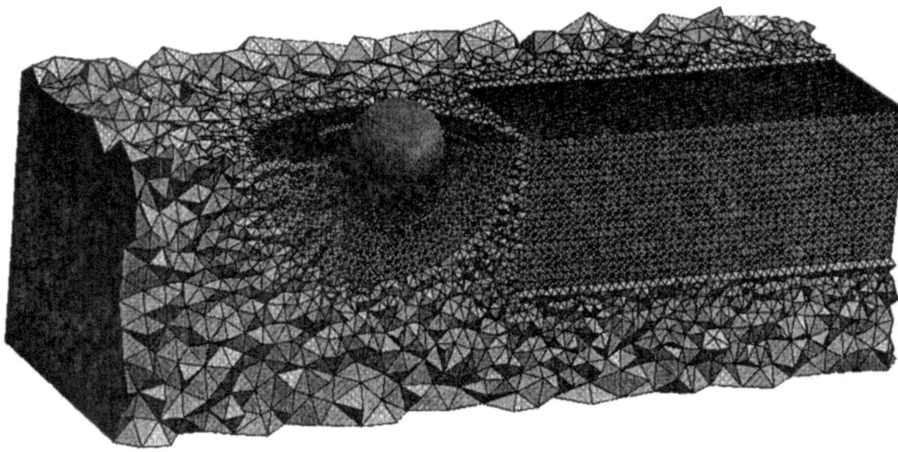

Figure 11 Broken partition of a hybrid grid for the computation of flows past a sphere. Mesh consisting of prisms, Cartesian hexahedra, pyramids and tetrahedra.

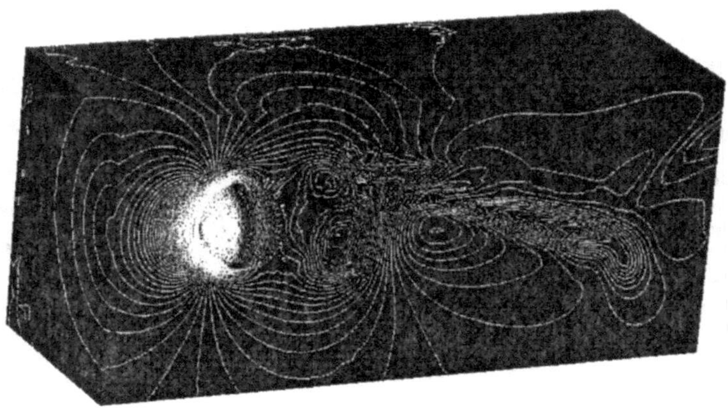

Figure 12 Lines of constant pressure for a laminar, time dependent compressible flow at $Re = 1000$ and $Ma_\infty = 0.2$, computed on the mesh above.

the most useful elements. Hybrid grids may therefore consist mainly of these two element types. Pyramids are required as intermediate elements between quadrilateral and triangular faces of other elements. Tetrahedra are still required, to provide maximum geometric flexibility.

The zonal approach
One possibility to generate geometrically complex grids with bilinear elements is the construction of structured or semi structured blocks, which are then connected to a whole mesh by unstructured mesh techniques. A simple example of such a grid, employed for the flow computation past a sphere is shown in Fig. 11. The mesh consists of prismatic layers around the triangulated sphere and a Cartesian block covered by pyramids in the wake region, while the "empty spaces" are filled by tetrahedra. A corresponding solution for a laminar flow is presented in Fig 12. A similar approach can also be found in [16].

The vast disadvantage of the method is, that it violates the demands, stated in sub-chapter 7. In fact it is inflexible (consider a bent object) and adaptive modifications are restricted. A fully automatic generation is however possible and the quality is considered good.

The seed and crystallization strategy Subject of the ongoing investigation is the trial to generate hybrid grids starting at fully unstructured, tetrahedral ones. Upon local transformations, the desired elements can be created within the tetrahedral mesh. It is thus possible, to include such modification tools to the mesh optimization algorithms for simplex grids, yielding automatic hybrid generators. Such methods open up a large number of possible strategies. Among these, a seed and crystallization strategy seems most promising. The corresponding development is currently restricted to 2-D, however the method is transferable to 3-D. It outlines as follows:

a) Create a first priming bilinear element: A good location to do so is typically a segment (face in 3-D) on a boundary line or surface of an object, as it is then automatically aligned.

b) Crystallization: A growth mechanism in the directions of the eigenvalues of the stretching matrix. That is, typically a grow mechanism aligned and normal to layers.

c) Squeeze simplex elements: Seed, crystallization and grow produce additional nodes, thus the mesh density is increased. The smoother is now employed for a redistribution, yielding a compression of surrounding simplex elements. The mesh density in these regions is then reduced by the edge-shrinking routine, thus deleting simplex elements.

Upon the above discussion on the dependence of the suitable element type according to a local stretching of the mesh, the creation and orientation of the bilinear elements is directly related to a local stretching matrix. Bilinear elements are created, if the aspect ratio QS exceeds a limiting value QS_{min} and the alignment is made according to the eigenvectors. This means, that mesh regions with low anisotropy are not converted. An implementation of this method for 2-D meshes is discussed in the next sub-chapter.

7.4 Implementation of a hybridization tool in 2-D

The 2-D mesh-generator, currently in development, uses a full elliptic approach for the generation of simplex meshes. Generation of quadrangular elements is done with the seed and crystallization strategy outlined above. Pentagonal elements or elements with even more nodes are not used at present, since the corresponding methods would not be extendeable to 3-D mesh generation and since such elements lead to big errors in case of upwind discretizations.

Local Strategies for Handling of Quadrilateral Elements

To implement this strategy, local methods for the creation and destruction of quadrangles are needed. At present the system is limited to one method for creation and one for destruction.

Creation of Quadrangles

The creation-algorithm needs an edge, on which it is able to grow a quadrangle. In the example shown besides, this is the edge AB which is part of the triangle ABC. The first step is to add a new point D inside the triangle ABC. This is the same method as it is used in the simplex-mesh-generation. Further, this point is split up into two points D' and D''. This splitting causes the edge CD to turn into the triangle $CD'D''$ and the triangle ABD into the quadrangle $ABD'D''$.

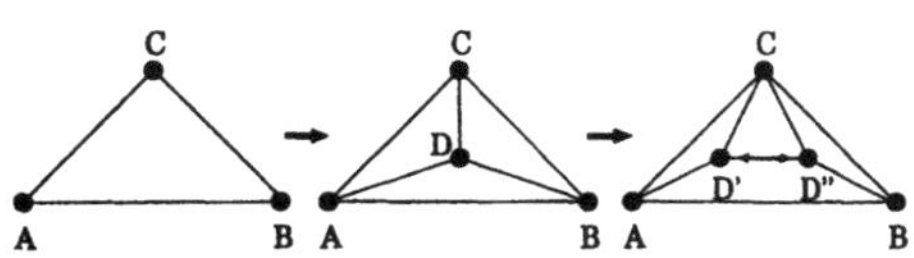

Figure 13 Generation of quadrangular elements

Destruction of Quadrangles

There is no direct destruction algorithm for quadrangles included in the mesh-generator. Quadrangles which have to be removed, are split into two triangles. Splitting is performed, starting with the node of the quadrangle, which has the largest inner angle. Irregular elements are avoided this way, when splitting convex elements. The picture on the right shows two different configurations, where this splitting is applied.

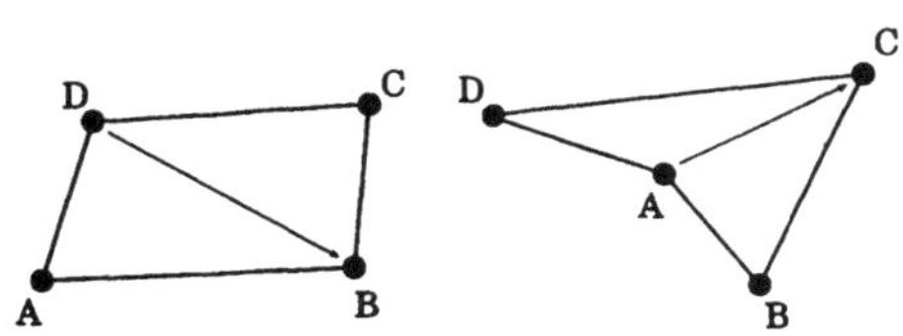

Figure 14 Splitting of quadrangular elements

Smoothing of Quadrangles

Smoothing for hybrid meshes is performed in the same manner as it has already been for simplex meshes. Every point is moved into an optimized position, by minimizing a local mesh error, which can be computed with the knowledge of the neighbouring elements. Only the formulation of this error is different for quadrangular regions. For a given quadrangle with the radius vectors $\vec{x}_1, \vec{x}_2, \vec{x}_3, \vec{x}_4$ the error will be computed as follows. Around the point $\vec{x}_m$ an ideal quadrangle is developed according to the eigenvectors $\vec{v}(\lambda_1), \vec{v}(\lambda_2)$ of the stretching matrix. $\vec{v}(\lambda_1)$ and $\vec{v}(\lambda_2)$ are normalized with the local net-density ρ_n and the eigenvalues.

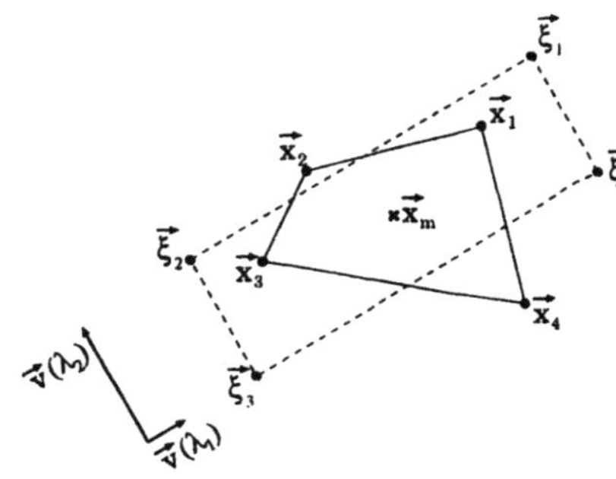

Figure 15 Quadrangular element and ideal shape

$$\vec{v}_n(\lambda) = \frac{1}{|\vec{v}(\lambda)|\rho_n}\vec{v}(\lambda). \tag{7.9}$$

The coordinates of the ideal shaped element are the following:

$$\vec{\xi}_1 = \vec{x}_m + \tfrac{1}{2}(\vec{v}_n(\lambda_1) + \vec{v}_n(\lambda_2)) \quad , \quad \vec{\xi}_2 = \vec{x}_m + \tfrac{1}{2}(-\vec{v}_n(\lambda_1) + \vec{v}_n(\lambda_2))$$
$$\vec{\xi}_3 = \vec{x}_m + \tfrac{1}{2}(-\vec{v}_n(\lambda_1) - \vec{v}_n(\lambda_2)) \quad , \quad \vec{\xi}_3 = \vec{x}_m + \tfrac{1}{2}(\vec{v}_n(\lambda_1) - \vec{v}_n(\lambda_2)).$$

For a single quadrangle the error is:

$$err_{quad} = \sum_{i=1}^{4}(\rho_n|\vec{xi}_i - \vec{x}_i|)^{\alpha}. \tag{7.10}$$

For a point, being a vertex of N quadrangles, the total error becomes:

$$err_{point} = \frac{1}{N}\sum_{i=1}^{N} err_{quad_i}. \tag{7.11}$$

Unfortunately it is not possible, to assign every point it's corresponding ideal partner uniquely. Therefore all four permutations are computed and the minimal error is taken. For $N = 0$ the error will be computed like in a pure simplex mesh-generator. In case of $N > 0$ no influence of the connected triangles will be respected. α can be used to adjust the behaviour of the smoothing algorithm. If $\alpha > 1$ a compromise between all connected quadrangles will be found, and for $\alpha < 1$ good elements will be favoured. $\alpha < 1$ only makes sense if an efficient quad-removal algorithm is available, which does not destroy an already established structure. At present $\alpha > 1$ is used.

Fig.16 exemplarily shows the growing of quadrangles. The result of a first automatic generation process is presented in Fig.17.

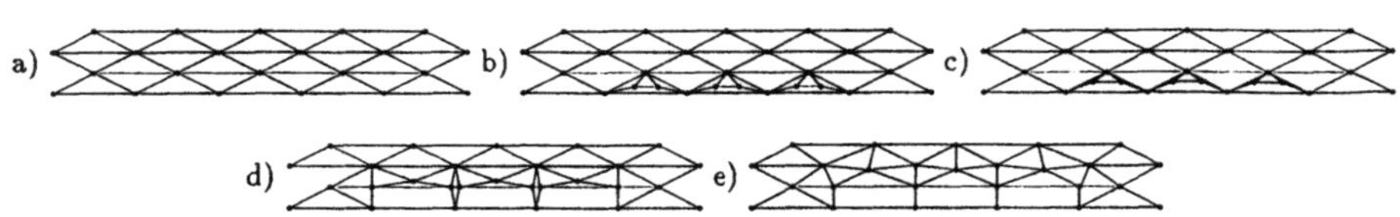

Figure 16 Growing of a Quadrangular Layer a) initial stretched simplex mesh b) insertion of quadrangular elements c) first smoothing of quadrangles (further improvement is prevented by the surrounding triangular elements) d) after reconnecting and another smoothing e) quadrangles grew together due to the elimination of flat triangles

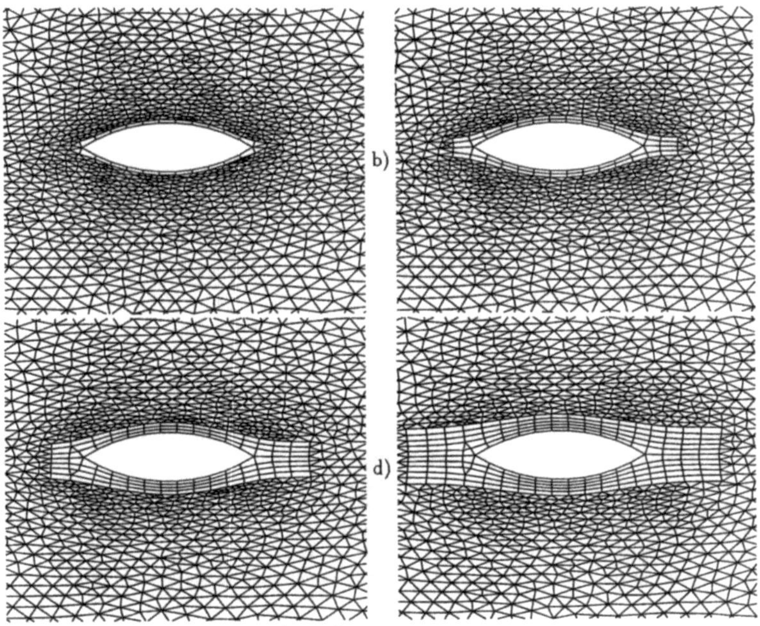

Figure 17 Several stages of an automatic growing process

7.5 A possible extension to 3-D via element conversion

The 2-D version of the method showed already to be very difficult. Nevertheless, a 3-D version is possible and is thus intended. Upon first thoughts, element conversion seems promising for the seed and crystallization process.
At first, we may restrict the hybridization to the generation of prisms. A further extension towards hexahedra will be developed, if the prismatic generation is successful. The hybridization method may be constructed upon the following conversion processes.

Triangular face to tetrahedron
Creating an additional edge, an existing triangle may be split up and converted to a tetrahedron. This operation will probably not be desired, but it appears as a side effect, of other conversions to ensure consistency. for a single triangular face, three possibilities exist.

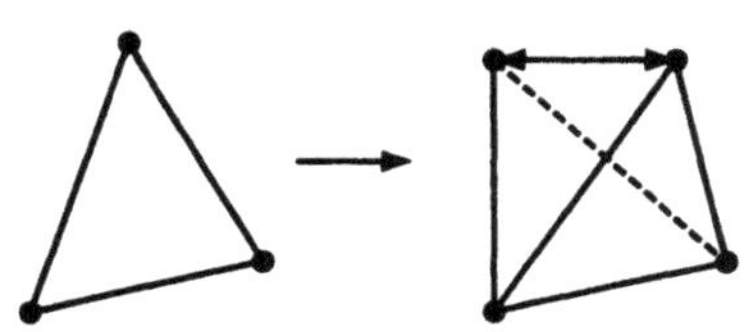

Tetrahedron to pyramid
The creation of an edge may produce a pyramid out of a tetrahedron. For this operation, 12 possibilities exist.

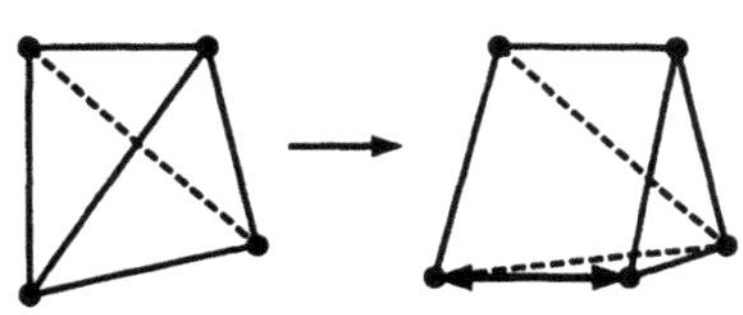

Pyramid to prism

A pyramid may be converted into a prism, again be creating an additional edge. The number of possible choices is two.

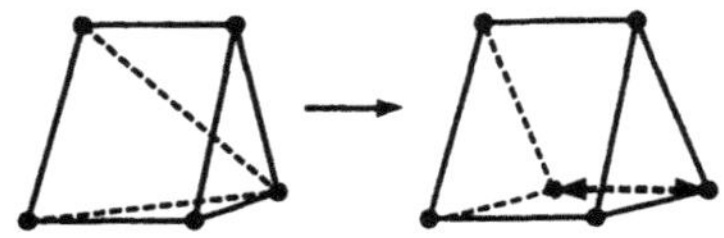

Set of tetrahedra to set of triangular faces

Deleting an edge surrounded by tetrahedra, destroys them all, yielding a set of triangular faces.

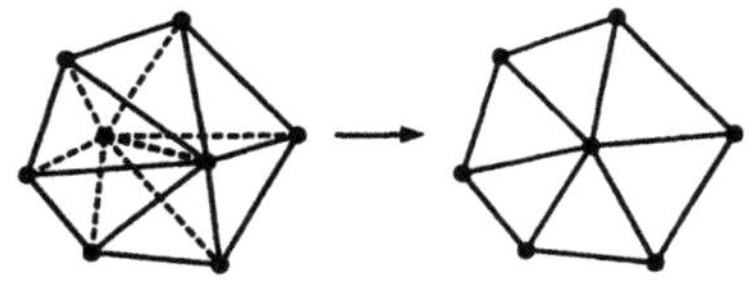

A seed process for the production of priming elements could thus look as sketched in figure 18. A triangular face between two neighbouring tetrahedra is split and a new tetrahedron is produced and directly converted to a pyramid which itself is converted to the first prism. In total, this process produces one prism, three adjacent pyramids and some new tetrahedra in between the new faces of the pyramids. The crystallization in the layer direction proceeds as following. An adjacent pyramid is converted into a prism. Since two of its triangular will then get quad-faces, the neighbouring elements will be converted to the "next higher" element type, figure 19. The process continues at the next prism.

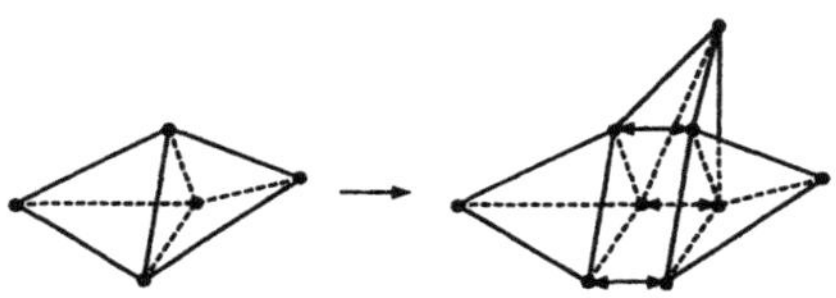

Figure 18: Seed process for the creation of a priming prism.

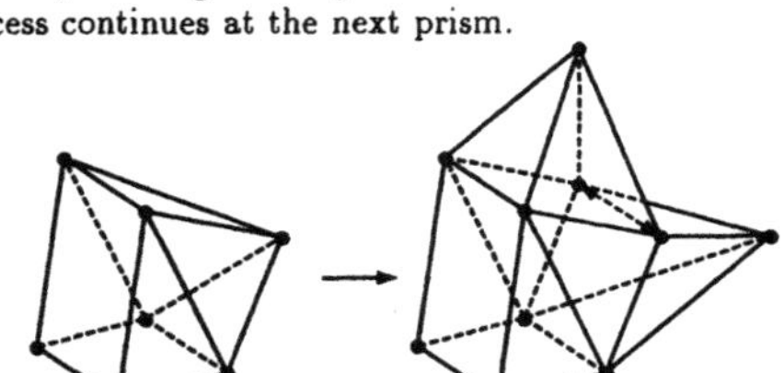

Figure 19: Crystallization process via element conversion.

It is interesting to note, that all these operations are based or on edge-creation or edge-deletion. Remembering the above discussion on the uniformity of data structures, this is obvious, since the connectivity of every mesh might be described by its set of edges. Therefore, modifying the set of edges accordingly may produce the desired mesh with any element types.

At present, all mesh generation algorithms are based on element structures. For the above conversion tools, the use of edge based or node-neighbournode structures seems useful. However, the sole use of these structures is not sufficient. At least hidden temporary element structures are additionally required.

References

[1] A. D. Robison, P. F. Dubois: C++ Gets Faster for Scientific Computing. In: Computers in Physics, Vol 10, pp 458-462, American Inst. of Physics (1996).

[2] J.A. Désidéri, P.W. Hemker: "Convergence Analysis of the Defect-Correction Iteration for Hyperbolic Problems," SIAM J. Sci. Comput., vol. 16, pp 88-118, (1995).

[3] R. Vilsmeier and D. Hänel: Adaptive Solutions for Unsteady Laminar Flow on Unstructured Grids. Int. Journal for Numerical Methods in Fluids, Vol 22, 85-101 (1995).

[4] M. Breuer, D. Hänel: A Dual Time-Stepping Method for 3-D, Viscous, Incompressible Vortex Flow. Computer & Fluids, vol. 22, pp. 467-484, (1993).

[5] Meinke M., Hänel D.: Time Accurate Multigrid Solutions of the Navier-Stokes Equations. Intern. Series of Numerical Mathematics, vol. 98, pp. 289-300, Birkhauser, Basel, (1991).

[6] M. H. Lallemand, H. Steve, A. Dervieux: Unstructured multigridding by volume agglomeration: current status. In: Computers and Fluids, Vol. 21, No. 3, pp. 397-443, 1992.

[7] P. L. Roe: Approximate Riemann Solvers, Parameter Vectors, and Difference Schemes, JCP, vol. 43,pp 357 - 372, (1981).

[8] E. Turkel: Preconditioned Methods for Solving the Incompressible and Low Speed Compressible Equations, JCP,vol.72, pp 277 - 298, (1987).

[9] M.S. Liou, C.J. Steffen: A New Flux Splitting Scheme: JCP, vol. 107, pp 23-39, (1993).

[10] R. Schwane, D. Hänel: An Implicit Flux Vector Splitting Scheme for Viscous Hypersonic Flow. AIAA paper, No. 89-0274, (1989).

[11] H. Guillard and C. Viozat: On the behavior of Upwind Schemes in the Low Mach Number Limit. Rapport INRIA no 3160,(1997). (To appear in Computers and Fluids).

[12] O.Gloth, R. Vilsmeier, D. Hänel: Object Oriented Programming for Computational Fluid Dynamics. In: M. Bubak, J. Mościński: Proceedings of HiPer'97, ISBN 83-902363-6-2 (1997).

[13] John R. Cary and Svetlana G. Shasharina: Comparison of C++ and Fortran 90 for Object-Oriented Scientific Programming. To appear in Computer Physics Communications, Available from Los Alamos National Laboratory as Report No. LA-UR-96-4064. (1996).

[14] D.J. Mavriplis: Unstructured Grid Techniques. Ann. Rev. Fluid Mech., vol. 29, pp. 473-514, (1997).

[15] Vilsmeier R. and Hänel D.: A Field Method for 3-D Tetrahedral Mesh Generation and Adaption. Proc. of 14th Int. Conf. on Num. Meth. in Fluid Dynamics, Bangalore, India 1994.

[16] A. Khawaja, H. McMorris, and Y. Kallinderis: "Hybrid Grids for Viscous Flows around Complex 3-D Geometries including Multiple Bodies," AIAA Paper 95-1685-CP, San Diego,CA, June (1995).

II. CRYSTAL GROWTH AND MELTS

High Performance Computer Codes and their Application to Optimize Crystal Growth Processes

P. Droll[2], M. El Ganaoui[4], L. Kadinski[1], M. Kurz[3], A. Lamazouade[4], O. Louchart[4], D. Morvan[4], M. Naamoune[4], A. Pusztai[3], I. Raspo[4], P. Bontoux[4], F. Durst[1], G. Müller[3], J. Ouazzani[5], M. Schäfer[2]

[1] Lehrstuhl für Strömungsmechanik, Universität Erlangen-Nürnberg
Cauerstr. 4, D-91058 Erlangen
[2] Fachgebiet Numerische Berechnungsverfahren im Maschinenbau,
Technische Universität Darmstadt
Petersenstr. 30, D-64287 Darmstadt
[3] Kristallabor am Institut für Werkstoffwissenschaften
Lehrstuhl Werkstoffe der Elektrotechnik, Universität Erlangen-Nürnberg
Martensstr. 7, D-91058 Erlangen
[4] Dpt. de Modélisation Numérique IRPHE - Réseau MFN
Université d'Aix-Marseille II
Technopôle de Chateau-Gombert, 13451 Marseille Cedex 20
[5] ArcoFluid Les Bureaux de l'Arche
5, rue des Allumettes
13086 Aix-en-Provence Cedex 02

Summary

The paper deals with the development of high performance computer codes and their application to modelling of crystal growth processes. The mathematical model is based on the conservation equations for momentum and heat transfer combined with mass transfer including chemical reactions. The thermal radiation analysis assumes a non-participating medium and semi-transparent quartz walls. The radiation heat transfer is coupled with convection and conduction. The heat conduction includes thermal solid/fluid interactions between the gas and solid parts of the computational domain. The models are implemented in a finite volume and a spectral and coupled finite volume/spectral numerical solution procedures on block-structured non-orthogonal grids.
The capability of the developed methods is demonstrated for the different practically used crystal growth processes. The calculations are done for Chemical Vapor Deposition, Vertical Gradient-Freeze and Czochralski processes at the practically used growth conditions.

1 Introduction

The goal of the collaborative work reported in this paper resides in the improvement of crystal growth systems and their control. In particular, the industrially widely employed growth techniques from the vapour phase by chemical vapour deposition (CVD) and from the melt phase by solidification (Czochralski, Bridgman) are considered. The objectives are sought via the modelling and via the numerical simulation. The elaboration of high-technology materials is known to generate complex fluid dynamics and heat transfer due to a number of interactive physical mechanisms. In particular, we have to account for very stiff concentration layers in vapour reactor flows that requires a high resolution. Transport in vapour and melt phases can give rise to instability, time-dependent and turbulent flows. In the case of solidification from a melt the constitution of the materials and the shapes of the successive interfaces during the process are of importance for the crystal growth. In the global approach of technological devices, the simulation simultaneously has to take into account the conduction in the walls of the set-up and the radiation between the solid structures in addition to the fluid flows in the vapour or in the liquid phases.

A flexible and efficient solution method for the global approach and for the prediction of the fully turbulent flows is the finite volume technique with multi-blocks. The description of the interface change for no-flow situations is also treated alternatively with unstructured meshes and via a homogenization approach based on an enthalpy-porosity technique and applied to interactive oscillatory melts and interface situations. For the investigation of stiff shear layers and of time-dependent instabilities, a spectral approach appears to be more advantageous.

The elaboration of complementary and interactive codes has been developed towards several directions : (i) the extension of the multi-block structured and unstructured code ; (ii) the extension of the spectral approximation to domain decomposition using direct (influence matrix) [1] and iterative solvers [2] together with a projection technique ; (iii) an original combined finite volume-spectral approach has been initiated on the basis of the multiblock code and of the spectral projection solver.

We focus on a limited number of aspects concerning the algorithms and the physics: the combined finite-volume spectral technique is presented and its capabilities are illustrated, the enthalpy method for the simulation of the interaction of an oscillatory melt with a solidification front during a Bridgman growth is described, the application of the finite volume multi-block code for investigating the physics of the flow and transfer during CVD growth.

2 Mathematical formulation

The mathematical model used in the present study is based on the solution of coupled flow, heat transfer and mass transport equations including multicomponent diffusion and chemical reactions [3, 4].
Continuity equation

$$\frac{\partial \rho}{\partial t} + \nabla \cdot (\rho \vec{v}) = 0 \qquad (2.1)$$

with: ρ - the density and $\vec{v}$ - the velocity vector.
Momentum equations (Navier-Stokes):

$$\frac{\partial}{\partial t}(\rho \vec{v}) + \nabla \cdot (\rho \vec{v} \vec{v}) = -\nabla p + \vec{f} + 2\nabla \cdot (\eta \dot{S}) - \frac{2}{3}\nabla \cdot (\eta \nabla \cdot \vec{v}) \qquad (2.2)$$

with: η - the dynamic viscosity coefficient, p - the dynamic (excess) pressure, $\vec{f}$ - the gravitational acceleration, $\dot{S}$ - the deformation rate tensor.
Energy equation:

$$c_p \frac{\partial}{\partial t}(\rho T) + \nabla \cdot (-\lambda \nabla T) + c_p \nabla \cdot (\rho T \vec{v}) - s_T = 0 \qquad (2.3)$$

with: c_p - mass specific heat, λ - thermal conductivity, s_T - density of heat source.
Heat transfer by radiation:
In the radiation model an enclosure with opaque or semi-transparent grey-diffuse radiating walls and non-participating gas inside and ambient outside are assumed. In this case the energy flux to the inside wall is composed of two contributions: radiative transfer from surfaces facing the wall and conduction by the gas. Energy is lost from the wall to the ambient by radiation and cooling by the surrounding (ambient) gas. Thus, the wall boundary conditions take the form:
Flow–Wall:

$$\mathbf{n} \cdot \Big(-\lambda \nabla T\Big)_f = \mathbf{n} \cdot \Big(-\lambda \nabla T\Big)_w + q_i^{in} - q_o^{in} \qquad (2.4)$$

Wall–Ambient:

$$\mathbf{n} \cdot \Big(-\lambda \nabla T\Big)_w = h(T_w - T_a) - q_i^{out} + q_o^{out} \qquad (2.5)$$

where the indices "'w"' and "'g"' are related to the wall and to the gas, respectively, T_a is the temperature of the ambient, h is the overall heat transfer coefficient, q^{out} and q^{in} are the radiation fluxes to the wall (index i) and from the wall (index o) on the outer and inner side, respectively, and $\mathbf{n}$ is the unit vector orthogonal to the wall and directed outwards. In the case of thin walls, the boundary conditions at the wall take the form:

$$\mathbf{n} \cdot \Big(-\lambda \nabla T\Big) = h(T - T_a) - \underbrace{(q_i^{out} - q_o^{out} + q_i^{in} - q_o^{in})}_{q_{net}}, \qquad (2.6)$$

where T is the temperature of the gas on the inner side of the wall. To close the model equations the radiation fluxes have to be specified. If radiative properties do not depend on wavelength, the net radiation flux q_{net} for the k–th surface element becomes:

$$q_{net,k} = E_k(q_{i,k}^{in} - s_k\sigma T_k^4 + (s_k - 1)\sigma T_a^4) \tag{2.7}$$

where $s_k = 1$ for opaque and $s_k = 2$ for semi-transparent walls. The radiation heat flux $q_{i,k}^{in}$ is defined by the system of equations:

$$q_{i,k}^{in} = \sum_{j=1}^{N} q_{o,j}^{in} F_{k-j} \tag{2.8}$$

$$q_{o,k}^{in} = \sigma E_k T_k^4 + R_k q_{i,k}^{in} + \sigma \mathrm{Tr}_k T_a^4, \tag{2.9}$$

providing the heat exchange relationship between the radiating boundaries.
Conservation of chemical species:

$$\frac{\partial}{\partial t}(\rho m_l) + \nabla \cdot (\rho \vec{v} m_l) = \nabla \cdot \left(\rho D_l \left(\nabla m_l + \alpha_l m_l \frac{\nabla T}{T}\right)\right) + R_l \tag{2.10}$$

with: m_l - mass fraction of species l, D_l - diffusion coefficient of species l, α_l - thermal diffusion coefficient of species l, R_l - rate of generation or consumption of chemical species per unit volume.

For gases the equation of state is valid:

$$\rho = \frac{\mathrm{P_0}}{\mathrm{RT}} \sum_{\mathrm{i=1}}^{\mathrm{N}} \mathrm{x_i M_i}, \tag{2.11}$$

with: x_i - molar fraction, M_i - molar mass P_0 - constant operating (thermodynamic) pressure.

3 Algorithmic Developments

The complexity of crystal growth process requires global simulations including a variety of governing physical and chemical phenomena. It is the aim of the present project to develop new efficient numerical techniques to carry out parameter studies for the performance of crystal growth configurations and utilizing it for equipment optimization.

3.1 Coupling of Finite-Volume Method and Spectral Method

Motivation

Spectral methods (SM) feature the property of being of high accuracy, provided the number of continuous derivatives of the solution to be approximated is large and the computational domain is a Cartesian one. Because of the higher accuracy of the spectral method, one gets the same accuracy as with a standard finite-volume method (FVM) for fewer degrees of freedom.

On the other hand, when the computational domain has a complex shape, finite-volume methods have the advantage of being more flexible in fitting the domain boundaries by arbitrarily complex grids. A coupling of the spectral and the finite-volume method in one CFD code enables the combination of the advantages of both methods:

- accurate solution with less degrees of freedom,
- geometric flexibility of finite-volume codes.

Such a coupled CFD code can be useful for various applications:

- Crystal growth simulation at high Rayleigh numbers also for complex geometries, fluid domain with SM and heat transport in solid parts with FVM.
- Problems, which need locally a high accuracy, e.g. direct numerical simulations or large eddy simulations. Areas with high velocity gradients are discretized with the SM, areas with low velocity gradients are discretized with the FVM.
- Time dependent CFD problems for complex geometries. Due to the fewer degrees of freedom (in comparison to the FVM) for the same accuracy, transient problems in complex geometries can be simulated with less computational costs.

In the following the various components of the coupled scheme are described, where, for simplicity, we restrict ourselves to two-dimensional incompressible flows with heat transfer and with constant material properties.

Finite-Volume Program FASTEST

The coupled code is based on the finite-volume code FASTEST-2d, which employs a fully conservative FVM for the solution of the flow equations. The main features of the implemented method are (DURST, SCHÄFER, [5]): non-orthogonal boundary fitted block-structured numerical grids, a collocated (non-staggered) arrangement of dependent variables, for the coupled system of equations a pressure-correction approach of SIMPLE type [6] , a nonlinear multigrid scheme for convergence acceleration and for the treatment of complex geometries the concept of block-structured grids is used, which forms also the base for the parallelization of the computations by grid partitioning. The solution method is formally second order accurate, since all approximations are performed in a central-difference manner. For discretization with respect to time a second-order fully implicit scheme is employed.

Chebyshev-Collocation Method

The spatial discretization of a function ϕ is based on the ansatz:

$$\phi(x,y,t) \approx \sum_{k=0}^{N} \sum_{l=0}^{M} a_{kl}(t) T_k(x) T_l(y), \tag{3.12}$$

where a_{kl} are the unknown coefficients and the Chebyshev polynomials $T_k(x)$, $T_l(y)$, $x,y \in [-1,1]$ are evaluated at the Gauss-Lobatto points. The temporal discretization of the time derivative uses a second-order fully implicit scheme. The diffusive terms are approximated by

$$\left.\frac{\partial^2 \phi}{\partial x^2}\right|_{x_i,y_j} \approx \sum_{k=0}^{N} d_{ik}^{(2,0)} \phi_{kj}^{n+1}, \qquad \left.\frac{\partial^2 \phi}{\partial y^2}\right|_{x_i,y_j} \approx \sum_{l=0}^{M} d_{lj}^{(0,2)} \phi_{il}^{n+1}, \tag{3.13}$$

where $d_{ik}^{(2,0)}$ and $d_{lj}^{(0,2)}$ denote the coefficients of the spatial differentiation matrix with respect to x resp. y. ϕ_{kj}^{n+1} and ϕ_{il}^{n+1} are the function values for the logical coordinates (k,j) resp. (i,l) at the time $n+1$. The convective term in the momentum and energy equation is treated explicitly with an Adams-Bashforth evaluation:

$$u_j \frac{\partial \phi}{\partial x_j} \approx 2\, u_j^n \frac{\partial \phi^n}{\partial x_j} - u_j^{n-1} \frac{\partial \phi^{n-1}}{\partial x_j}. \tag{3.14}$$

Due to this explicit treatment the overall scheme is only semi-implicit. Therefore, a restriction to the time step size exists. The above scheme was first proposed by VANEL et al. [7] and its stability was discussed by OUAZZANI et al. [8].

Solution Algorithms

Pressure Correction Scheme
To fulfil the continuity equation a pressure-correction scheme is used, which was first proposed by GODA [9] and implemented for the collocation Chebyshev approximation by RASPO [1], HUGUES and RANDRIAMAMPIANINA [10]. The global iteration process is as following:

1. Initialize the field values (u_i, p, T) with some initial values.

2. Solve the momentum equations for the (provisional) velocity components $\tilde{u}_i^{n+1}$:

$$\frac{\eta}{\rho}\Delta\tilde{u}_i^{n+1} - \frac{3}{2\Delta t}\tilde{u}_i^{n+1} = F_i^{n+1} \tag{3.15}$$

with

$$F_i^{n+1} = f_i(T^n) + \frac{1}{\rho}\frac{\partial p^n}{\partial x_i} - \frac{1}{2\Delta t}\left(4\tilde{u}_i^n - \tilde{u}_i^{n-1}\right) + 2\,u_j^n\,\frac{\partial u_i^n}{\partial x_j} - u_j^{n-1}\,\frac{\partial u_i^{n-1}}{\partial x_j}.$$

3. Solve pressure-correction equation with homogeneous Neumann boundary conditions:

$$\Delta\psi^n = \frac{3\rho}{2\Delta t}\left(\frac{\partial\tilde{u}_j^{n+1}}{\partial x_j}\right). \tag{3.16}$$

4. Correct velocities and calculate pressure:

$$u_i^{n+1} = \tilde{u}_i^{n+1} - \frac{2\Delta t}{3\rho}\frac{\partial\psi^n}{\partial x_i}, \tag{3.17}$$

$$p^{n+1} = p^n + \psi^n. \tag{3.18}$$

5. Solve the energy equation, if involved.

6. Return to step 2 and calculate the next time step.

Solver

All equations can be written in the form (see [10, 11, 12] for details):

$$A_{ik}\phi_{kj} + \phi_{il}B_{lj} - \sigma\phi_{ij} = F_{ij}. \tag{3.19}$$

These systems are solved by a complete diagonalization matrix technique.

Comparison of the Finite-Volume Method and the Spectral Method

To verify the efficiency and accuracy of the spectral method, the spectral code was first compared with the finite-volume method. A thermally driven cavity problem [13] was selected to investigate the two methods for steady and unsteady cases. A square with a side wall length L encloses a fluid initially at temperature $T_0 = 0K$. At time $t = 0$, the temperature of the left and right walls is raised to T_h and T_c, respectively. The horizontal walls are insulated.

Steady Problem
The Prandtl number is $Pr = 0.71$ and for the Rayleigh number the value $Ra = 10^5$ is considered. All parameters are summarized in Table 1. To compare the results, the error of the Nusselt number Nu relative to the respective extrapolated value of the FVM solution is calculated. The FVM results were calculated with the full multigrid and for the spectral code a nested grid technique was used.

Table 1 Fluid properties and dimension of the steady test problem

Ra	$\rho\,[\frac{kg}{m^3}]$	$\eta\,[\frac{kg}{ms}]$	$\beta\,[\frac{1}{K}]$	Pr	$g\,[\frac{m}{s^2}]$	$T_h\,[K]$	$T_c\,[K]$	$L\,[m]$
10^5	1.19	1.8*10^{-5}	0.00341	0.71	9.81	12	2	0.045841

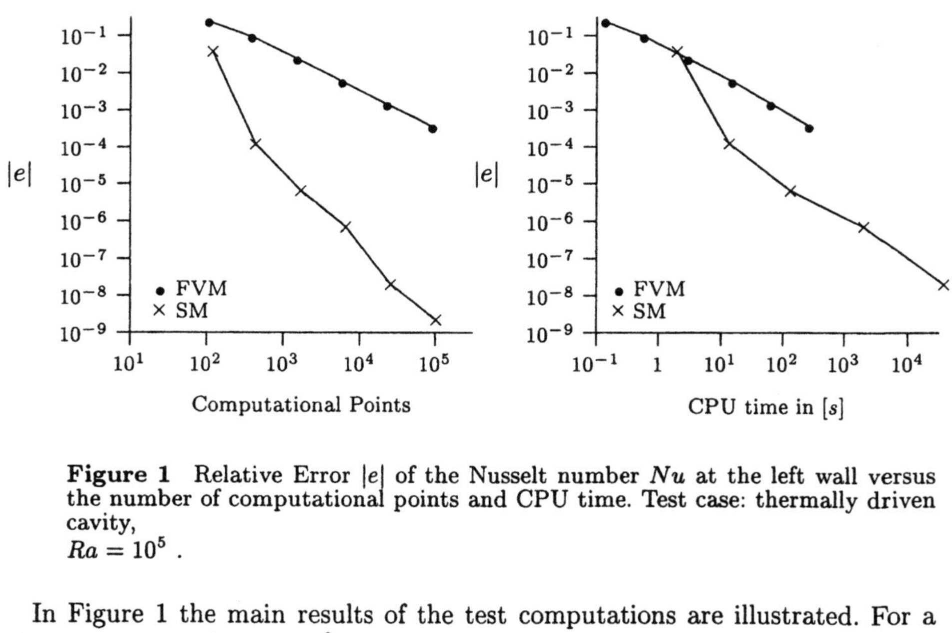

Figure 1 Relative Error $|e|$ of the Nusselt number Nu at the left wall versus the number of computational points and CPU time. Test case: thermally driven cavity,
$Ra = 10^5$.

In Figure 1 the main results of the test computations are illustrated. For a given accuracy of e.g. 10^{-3} the SM needs approximately 100 times less computational points resp. required memory than the finite-volume code (Figure 1 (left)). Due to the fixed convergence order of the FVM this advantage increases with an increasing number of computational points. Because the number of operations per computational point is higher for the spectral program, the benefit is not as big if one looks to the accuracy versus computational time (Figure 1 (right)). For a given accuracy of e.g. 10^{-3}, the FVM needs about 10 - 50 times more CPU time. In Table 2 the Nusselt number calculated with the SM and the FVM are summarized. The values correspond very well with the extrapolated

Table 2 Nusselt number Nu on the finest grid for $Ra = 10^5$ compared with the extrapolated values of PERIĆ (Finite-Volume code).

Nusselt number Nu on finest grid	SM	FVM	PERIĆ [13]
$Ra = 10^5$	4.521636	4.52308	4.52164

values of PERIĆ [13].

Unsteady Problem

The cavity problem was calculated with time dependent boundary conditions $T_h(t) = 5K\,(1 - \cos(t\,\pi))$, $T_c = 0K$ at the left wall. The material properties correspond to Table 1. The mean Nusselt number was calculated between $t = 150s$ and $t = 250s$.

The overall error $|e|$ is composed of the time discretization error $|e_{time}|$ and the space discretization error $|e_{space}|$. The time discretization error $|e_{time}|$ is computed with the respective time step independent value of each grid (which is obtained by Richardson extrapolation). The reference value of the space discretization error $|e_{space}|$ is the SM solution with a 81×81 grid and a time step size of $\Delta t = 0.003125s$.

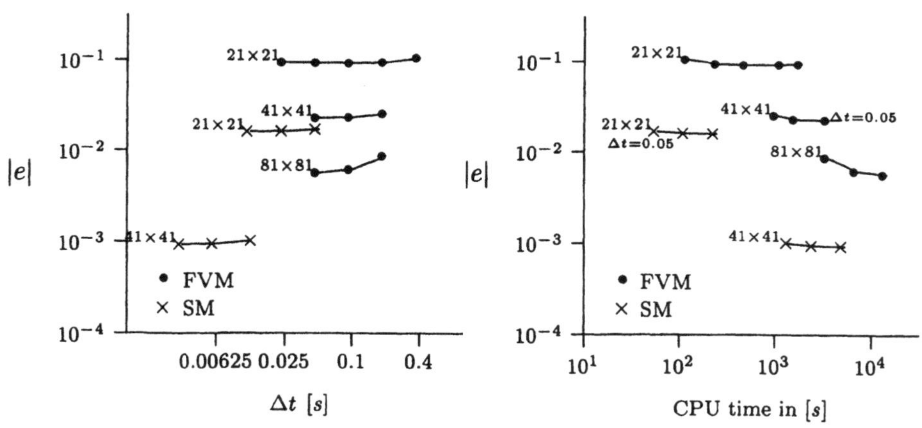

Figure 2 Relative overall error $|e| = |e_{time}| + |e_{space}|$ of the mean Nusselt number Nu at the right wall versus time step size Δt and CPU time. Test case: thermally driven cavity with time dependent boundary conditions.

For both methods, the relative overall error is dominated by the space discretization error (Figure 2, (left)). As expected, it is significantly smaller for a fixed number of computational points for the SM. The error versus CPU time has

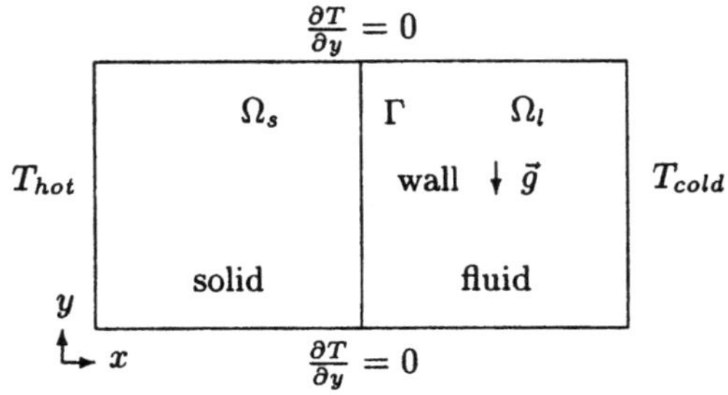

	Ω_s	Ω_l
Density	ρ_s	ρ_l
Thermal conductivity	λ_s	λ_l
Specific heat	c_{p_s}	c_{p_l}
Dynamic viscosity		η
Volumetric expansion		β

Figure 3 Coupling of a solid subdomain Ω_s and a fluid subdomain Ω_l.

to be discussed in relation to the time step size. For a given time step size (e.g. $\Delta t = 0.05s$) the spectral code is approximately 60 times faster than the finite-volume code for the same accuracy (Figure 2, (right)). For a freely selectable time step size (according to temporal accuracy requirements), the benefit is less. Here, the finite-volume code gains from the fully implicit formulation and can cope with a greater time step size. Due to the semi-implicit formulation of the spectral code, the time step size is restricted, the critical values decreases with an increasing number of computational points according to $\Delta t \sim O(N^{-2})$.

Mathematical Formulation of the SM-FVM-Coupling

For the coupling of the SM and FVM we restrict ourselves to the case where the fluid parts are calculated fully with SM and the heat conduction in solid and fluid parts with the coupled SM-FVM formulation. For such cases, a coupling of the two methods within the energy equation is sufficient.

In the solid domain Ω_s the energy equation describes the diffusive heat transport. The equation of mass, momentum and energy govern the fluid flow and heat transport in the fluid domain Ω_l (Figure 3).

At the solid-fluid interface Γ we have continuity constraints:

$$T^{\Omega_s}\big|_\Gamma = T^{\Omega_l}\big|_\Gamma, \qquad \lambda_s \frac{\partial T^{\Omega_s}}{\partial x}\bigg|_\Gamma = \lambda_l \frac{\partial T^{\Omega_l}}{\partial x}\bigg|_\Gamma .$$

The FVM code FASTEST uses the concept of block-structured grids. For the efficient handling of the coupling of the blocks auxiliary control volumes containing the corresponding boundary values of the neighbouring blocks are introduced along the block interfaces (Figure 4 (left)). The coupling of the blocks is then ensured by the interchange of these boundary values. To preserve this structure also in the coupled SM-FVM code, an overlapping grid partitioning technique is used for the SM-FVM coupling. The overlapping is constructed by enlarging the finite-volume subdomain into the spectral subdomain. Therefore the control volumes at an interface have to be displaced relative to the interface (Figure 4 (right)).

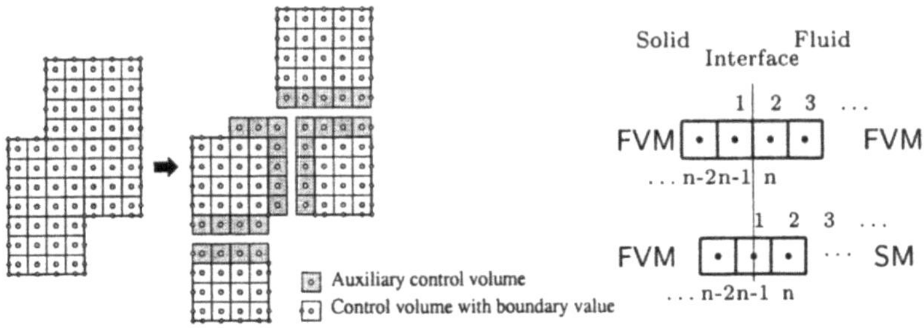

Figure 4 Block-structured grid and computational interface handling used in the pure finite-volume code (FVM) and in the coupled code (SM-FVM).

The exchange of the boundary values at the SM-FVM interfaces requires the following steps:

Step 1. Solving with FVM for all finite-volume blocks

Step 2. Exchange of boundary values FVM $\overset{\frac{\partial T}{\partial x}}{\rightarrow}$ SM (Neumann boundary condition)

$$\left.\frac{\partial T^{SM}}{\partial x}\right|_{interface} = \frac{T_n^{FVM} - T_{n-1}^{FVM}}{x_n - x_{n-1}}$$

Step 3. Solving with SM for all spectral blocks

Step 4. Exchange of boundary values SM $\overset{T}{\rightarrow}$ FVM (Dirichlet boundary condition)

$$\left.T_n^{FVM}\right|_{bc} = T_1^{SM} + \left.\frac{\partial T}{\partial x}\right|_{interface}(x_n - x_{n-1})$$

Numerical Comparison

A thermally driven cavity problem combined with a solid block (Figure 3) was calculated with the FVM and the coupled SM-FVM code. The material properties and boundary conditions correspond to that in Table 1. The FVM results were calculated with the full multigrid (MG) and the single grid algorithm (SG). The Nusselt number Nu calculated at the interface wall Γ is used as an indicator for the convergence and accuracy. The error is computed with the extrapolated Nusselt number on the finest grids (FVM).

In Figure 5, left, the most important property of the coupled code is shown: also the solution of the coupled code is of high accuracy. For a given accuracy, the finite-volume code needs up to 100 times more computational points than the coupled program.

The error of the Nusselt number calculated by the FVM decreases only with first order, although all discretization schemes are of second order. The dominant first

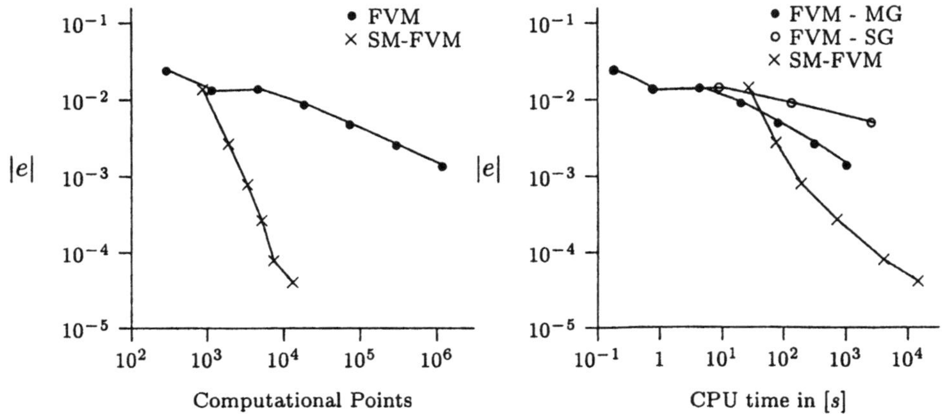

Figure 5 Relative Error $|e|$ of the Nusselt number Nu at the interface versus the number of Computational Points and CPU time. The extrapolated Nusselt number $Nu_{ex} = 2.3956$ is computed with the solutions of the FVM on the finest grids.

order term in the truncation error does not vanish, because the second derivative of the temperature is not smooth at the interface Γ.
Thus, already for the calculated steady problem, the coupled code reaches the highest accuracy with significantly less computational cost (Figure 5, right). Also the computations accelerated with multigrid (FVM - MG) need distinct more computing time. Due to the formulation of the SM, for time dependent problems the coupled code is expected to be even more advantageous compared with a pure FVM treatment.

3.2 Enthalpy-Porosity Method and Finite Volume Approximation

For solidification problems the front tracking methods deal with a weak formulation. The energy equation is formulated with the temperature field. Each solid and liquid phases are solved separately. The interface is treated with both thermal Neumann and Dirichlet boundary conditions. The need of tracking the interface induces a remeshing of both domains at each time step, which is expensive in computing time.

The enthalpy method [14, 15, 16] gives a way of reformulating the problem using an enthalpy function and a porous model in the momentum equation. The transmission conditions at the interface are implicitly bound up in a new form of the equations, which applies over the whole of a fixed domain reducing the computation time.

For time-dependent problems, as the crystal growth applications, this approach has been validated to represent properly the flow pattern and the heat transfer and to determine realistic sections of solidified crystals [16].

Mathematical Formulation

The phase change problems are defined by the averaged mass, momentum and energy conservation equations. For binary alloys, it should be added species equations. Theses ones are respectively written in their cartesian form, where the indices s, l denote solid resp. liquid :

$$\frac{\partial \rho}{\partial t} + \nabla \cdot \rho \vec{u} = 0, \tag{3.20}$$

$$\frac{\partial}{\partial t}(\rho \vec{u}) + \nabla \cdot (\rho \vec{u}\vec{u}) = \nabla \cdot \left(\eta_l \frac{\rho}{\rho_l} \nabla \vec{u}\right) - \nabla p + \vec{f}(T) - \frac{\eta_l}{K}\frac{\rho}{\rho_l}(\vec{u} - \vec{u}^s), \tag{3.21}$$

$$\frac{\partial}{\partial t}(\rho h) + \nabla \cdot (\rho \vec{u} h) = \nabla \cdot \left(\frac{\lambda}{c_{p_s}} \nabla h_s\right) - \nabla\left(\rho (h_l - h)(\vec{u} - \vec{u}^s)\right), \tag{3.22}$$

$$\frac{\partial}{\partial t}(\rho f^\alpha) + \nabla \cdot (\rho \vec{u} f^\alpha) = \nabla \cdot (\rho D_l \nabla f^\alpha) - \nabla\left(\rho (f_l^\alpha - f^\alpha)(\vec{u} - \vec{u}^s)\right). \tag{3.23}$$

The average density, velocity vector components, enthalpy species and thermal conductivity are defined as following:

$$\begin{aligned} \rho = &\quad g_s \rho_s + (1 - g_s)\rho_l, \quad \vec{u} = \quad f_s \vec{u}^s + (1 - f_s)\vec{u}^l, \quad h = \quad f_s h_s + (1 - f_s) h_l \\ f^\alpha = &\quad f_s f_s^\alpha + (1 - f_s) f_l^\alpha, \quad \lambda = \quad g_s \lambda_s + (1 - g_s)\lambda_l. \end{aligned}$$

For the permeability that models the liquid-solid interaction forces in the momentum equation, we use the Carman-Kozeny relation: $K = K_0 \frac{g_l^3}{(1-g_l)^2}$ The energy equation is formulated with the enthalpy variable which is the sum of the specific heat and the solidification latent energy. A source term is added in this equation, taking into account the latent heat jump. We consider that the exchanges of latent energy are continuous phenomena spreaded through a wide range of temperature to define a 'mushy zone' considered as a porous medium. The interaction between phases can be modelled by Darcy's law in the region where the solid fraction predominates. This Darcy's term is proportional to relative phase velocities. It corresponds to a damping force, and depends also on the anisotropic permeabilty of the mushy zone. We also assume that the solid phase is free of internal stress, translates at a fixed velocity, and that viscous stresses due to local density gradients are negligible. We also suppose the validity of the Boussinesq approximation, and that the fluid is Newtonian in the liquid phase. For the conservation equations, we suppose there is no solute diffusion in the solid phase.

Equilibrium phase diagrams are used to close the system. These diagrams allow to calculate the different mass fractions of constituents in each phase. The solidus and liquidus curvatures can be linearised around the melting temperature because of the low dopant concentration considered. We obtain then direct linear relations between mass fractions in each phase.

Numerical Method

The spatial discretization of the equations uses a finite volume method. The convective terms are evaluated using a Quadratic Upwind Interpolation for Convective Kinematics scheme (QUICK). It uses a quadratic interpolation between two upstream and one downstream neighbours in order to estimate the variable considered at any control-volume interface. The formulation of this interpolation depends on the sign of the convective flux on the interface. The QUICK scheme assumes a parabolic shape profile for conducting the interpolation procedure. This scheme preserves the pentadiagonal form of the linear system, which can be then solved with a classical Tridiagonal Matrix Algorithm (TDMA). It is stable as long as the diagonal dominance of the matrix is ensured. It is also less sensitive to the consistency of boundary treatments than the second order Upwind scheme. The time discretization uses a second order Euler scheme. The pressure velocity coupling is ensured by a PISO algorithm.

4 Applications to CVD

Chemical Vapour Deposition (CVD) is a technique that is widely used, particularly in the semi-conductor industry. There is a continuous demand for scale-up reactors to produce larger wafers. At the same time the feature sizes are reducing, bringing ever increasing demands on film quality and process performance. As a result, the life cycle of CVD equipment is short and new equipment has to be developed every few years. Conventionally, such requirements have been met by empirical trial-and-error methods, straightforward scale-up of existing equipment and ad hoc modifications to existing processes. However, computer simulation offers the possibility of a more fundamental understanding of the essential physics and chemistry, leading to a systematic approach to reactor and process design.

Our main objectives were the compilation of modelling equations for the transport, heat transfer and chemical phenomena important in CVD processes, together with the provision of a practical engineering tool to permit the simulation of 3D reactors with rotating circular substrates, incorporation of the models in a finite-volume flow simulation codes (FASTEST-3D and CFD2000/STORM) and their validation against previously existing validated data.

4.1 Transport mechanisms

Various basic assumptions are made about the nature of the gas flow in the reactor. The first is that the gas mixture will behave as a continuum, i.e. that the mean free path for the molecules is much smaller than the characteristic dimensions of the reactor. This effectively imposes a lower limit on the operating pressure of, typically, about 30 Pa. The assumption is also made that ideal gas behaviour is a suitable approximation. Viscous heating due to dissipation is also neglected. All these limitations are entirely compatible with most CVD applications.

The fundamental equations to be solved are those representing conservation of mass, momentum and energy (eqs. (2.1), (2.2), (2.3)), coupled with a similar equation for each of the individual species (2.10).

Besides ordinary diffusion, thermal diffusion has been incorporated in the model. The chemistry has been treated through the incorporation of linearized source terms. The transport properties of the gas mixture have all been taken as a function of the local temperature. Due to the large gradients of temperature thermal (Soret) diffusion has been included in the model.

4.2 Geometrical considerations

To represent a circular substrate in a rectangular reactor (Figure 6) we have used a Body Fitted Coordinates (BFC) grid as shown in Figure 7. We have considered 4 configurations noted A, B, C, D and corresponding to different aspect ratios of the section (width-to-height, l/H) and of the diameter of the circular substrate to the width (D/l). Geometrical characteristics of the reactors of same width l=10cm and length L=20 cm, are given in 4.2. The hot and cold temperatures inside the device are 1323K and 300K. For case A, 40x16x32 curvilinear mesh is used in x, y and z directions.

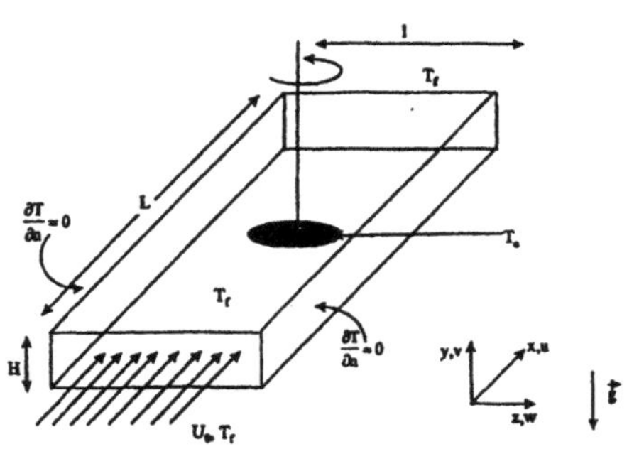

Figure 6 Sketch of a CVD reactor with rotating substrate: geometry, flow and thermal boundary conditions

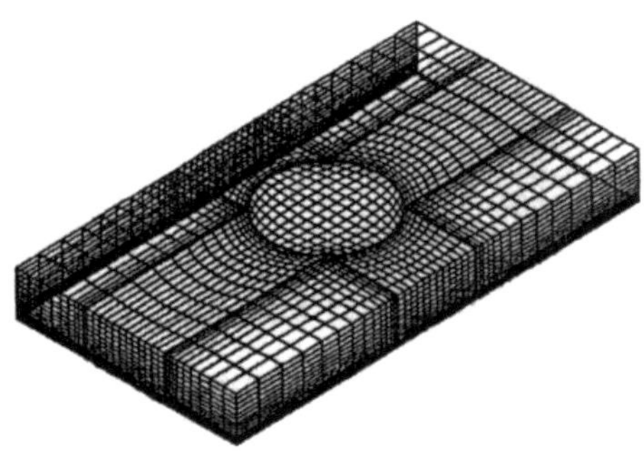

Figure 7 Structured grid of the CVD reactor with rotating substrate, 40×16×32 CVs

The viscosity of the gas mixture is increased by a ratio of 10 when the gas goes from the cold entrance section to the hot substrate. It results in large variation of the local Reynolds number while the Prandtl number stays nearly constant at about 0.75 to 0.70. The typical rotation rate in CVD reactor is up to 10 rad/sec. The flow in the reactor is laminar and we consider the steady-state equations.

Table 3 Details of the geometrical configuration

Configuration	H	D	l/H	D/l
A	2cm	5cm	5	0.5
B	2cm	8cm	5	0.5
C	1cm	5cm	10	0.8
A	1cm	8cm	10	0.8

4.3 Chemical Reactions - Si deposition from SiH_4

We have studied the deposition of Si from SiH_4 with H_2 as carrier gas. The validation of the chemistry model has been done in 2D and 3D against published data of Ouazzani et al. [17, 18] and the references therein.

We consider homogeneous gas-phase reactions. The conditions correspond to the experiments of Eversteyn et al. [19]. The kinetics of silane decomposition has been extensively studied. Based on these studies and their own investigations, Coltrin et al [20, 21] have introduced a path of chemical reactions containing twenty steps.

However, since the formation of Si from SiH_2 is kinetically favoured over that from the other Si-containing species, one can well approximate the system by using only the two chemical reactions :
$SiH_4 \rightleftarrows SiH_2 + H_2$, $Si_2H_6 \rightleftarrows SiH_4 + SiH_2$
The rates of these two reactions fit the Arrhenius form : $k = AT^{\beta}\exp(-E_a/RT)$ and they are considered in each of the 4 species equations with adapted values of coefficients (see [22]). The heat from these reactions is neglected in the energy equations.

4.4 Surface reaction

Following Moffat and Jensen [23, 24] we have allowed for Si formation from SiH_4 , Si_2H_6 and SiH_2 . As sticking coefficients g_j, we used :
- unity for SiH_2 and Si_2H_6 ,
- and $g_j = 5.37 \cdot 10^{-2}\exp(-18680/RT)$, for SiH_4 .

corresponding to a Damköhler coefficient less than unity.

4.5 Finite volume code

The computations have been performed using finite volume methods, (CFD2000 [25] and FASTEST-3D [26] software package). The main features for the first one are the use of the PISO algorithm, of collocated finite volumes with 2nd and 3rd order schemes, the solution of the coupled differential equations in the predictor and corrector steps of the PISO algorithm. The rotation boundary condition has been discretized using a second-order finite difference scheme. The dependent variables are the velocity, the pressure, the temperature distribution, the species concentration in 3D and 2D geometries and for steady stat. The chemistry has been solved with appropriate linearization. A module has been developed which automatically linearize any number of chemical reactions. This method has been compared also to the method of solving using an ODE solver (LSODA, DVODE etc....). The surface chemistry has been modeled through a term source in the boundary conditions. For the rotation of the substrate, we have introduced a momentum source which has been defined over the substrate surface.

4.6 Results

We have validated first the finite volume code CFD2000 on an experimental benchmark.Then we have been able to treat a realistic industrial CVD reactor with rotating substrate and chemical reactions. We have shown that using a non uniform mesh and second order convective schemes we can resolve the steep gradients of concentrations occurring at the surface substrate.

We have considered the configurations described in Table 5.1 with various flow rate at the entrance of the reactor, from 0.05 m/sec to 0.25 m/sec, and with various rotation rate of the substrate, from 0 to 10 rad/sec.

The goal of the study is to understand the interplay between the three different convective mechanisms (rotation, forced flow, buoyancy), their effect in the region of the substrate and then on layer uniformity. The basic flow at low rotation and mass-flux corresponds to a closed roll driven par the horizontal temperature gradient at the border of the disk and of the bottom wall. The effect of rotation is illustrated in Figure 8 at low mass flux and high rotation rate. The particle tracks exhibits a pumping effect near the rotating substrate and large complex vertical structures in the height above and downstream the disk.

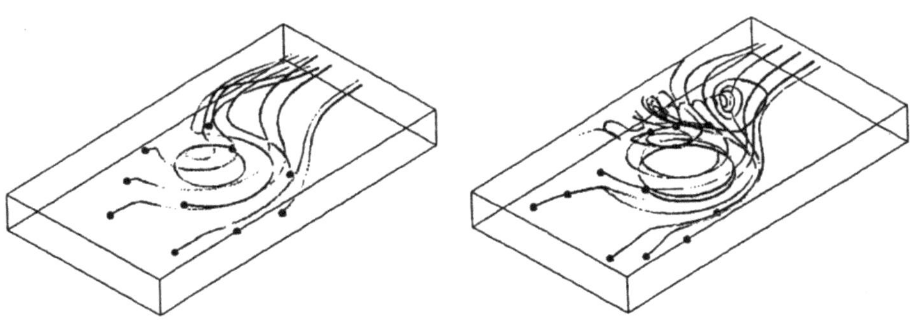

Figure 8 Typical particle paths near the rotating substrate at $\Omega = 10$ rad/s for two inflow rates: $U_{inflow} = 0.02m/s$ (left) and $U_{inflow} = 0.05m/s$ (right)

The rotation of the substrate can affect significantly the rate of deposition.

The modification of the growth rate is shown for 0.05 m/sec when increasing the rotation from 0 to 10 rad/sec (Figure 9).

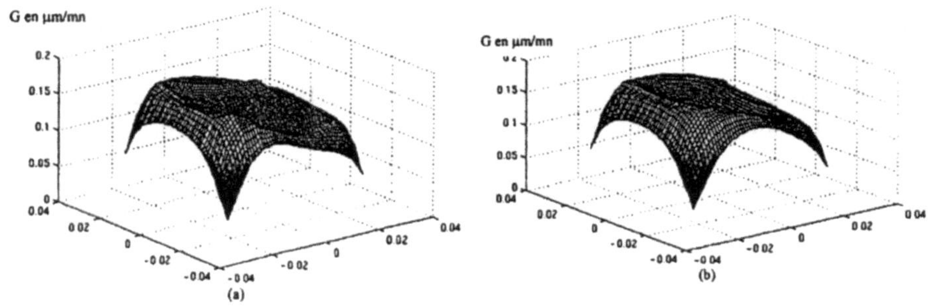

Figure 9 Typical growth rates at $U_{inflow} = 0.05m/s$ on the rotated substrate (Ω = 10 rad/s) (left) and on the static substrate (right)

In the next part calculations using FASTEST 3D have been performed and compared to the results given by the first code.
For our calculations the Configuration B has been taken for two different inflow rates: 0.25 m/s and 0.02 m/s at the rotation speed Ω = 10 rad/s.
The results of calculations are shown in Figure 10 for $U_{inflow} = 0.25m/s$ and in Figures 11, 12 for U_{inflow} = 0.02 m/s. The results of calculations agree with the results in [22] for that two completely different flow regimes.
The calculations on three grids (the finest with 128 x 80 x 32 = 327680 CV) on a workstation SUN ULTRA2 were obtained in approximately 5 hours.

5 Applications to bulk crystal growth

In this application we present two aspects of using numerical methods based on finite volume approximation in the investigation of Vertical Bridgman solidification. The global method allows the consideration of realistic configurations, while the enthalpy method deals with complex physical phenomena like interaction between unsteady melt and its interface.

Using the Bridgman technique on can obtain GaAs and InP crystals with low dislocation density, defined by etch pitch density (EDP), as needed by the electronic and optoelectronic industry (see [27, 28]).

5.1 Global simulation

In global simulation one wishes to solve one of two typical problems: a) The direct problem: given the heating powers in the heaters, compute the temperature (or any other quantity) distribution in the whole setup. b) The inverse problem:

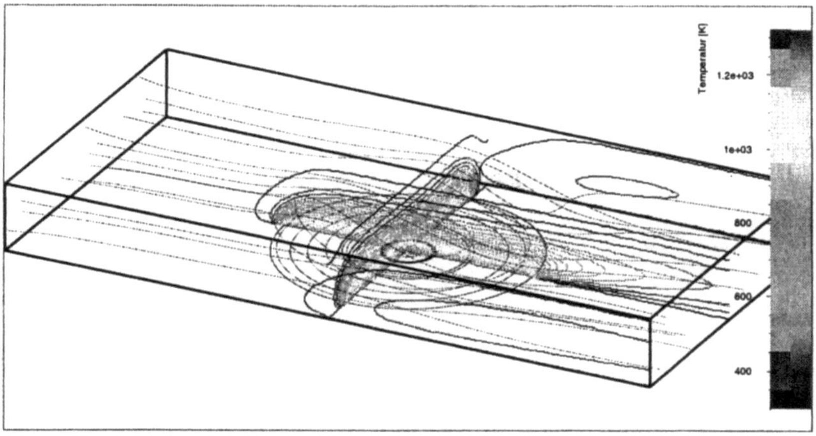

Figure 10 The temperature distribution over the susceptor in two opposite directions; along and across the susceptor including isotherms on the side walls (different gray lines); and traces (gray thin lines) corresponding to the flow in the reactor. $U_{inflow} = 0.25m/s$, $\Omega = 10$ rad/s

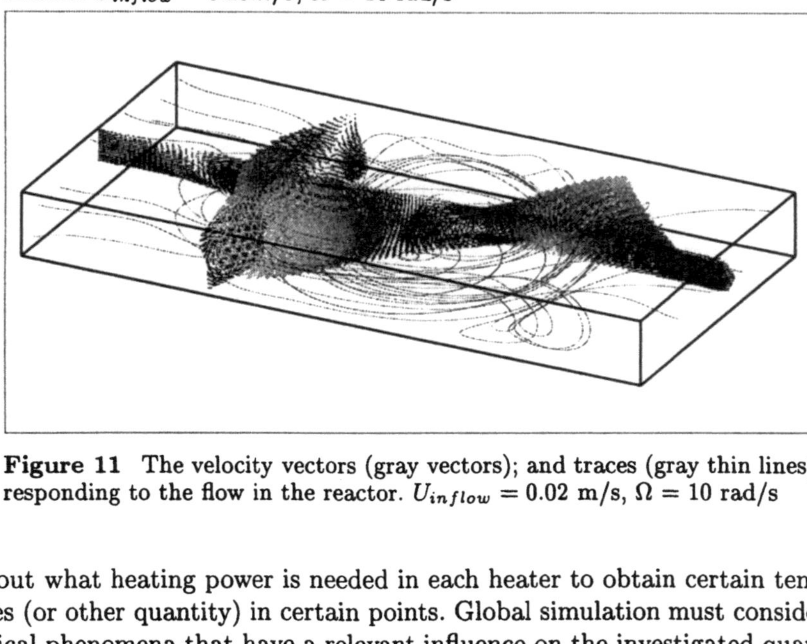

Figure 11 The velocity vectors (gray vectors); and traces (gray thin lines) corresponding to the flow in the reactor. $U_{inflow} = 0.02$ m/s, $\Omega = 10$ rad/s

find out what heating power is needed in each heater to obtain certain temperatures (or other quantity) in certain points. Global simulation must consider all physical phenomena that have a relevant influence on the investigated quantity. Currently only setups with axial or translational symmetric geometries can be modelled globally, because of the large systems of equations that would result in 3D modelling. At the Institute for Material Sciences in Erlangen we have developed a computer program for global numerical simulation of processes in real crystal growth furnaces with axial symmetry. The numerical method used is based on the finite volume method and is applied in conjunction with an unstructured grid. In our model, nonlinear conductive (eq. 2.3) and radiative

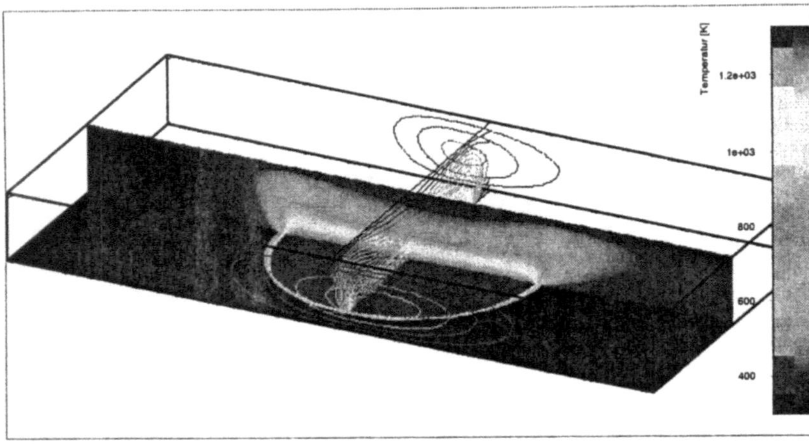

Figure 12 The temperature distribution over the susceptor in two opposite directions; along and across the susceptor including isotherms on the side walls (different gray levels); $U_{inflow} = 0.02$ m/s, $\Omega = 10$ rad/s

(eq. 2.8) heat transfer is handled (see [29]) with respect to latent heat of the solidification interface. Usually the so called inverse problem can be solved in other codes only for one heater. But in our code one can chose as many heaters as needed, provided that each of the heaters has enough influence on at least one of the points at which temperature is imposed. "Enough influence" means that a change of the heating power is reflected in the next outer iteration on the temperature on at least one control point. The aim of this work is to verify the code for global simulation of heat transfer by conduction and radiation in crystal growth. We have chosen for our simulation a Vertical Gradient-Freeze (VGF) furnace for which we have experimental data from a real growth process. The furnace is composed of nine cylindrical heating zones, each of them independently controllable. The heaters are heating a ceramic pipe, which contains the quartz ampoule, as shown in Figure 13 (left). Inside the quartz ampoule the pBN crucible is supported by a quartz cylinder. The crucible itself consists in the main part of a cylinder, at the lower end of a conical zone and a small channel for the seed.

For each heater, there is a thermocouple on the exterior side of the ceramic pipe. This thermocouple can be used to adjust the power of the heater in order to obtain a certain temperature in that point. The heaters are isolated from the exterior medium through an good isolating material. In the experiment a time dependent temperature profile along the ceramic pipe was imposed using the controlling mechanism for 8 of the thermocouples at the ceramic pipe and the corresponding 8 heaters. Two additional other thermocouples near the seeding channel of the ampoule were used to find the thermal conditions for the seeding process. We have picked up one set of temperatures at the thermocouples at an early stage of the growth process, at a time at which it is supposed that the solidification front was in the conical part of the crucible, and we have

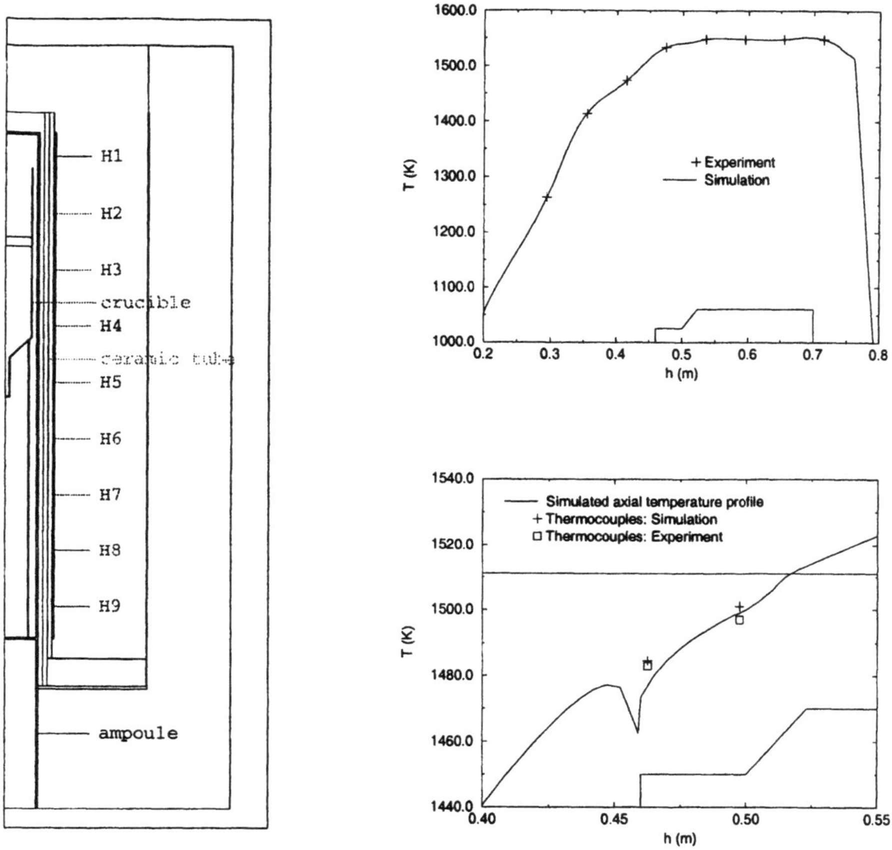

Figure 13 **Left:** The furnace and the main components. **Upper right:** Temperature profile along the wall of the ceramic pipe: experiment and simulation. **Lower right:** Temperature profile along the symmetry axis obtained in the simulation and the temperatures at the two thermocouples near the seeding channel.

used the controlling mechanism in our program to adjust the heating power in order to obtain the same temperatures as in the experiment at the ceramic pipe. The temperature distributions along the ceramic pipe is shown in Figure 13 (right, above). As one can see, the profile is in good agreement with the experimental values, the difference is less than 0.1K. The two temperatures at the two thermocouples near the seeding channel are also in good agreement, see Figure 13 (right, below): the difference between measured and simulated values is smaller than 3K in both cases. The VGF furnace was chosen because we expect that the convection plays no important role in the energy flow, and therefore expect our model to be a good approximation for this case. The interface shape

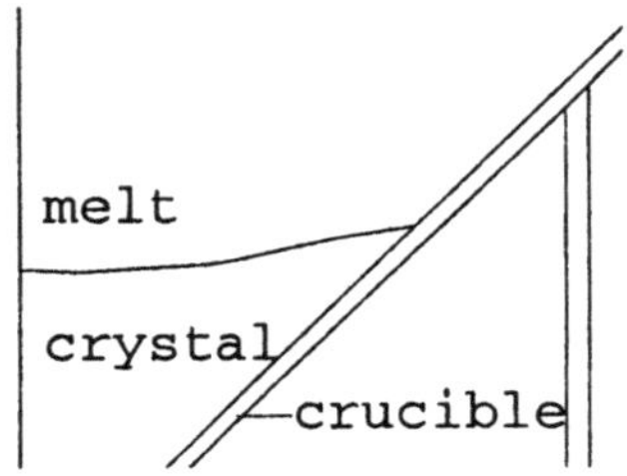

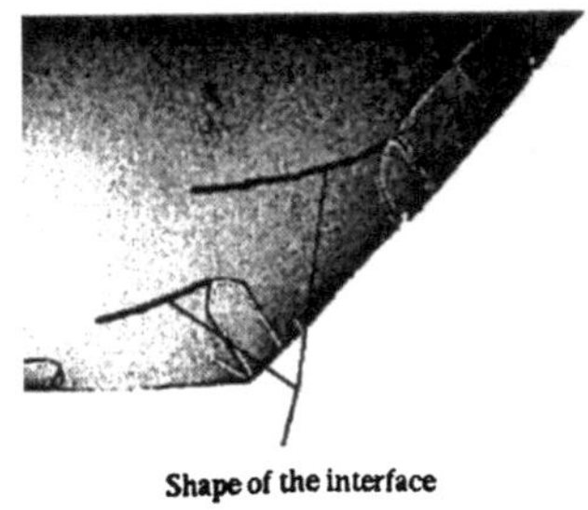

Figure 14 **Left:** the calculated position of the solid-liquid interface conform to our global simulation; latent heat was considered for a growth rate of 2.5mm/h. **Right:** A transversal section through the conical part of the crystal. The shape of the solidification interface can be observed for two positions.

in the conical part is used to verify our model as for this part of the crystal experimental data were available from a preparation of a longitudinal section along the crystal axis. An image of a transmitted infrared radiation was recorded for a this thin plate, see Figure 14 (right). Along the crucible walls one can recognise the growth of facets structures. But the main part of the interface follows the melting point isotherm, and one can see that the shape is matching with the results of our simulation (left hand-side in Figure 14).

5.2 Complex dynamic fields and interface shape by enthalpy method

Unsteady axisymmetric melt flow and interaction with the solidification interface

For this study we have used inverted Bridgman configuration heated from below (IVB) (Figure 15 a). Comparison with Brown's results performed in the case of steady state solution have shown a very good agreement qualitatively and quantitatively [16]. Interface remains quasi-planar in all the steady state configurations. The simulations were carried out above the threshold of oscillatory convection. The first transition at $Ra = 10^6$ corresponds to a periodic flow in the melt which does not bring severe modification to the interface shape. It remains quasi-steady but a slightly increased convex shape. Ra is increased up to $Ra = 10^7$, the melt flow structure becomes very complex and the motion is aperiodic. The driving cell moves progressively with increasing Ra to the bottom of the crucible, with reducing size and increasing strength. The neighbour melt is submitted to the oscillating cell which drives a succession of moving cells which rise up to the top of the melt and affect the interface curvature. Secondary cells are shown at three different stages of the process, to steeply modify the shape of the interface Figure 17, following the time-dependent patterns.

Asymmetric melt convection and melt interface change

We have represented in Figure 15 the axisymmetric model with two-dimensional model (in cartesian coordinate and corresponding to a diameter section of the Bridgman ampoule) that we consider in this section. A two-dimensional model with a planar interface and without solid phase is also considered in order to qualify the solution with respect to time dependent behaviour (and to spectral solutions existing in this configuration). This basic cavity is square $L/2$, and the temperature gradient is $\Delta T/2$. In the case of solid liquid model, the length is L and the temperature gradient is ΔT. In the axisymmetric configuration the height is still L but the width is $L/4$. The relevant Rayleigh number are deduced from model $M_{\alpha\beta}$ with characteristic length scale $(\beta L/2)$ and temperature difference $(\alpha\Delta T/2)$ as: $Ra_{\alpha\beta} = (\alpha\beta^3)Ra_{11}$.
With this notation Ra_{22} is the Rayleigh number used for the axisymmetrical case. Ra_{21} for the 2D-2 phase case and Ra_{11} for the 2D-1 phase case.

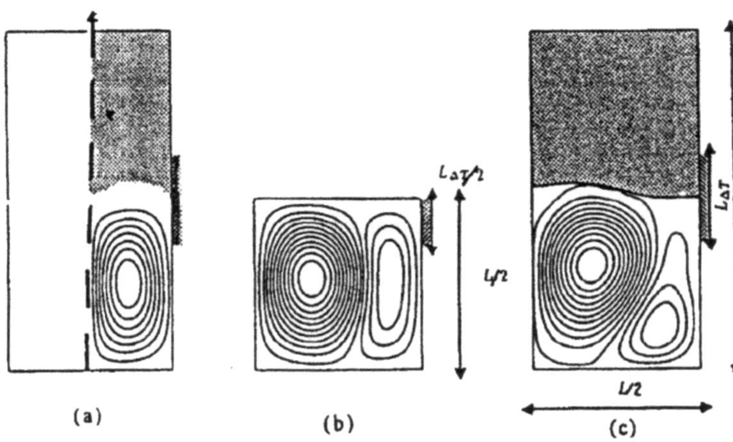

Figure 15 Models configuration used in unsteady study. (a) Axisymmetric model (M_{22}), (b) Basic 2D fluid phase model (M_{11}), (c) 2D-2 phases model (M_{21}).

Ra_{11} is used as a common reference of Rayleigh number for the 3 models. The computation exhibits steady symmetrical solutions up to about $Ra_{11} = 5000$, transition to asymmetrical solutions at $10000 < Ra_{11} < 15000$. The asymmetrical instability is shown to become time-dependent at $25000 < Ra_{11} < 30000$.
The ability of the finite-volume solution to predict such behaviours is assessed in the case of the simplified one-fluid-phase model for which we obtained highly accurate spectral solutions [30]. The finite volume solution exhibits the same flow features : steady symmetrical - asymmetrical - time dependent. The two associate critical Ra ranges are of the same order of magnitude. The first asymmetrical transition occurs at $5000 < Ra_{11} < 7000$ and the time-dependent solution arises

at $30000 < Ra_{11} < 40000$. Respectively, the spectral solution gives $2500 < Ra_{11} < 3500$ and $10000 < Ra_{11} < 20000$. The comparisons are shown in table 4.

Table 4 Regimes and transitions for the different models

Regimes	Finite-volume			Spectral
	Axis	2 Phases	1 Phase	1 Phase
Steady	6250	5000	3000	1000
		Sym → *Asym* 10000 – 15000	*Sym* → *Asym* 5000 – 7000	*Sym* → *Asym* 2500 – 3000
		Asym 25000	*Asym* 30000	*Asym* 10000
Unsteady	62500	30000	40000	20000

Some insight on crystal constitution

Crystal constitution process is simulated by considering the successive interfaces during a characteristic growth time scale and for the two axisymmetrical and two-dimensional models. The crystal characterised by its shapes is represented in (Figure 16). The figure exhibits irregularities which shows similarities with the ones exhibited in experimental result [31]. In this case the occurrence of an asymmetric mode of convection is suggested and the two-dimensional solution exhibits very similar features.

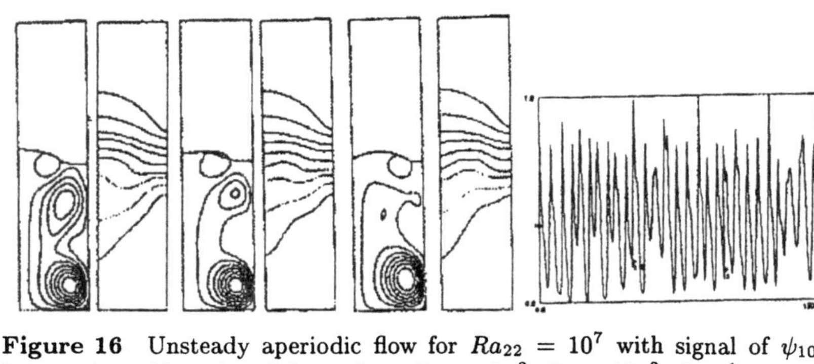

Figure 16 Unsteady aperiodic flow for $Ra_{22} = 10^7$ with signal of $\psi_{10,10}(t)$. Interaction with melt-solid interface. $Pr = 10^{-2}$, $Pe = 10^{-2}$, $Ste^{-1} = 1$.

Study of solutal convection

We study solutal convection during directional solidification of a lead-30% thallium alloy. The crucible which walls are heated at a temperature greater than

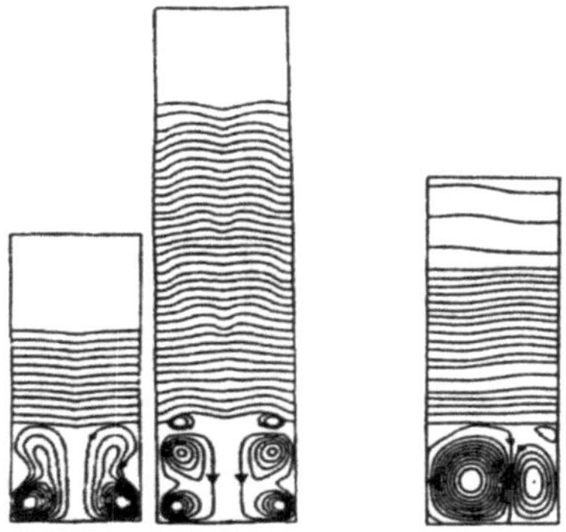

Figure 17 (a) crystal constitution and streamlines patterns for steady and time-dependent axisymmetric regimes: $Pr = 10^{-2}$, $Pe = 10^{-2}$, $Ste^{-1} = 1$ and $Ra_{22} = 10^6$, 10^7. (b) symmetric regime $Pr = 10^{-2}$, $Pe = 5 \times 10^{-3}$, $Ste^{-1} = 0$ and $Ra_{21} = 80000$.

the fusion temperature initially contains a melt pool at uniform temperature. The aspect ratio of the rectangular cavity is 9 by 20. In our case, the phase diagram gives a segregation coefficient $k = 1.1$. The solution (thallium) solidifies more easily than the solvent (lead). Nearby the interface the melt is less dense which leads to a solutal convective instability similar to the Rayleigh-Benard thermal instability [16]. In this configuration, the vertical temperature gradient has a stabilising effect, the radial gradient is only destabilising. Nevertheless, the contribution of thermal buoyancy to hydrodynamics in the melt pool is much weaker than the solutal one, and thus it can be neglected. A simulation of thermo-solutal convection has validated this assumption. The thermal effects determine the initial interface shape which then affects the structure of the convection. The velocity and the length are made dimensionless with respectively the drawing velocity V_0 and the solutal length $l_s = V_0/D$, D solutal diffusivity. The main parameters are the solutal Rayleigh number $Ra_S = \beta_S g \Delta f^\alpha / \eta_l D l_s^3$, and the Schmidt number $S_c = \eta_l / D$ We take $Sc = 121.5$ and $1000 < Ra_S < 8000$ (which correspond to growth velocities from 14 to $28 \mu m/s$ and solutal lengths from 0.7 to $1.4 mm$). The calculation grid has been refined in the solidification zone and uses 40×105 points respectively in x and y directions. Figure 18 shows the streamlines and the concentration patterns for $Ra_S = 4000$ and 8000. The solutions are composed of two symmetric convective cells in the melt zone. The flow goes upward along the walls and downward in the middle of the ampoule. Other studies have stated that asymmetric single-cell solutions can develop in that kind of configuration with no phase change [32]. The increase of convection in the liquid part leads to a longitudinal extension of a recirculation cells which feel 1/3 of the cavity for $R_{aS} = 4000$ and 2/3 of the cavity for $R_{aS} = 8000$. The influence of convection on the solute transport is especially noticeable along the walls as shown by the concentration pattern on Figure 18. This leads to a

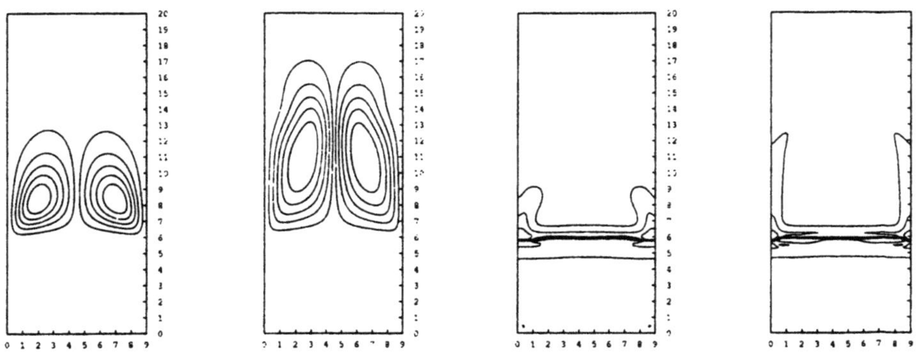

Figure 18 Streamlines and concentration isovalues for solutal Rayleigh $Ra_S =$ 4000 and 8000.

degeneration of solute on the periphery of the crystal nearby the walls and an enrichment in the central zone. The inhomogeneity of solute along the interface, as shown in Figure 19 tends to be increased with Ra_s. The difference of radial concentration in the solidified zone goes from 3.5% for Rayleigh 1000 to 15% for Rayleigh 8000. The streamlines nearby the solidification zone (see Figure 18) exhibit a slightly incurved interface.

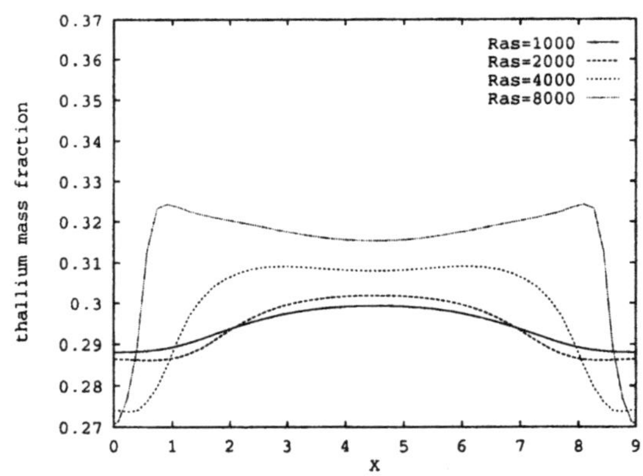

Figure 19 Solute distribution in the solidified zone for solutal Rayleigh values $Ra_S = 1000 - 8000$.

6 Conclusions

The outstanding properties of the spectral method with respect to accuracy and efficiency for the simulation of fluid flow in simple geometries can be preserved also in a coupled SM-FVM code. The accuracy of the coupled code for a given number of computational points is significantly higher than the accuracy of the finite-volume code. If a high accuracy is required, the coupled code outperforms the finite-volume code significantly. The coupled code reaches the highest accuracy with less computational cost (Figure 5) and memory requirement.

The developed CVD simulation models have successfully been applied to SiC-CVD process at largely varying process conditions. The interaction between the three different convective mechanisms (rotation, forced flow, buoyancy) and their effect on layer uniformity have been studied using two different 3-D finite volume codes. The results of computation are compared favourably and shown that the both codes can be effective used as prediction tools.

A real Bridgman configuration was investigated using a model for global simulations that takes into account conductive and radiative heat transfer, and a good agreement with the experimental results is found.

The interaction between unsteady melt and the solidification interface is studied by an enthalpy model and we can describe for aperiodic flow strong effects on the shape of the interface. Various symmetric, asymmetric, steady and time-dependent flow solutions are considered. Transitions between regimes are analysed and we have determined realistic sections of solidified crystals.

References

[1] I. Raspo, J. Ouazzani, R. Peyret. A spectral multidomain technique for the computation of the Czochralski melt configuration. *Int. J. Num. Meth. Heat Fluid Flow*, vol. 6, 31-58, 1996.

[2] E. Serre, S. Hugues, E. Crespo del Arco, A. Randriamampianina, P. Bontoux. Spiral Patterns in an Ekman Boundary Layer Flow. *Phys. Fluids*, to appear 1998.

[3] F. Durst, L. Kadinski, and M. Schäfer. A multigrid solver for fluid flow and mass transfer coupled with grey-body surface radiation for the numerical simulation of CVD processes. *J. Crystal Growth*, 146:202–208, 1995.

[4] L. Kadinski. *Mathematische Modellierung und numerische Simulation von CVD-Prozessen in der Halbleitertechnik.* PhD thesis, Friedrich-Alexander-Universität zu Erlangen, 1996.

[5] F. Durst, M. Schäfer. A Parallel Blockstructured Multigrid Method for the Prediction of Incompressible Flows. *Int. J. for Num. Meth. in Fluids*, Vol. 22, p. 549-565, 1996.

[6] S. Patankar, D. Spalding. A Calculation Procedure for Heat, Mass and Momentum Transfer in Three-Dimensional Parabolic Flows. *Int. Journal of Heat and Mass Transfer*,p. 1787:1806., 1972.

[7] J. Ouazzani, R. Peyret, and A. Zakaria. Stability of collocation-chebyshev schemes with application to the Navier-Stokes equations. In D. Rues and W. Kordulla, editors, *Sixth GAMM-Conf. Numer. Meth. Fluid Mech.*, 1986.

[8] J. M. Vanel, R. Peyret, and P. Bontoux. A pseudospectral solution of vorticity-stream function equations using the influence matrix technique. *Numerical Methods for Fluid Dynamics*, pages 463–475, 1986.

[9] K. Goda. A multistep technique with implicit difference schemes for calculating two- and three-dimensional cavity flows. *Journal of Computational Physics*, 30:76–95, 1979.

[10] S. Hugues, A. Randriamampianina. An improved projection scheme applied to pseudospectral methods for the incompressible Navier-Stokes equations. *Int. J. for Num. Meth. in Fluids*, in print 1998.

[11] D. B. Haidvogel and T. A. Zang. The accurate solution of Poisson's equation by expansion in chebyshev polynomials. *Journal of Computational Physics*, 30:137–180, 1979.

[12] P. Haldenwang, G. Labrosse, S. Abboudi, and M. Deville. Chebyshev 3d spectral and 2d pseudospectral solvers for the Helmholtz equation. *Journal of Computational Physics*, 55:115–128, 1984.

[13] M. Hortmann, M. Perić. Finite volume multigrid prediction of laminar natural convection: Bench-mark solutions. *Int. J. for Num. Meth. in Fluids*, 11:189–207, 1990.

[14] W.D. Bennon and F.P. Incropera. A continuum model for momentum, heat and species transport in binary solid-liquid phase change systems: 1 model formulation. *Int. J. Heat Mass Transfer*, 30(10):2161–2170, 1987. See also 30(10):2171–2187, 1987.

[15] D. Morvan, M. El Ganaoui, and P. Bontoux. Numerical simulation of 2D crystal growth problems in a vertical Bridgman-Stockbarger furnace, latent heat effects and crystal-melt in interface morphology. *Int. J. Heat Mass Transfer* in print, 1998.

[16] M. El Ganaoui Modélisation de la convection instationnaire en présence d'un front de solidification déformable. *Phd Thesis* Université de la Méditerranée Aix-Marseille II, Octobre 1997.

[17] J. Ouazzani, K.-C. Chiu, and F. Rosenberger. On the 2d modelling of horizontal CVD reactors and its limitations. *J. Crystal Growth*, 91:497–508, 1988.

[18] J. Ouazzani and F. Rosenberger. Three-dimensional modelling of horizontal chemical vapor deposition I. MOCVD at atmospheric pressure. *J. Crystal Growth*, 100:545–576, 1990.

[19] F. C. Eversteyn, P. J. Severin, C. H. van der Brekel and H. L. Peek newblock *J. Electrochem. Soc.*, 117:925, 1970.

[20] M. E. Coltrin, R. J. Kee, and J. A. Miller. A mathematical model of silicon vapor deposition: further refinements and the effects of thermal diffusion. *J. Electrochem. Soc.*, 133:1206–1213, 1986.

[21] M. E. Coltrin, R. J. Kee, and J. A. Miller. A mathematical model of the coupled fluid mechanics and chemical kinetics in a chemical vapor deposition reactor. *J. Electrochem. Soc.*, 131:425., 1984.

[22] M. Naamoune. *Modelisation et optimisation numerique de reacteurs CVD.* PhD thesis, L'Universite D'Aix–Marseille II, l'Institut de Recherche sur les Phénomènes Hors Equilibre, 1997.

[23] H. Moffat and K. F. Jensen. Complex flow phenomena in MOCVD reactors. *J. Crystal Growth*, 77:108–119, 1986.

[24] H. K. Moffat and K. F. Jensen. Three-dimensional flow effects in silicon CVD in horizontal reactors. *J. Electrochem. Soc.*, 135:459–471, 1988.

[25] CFD2000, User Manual, Adaptive Research, Huntsville, Alabama, USA

[26] F. Durst, M. Schäfer, K. Wechsler. *Efficient Simulation of Incompressible Viscous Flows on Parallel Computers,* DFG Priority Research Prog. results, ed. E.H. Hirschel, Notes on Num. Fluid Mech., vol. 52, pp. 87 – 101, 1996.

[27] G. Müller, G. Hirt, D. Hofmann: *Current issues in bulkgrowth of s.i. III-V materials,* 7th Int. Conf. on Semi-insulating III-V Materials 1992, ed. by C.J. Miner et al., Inst. of Physics Publishing, Bristol 1993, p. 73.

[28] J. Amon, *Züchtung von Versetzungsarmen Si-dotiertem GaAs mit dem Vertikalen Gradient-Freeze Verfahren,* Dissertationsschrift, Technische Fakultät, Universität Erlangen-Nürnberg, 1998.

[29] M. Kurz and A. Pusztai, Presentation of a gentle discretisation scheme for the numerical treatment of nonlinear heat conduction on unstructured grids in finite volume technique, Comput. Journal of Mass and Heat Transfer, to be published.

[30] P. Larroudé, J. Ouazzani, J.I.D. Alexander and P. Bontoux, Symmetry breaking flow transitions and oscillatory flows in a 2D directional solidification model, European Journal of Mechanics/B, 13(3), pages 353-381, 1994.

[31] H. Jamgotchian, B. Billia and L. Capella. *J. Crystal Growth,* 85:318, 1987.

[32] C. Le Marec, R. Guerin and P. Haldenwang. Pattern study in the 2-d solutal convectiom above a bridgman-type solidification front. *Physics of Fluids,* 9(11):3149-3161,1997.

Multigrid Methods for Two Phase Flows

I. Ginzbourg and G. Wittum
Institute for Computer Applications
University of Stuttgart, Pfaffenwaldring 27, 70569 Stuttgart
Germany

Summary

In this paper we present new Volume of Fluid two phase model with surface tension. The model is based on the staggered, implicit in time Finite Volume discretization of basic equations, using so called rotated elements. Interface adaptive and/or interface aligned deformable grids are reconstructed at each time step with help of the Piecewise Linear Interface Calculation Method. First and second order Langrangian propagation of one fluid with respect to regular mesh is introduced as alternative to the Eulerian split advection algorithm. The simulations of buoyant bubbles are compared with the theoretical predictions and the SURFER code numerical computations.

1 Introduction

Multigrid approach is known to be as one of the most efficient computational techniques for solving a wide range of technical applications. Recently, it is successfully applied for solving two-phase problems with high density ratios. The difficulty of two-phase problems arrises from the presence of sharp moving fronts where pressure and velocity derivatives may have a jump. Although conservative numerical schemes do not require the solution to be differentiable, it is known for a long time that low order approximations of governing equations destroy the sharpness of the front due to excessive numerical diffusion whereas high order schemes provoke numerical oscillations near the interface. Regular Eulerian grids provide good approximations for pressure and velocity solutions. When widely used projection methods are applied for MAC staggered grids, multigrid method is used to solve the resulting Poisson equation for pressure ([14], [19], [25]). By semi-implicit time discretization of the viscous term combined with projection methodology, multigrid based techniques are also used as viscous solvers ([19], [25]); in the same time, the majority of the investigators treat the convection term explicitly. Numerous methods to describe the interface and its propagation are often based on similar discretizations of basic conservation laws. Two main directions, *surface tracking* and *surface capturing*, can be found in literature (see [8],

Surface tracking methods explicitly treat the interface as a discontinuity; they specify it by an ordered set of marker points, connected by some interpolation curve; the markers are advanced in the Langrangian manner and then redistributed to obtain the best resolution of the interface. Cubic splines ([5], [16], [17]) are found to be smooth enough to ensure the correct discretization of the surface tension force. However, it is underlined in [17] that an accurate representation of the pressure jump across the interface requires not only a precise description of the surface tension force but also taking into account the interface position while discretizing the pressure gradients. This is difficult until the grid is fixed. Moreover, deformable grids with cell boundaries aligned with the reconstructed interface allow to impose on it the conjunction conditions exactly. Unstructured grids induce the development of multiphase finite volume schemes based on finite-element approximations (see [15], for instance). Free *Langrangian* methods do not treat the basic equations and the interface propagation separately but advect the calculation grid itself by the Langrangian equations.

Surface capturing methods are implicit with respect to the interface: using the special functions to describe the multiphase distribution, they do not need the exact interface position to discretize the governing equations. Well known are the Volume of Fluid (VOF) method introduced by Hirt and Nichols ([11]) and the Level Set approach due to Osher and Sethian ([18]). These methods handle complicated interfaces, including their reconnection, more easily than the surface tracking methods. Considerable efforts have been devoted to create accurate advection schemes for the VOF method. More often, volume fractions are advanced by fractional step, flux based algorithms: first in one space direction, then in another; the fully coupled advection methods are more difficult to implement (see in [19]). Different methods are available to reconstruct the interface from the volume fraction distribution; the recent schemes ([2],[14],[19],[22]) take into account the interface propagation while computing the fluxes across the cell boundaries.

Our purpose is the construction and justification of approximation schemes and solution methods for interfacial flows. The general frame of our work is as follows. First, we have combined a VOF interface representation with a collocated, Finite Volume discretization of the Navier-Stokes equations on both uniform and non-uniform grids (see [20], [24] for discretization). With a collocated discretization, the stabilizing pressure term should be introduced into the basic relation for mass conservation. Therefore, the resulting velocity field violates the continuity equation, especially near the interface and affects the mass conservation during interface advection noticeably. Besides that, in case of the aligned grid, the presence of the pressure discretization points on the interface does not allow to describe a pressure jump correctly, even if the surface tension force is imposed precisely. Because of these reasons, we have introduced a new, *staggered* discretization as an alternative to the previous nodal description. Both discretizations are discussed and numerically compared in [7].

In the present work we only use a staggered discretization, also applicable on unstructured grids due to the development of finite-element approximations with respect to *rotated* elements. The discretization is based on the implicit in time and Finite Volume schemes. The surface tension is treated in a momentum conserving way following the idea of our French partners, S. Zaleski and cooperators ([12]). Three different types of calculation grids are available to track the pressure and velocity solution: regular, adaptive and interface aligned. The aligned grid separates the fluids in such a way that no mixture of phases occurs in one element; the Finite Volume method provides then the exact conjunction of the viscous stresses on the reconstructed interfaces. The interface is approximated by the Piecewise Linear Interface Calculation method (PLIC), introduced by J.Li and S. Zaleski ([13], [14], [29]). The PLIC method allows to reconstruct free surfaces of arbitrary geometry locally in each cell of the *uniform grid*; however, the locality of the method results in discontinuity of the interface on the cell edges. Three different advection techniques are implemented: first and second order Langrangian moving are introduced as an alternative to the fractional, PLIC based advection algorithm ([13], [14], [22], [29]). We do not understand the Langrangian moving in sense of free Langrangian methods. Indeed, we advance the part of the aligned grid, occupied by one phase, in Langrangian manner; the new volume fraction distribution is derived from the new grid position; the calculation grid is then aligned with new *reconstructed* interface. A similar approach, the so called hybrid finite-element-volume-of-fluid method, has been recently described by Mashayek and Ashgriz ([15]). They align the collocated grid with the interface which consists of connected line segments; the segments cut the cell boundaries while the junction takes place somewhere inside a cell. Advection is performed with a flux based FLAIR VOF method (see [1], [2], [15]), without any Langrangian moving of the calculation grid.

The current paper does not focus on solvers. When the convective term is treated implicitly, the arising nonlinear problem is solved by the Newton method. The multigrid method with BiCGstab acceleration (see [23]) is used to solve the linear saddle-point problem with respect to all, pressure and velocity, unknowns on the grid. Different iterative smoothing procedures are adapted: Transforming smoothers of Wittum ([28]) have been generalized; *collective* Gauss-Seidel iterative smoothing technique was introduced as an extension of Vanka ([27]) for staggered non-uniform grids.

Thus, the paper is organized as follows. In Section 2 the basic conservation relations are given. In Section 3.1 we reconstruct the interface, determine the criterion for the adaptive refinement and describe the construction of the aligned calculation grid. Space and time discretizations on a non-uniform staggered grid are introduced in Section 3.2; different transgrid interpolations are discussed. Advection techniques are described in Section 4. The measurements of Laplace law for a stationary bubble are given in Section 5. Section 6 describes the simulations of bubble rising in a confined box.

2 Basic equations

Assume that two immiscible phases (fluid-fluid or fluid-gas) occupy a 2D rectangular domain Ω. In bulk, the velocity field $\vec{u} = (u_x, u_y)$ and pressure p of each phase obey the continuity equation

$$\nabla \cdot \vec{u} = 0 \tag{2.1}$$

and the Navier-Stokes equations

$$\frac{\partial \rho \vec{u}}{\partial t} + \nabla \cdot (\rho \vec{u} \otimes \vec{u}) = -\nabla p + \rho \vec{g} + \nabla \cdot (2\mu \mathbf{D}), \quad \mathbf{D} = \frac{1}{2}(\nabla \vec{u} + \nabla^t \vec{u}) \ . \tag{2.2}$$

Here $\rho = \{\rho_1, \rho_2\}$ and $\mu = \{\mu_1, \mu_2\}$ are density and viscosity of phase 1 or 2, respectively. Note that eq. (2.2) is considered in fully conservative form. The surface tension force $\vec{F}_s$ acts on the interface S between the fluids. When the surface tension coefficient σ is constant, the surface tension force per unit area $\vec{F}_s(\vec{r})$ is defined as

$$\vec{F}_s(\vec{r}) = \sigma \kappa(\vec{r}) \vec{n}(\vec{r}) \ , \qquad \forall \vec{r} \in S \ . \tag{2.3}$$

Here $\kappa(\vec{r}) = -(\nabla \cdot \vec{n})|_{\vec{r}}$ is a local curvature, taken positive if the center of curvature is in fluid 1; $\vec{n}(\vec{r})$ is the unit normal to S at $\vec{r}$, directed from fluid 2 into fluid 1. Precise numerical computations of the curvature are difficult until the interface is not represented by a sufficiently smooth curve, e.g. continuous together with its first and second derivatives. Lafaurie et al. ([12]) have shown, that the surface tension force can be included into eq. (2.2) in the conservative form

$$\vec{F}_s = -\nabla \cdot \mathbf{T} \ , \qquad \mathbf{T} = \sigma(\mathbf{I} - \vec{n} \otimes \vec{n})\delta_S \ . \tag{2.4}$$

Tensor $\mathbf{T}$ is tangential with the interface ($\mathbf{T} \cdot \vec{n} = 0$); the function δ_S is concentrated on it. The velocity is assumed to be continuous across the interface: $\vec{u}]_S = 0$. Here and below, $[\psi]_S = \psi_2 - \psi_1$ denotes the jump of a scalar or vector field ψ across the interface S. Momentum conservation across the interface supplies the additional interface conjunction condition

$$[2\mu \mathbf{D} \cdot \vec{n} - p\vec{n}]_S = \vec{F}_s \ . \tag{2.5}$$

At the boundary of the domain Ω, we impose either noslip or freeslip conditions.

3 Discrete formulation of the two phase 2D model

3.1 Calculation grids

Let us refer to the finest grid where mass and momentum conservation equations are discretized as *calculation* grid. Below, we discuss three different types of the calculation grids. First, we consider a uniform square grid on which we discretize the basic equations (2.1) and (2.2) without any reconstruction of the interface between the phases. Second, we apply successively some refinement criterion in the vicinity of the interface to obtain the hierarchy of non-uniform *adaptively refined* grids. Third, we introduce an interface *aligned* grid which has no mixture of phases inside its elements. Finally, we combine the two last approaches and construct an aligned grid on the finest adaptive grid. The construction of the calculation grids is based on the Piecewise Linear Interface Calculation (PLIC) method (see [13], [14], [29]).

3.1.1 Construction of the interface

Introduce on a 2D rectangular domain Ω the uniform rectangular mesh with constant space step h. Denote its elements as (i, j). With the well known Volume of Fluid (VOF) approach of Hirt and Nicholls ([11]), the distribution of phases is represented by *volume fraction* $C_{i,j}$ of fluid 1 in cell (i, j). Then $C_{i,j} = 1(C_{i,j} = 0)$ in the part of Ω occupied by fluid 1 (respectively, by fluid 2), and $0 < C_{i,j} < 1$ in cells intersected by an interface. The PLIC method uniquely defines the interface in each cell with $0 < C_{ij} < 1$ by a slope segment, which is perpendicular to a given normal $\vec{n}^{i,j} = (n_x^{i,j}, n_y^{i,j})$ and cuts exactly the fraction $C_{i,j}$. The local normal $\vec{n}^{i,j}$ to the interface is approximated from the phase distribution C as

$$\vec{n}^{i,j} \approx \nabla C|_{i,j} \,. \tag{3.6}$$

It can be computed, for instance, with a 8 point finite difference scheme

$$\begin{aligned} n_x^{i,j} &= \frac{1}{4}(2\nabla_x C_{i,j} + \nabla_x C_{i,j-1} + \nabla_x C_{i,j+1}) \,, \\ n_y^{i,j} &= \frac{1}{4}(2\nabla_y C_{i,j} + \nabla_y C_{i-1,j} + \nabla_y C_{i+1,j}) \,, \end{aligned} \tag{3.7}$$

where $\nabla_x C_{i,j}$ and $\nabla_y C_{i,j}$ are central difference approximations of the first derivatives on uniform grid with mesh size h

$$\nabla_x C_{i,j} = \frac{C_{i+1,j} - C_{i-1,j}}{2h} \,, \ \nabla_y C_{i,j} = \frac{C_{i,j+1} - C_{i,j-1}}{2h} \,. \tag{3.8}$$

The construction of the interface with PLIC is done locally, cell by cell; it does not need any assumption about phase distribution. The deficiency of this locality manifests itself in the discontinuity of the interfaces at the boundaries of the cells.

3.1.2 Adaptive refinement near the interface

Assume that the phase distribution C is defined on a regular mesh. Let corners of the element (i, j) be numbered as $(i \pm \frac{1}{2}, j \pm \frac{1}{2})$. The normal $\vec{n}^{i\pm\frac{1}{2}, j\pm\frac{1}{2}}$ to the interface at the corners can be approximated with four point finite-differences similar with (3.7), (3.8). Let $|| \cdot ||$ denote the Euclidean norm of a vector. We are able now to formulate our refinement criterion: if $|| \vec{n}(\vec{r}) ||$ is different from zero for some node $\vec{r}$ of the given element (i, j), we mark for regular refinement the element (i, j) itself and three its neighbours which contain the given node $\vec{r}$. Regular refinement means that the square element will be subdivided into four square elements. When all regular refinements are done, some irregular refined elements may appear in their neighbourhood to guarantee the closure of the grid (see Fig. 1a and b). Once the adaptive grid is constructed, we have to determine the volume fractions on it. Volume fraction of the finer element is defined as the square of its intersection with PLIC polygons constructed inside a father element. Since the vicinity of any interfacial node is regularly refined, PLIC method can be applied on the adaptive grid itself. Repeating the refinement procedure l_{adap} times, we obtain a hierarchy consisting of l_{adap} adaptively refined grids. The finest grid of the hierarchy is considered as a calculation grid where basic equations are discretized and solved.

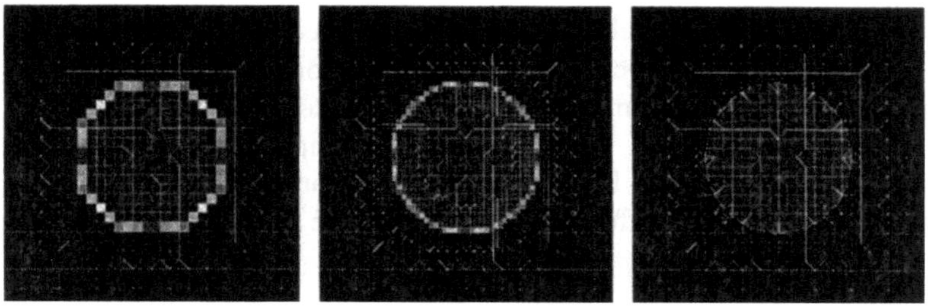

a: One adaptive level. b: Two adaptive levels. c: Aligned grid on one level.

Figure 1: Calculation grids corresponding to a circle $R = 0.25$.

3.1.3 Interface aligned grid

Assume that we are able to separate completely the phases on the calculation grid by the boundaries of its geometrical elements; then no approximation of the physical parameters (density and viscosity, for instance) will needed. This gives us the idea of the *interface aligned* calculation grid. In order to construct it, we proceed in three steps. First, piecewise-linear interface is reconstructed on the finest, uniformly or adaptively refined, grid. Second, irregular refining rules are applied to interfacial cells in such a way that any interface segment will represent a boundary of some son element as illustrated on Fig. 2a. Let us refer to the obtained set of polygons as *moving grid* since it is used to advect the interface with our *Langrangian* approach (see in section 4). The moving grid separates the fluids but it can not be used as a calculation grid because of the discontinuity of the piecewise interface at the boundaries of the cells. Nevertheless, one can convert the moving grid into the calculation grid with the help of the subsequent refinement procedure. Additional segments should connect the corners of the moving grid *inside each phase* when the discontinuity occurs as demonstrated on Fig. 2b. This completes the third step to construct the aligned grid. When the PLIC interface is smooth enough but discontinuous, the aligned grid contains the elements with bad aspect ratio, which can deteriorate the solving procedure. When it becomes necessary (for instance, when the iteration process does not converge on the aligned grid), we connect two interface segments in the middle of the discrepancy if the last one is

a). b).

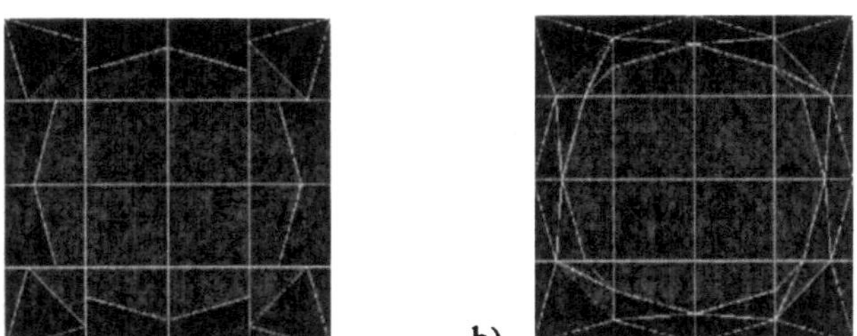

Figure 2: Moving grid (a) and the interface aligned grid (b).

smaller than a given small value ϵh. This is illustrated on Fig. 1c where the aligned grid is constructed on adaptively refined grid and smoothed with $\epsilon = 0.1$. The smoothing procedure does not guarantee the complete absence of the "bad elements"; in spite of that it turns out to be very helpful in practical computations. Indeed, although the grid smoothing deteriorates a little bit the volume balance of phases on the calculation grid, the moving grid is not affected. Consequently, the advected phase volumes are not changed.

3.2 Finite Volume Method

Let initial conditions $\vec{u}_0$ and p_0 be given at $t = t_0$. We look for the solution $\vec{u}^{t_i}$ and p^{t_i} at discrete time moments $t_1, t_2, \ldots, t_n$. Consider an arbitrary interval $[t, t + \Delta t]$ and assume that the distribution of phase field $C^\tau = C(t = \tau)$ is known at some time τ inside the time interval under consideration. Suppose the calculation grid corresponds to C^τ or is uniform. We present new staggered discretization on uniform or non-uniform calculation grid, consisting of convex quadrilateral and/or triangular elements. Similar to the well known MAC staggered discretization of Hirt and Nicholls ([11]), pressure unknown is located at the center of the element and the velocity variables are found at the middle of its edges: however, unlike MAC method, each edge contains *both* components of the velocity. This modification enables us to discretize the basic equations in the same manner on the uniform and non-uniform grids. With the Finite Volume

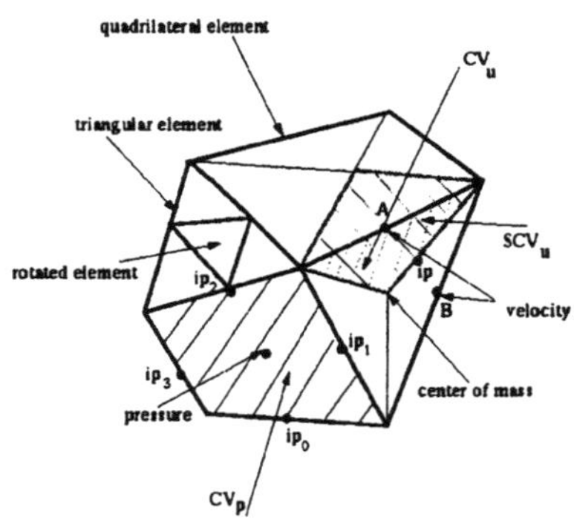

Figure 3: Staggered grid.

method, mass conservation eq. (2.1) is integrated over a control volume $CV_p(\vec{r})$ which coinsides with the geometrical element with the center $\vec{r}$. The Gauss theorem yields then

$$\oint_{\partial CV_p} (\vec{u} \cdot \vec{n}_p) ds = 0 \; . \tag{3.9}$$

Here and below, n_α denotes the outer unit normal to the boundary segment ∂CV_α; s_α equal n_α scaled with the length of the segment. Approximation of the boundary integral (3.9) with quadratures involves the velocity variables $\vec{u}(\vec{r}_i)$ lying on the boundary of the element. With *implicit* Euler time discretization, we represent (3.9) as

$$\sum_i (\vec{u}^{t+\Delta t} \cdot \vec{s}_p)|_{\vec{r}_i} = 0\,, \qquad \vec{r}_i \in \partial CV_p\,. \tag{3.10}$$

In order to discretize the Navier-Stokes eq.(2.2), we integrate them over control volumes CV_u constructed for each velocity discretization point; the control volume is subdivided into the subcontrol volumes SCV; each SCV belongs only to one geometrical element (see Fig. 3). With the help of the Gauss theorem we represent eq.(2.2) with surface tension force (2.4) as

$$\int_{CV_u} \frac{\partial \rho \vec{u}}{\partial t} + \oint_{\partial CV_u} \{(\rho \vec{u} \otimes \vec{u})) + p - 2\mu \mathbf{D}\} \vec{n}_u ds = \int_{CV_u} \rho \vec{g} - \oint_{\partial CV_u} \mathbf{T} \cdot \vec{n}_u ds\,. \tag{3.11}$$

Boundary integrals are approximated by the mean point rule at integration points ip. Density ρ^τ and viscosity μ^τ are computed at time $t = \tau$ as a volume averaging inside each geometrical element

$$\rho^\tau = C^\tau \rho_1 + (1 - C^\tau)\rho_2; \quad \mu^\tau = C^\tau \mu_1 + (1 - C^\tau)\mu_2\,. \tag{3.12}$$

They are assumed to be piecewise constant functions. When the grid is interface aligned, i.e. its each element belongs either to phase 1 or 2, no averaging occures in (3.12) so that (3.11) satisfies the tangential conjunction condition (2.5) exactly on the reconstructed interface.

• **Mass and gravity terms**. Taking into account that density can vary inside the control volume, the integration is subdivided into the summation over the subcontrol volumes. Mass term takes then a form

$$\int_{CV_u(\vec{r})} \frac{\partial \rho \vec{u}}{\partial t} \approx \frac{\vec{u}^{t+\Delta t}(\vec{r}) - \vec{u}^t(\vec{r})}{\Delta t} \sum_i \rho^\tau(SCV_i) \times area(SCV_i)\,. \tag{3.13}$$

In the same way we approximate the gravity term

$$\int_{CV_u} \rho^\tau \vec{g} = \vec{g} \sum_i \rho^\tau(SCV_i) \times area(SCV_i)\,. \tag{3.14}$$

• **The surface tension term** is completely defined by the phase distribution C^τ (cf. (2.4))

$$\oint_{\partial CV_u} (\mathbf{T} \cdot \vec{n}_u) ds \approx \sum_{ip_i} (\mathbf{T}^\tau \cdot \vec{s}_u)|_{ip_i}\,. \tag{3.15}$$

Let the finite element (f.e.) approximation of arbitrary scalar or vector field ϕ inside or outside a given triangular/quadrilateral element be written as

$$\phi^{f.e.}(\vec{r}) = \sum_{j=1}^{m} N_j(\vec{r})\phi_j\,. \tag{3.16}$$

Here N_j are standard linear/bilinear basis functions; m is the number of the element corners. In order to compute $\mathbf{T}^\tau(ip)$, we use (3.16) in two different ways. With the first technique, the normal $\vec{n}(ip)$ in an integration point is approximated with f.e. extension (3.16) from its nodal values; then $\mathbf{T}^{\tau}(ip)$ is computed using (2.4). In a second approach, tensor $\mathbf{T}^{\tau}$ itself is first calculated in the corners of the elements with (2.4) and then, with help of the relations (3.16), is prolongated into the integration points. With both techniques, concentrated on the interface function δ_S (cf. (2.4)) is approximated by $|\nabla C|$; relation (3.6) gives then: $\delta_s(\vec{r}) \approx\| \vec{n}(\vec{r}) \|$.

• **The pressure term** takes the form

$$\oint_{\partial CV_u} p\vec{n}_u \approx \sum_{ip_i} p\vec{s}_u|_{ip_i} , \qquad p = \alpha^+ p^{t+\Delta t} + \alpha^- p^t . \tag{3.17}$$

A piecewise constant approximation of the pressure is applied: $p(ip_i)$ is equal to the pressure value at the center of the element for any integration point lying inside a given element. The pressure is treated either with implicit Euler ($\alpha^+ = 1, \alpha^- = 0$) or with Crank-Nicolson time discretization scheme ($\alpha^+ = \alpha^- = \frac{1}{2}$).

• **Diffusion term**. The rate of deformation tensor $\mathbf{D}(\vec{u}^*)$ is discretized with implicit Euler, Crank-Nicolson and explicit Euler schemes ($\alpha^+ = 0, \alpha^- = 1$) as

$$\oint_{\partial CV_u} \mu\mathbf{D} \cdot \vec{n}_u \approx \sum_{ip_i} \mu^{\tau}(\mathbf{D}(\vec{u}^*) \cdot \vec{s}_u)|_{ip_i} , \; \vec{u}^* = \alpha^+ \vec{u}^{t+\Delta t} + \alpha^- \vec{u}^t . \tag{3.18}$$

Spatial velocity approximations are fulfilled with the finite element relations (3.16) on *rotated* elements: the corners of the rotated element coincide with the velocity discretization points (see Fig. 3). Substitution of the f.e. approximation over the rotated element

$$\nabla\vec{u}^*(ip) = \sum_{j=1}^{m} \nabla N_j(ip)\vec{u}_j^* \tag{3.19}$$

into (3.18) keeps the diffusion linear w.r.t. the velocity variables. In case of equal dynamical viscosities, and with use of continuity eq. (2.1), the approximation of the diffusion term can be done in Laplacian form, similar with the one phase model

$$2\oint_{\partial CV_u} \mu\mathbf{D} \cdot \vec{n}_u = \mu \sum_{ip_i} (\nabla\vec{u}^* \cdot \vec{s}_u)|_{ip_i} . \tag{3.20}$$

Note that contrary to this particular case (3.20), no diagonal dominance is guaranteed by the general form (3.18) for the diffusion part of the stiffness matrix.

• **Convection term**. By analogy with (3.18), let us represent the convection term as

$$\oint_{\partial CV_u} \rho\mathbf{V} \cdot \vec{n}_u \approx \sum_{ip_i} \rho^{\tau}([\alpha^+ \mathbf{V}(\vec{u}^{t+\Delta t}) + \alpha^- \mathbf{V}(\vec{u}^t)] \cdot \vec{s}_u)|_{ip_i}, \; \mathbf{V} = \vec{u} \otimes \vec{u} . \tag{3.21}$$

As for the diffusion term, implicit Euler, Crank-Nicolson and explicit Euler schemes are implemented. As above, rotated f.e. approximations (3.16) are called to relate the velocity $\vec{u}(ip)$ with the velocity variables. Implicit Euler and Crank-Nicolson schemes lead to the non-linear system of equations. Its solving with Newton method needs the computation of the Jacobian $\frac{\partial(\mathbf{V}\cdot\vec{s}_u)}{\partial\vec{u}}$ in each discretization point. Similar to the diffusion term, the convection part of the stiffness matrix yields no diagonal dominance unless the quasi-Newton linearization is used (see [20]). On the other hand, the convection term can be linearized with respect to time

$$(\vec{u}\otimes\vec{u})^{t+\Delta t} = (\vec{u}\otimes\vec{u})^{t} + \frac{\partial(\vec{u}\otimes\vec{u})}{\partial\vec{u}}^{t}\frac{\partial\vec{u}}{\partial t}\Delta t\,; \qquad \frac{\partial\vec{u}}{\partial t} \approx \frac{\vec{u}^{t+\Delta t} - \vec{u}^{t}}{\Delta t}\,. \tag{3.22}$$

The second order time linearization scheme (3.22) preserves stability properties of the implicit approach and avoids the necessity to solve the non-linear system of equations. Both approaches, with and without linearization (3.22), are implemented and compared. This completes the discretization of the basic equations on a single grid.

3.3 Transgrid solution interpolations

Once the velocity field $\vec{u}^{t+\Delta t}$ is computed, the phase field C^{τ} is advanced to new position. We do this either by Langrangian or Euler approach, described in Section 4. New phase distribution requires the construction of new calculation grid unless it is uniform. The discretization of the basic equations in the next time interval $(t+\Delta t, t+2\Delta t)$ needs the old solution $\vec{u}^{t+\Delta t}$ and, eventually, $p^{t+\Delta t}$ at the new discretization points. The old solution can be interpolated from the previous calculation grid or prolongated from some coarse grid, which is not altered between two time steps. We discuss below some different transgrid operators implemented for solution interpolations.

3.3.1 Velocity interpolation

• **Nodal interpolation.** The velocity is first extrapolated from the edges into the corners of the geometrical elements on the old calculation grid using f.e. relations (3.16) w.r.t the rotated element. The averaging is done over all elements to whom the node belongs. These nodal values are then interpolated into new discretization points with f.e. relations (3.16) over old geometrical elements. In general, the interpolated solution does not satisfy exactly the mass conservation relation (3.10).

• **Edge interpolation.** Velocity is directly propagated from the old discretization points to new ones with f.e. interpolations (3.16) w.r.t the old rotated elements; the averaging is fulfilled over all old geometrical elements to whom new discretization points belongs. As above, mass conservation is not preserved by this rotated interpolation.

• **Divergence free interpolation.** Let us remind here that by the construction any son element on the finer grid is either a copy of its coarse father or its coarse father represents a regular element. Then non-trivial transgrid interpolations occur only between a regular father and its son elements. Consider a regular father element and the set of its son elements, obtained with regular or irregular refinement rules. Velocity in father discretization points can be restricted from sons elements with nodal or edge approach discussed above. We are interested, however, in divergence free solution restriction. Assume that the mass conservation relation (3.10) is satisfied for each son element and consider the sum of these relations over all sons of one father. Since any internal, w.r.t the father element, velocity discretization point belongs to two son elements, the sum provides the following relation

$$\sum_k (\vec{u} \cdot \vec{s}_p)|_{\vec{r}_k} = 0 \, , \qquad \vec{r}_k \in \partial CV_p \, . \tag{3.23}$$

Here, summation goes only over *sons* discretization points lying on the boundary ∂CV_p of the father element. Let us put the velocity $\vec{u}$, defined at the edges of father element, be equal to the weighted average of the velocity at sons discretization ponts $\vec{r}_k$ lying on the same edge

$$\vec{u} = \frac{1}{\sum_k \parallel \vec{s}_p(\vec{r}_k) \parallel} \sum_k (\parallel \vec{s}_p \parallel \vec{u})|_{\vec{r}_k} \, . \tag{3.24}$$

Due to relation (3.23), the velocity (3.24) is divergence free in sense (3.10).

Starting from finest level and proceeding in this way, one can construct divergence free fields on all coarser levels. Projected solution can be used, first, to compute the stiffness matrix on coarse grids required for multigrid technique ; second, to advance the phase field on the regular mesh with split advection algorithm (see in Section 4.3); finally, to interpolate the solution onto the hierarchy of new grids. In order to prolongate the solution from coarser to finer grid, assume that the velocity $\vec{u}$, defined on the edges of $(h \times h)$ element (i, j), obeys (3.10). Its divergence free prolongation can be done with the linear interpolations

$$\begin{aligned} u_x(x,y) &= (1-x)u_{i-\frac{1}{2},j} + xu_{i+\frac{1}{2},j} \, , \\ u_y(x,y) &= (1-y)u_{i,j-\frac{1}{2}} + yu_{i,j+\frac{1}{2}} \, . \end{aligned} \tag{3.25}$$

The resulting solution will satisfy (3.10) for any son element. Together, linear interpolations (3.24)-(3.25) supply divergence free finer-coarse and coarse-finer approximations.

3.3.2 Pressure interpolation

We need the restriction of pressure solution to coarser levels for multigrid method as well as its expansion to new discretization points. Projection of the pressure is simple: the arithmetical mean pressure value over all son elements is appropriated to their father. In order to approximate the pressure in new discretization points, we proceed as follows. First, we compute the pressure solution at the nodes of the calculation grid as an arithmetical mean of the pressure values over all elements whom belongs the given node. Then we project this nodal solution to all coarser levels. Once a hierarchy of new grids is constructed over the finest regular level, its nodal solution is interpolated to the sons nodes with f.e. relations (3.16). The solution in new discretization points, i.e. in the centers of new elements, is computed as an arithmetical mean value of the nodal solution. The procedure is then successively repeated for next finer levels.

4 Propagation of the interface

When the velocity field $\vec{u}$ is divergence free, the conservation of volume of each phase can be expressed by one cinematic equation

$$\frac{\partial C}{\partial t} + \nabla \cdot (\vec{u}C) = 0 \,. \tag{4.26}$$

We have implemented two different approaches to solve the advection equation (4.26): *Langrangian moving* and *Euler* split advection algorithm. Once phase distribution C^t, $t = t_1, t_2, ..., t_n$, is found, new calculation grid is constructed and basic equations are discretized at next time interval. The first time step is a bit different from the others. Using the initial distribution $C(t = 0)$, the first approximation to solution $\vec{u}^{\Delta t}$ is computed. Then, with the mean solution $\frac{1}{2}(\vec{u}_0 + \vec{u}^{\Delta t})$, the advection equation (4.26) is solved for $C(t = \frac{1}{2}\Delta t)$. The distribution $C(t = \frac{1}{2}\Delta t)$ is used to recalculate anew $\vec{u}^{\Delta t}$. Starting from this moment, the solution $C(t + \frac{1}{2}\Delta t)$ in the middle of the next time interval $(t, t + \Delta t)$ is obtained always similar: namely, the field $C(t - \frac{1}{2}\Delta t)$ is advanced with the velocity $\vec{u}^t$. In its turn, we set C^τ (cf. (3.12)) be equal to $C(t + \frac{1}{2}\Delta t)$ in order to obtain the next solutions $\vec{u}^{t+\Delta t}$ and $p^{t+\Delta t}$.

4.1 First order Langrangian method

The construction of the interface with the PLIC method enables us to transport the fluid 1, i.e to solve eq.(4.26), in *Langrangian manner*. It is understand in the following way here. Let (i, j) be regular cells lying on the finest uniform grid where phase distribution is initially defined. Once the interface is reconstructed with the PLIC method on the finest uniform mesh, or on the finest adaptively refined grid, each phase is represented by a set of polygons which we have refered as *moving grid* (cf. Fig. 2a). Assume that the velocity $\vec{u}$ is known at the nodes of the moving grid. If we shift them on the distance $\vec{u}\Delta t$, the total area of the intersection of a cell (i, j) by advanced polygons determines new phase distribution $C^\star_{i,j}$. In practice, in order to avoid an appearance of non-convex polygons, we triangulate the polygons before moving. Proceeding in this way, we advance all cells occupied completely or partially by fluid 1. If necessary, we adjust the values $C^\star$ to be in the interval $(0, 1)$ similar with [29]

$$C^{new}_{i,j} = \min[\max(C^\star_{i,j}, 0), 1] \,. \tag{4.27}$$

Advection with the given velocity is explicit in time and only first order accurate in space since the variation of the velocity during the moving is neglected. One can easily show, using any non-constant divergence free velocity field as an example, that such a moving does not preserve the volume even if the velocity is a linear function of coordinates. In order to improve the volume or *mass* conservation, we have introduced a second order Langrangian method described in the next section. It is important to emphasize here that the mass conservation depends on the error with which the velocity in the nodes of the moving grid is obtained. Three types of solution interpolations between the finer-finer and the coarse-finer grids, introduced in Section 3.3, are implemented and compared in Section 6.

4.2 Second order Langrangian method

Consider some node $\vec{r} = (x, y)$ of the moving grid. Using above approach, its new position is defined as $\vec{r}^{new} = \vec{r} + \Delta\vec{r}$, $\Delta\vec{r} = \vec{u}\Delta t$. It is much more accurate to compute the displacement $\Delta\vec{r} = (\Delta x, \Delta y)$ as

$$\Delta\vec{r} = \frac{1}{2}(\vec{u}(\vec{r}) + \vec{u}(\vec{r}^{new}))\Delta t\,. \tag{4.28}$$

Substitution of the second order Taylor expansion for $\vec{u}(\vec{r}^{new})$

$$\vec{u}(\vec{r}^{new}) = \vec{u}(\vec{r}) + \frac{\partial\vec{u}}{\partial\vec{r}}\Delta\vec{r} + O(\|\Delta\vec{r}\|)^2) \tag{4.29}$$

into equation (4.28) results in linear system of two equations with respect to $(\Delta x, \Delta y)$. The determinant of its matrix

$$\mathbf{D}^\star = \begin{pmatrix} \frac{2}{\Delta t} - \frac{\partial u_x}{\partial x} & -\frac{\partial u_x}{\partial y} \\ -\frac{\partial u_y}{\partial x} & \frac{2}{\Delta t} - \frac{\partial u_y}{\partial y} \end{pmatrix} \tag{4.30}$$

depends on the time step and is in general nonzero. The velocity gradients at the nodes of the moving grid are computed with f.e. relations (3.19) from the nodal velocity. When $\Delta\vec{r}$ is derived from (4.28)-(4.29), the mass is conserved exactly at least when each velocity component is a linear function of coordinates. To avoid the computation of the derivatives, one can proceed a bit less accurately, computing $\Delta\vec{r}$ directly from (4.28) with $\vec{u}(\vec{r}^{new})$ replaced by $\vec{u}(\vec{r} + \vec{u}\Delta t)$.

4.3 Split advection algorithm with PLIC method

In this section, we follow the split advection algorithm ([13],[14],[22],[29]); its realization is due to an acquaintance with the SURFER code of our French co-workers. Consider a square mesh with elements (i, j). To simplify the algebra, let us rescale the problem and put, locally in this section, $h = 1$, $\Delta t = 1$ and $\vec{u}^\star = \frac{\Delta t}{h}\vec{u}^{t+\frac{1}{2}\Delta t}$. Assume the velocity $\vec{u}^\star$ obeys the continuity relation (3.10)

$$u^\star_{i+\frac{1}{2},j} - u^\star_{i-\frac{1}{2},j} + u^\star_{i,j+\frac{1}{2}} - u^\star_{i,j-\frac{1}{2}} = 0\,. \tag{4.31}$$

The algorithm combines the Eulerian method, which is a direct integration of the advection equation (4.26) over the element surface, with the Langrangian moving of PLIC segments while computing the fluxes. In order to simplify the programming, the moving is done with fractional step or *split* strategy, first in $x-$ and then in $y-$ directions. The new fraction $C^{t+\Delta t}_{i,j}$ is obtained with a two step volume conservative scheme

$$C^\star_{i,j} = C^t_{i,j} + [F_{i-\frac{1}{2},j} - F_{i+\frac{1}{2},j} + C^t_{i,j}div_x u_x]\,; \tag{4.32}$$

$$C^{t+\Delta t}_{i,j} = C^\star_{i,j} + [F_{i,j-\frac{1}{2}} - F_{i,j+\frac{1}{2}} + C^t_{i,j}div_y u_y]\,. \tag{4.33}$$

Here, $F_{l,m}$ denotes phase flux during time interval Δt through $h \times h$ face, centered at (l, m). The terms $div_x u_x$ and $div_y u_y$ correspond to $x-$ and $y-$ parts, accordingly, of the continuity equation (4.31): $div_x u_x + div_y u_y = 0$. As discussed in Puckett et al. (see [19]), addition of these terms in split advection scheme allows to maintain better mass conservation. The principal problem of Eulerian approach is to determine accurately the fluxes $F_{l,m}$. Well known donor-acceptor flux approximation, employed in SOLA-VOF (see [11]) is only first order accurate. More precise methods use locally reconstructed interfaces to estimate the fluxes. With PLIC method, we calculate fluxes from new interface position as follows.

Assume equation of the interface segment inside the cell (i, j) reads as $n_x x + n_y y = \alpha$. After moving in $x-$ direction with the linear interpolated velocity (3.25), new equation of the interface segment reads

$$n_x^{new} x + n_y y = \alpha^{new}; \; n_x^{new} = \frac{n_x}{1 - u^\star_{i-\frac{1}{2},j} + u^\star_{i+\frac{1}{2},j}}; \; \alpha^{new} = \alpha + n_x^{new} u^\star_{i-\frac{1}{2},j} . \tag{4.34}$$

Assume that $u^\star_{i+\frac{1}{2},j}$ is positive, i.e. the cell (i, j) is upstream. Then $F_{i+\frac{1}{2},j}$ is set equal to the amount of the fluid contained in the rectangle $[1, 1 + u^\star_{i+\frac{1}{2},j}] \times [0, 1]$, computed from the new interface position (4.34). When $u^\star_{i+\frac{1}{2},j}$ is negative, $F_{i+\frac{1}{2},j}$ is also negative and its absolute value is equal to the quantity of fluid contained in rectangle $[1 + u^\star_{i+\frac{1}{2},j}, 1] \times [0, 1]$. This quantity is also computed from the new interface position but now the interface is shifted from the upstream cell $(i + 1, j)$. The computation of the flux through left face is done by analogy. In order to compute the fluxes through horizontal faces, the interface is advected in $y-$ direction following the same lines as (4.34). When all fluxes are computed, a new phase distribution is obtained with a help of the relations (4.32)-(4.33). Theoretically, when the CFL criterion

$$max(\| \vec{u} \|)\frac{\Delta t}{h} < \frac{1}{2} \tag{4.35}$$

is satisfied, the algorithm is stable and conserves the mass. Numerically, the necessity to adjust volume fractions to interval $(0, 1)$ (cf. (4.27)) leads to mass violation. As an example, we have advanced the circle with the given velocity $(1, 1)$ and $\Delta t = \frac{h}{4}$, $h = \frac{1}{16}$. With the Langrangian approach, mass is completely conserved and the center of the circle is found in the prescribed point. However, such a "diagonal" moving is difficult for split advection algorithm: the loss of mass relatively to its initial value is about -0.4% after only eight time steps and the circle is elongated in the velocity direction. Nevertheless, when the moving is generally in one direction but velocity distribution is far from linear, as in the case of rising bubbles considered in this paper, the situation with mass balance is inverse.

5 Laplace law

Validation of the Laplace law for stationary bubble represents a well known test of surface tension methods. At equilibrium, the pressure jump across the interface is related with its radius R according to (2.5)

$$p_{in} - p_{out} = \frac{\sigma}{R} \, . \tag{5.36}$$

Here, p_{in} and p_{out} are pressure values inside and outside the bubble. With a piecewise constant, finite volume discretization (3.17), the pressure gradient across the interface is correctly described only on the aligned grid where phase separation inside a control volume is done by the boundary between two elements. As fas as we know, no way to reconstruct a smooth enough, continuous interface without *volume violation* is proposed for VOF models. When neither equation of the interface nor its curvature is known, the surface tension force can be applied with the approximation (2.4). This results in well-known numerical artifacts, called in literature "anomalous" or "spurious" currents around a macroscopically static bubble. In order to demonstrate that the anomalous currents originates from bad approximation of the surface tension, let us consider the following experiment.

Table 1: Amplitude of the anomalous currents on the aligned grid is shown for continuous interface. Case a) addresses the computations of $\vec{n}(ip)$ with f.e. relations (3.16) whereas case b) corresponds to f.e. approximation of tensor $\mathbf{T}(ip)$ itself.

$R\backslash h$	$\vec{F}_s = \sigma\kappa\vec{n}$	$\vec{F}_s = \sigma\nabla\cdot(\mathbf{I} - \vec{n}\otimes\vec{n})\delta_S$	
		a).	**b).**
4	$6.7 \times 10{-}9$	0.033	0.034
8	$8.5 \times 10{-}9$	0.023	0.024
16	$1.2 \times 10{-}8$	0.024	0.027

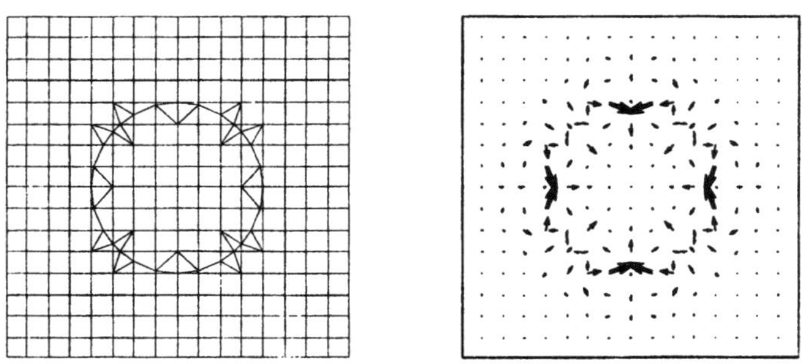

Figure 4: Anomalous currents on the grid aligned with the continuous interface. The surface tension force is computed in conservative form (3.15).

We approximate circle of the radius $R = 0.25$, centered in the computational domain $[0, 1] \times [0, 1]$, by a *continuous linear* interface and align the calculation grid with it (see Fig. 4a). Both fluids have the same densities and viscosities equal to one; the diffusion term in the Laplacian form (3.20) is assembled. The surface tension force with $\sigma = 1$ is applied in both forms: exact (2.3) and conservative (2.4), (3.15). In former case, $\int_{CV_u} \vec{F}_s$ reduces to an integration over the interface curve lying inside of the control volume; the integration is done exactly. In the second case, two finite element techniques are implemented as described in Section 3.2. The amplitude of the anomalous currents, $\max(\| \vec{u} \|)$, is given in Table 1. As we could expect, anomalous currents almost vanish when the exact surface tension force is applied on the continuous interface. We believe that the small non-zero amplitude of order 10^{-8} is related with the *linearization* of the interface. The results obtained with (3.15) are similar for both finite element approximations: the amplitude of currents varies weakly with space resolution and can even increase with it when the space resolution is still too coarse, similar to the results [12], [17] obtained on uniform staggered grid. We set inside pressure $\bar{p}_{in}$ and outside pressure $\bar{p}_{out}$ be equal to a weighted mean value computed over all elements lying completely inside or outside, accordingly, the bubble; the weight coefficient is equal to the element surface. Remarkably then that exact application of the surface force on the aligned grid provides an exact pressure distribution (5.36) up to machine accuracy while the precision of Laplace Law in case (3.15) varies from 0.961 to 0.992 for computations shown in Table 1.

6 Rising bubbles

In case of free motion in infinite media, the bubble can be characterized in terms of Eötvös number E_0, Morton number M and Reynolds number Re (cf.[4]) :

$$E_0 = g\Delta\rho\frac{d^2}{\sigma}\ ; \qquad M = g\mu^4\frac{\Delta\rho}{\rho^2\sigma^3}\ ; \qquad Re = \rho d\frac{U_\infty}{\mu}\ . \tag{6.37}$$

Here, U_∞ is a terminal bubble velocity with respect to the fluid; d is its effective diameter; $\rho(\rho_b)$ and $\mu(\mu_b)$ are density and viscosity of the fluid (or bubble, respectively); $\Delta\rho = \rho_b - \rho$. In fact, the Eötvös number is a dimensionless size of a bubble; it belongs to the interval $(10^{-2}, 10^3)$. The Morton number depends only on the properties of the fluids, it varies strongly depending on the viscosity of the surrounding phase, from 10^{-14} for liquid metals up to 10^8 in viscous oils.

• **First experiment.** We deal here with a relatively large bubble ($E_0 = 40.1$, $\sigma = 0.026$, $\rho = 1$, $\rho_b = \rho/40$) traveling in the closed 2D solid box. Two main cases are considered throughout this experiment: simulations with equal dynamical viscosities ($\mu_b = \mu$) and with different viscosities ($\mu_b = \mu/88$). Variation of M is only due to fluid viscosity: we consider $M = 3170$ ($\mu \approx 0.28$) and $M = 125$ ($\mu \approx 0.12477$) in each case. In infinite media, the terminal velocity of a large bubble ($E_0 > 40$) satisfies the following equation (see [6])

$$2Re^2 + 6Re\frac{2+3\kappa}{1+\kappa} - E_0^{\frac{3}{2}}M^{-\frac{1}{2}} = 0\ ,\ \kappa = \frac{\mu_b}{\mu}\ , \quad M \geq 100\ ,\ \forall Re. \tag{6.38}$$

When the bubble travels in a confined box, boundary conditions at solid walls affect its terminal velocity. Let D be the diameter of a tube and L be its length. Denote the actual velocity as U_w in order to distinguish from the velocity U_∞

$$U_w \approx \frac{U_\infty}{K}\,. \tag{6.39}$$

For creeping flow, the approximate solution [10] for wall correction $K(\kappa, \lambda)$, $\lambda = \frac{d}{D}$ yields

$$K = \frac{1 + 2.2757\lambda^5 \frac{1-\kappa}{2+3\kappa}}{1 - 0.7017\frac{2+3\kappa}{1+\kappa}\lambda + 2.0865\frac{\kappa}{1+\kappa}\lambda^3 + 0.5689\frac{2-3\kappa}{1+\kappa}\lambda^5 - 0.72603\frac{1-\kappa}{1+\kappa}\lambda^6}\,. \tag{6.40}$$

The parameters of the numerical experiment are chosen in such a way that the ends

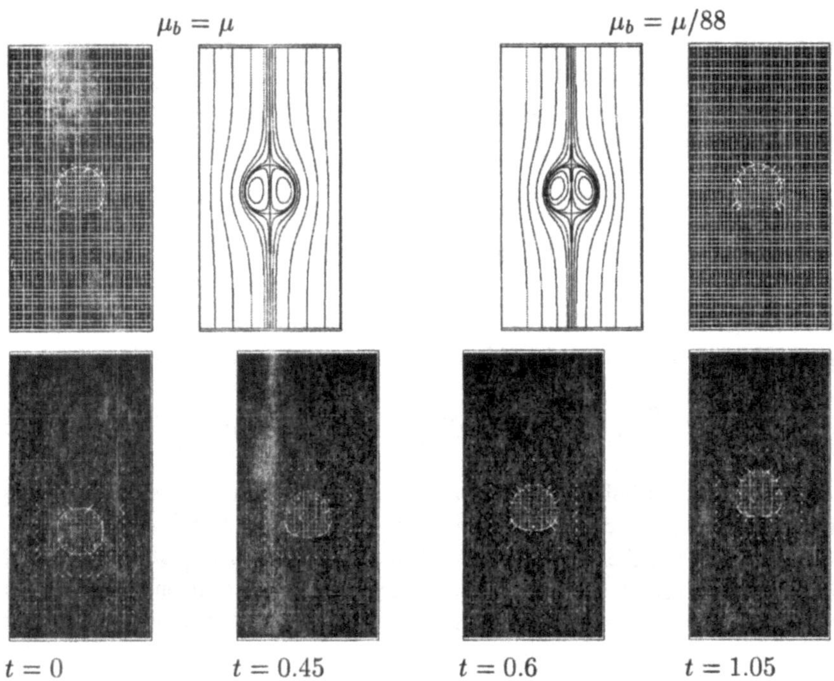

Figure 5: Bubble shapes and stream lines are shown in the middle of the box for $M = 125$. Evolution of rising bubble in the bottom row corresponds to $\mu_b = \mu/88$.

effects are negligible and the wall correction (6.40) is still valid: $D = 1$, $L = 2$, $d = \frac{1}{3}$, initial position $y = \frac{3}{4}$ of the bubble center. We show the obtained bubble shapes and stream lines on Fig. 5: the bubbles attain a steady state with flat or a bit dimpled at the rare circular shape. This initial transition from spherical to oblate-ellipsoidal form agrees with the theoretical predictions and the numerical results [21] for large bubbles at low Re ($Re < 0.5$) and low Weber numbers We ($We = \frac{\rho U_\infty^2 d}{\sigma}$, $We < 2$). The stream lines $\psi = const$ ($u_x = -\frac{\partial\psi}{\partial y}$, $u_y = \frac{\partial\psi}{\partial x}$) are plotted in a frame of reference moving with the bubble. We confirm that the surrounding flow is close to Stokes flow and that ψ is nearly constant along the bubble surface, according to the theory of the axially symmetric steady motion (see in [9]). Interface aligned grids, constructed on one adaptive level, are shown in time in the bottom row of Fig. 5. The computed ter-

Table 2: Comparison of the terminal velocity with the theoretical predictions U_w^{th} (6.38), (6.40). Adaptively refined and aligned grids are constructed on the regular mesh 32 × 64.

$\mu_b = \mu$					
M	U_w^{th}	**Regular grid**		**Adaptive grid**	**Aligned grid**
		32 × 64	64 × 128		
3170	0.11	0.12	0.122	0.1207	0.121
125	0.219	0.249	0.253	0.2502	0.2512
$\mu_b = \mu/88$					
M	U_w^{th}	**Regular grid**		**Adaptive grid**	**Aligned grid**
		32 × 64	64 × 128		
3170	0.1578	0.1637	0.1626	0.162	0.1614
125	0.3	0.3343	0.33	0.3317	0.338

minal velocities are compared with 3D theoretical predictions (6.38), (6.40) in Table 2. We stop the computation when the bubble center reaches the middle of the box in order to avoid the influence of the box ends. The local rising velocity is computed from the change of the position of bubble center at each time step; its arithmetical mean value during the calculation is considered as terminal velocity. Time step $\Delta t = 0.005$ satisfies (4.35) and the restriction coming from the stability condition at capillary level (see [3]). The results gathered in Table 2 are obtained by the split advection algorithm described in Section 4.3. We observe that the terminal velocity computed for different grids regularly exceeds the theoretical predictions but differs not so much from each other. Langrangian moving provides practically the same results; as an example, we show in Table 3 the terminal velocity obtained with the split advection algorithm and with Langrangian approach when the calculations have been performed on the adaptively refined grid. Rising velocities are found to be in good agreement except of the case C, corresponding to first order Langrangian moving with direct rotated interpolation from the calculation grid into the nodes of the moving grid. In fact, this scheme conserves the mass worser than other advection algorithms, what results in less steady flow solution. We measure during the computations the change of the total mass $M(t)$; its relative value $M^{rel.}(t) = \frac{|M(t)-M(t=0)|}{M(t=0)}$ is plotted on Fig. 6.A for computations shown in Table 3. The mass loss in split advection algorithm is due to physical constraint (4.27); usually it is relatively small, less 10^{-6} per time step. We see that after 300 time steps the total mass violation with the split algorithm is about 10^3 times smaller than with the Langrangian moving. The first order Langrangian algorithm with the linear interpolations (3.25) provides nearly the same mass violation as the second order

Table 3: Comparison of the terminal velocity obtained with different advection techniques in case $M = 3170$ and $\mu_b = \mu/88$. Calculations are performed on the adaptively refined grid, constructed on the regular mesh 32 × 64. A: split advection algorithm, B: first order Langrangian moving with linear interpolation (3.25) from regular to moving grid; C: first order Langrangian moving with rotated interpolation from calculation to moving grid; D: second order Langrangian moving with rotated interpolation.

U_w^{th}	A	B	C	D
0.1578	0.162	0.1622	0.1576	0.1619

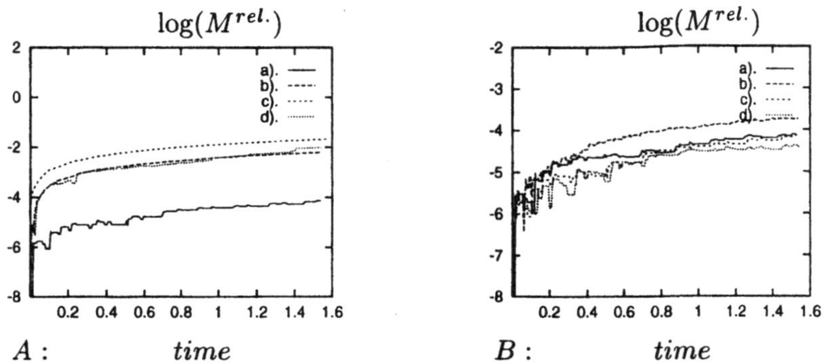

Figure 6: Relative mass loss $M^{rel.}(t) = \frac{|M(t)-M(t=0)|}{M(t=0)}$ is plotted in log. scale as function of time. Computations correspond to $M = 3170$ and $\mu_b = \mu/88$; $\Delta t = 0.005$. A: The same caption as in Table 3. B: Split advection algorithm. a) regular grid 32×64; b) regular grid 64×128; c) adaptively refined grid; d) aligned grid.

Langrangian moving with rotated f.e. approximations. In its turn, the later scheme preserves the total mass twice better than the first order Langrangian scheme when the rotated approximations onto the moving grids are applied in both methods. In whole, the moving with the velocity interpolated from the aligned grid has the same mass conservation properties as for the adaptively refined grid. Usually, the Langrangian approach preserves the mass better when the velocity is interpolated from the adaptively refined mesh than from the aligned grids. Let us compare now mass conservation provided by split advection algorithm for computations done on regular, adaptively refined and aligned grids (see Fig. 6B). First, we observe that mass is preserved better on 32 × 64 regular grid comparing with finer, 64 × 128, mesh. There is no contradiction in this since equal time steps are used in both computations, resulting more probably in violation of the restriction $0 \le C \le 1$ on the finer grid. In case of deformable calculation grids, relation (3.24) provides a divergence free velocity field on the finest underlying regular mesh where moving is done. Unlike with Langrangian approach, the smallest mass loss is observed in case of the aligned calculation grid.

• **Second experiment**. Buoyant bubbles with $E_0 = 10$, modelled in [26] and [14], are found to be in good qualitative agreement with each of the Morton number M,

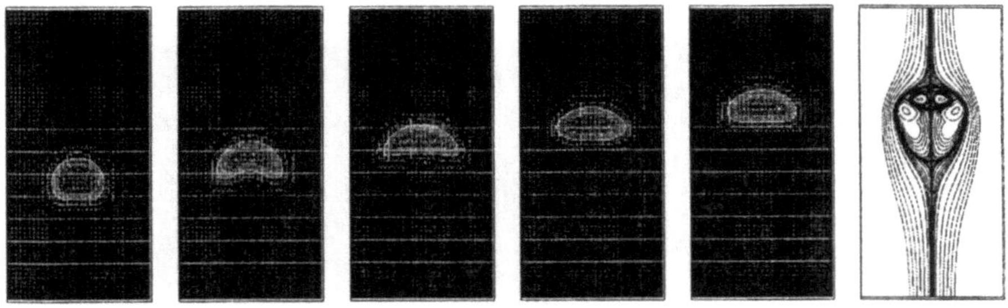

$t = 0.18$ $t = 0.36$ $t = 0.54$ $t = 0.72$ $t = 0.9$ $t = 0.9$

Figure 7: Computations are done for $E_0 = 10$, $M = 0.01$ with two levels of the adaptive refinement. Time step $\Delta t = 5 \times 10^{-4}$; split advection algorithm is applied on finest regular grid. Relative violation of total mass after 1800 time steps, at $t = 0.9$, is about 1.5×10^{-5}.

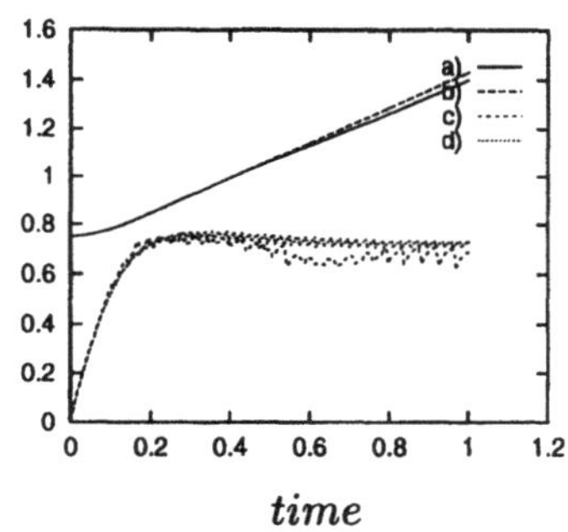

Figure 8: Position of bubble center (a,b) and rising velocity (c,d) are plotted as function of time for $M = 0.01$. Computations are done on uniform 64×128 grids with $\Delta t = 5 \times 10^{-4}$. a), c) present method; b), d) SURFER method.

$M = 10^{-3}, 10^{-2}, 10^{-1}$. The explicit finite difference MAC discretization, based on a first order projection method, is used in both works. We compare our results with the solutions obtained directly with 2D version of the SURFER code of our French co-workers (see [12], [22]) which was used in [14]. Split advection algorithm provides the moving of the interface in both models; boundary conditions are freeslip and the initial phase distribution is as above. We decrease M with the fluid viscosity μ;. μ_b changes accordingly ($\mu_b = \mu/88$) whereas all other parameters are kept constant: $\sigma = 0.1089$, $\rho/\rho_b = 40$, $\rho = 1$. When $M \geq 0.1$, no significant difference either in solutions behavior or interfacial shapes is detected between the two models: bubble shape has nearly flat back and semicircular front; the internal circulation consists only of so-called main vortices. At $M = 0.01$ ($\mu \approx 0.0339$), we observe that bubble shape varies from spherical, through oblate-ellipsoidal, to ellipsoidal cap (see Fig. 7), what agrees with the results in [21] in this range of E, M and Re numbers. Bubble shapes obtained with the SURFER code are similar but less indented at the rare during the motion. In agreement with the qualitative analysis [4], the separation occurs at the rim of the ellipsoid and

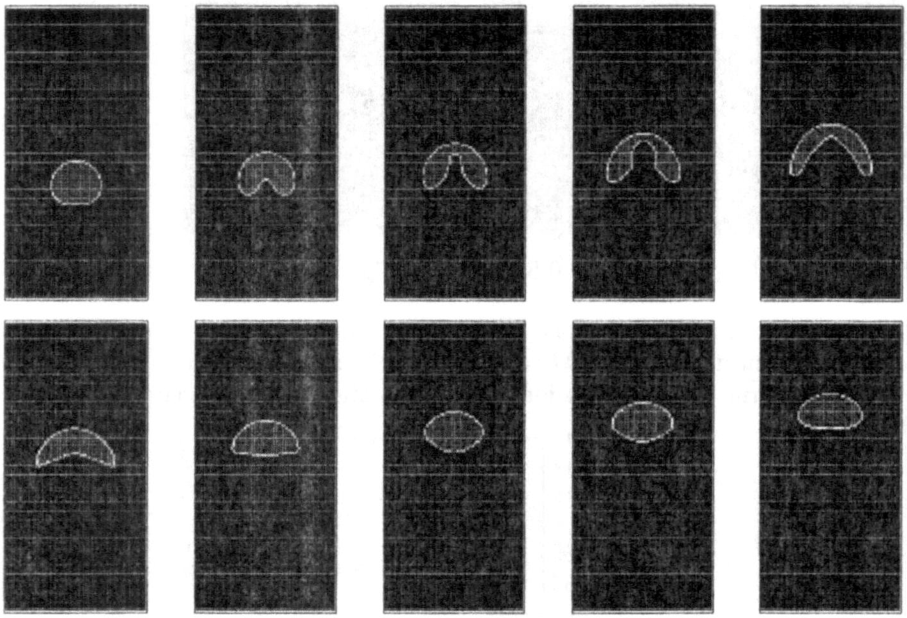

Figure 9: Bubble shapes are shown for $E_0 = 10$, $M = 0.001$; computations are done on uniform 64×128 grid. Time step $\Delta t = 5 \times 10^{-4}$. Relative mass violation at $t = 0.9$ is 7.87×10^{-5}.

"egg-shaped" closed wake appears behind it. The volume of the wake seems to be greater comparing with the results [26] and [14], confirming that our bubble undergoes stronger shape oscillations during the rising. Comparison of the position of the bubble center and its current rising velocity is drawn on Fig. 8. We observe, first, that the rising velocity corresponding to the SURFER model is less oscillated and, second, that it is greater than our solution. Stronger fluctuations of rising velocity are related probably to the variety of transition interfacial shapes which suffers the bubble during the motion in our experiment.

Ryskin and Leal ([21]) refer to the liquids with $M \geq 4 \times 10^{-3}$ as high-M liquids in difference with low-M liquids. For low-M liquids the transition from spherical to oblate-ellipsoidal shapes at low Re "does not pass through a sequence of steady shapes as in the high-M case"([21], p.21). Indeed, we observe that the transition behavior is quite different for $M = 10^{-2}$ and $M = 10^{-3}$ ($\mu \approx 0.019$). This is demonstrated on Fig. 9: very strong indentation develops first at the rare but, with further ascent, it completely disappears and the bubble looks similar to an oblate-ellipsoid. This shape transformation is accompanied by a very rough changing in rising velocity. In the same time, the solution obtained with the SURFER method remains sufficiently smooth, both for bubble shape and its rising velocity. We suppose that smooth behavior of bubbles computed with the explicit scheme is related with the inherent to these models excessive numerical diffusion. On the other hand, approximation and stability properties of our model at low Morton numbers are currently under study while comparing the results obtained on uniform, adaptively refined and aligned grids.

7 Conclusion

The computations done on the interface aligned, adaptively refined or uniform grids are shown to be in quite good agreement. Due to separation of fluids by element boundaries, the aligned grid enables us to model accurately the conjunction condition for the viscous stresses and pressure jump across the interface. However, the presence of the elements with bad shapes near the interface can decrease considerably the accuracy of the solution. Comparison with the exact solutions in simple tests, as linear flow with different viscosities or Laplace low, confirms increasing of the solution accuracy on the aligned grids; no precise results is obtained up to now for more complicated flows with high viscosity ratio and/or strong surface tension coefficients. The problem of bad elements can be solved, even for PLIC discontinuous interfaces, by using another refinement procedure, which subdivides not only interfacial cells but also the rest of the grid in order to obtain the aligned grid with appropriate triangulation. A much more important problem, however, is to apply correctly the surface tension force on the interface itself rather than at its neighbourhood. This requires a rather smooth representation of the interface. The construction of a continuous interface without violation of volume fraction distribution is a difficult task, not solved up to now. Our idea is to reduce the problem to a construction of a cubic spline interpolant which cuts exactly the given fraction in each interfacial cell. We see this as a primary purpose of the coming work. We expect that the aligned with the continuous interface, smooth enough grid, enables us to improve significantly the solution accuracy and stability for problems with discontinuous coefficients and surface tension effects.

The restrictions of our current model with respect to the viscosity and density ratio, as well as in terms of Re and M numbers, are not yet established. Indeed, available now the computational grids are not sufficiently fine to merit extension of the current computations to larger Re or smaller M while the use of finer grids is currently too expensive. Therefore, we see the remedy in adaptive refinement. The convergence of the multigrid in its current version we use, deteriorates on the unstructured grids. We do not practically feel this in case of the adaptive refinement but it is what we have in case of the aligned grids. In our case, the problem becomes more complex because of the jumps in the coefficients, especially unfavourable when thin zones, or fractions, of one fluid appear inside another one. We see the development of special tuning of the solution techniques to interfacial problems as another important task of our future work.

References

[1] N. ASHGRIZ AND J. Y. POO, *A Computational Method for Determining Curvatures.* J.Comp.Phys., vol. 84, 1989, pp.483-491.

[2] N. ASHGRIZ AND J. Y. POO, *FLAIR: Flux Line-Segment Model for Advection and Interface Reconstruction.* J.Comp.Phys., vol. 93, 1991, pp.449-468.

[3] J. U. BRACKBILL, D. B. KOTHE AND C. ZEMACH, *A continuum method for modeling surface tension.* J.Comp. Phys., vol. 100, 1992, pp. 335-354.

[4] R. CLIFT, J. R. GRACE AND M. E. WEBER, *Bubles, Drops, and Particles.* Academic Press, 1978.

[5] B. J. DALY, *A technique for Including Surface Tension Effects in Hydrodynamic Calculations.* J.Comp. Phys., vol.4, 1969, pp.97-117.

[6] R. C. DARTON AND D. HARRISON, trans.Inst.Chem.Eng., vol.52, 1974, pp.301-306.

[7] I.GINZBOURG AND G.WITTUM, *Multigrid methods for interfacial flow.* Report in the frame of the German-French programme "Numerische Strömungssimulation - Simulation Numérique d'Ecoulements"., August 1997.

[8] J. M. HYMAN, *Numerical methods for tracking interfaces.* Physica 12D, 1984, pp.396-407.

[9] J. HAPPEL AND H. BRENNER, *Low Reynolds number hydrodynamics.* Prentice Hall, Englewood Cliffs, NY, 1965.

[10] W. L. HABERMAN AND R. M. SAURE, David Taylor Model Basin Rep.no.1143, 1958.

[11] C. W. HIRT AND B. D. NICHOLLS, *Volume of fluid (VOF) method for the dynamics of free boundaries.* J.Comp.Phys., vol.39, 1981, pp.201-225.

[12] B. LAFAURIE, C. NARDONE, R. SCARDOVELLI, S. ZALESKI, *Modeling Merging and Fragmentation in Multiphase Flows with SURFER.* J.Comp.Phys., vol.113, 1994, pp.134-147.

[13] JIE LI, *Calcul d'Interface Affine par Morceaux.* C.R.Acad.Sci.Paris, vol.t.320, série IIb, 1995, pp.391-396.

[14] JIE LI, *Résolution numérique de l'équation de Navier-Stokes avec reconnexion d'interfaces. Méthode de suivi de volume et application à l'atomisation.* PhD PARIS VI, 1996.

[15] F. MASHAYEK AND N. ASHGRIZ, *A hybrid finite-element-volume-of-fluid method for simulating free surface flows and interfaces.* Int. J. for Numerical Methods in Fluids, vol. 20, 1995, pp.1363-1380.

[16] S. POPINET, *Mémoire de DEA.* LMM, Université Pierre et Marie Curie, 1995.

[17] S. POPINET, S.ZALESKI, *A front-tracking algorithm for the accurate representation of surface tension.* private communication.

[18] S. OSHER, J. A. SETHIAN, *Fronts Propagating with Curvature-Dependent Speed: Algorithms Based on hamilton-Jacobi Formulations.* J.Comp.Phys., vol.130, 1997, pp.269-282.

[19] E. G. PUCKETT, A. S. ALMGREN, J. B. BELL, D. L. MARCUS AND W. J. RIDER, *A High-Order Projection Method for Tracking Fluid Interfaces in Variable Density Incompressible Flows.* J.Comp.Phys., vol.79, 1988, pp.12-49.

[20] H. RENTZ-REICHERT, *Robuste Mehrgitterverfahren zur Lösung der inkompressiblen Navier-Stokes Gleichung: Ein Vergleich.* PHD 1996, Institut für Computeranwendungen der Universitat Stuttgart.

[21] G. RYSKIN AND L. G. LEAL, *Numerical solution of free-boundary problems in fluid mechanics. Part 2. Buoyancy-driven motion of a gas bubble through a quiescent liquid.* J.Fluid.Mech., vol.148, 1984, pp.19-35.

[22] D.GUEYFFIER, J.LIE, R.SCARDOVELLI, S.ZALESKI, *Volume of fluid interface tracking with smoothed surface stress methods for three-dimensional flows applied to pinching threads.* private communication.

[23] G. L. SLEIJPEN, H. A. VAN DER VORST AND D. R. FOKKEMA, *BiCGstab(l) and other hybrid BI-CG methods.* Numerical Algorithms, vol.7, 1994, pp.75-109.

[24] G. E. SCHNEIDER AND M. J. RAW, *Control volume finite element method for heat transfer and fluid flow using collocated variables.* Numer. Heat Transfer, vol.11, 1987, pp.363-390.

[25] M. SUSSMANN, A. S. ALMGREN, J. B. BELL, P. COLLELA, L. HOWELL, M. WELCOME, *An adaptive Level Set Approach For Incompressible Two-Phase Flows.* May be obtained from //www.nersc.gov/research/CCCE/publications/pub.html.

[26] S. H. UNVERDI AND G. TRYGGVASON, *A Front-Tracking Method for Viscous, Incompressible, Multi-fluid Flows.* J.Comput.Phys., vol.100, 1992, pp.25-37.

[27] S. P. VANKA, *Block-Implicit Multigrid Solution of Navier-Stokes Equations in Primitive Variables.* J.Comp.Phys., vol. 65, 1986, pp.138-158.

[28] G. WITTUM, *Multi-Grid Methods for Stokes and Navier-Stokes Equations. Transforming Smoothers: Algorithms and Numerical Results.* Numer.Math.54, 1989, pp.543-563.

[29] S. ZALESKI, *Méthodes de Simulation d'Interfaces Libres entre Fluids.* Cours de l'ecole d'ete, Carcans-Maubuisson, 1995.

Modeling of Free Surfaces in Casting Processes

Jürgen Neises[1], Ingo Steinbach[1], Yves Delannoy[2]
[1]ACCESS e.V./ Intzestr. 5, D-52072 Aachen, Germany
[2]EPM / MADYLAM CNRS UPR 9033 BP 95, F-38402 St Martin d'Heres Cedex, France

Summary

The modeling of free surfaces is an important aspect of casting process simulation, due to the influence of mold-filling on the whole casting process. A new stabilization of the VoF method without reconstruction is presented for a control volume method on unstructured grids. This method has been inspired by level set methods. The stabilization transforms the hyperbolic equation of transport into a modified parabolic equation. Tests on simple problems and industrial applications are presented.

1 Introduction

In the 1980s, extensive research was dedicated to the development of modeling and simulation tools for casting processes. This is partly due to the wide range of physical phenomena present which work within different length scales. These phenomena range from fluid flow and free surface dynamics, which govern the mold-filling process, to heat transfer and phase-change process (important during cooling and solidification). Grain growth and material response (e.g., residual stress) start as the part continues to cool. Mold-filling as the initial part of the casting process affects the subsequent casting process and the quality of the part. Analogously, modeling mold-filling affects the simulation of the complete casting process.

Mold-filling is currently mostly carried out using VoF (Volume-of-Fluid) [1, 2] and ALE (Arbitrary Lagrangian Eulerian) [3a, 3b] techniques. ALE techniques are usually based on finite element (FE) methods and may utilize fixed meshes, using a moving refined one at the transition zone. Remeshing is thus reduced to switching between the original and the refined mesh, which may be adapted if necessary. Due to the complexity of the implementation of an ALE method, this paper only deals with the VoF method.

The accuracy of the VoF method, based on the control volume (CV) method and structured grids, has been improved since it was first developed. Currently, some enhancements to VoF methods for unstructured grids are being designed [4a, 4b], for example, the PLIC (Piecewise Linear Interface Calculation) algorithm.

These enhancements are all based on reconstructions of the front. This achieves a resolution at subgrid level. The main disadvantages of these schemes are their 3-dimensional complexity and the limitations of the time stepping due to their explicit form.

A new VoF method, applying the concept of interfacial energy and requiring no reconstruction, is now presented. Instead of subgrid resolution techniques, a simple stabilization, achieved by adding a surface potential, has been developed.

Section 2 of this paper outlines the VoF method and enhancements. Following the development of the new VoF method (FF-VoF), a hybrid (FE/CV) approach and its implementation for the FF-VoF method is described. The efficiency of this stabilization is demonstrated by tests applied to simple gemetries in 2D and industrial examples.

2 Stabilization of the Volume-of-Fluid Method

There are several existing approaches to solving problems with free surfaces [4b]. The VoF method is a commonly used algorithm which has proven to be a useful and robust tool since its development over two decades ago. It has therefore become a frequent choice in Eulerian models of interfacial flows, especially where interfaces undergo topological changes. The essential features of the VoF method are as follows:

- The front is modeled as a transition zone of values of the filling function F between 0 (empty) and 1 (full). Usually, the isosurface F=0.5 is taken to be a description of the actual front.
- Firstly, the initial interface geometry is used to compute fluid volume fractions F in each computational cell (control volume). This task amounts to computing the volume truncated by the fluid interface in each cell containing an interface.
- Accurate interface information is discarded in favor of discrete volume fraction data. Interfaces are subsequently tracked by evolving fluid volumes in time with the solution of a standard convection equation $\frac{\partial F}{\partial t} + \vec{u}\nabla F = q$
- Volume fractions F of fluid in a cell result from normalization of fluid volumes relative to cell control volume. The exact interface location is unknown. Location of the interface can be reconstructed from local volume fraction data, thus allowing tracking of the interface.

The reconstructed locations then are applied to compute the fluxes necessary for the convective term in the transport equation. Without reconstruction of the interface, the thickness of the interface will not remain compact. The interface spreads over the domain as a result of numerical diffusion. This diffusion can be reduced in certain methods to a determined width (e.g. superbee limiter [5], steepened PPM [6]), but these high order advection schemes have proved to be more expensive than methods with reconstruction.

Another limitation of the VoF method is the limitation of time stepping by the Courant-Friedrich-Levy condition due to the explicit calculations [5]. This either intensifies spreading of the front or leads to schemes which yield a subgrid resolution.

A description of the front was obtained by implementing various reconstruction methods, such as SLIC (Simple Line Interface Calculations) and PLIC (Piecewise Linear Interface Calculations). Some enhancements to VoF methods for unstructured grids are currently being designed [4b], e.g. the PLIC method. In the PLIC method, the interface is reconstructed in a piecewise linear manner, taking curvature (and surface tension) into account, and is positioned in the cell according to fluid volume. Finally, the front is advected in a conservative way. The PLIC method is implemented for arbitrary hexahedra and will be extended to 3D unstructured meshes [7]. The PLIC algorithm is based on geometrical descriptions of control volumes and benefits from possible high resolution methods. Difficulties occur in implementing a 3D control volume method for unstructured meshes, in calculating the normal of the interfaces accurately, and in constructing an accurate 3D advection scheme. Time steps are limited by the Courant-Friedrich-Levy condition [5], as PLIC still is an explicit method.

Another possible option is a stabilization, which gives resolution only on cell level, but also keeps the transition region compact and allows larger time steps by implicit calculations. A new enhancement to the VoF method named FF-VoF (Filuid Field F-VoF) is inspired by the phase field or level set method. Analogous to a level set method, the volume fraction of the VoF method describes the transition between liquid and gas/vacuum. Using this approach the reconstruction on subgrid level is avoided.

The concept of the new enhancement (FF-VoF) is the description of the interface free energy as a model for the liquid-gas front. The phase-field concept is interpreted as level set or volume fraction F. The microscopic phases liquid and solid are now referred to as the macroscopic liquid and gas phases. This way the volume fraction is treated as a kind of fluid field. Analogous to the concepts of the phase-field method, the energy at the interface may be described by diffusion and molecular forces, which tend to cancel each other out in a state of equilibrium. In non- equilibrium, both forces control the shape and width of the interfacial region [8a, 8b]. Thus, the additional consideration of these forces at the transition region gives a consistent perturbation of the transport equation by a surface potential:

$$\frac{\partial F}{\partial t} + \vec{u}\nabla F = q + \epsilon^2\nabla^2 F - \beta F(1-F)(\frac{1}{2} - F). \tag{2.1}$$

A similar enhancement is derived by the model of continuous surface force [4b] for the PLIC method.

The added potential models both the diffusion forces by the Laplacian $\epsilon^2\nabla^2 F$ and the molecular forces by the polynomial source term $P(F) = \beta F(1-F)(\frac{1}{2}-F)$ at the interface. In equilibrium these forces cancel each other out. As this surface is not usually in equilibrium, a perturbation modeling the interfacial energy is added to the transport equation.

The interfacial potential controls the shape and the width of the interface. The surface potential in equilibrium gives

$$0 = \epsilon^2\nabla^2 F - \beta F(1-F)(\frac{1}{2}-F). \tag{2.2}$$

With boundary conditions $F \to 1$ as $x \to -\infty$ and $F \to 0$ as $x \to \infty$ the solution of this equation in 1 dimension is found to be

$$F(x) = \frac{1}{2}\left[1 + \tanh\left(\frac{x}{2\epsilon\sqrt{2\beta^{-1}}}\right)\right]. \tag{2.3}$$

Thus, the shape of the interface is determined by this tanh profile. If the transition is assumed in the range of $0.05 < F < 0.95$, the width δ of the interface is given by

$$\delta = 6\epsilon\sqrt{\frac{2}{\beta}}. \tag{2.4}$$

Therefore, the width of the interface is determined by ϵ and β.

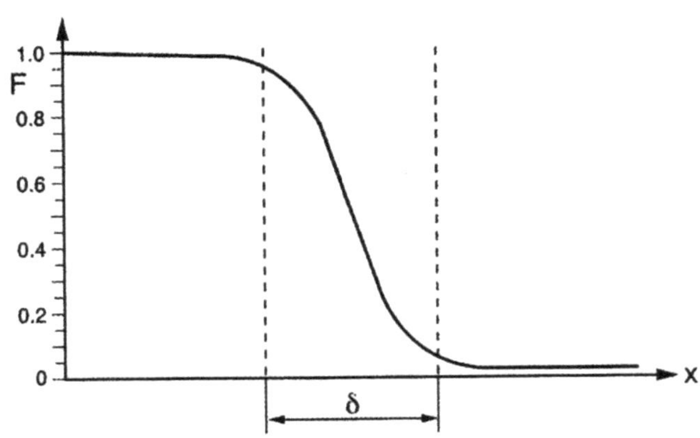

Figure 1 Numerical profile of the interface in one dimension. The degree of filling is described by the variation of c from 1 to 0. The interfacial thickness is given by $\delta = 6\epsilon\sqrt{\frac{2}{\beta}}$.

It is clear that the original VoF method is included in FF-VoF. In an exact calculation and with $\epsilon = 0$ the transition zone should be zero and a solution with jumps should occur [5]. Therefore, with ϵ=0 the numerical diffusion of the method can be estimated by the size of the transition zone. If this numerical diffusion is taken into account, the correction of the fluid-field by the source term only, i.e. $\epsilon = 0$ and $\beta \neq 0$ is a simple variant of FF-VoF.

By varying the parameters ϵ and β locally, the interface region can be adapted to a distorted mesh. The parameters are controlled locally by the size of the elements. Thus, a fixed transition zone of the thickness of a few (usually 3) elements may be achieved.

The modified differential equation is parabolic rather than hyperbolic. The perturbation by adding the surface potential smoothes the problem. Numerical difficulties, such as the need for sophisticated advection schemes, no longer exist. Furthermore, no reconstruction of the interface is performed. Implicit time stepping may be implemented in a straightforward manner.

The VoF method developed is to be implemented in the context of mold-filling. Thus, the implementation of the VoF method is part of a coupled calculation of heat transfer, fluid flow and free surfaces in the simulation of casting processes. Such simulations require the handling of geometrically complex parts. Simulation tools must capture complicated 3D geometries, which leads to the use of unstructured meshes.

There are several kinds of control volumes [9, 2], each having their respective advantages. Very close to the classical volumes used in structured meshes is the cell centered approach of subdividing the FE mesh both in the centers of gravity of the elements and at the centers of the links in order to construct control volumes around the nodes (Fig. 2). Different control volumes are constructed for different types of meshes and elements in this way. A Cartesian grid results in control volumes just as in classical cell centered approaches. More refined meshes result in more complicated cells (Fig. 2). Using this approach, all techniques using high resolution schemes as PLIC can be implemented. The implementation of this in 3D is under development at MADYLAM.

The collaboration between ACCESS and MADYLAM now allows the joint development of both approaches, i.e. resolution on subgrid level by the PLIC method and stabilisation by a surface potential. Thus, a comparison of the numerical properties of the different approaches becomes possible. This task cannot be managed by one group working on itself.

Due to the complexity of implementing the PLIC method for 3D unstructured control volume methods, only the FF-VoF method is implemented and examined in numerical examples.

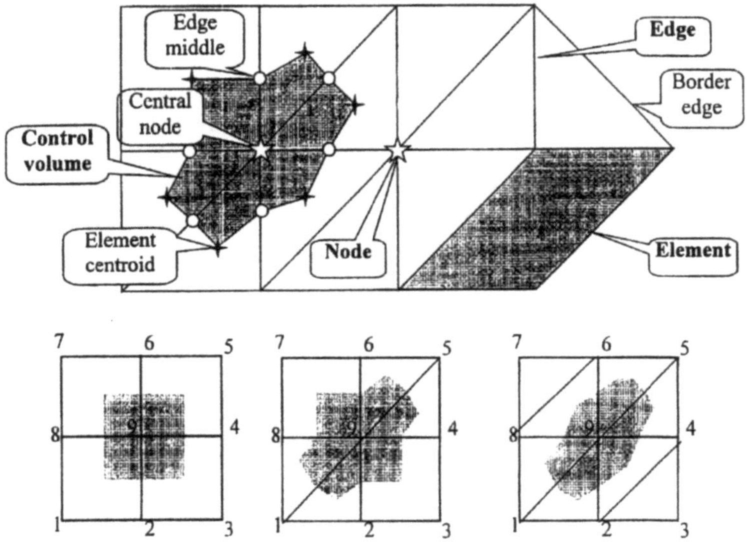

Figure 2 Construction of a control volume by subdividing an FE mesh. The mesh is subdivided at the centers of gravity of the elements and at the centers of the connecting links between the nodes. Cartesian meshes result in the classical cell-centered scheme, which is reproduced by the letfmost example in the lower line. The other examples reveal the complicated structures, which may occur.

3 Implementation

At ACCESS, the finite-element-based program CASTS (Computer Aided Solidification TechnologieS)[10a, 10b, 10c] is an integrated tool to simulate 3D solidification problems. Currently, the program covers heat transfer including radiosity heating, stress / strain analysis and convection / fluid flow. An important development of recent years has been the change from a classical finite element formulation to a hybrid finite-element-based control volume (FE/CV) method and the implementation of thermohydrodynamic simulations.

The FE/CV method is based on similarities between finite element and control volume methods [2]. CV-like stencils can be reproduced by FE methods and vice versa. Using Cartesian grids, a 7-point stencil in 3 dimensions is established by a CV method. This can be reproduced by linear shape functions and numerical integration by the trapezoidal rule in an FE method [11]. The only difference between CV methods and FE methods is the evaluation of physical parameters for nonlinear problems. In CV methods, these parameters are evaluated at the faces of the cells, whereas in FE methods they are integrated over the whole element.

In the hybrid FE/CV method, the finite element integrals are reordered by the links between the nodes. The FE integrals attributed to a pair of nodes are summed for the advective and the diffusive operator respectively. This results in a discrete operator on each link. As in classical CV methods, only these links are used in calculation. The discrete operators may be interpreted in a geometrical way as cell face area divided by node distance.

The reordering of an FE discretization to a CV method can be outlined, e.g. for the Laplacian $-(\dot{\nabla}(\lambda(T)\nabla T) = 0$. The discretization at node i for an FE method with piecewise linear shape functions is ϕ_j

$$\sum_j T_j \int_\Omega \lambda(T)\nabla\phi_i\nabla\phi_j = 0 \,. \tag{3.5}$$

The nonlinear thermophysical parameter $\lambda(T)$ may be approximated by a locally constant λ_{ij}. Thus, the equation (3.5) can be approximated as

$$\sum_j \lambda_{ij}T_j \int_\Omega \nabla\phi_i\nabla\phi_j = 0 \,. \tag{3.6}$$

The ingredients of the FE/CV method can now be demonstrated. The integral is the sum of the integrals of the elements to which the nodes i and j belong. Only within those elements is $\nabla\phi_i\nabla\phi_j \neq 0$. The approximation of the physical parameter λ_{ij} only depends on T_i and T_j. In CV methods, there are two common ways of defining λ_{ij}: the arithmetic or the geometric mean of $\lambda_i = \lambda(T_i)$ and $\lambda_j = \lambda(T_j)$ can be selected.

In CV methods, the fluxes between the neighboring control volumes are balanced. Thus, the matrix assembly of the hybrid methods is completed by computing the elements at the offdiagonals, of which the negative sum defines the diagonal. In this way, a CV method is implemented, but no geometric definition of a control volume is necessary.

As only integrals of shape functions remain to be evaluated, integration can be performed once at the beginning of a transient analysis. No further integration is necessary during the following time steps. Mass lumping techniques can be used, according to the control volume framework.

The implementation of fluid flow follows the principles outlined. With an FE approximation of first order velocities and zeroth order pressure elements, a kind of staggered grid CV method is derived. The incompressibility is modeled by artificial compressibility and the iteration of pressure and velocities until incompressibility is almost reached [12, 13].

To integrate the FF-VoF method into CASTS, the differential equation, which includes mass transport and free energy of the interface, has to be discretized in the hybrid FE/CV framework. The parabolic nature of the stabilized FF-VoF equation allows reuse of the existing implementation of convection diffusion problems. The polynomial source term is treated in an explicit way to keep the resulting equation system linear. Thus, the polynomial expression can be evaluated once per time step at the nodes (control volumes). The FF-VoF method is realized in a straightforward manner using this procedure.

Since for cell centered CV methods the cells are identified with the nodes, the calculation of the volume of fluid is performed at the nodes. The interface is modeled as the region with $0.95 > F > 0.05$, and cells outside this range may be treated as empty or filled. This determines the domains of computation for the FF-VoF method (fig 3).

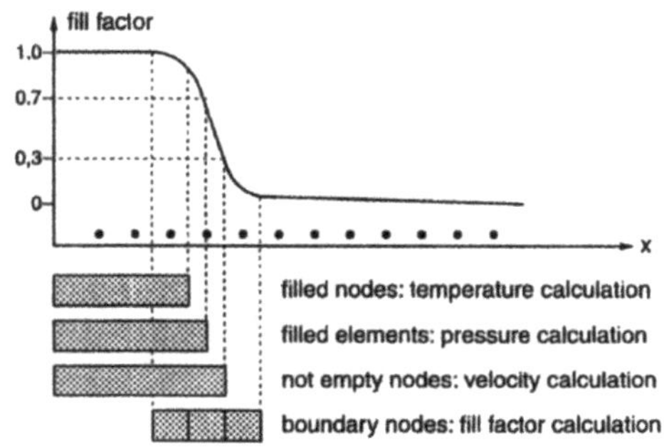

Figure 3 Domains of Calculation within FF-VoF. For filled nodes - $F > 0.95$ - the temperature field is calculated. The pressure field is solved on filled elements. It is set to 0 at the front. If nodes are not empty $F > 0.05$, the velocity is solved. In the transition region and at the front, the fill function is propagated.

Temperature and velocity is calculated for filled nodes, which are nodes (control volumes) with $F > 0.95$. Pressure is calculated for elements consisting of at least one filled node. In the transition region, which is three elements wide and is determined by $0.95 > F > 0.05$, zero pressure is assumed. This is according to a vacuum process. Velocity is also calculated at these nodes. The volume fraction F is calculated in the transition region and at the front. If the fill function F exceeds the threshold 0.05, the node is added to the transition region and its velocity is extrapolated from nodes in the upwind direction.

Therefore, the FF-VoF method is integrated into the FE/CV framework of CASTS in a simple and efficient manner. Implicit methods are realized straightforwardly.

4 Numerical Examples

First simulations with the implemented FF-VoF method were performed for academic test configurations in two dimensions using an equidistant Cartesian grid [14]. These tests include the original VoF method with and without upwind schemes and different in-flow conditions. These tests demonstrate the capability of the new FF-VoF method for stabilizing interfacial flows. Numerical diffusion is diminished without reconstructing the interface. The results encouraged the implementation of an FF-VoF method in the hybrid FE/CV framework of the program CASTS to simulate mold-filling. Several tests were completed on actual industrial components.

4.1 Benchmark-Tests on Simple Geometries

To validate the method developed, benchmark-tests on simple geometries in 2D were performed. A rectangular domain $[0,1]^2$ with an equidistant Cartesian grid was used and a constant velocity $\vec{u}$ was set. In-flow conditions were set at the origin. Within the figures, 4 different time steps were plotted to show progress of the fluid. The transition is shaded from gray to dark. The filled region is shaded light gray.

One weakness of the original VoF method is numerical diffusion perpendicular to flow direction. The front leaks out and the fluid is spread over a widening area. Without an upwind scheme, the original method results in oscillations in the direction of the flow (Fig. 4) which is usually observed with centered schemes applied to hyperbolic differential equations. These oscillations do not occur if an upwind scheme is used (Fig. 5).

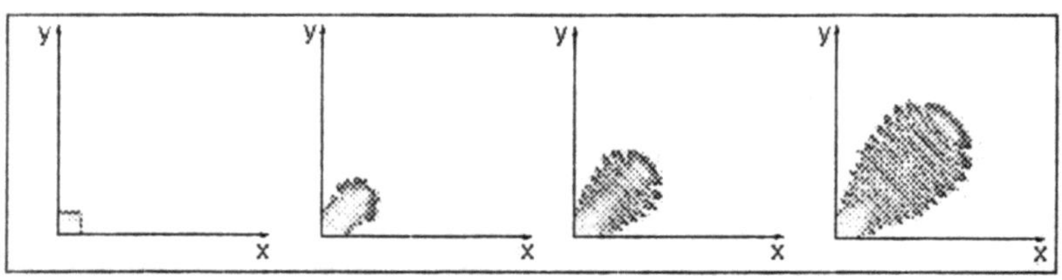

Figure 4 Propagation of the interface with the original VoF method and a constant velocity field $\vec{u} = (1,1)$. The central scheme results in oscillations in the direction of velocity. Numerical diffusion in the direction normal to $\vec{u}$ is observed.

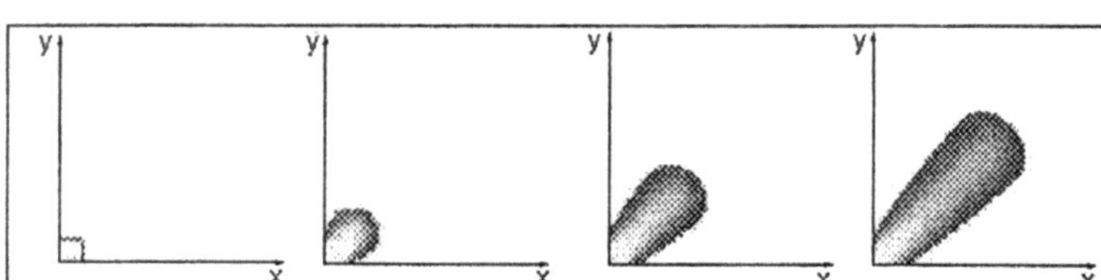

Figure 5 Propagation of the interface using the original VoF method and a constant velocity field $\vec{u} = (1,1)$. An upwind scheme suppresses the oscillations.

Because the modified approach leads to a parabolic differential equation, the problem is smoothed. Thus, no upwind scheme is needed and no oscillations occur (Fig. 6). The numerical diffusion is corrected through the added perturbation using $\epsilon = 1$ and $\beta = 1$. The front is kept compact in this way and the transition is limited to a few cells with no observable widening. The fluid progresses only in the direction of $\vec{u}$.

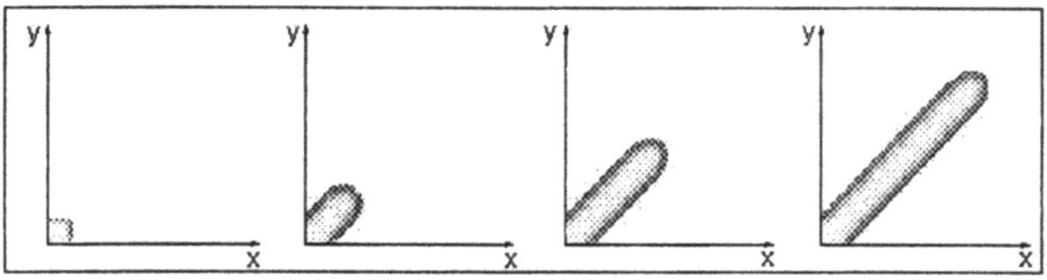

Figure 6 Propagation of the interface using the FF-VoF method and a velocity of $\vec{u} = (1,1)$. With no upwind scheme, no oscillations occur because the stabilized differential equation is parabolic. Also numerical diffusion is diminished, and transition remains limited to a few cells.

To test the sensitivity of the method, a calculation with the same correction was performed with velocity $\vec{u} = (2,1)$ (Fig. 7). The angles between flow direction and grid lines are changed in comparison to (Fig. 6). Within this calculation, the front also remained limited to a few cells, and the area of filled cells also progresses only in flow direction. This demonstrates the independence of the relative orientation of the grid, because the different angle of flow direction does not affect the flow pattern.

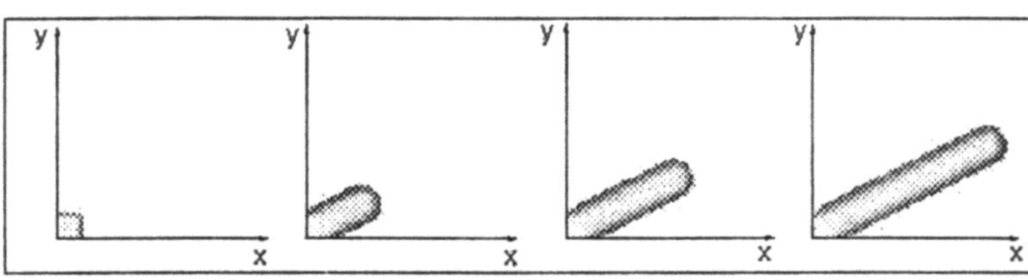

Figure 7 Propagation of the interface using the FF-VoF method and a velocity of $\vec{u} = (2,1)$. The same correction as in Fig. 6 is used. The solution is independent of the relative orientation of the grid, because the different angle of flow direction does not affect the flow pattern.

Efficient stabilization via the surface potential can be observed if two neighboring in-flow conditions with the same velocity $\vec{u} = (1,1)$ are set. Using the original VoF method, numerical diffusion causes a spreading of the fluid. Thus, both areas of fluid overlap (Fig. 8) and mix. In contrast, in the FF-VoF method, the areas of both fluids remain separated. The transition regions of each fluid are limited to just a few elements. Thus, two separate areas progress in the direction of flow (Fig. 9).

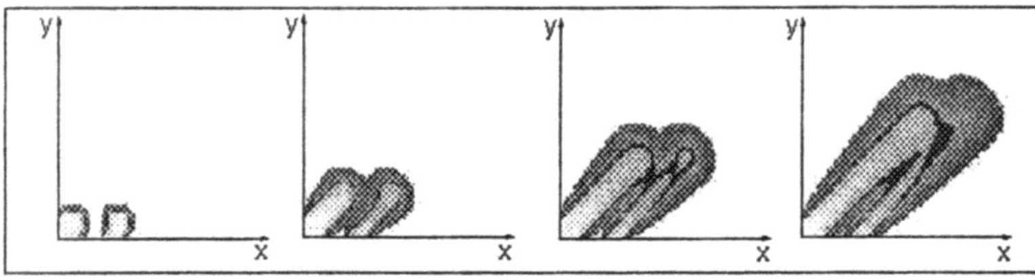

Figure 8 Propagation of the interface using the original VoF method and a constant velocity field $\vec{u} = (1,1)$. Two in-flow conditions were set at the origin. The smearing of the front causes the mixing of the two areas of fluid. The back color represents the isoline F=0.5.

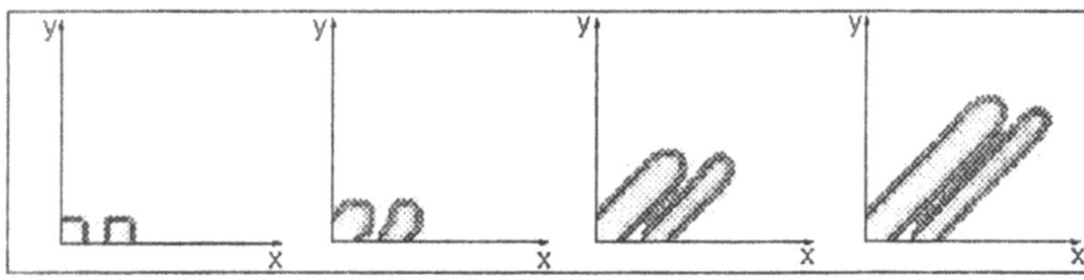

Figure 9 Propagation of the interface with the FF-VoF method and a constant velocity field $\vec{u} = (1,1)$. Two in-flow conditions were set at the origin. The transition regions are limited to a few elements, thus the areas of fluid remain separated.

These tests on academic problems demonstrate efficient stabilization using the FF-VoF method:

- No upwind scheme is needed, since the added perturbation results in a parabolic equation.
- The transition region is limited constantly to a few cells.
- The stabilization is independent of the orientation of the grid.

Therefore, an efficient and easily implemented stabilization, at least for regular grids, is available.

4.2 Tests on real components

The results of the simulations with Cartesian grids led on to testing the FF-VoF method in real casting processes. The processes investigated cover various aspects of casting:

- centrifugal casting of an automotive valve,

• casting of a thin plate as benchmark.

Each example presents different difficulties. Filling times are short and flow rates are high. Therefore, these problems require accurate and robust solvers. Due to the geometry selected, the benchmark example takes special account of the resolution of the flow field.

Centrifugal casting of Automotive Valves

Due to its low density and high-temperature strength, titanium aluminide is an excellent candidate for automotive exhaust valve applications. The cost of valves made of TiAl must, of course, be competitive. Existing production routes developed for aircraft industry applications are very complex and expensive, and hence not appropriate for economic mass production of TiAl valves. The cost-saving process consists of melting and alloying using a cold crucible furnace and centrifugal casting. The simulation of mold filling should give insight into this process. A serious problem which must be avoided is porosity affecting the strength of parts. Numerical simulation can be an aid in identifying possible problems and determining process parameters.

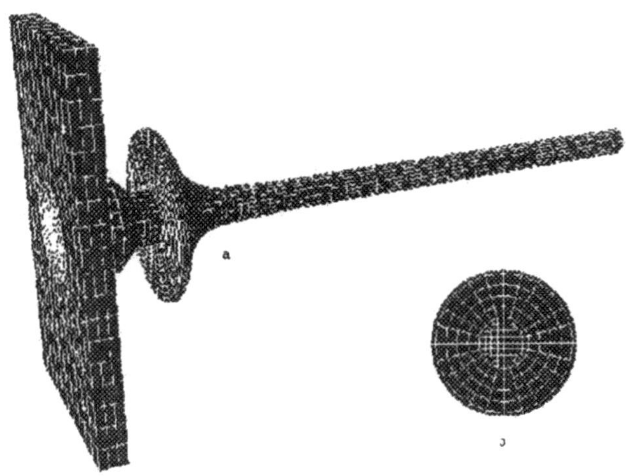

Figure 10 Solid model and cross-section of the FE-mesh of a valve with a plate used to simulate the in-flow. The mesh is consists of 15332 hexahedra and 16750 nodes. The plate is a part of the chill, and is used to model the boundary conditions. The lower part of the plate is assumed to be filled and a vertical velocity is set to simulate the rising of the melt.

The casting facility consists of a cylindrical chill with eight stacked disks, each having twelve molds for the valves. Before pouring, the chill is heated and the air is evacuated. The melt is poured into the center when the chill rotates at the desired

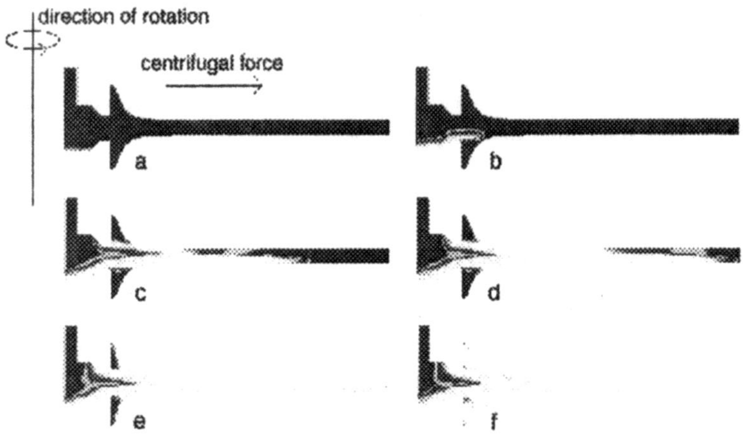

Figure 11 The figure shows the results of the simulation on a vertical cross-section of a valve. A correction of the filling front with $\epsilon = 0$ and $\beta = \frac{1}{\Delta t}$ was used. The chill rotates at 100 rpm and a vertical velocity of 15 cm/s was set at the in-flow. The initial condition of the fill function is shown in (a). The plate is filled partially. The melt enters the valve in (b) and flows into the stem directly driven by the centrifugal force (c), leaving the head empty. The stem is filled first (d,e) from end to head. Next, the valve stem is filled completely. Finally, the valve head is filled (e, f).

velocity. The centrifugal force pushes the melt into the molds. This part of the process was simulated using the FF-VoF method. For reasons of symmetry only a single valve was simulated. The mesh of the valve consists of 15332 hexahedra and 16750 nodes (Fig. 10). A core of regular hexahedra is covered by 5 radial layers. A thin layer is used to model the flow near the wall.

A plate at the inlet is used to model the flow of the melt up to the mold. The lower part is assumed to be filled and a vertical velocity is set as initial value. A fixed rotational velocity is assumed. A stabilized VoF method with $\epsilon = 0$ and $\beta = \frac{1}{\Delta t}$ is tested (Fig. 11 and Fig. 12). The number of rotations is set to 100 rpm, the velocity at the in- flow is set to 15 cm/s. The colors represent the volume of fluid from empty (black) to filled (white).

The vertical cross-section demonstrates the effect of the centrifugal force (Fig. 11). In the initial state, the plate is filled partially (a). The melt rises against the wall and enters the valve in (b). It is driven directly into the stem by the centrifugal force (c), leaving the sides of the head empty. The stem is filled first (d,e). The melt flows to the end of the stem, which is filled from the end to the valve head. After the valve stem is filled completely, the valve head is filled (e, f).

The effects of the Coriolis force are visualized in a horizontal cross-section (Fig. 12). The Coriolis force acts against the direction of rotation. Thus, the flow pattern

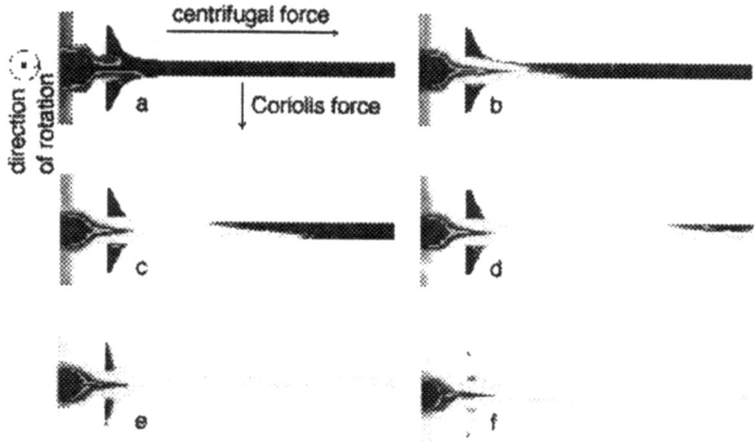

Figure 12 This figure is a horizontal cross-section through the valve. A stabilization with $\epsilon = 0$ and $\beta = \frac{1}{\Delta t}$ is employed as in figure 11. The melt rises at the sides of the valve (a). The centrifugal force drives the melt into the valve stem. Due to the Coriolis force, the melt progresses faster at the lower side of the valve (b to d). The valve head is filled last (e, f).

is not symmetric. The melt is pushed in the direction of the Coriolis force already as it enters the valve (a). Nevertheless the melt enters the valve stem leaving the valve head empty as in (Fig. 11). The melt fills the valve stem from the end to the valve head. Due to the Coriolis force the melt reaches the end of the stem at the "lower" side of the wall first. But the stem is filled completely.

Mold Filling Benchmark 1995

At the 7th Conference on Modeling of Casting, Welding and Advanced Solidification Processes [15] the results of a benchmark problem of mold-filling was presented. This benchmark was accompanied by a series of experiments to evaluate the simulations. This benchmark test is well-documented and reproducible.

The overall geometry of the casting is a simple plate, with a tall sprue and a bottom gated running system (Fig. 13). However, the filling system was less easy to model as a result of its delicate geometry and the necessity to model the gravity correctly. A tall sprue was chosen because sprues of this kind generate an impressive degree of turbulence in the runner and the gate.

The mold is discretized by a mesh of 10464 hexahedra and 7908 nodes. No-slip boundary conditions are imposed. The pouring basin is kept full and the melt is poured by free fall into the mold. The front is represented in the visualization by

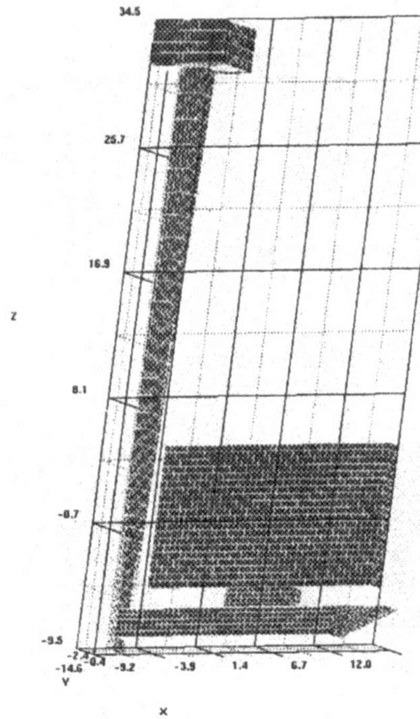

Figure 13 This figure shows the geometry and the FE-mesh of the plate. The basin on top of the sprue is kept full during the simulation. The mesh consists of 10464 hexahedra and 7908 nodes.

the isosurface of a volume of fluid of 0.5. The degree of filling is represented by the gray scale from empty (black) to full (white).

This example reveals problems which occur using VoF methods without reconstruction, when high velocities and mixing are present. Small pieces of material (called flotsam or jetsam) about the size of the mesh spacing are separated from the main body of the fluid [4a], and the front leaks out in spite of correction. Thus, some parts of the transition area are filled more than others. This starts after the melt leaves the sprue. The free fall of the melt induces high velocities. When the melt arrives at the bottom of the sprue, the pressure rises instantaneously. The modeling of the free fall and the pressure rise demand a robust solver of the flow field. The extrapolation of velocities at the front and the resolution of pressure seem to lead to a separation of drops (Fig. 14).

After the melt reaches the end of the runner, it swashes back and the waves mix. This phenomenon also demands an accurate and robust solution of the flow field. The approach of piecewise constant pressure and first order velocity (staggered grid)

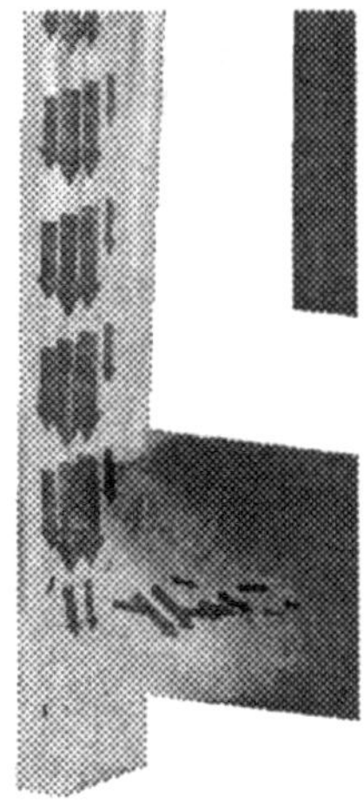

Figure 14 The melt reaches the bottom of the sprue. The high velocities during the free fall cause smearing of the front.

leads to technical difficulties in computing accurately the flow field of the interfacial region. The assumption of zero pressure works on the frontal element and affects not only the frontal nodes, but also the layer behind. This makes the calculation of the velocities at the front inaccurate. Clearly, this affects the development of the front, especially at high flow rates.

After the rear section of the runner is filled, the flow patterns become smooth and are well resolved. Thus, when the melt rises through the gate (Fig. 15), the shape of the front is analogous to the experimentally observed shape of a mushroom and the plate fills with the rising melt.

In regions of smooth flow patterns, the resolution of the stabilized method reproduce the experimental results. The experimental filling patterns are reproduced. However, the front tends to leak out if the resolution of the flow field is not sufficient. In regimes of high velocities at the front, the resolution and accuracy of the flow field is crucial to the development of the interface. Thus, the modeling of free surface flow will benefit much from an improvement of the robustness and accuracy of the Navier-Stokes Solver in such regimes.

5 Conclusions

A stabilization of the VoF method inspired by phase field or level set methods has been developed. The equation of mass transport is transformed from a hyperbolic

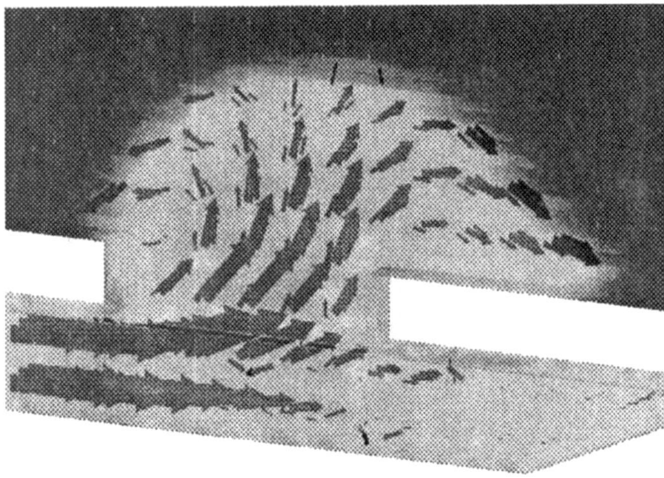

Figure 15 The melt rises through the gate. The flow is smooth and the front has the shape of a mushroom. This shape was observed in experiments. The velocities are well resolved.

differential equation into a parabolic equation by a consistent perturbation, which is a generalized form of surface tension. In this way, the shape and width of the interface may be controlled by a tanh profile. No reconstruction of the interface is needed. The modified equation can be implemented easily in a hybrid FE/CV context for unstructured meshes in 3D. Tests on Cartesian grids demonstrate the efficiency of this method. The smearing of the front is suppressed, and numerical diffusion and oscillations are diminished. The results are independent of grid orientation. This new method has been implemented in the program CASTS to simulate mold filling. It has been tested on industrial problems. The results demonstrate the ability of the FF-VoF method to handle such problems. In regimes of high velocities and mixing, the front smears over many cells, despite the correction terms of the FF-VoF method. This is mainly caused by a limited resolution of the flow field. Especially at the front, technical difficulties in approximating velocities and pressure arise from the staggered grid. Calculation of the frontal velocities is inaccurate due to the zero pressure assumption in the interface. Therefore, an approximation of velocity and pressure in the nodes is desired. A first order approximation of pressure and first order velocities enriched by bubble functions will be integrated into the hybrid FE/CV framework. The FF-VoF method will benefit from this. The more accurate flow field will give a more accurate transport and the consistent areas of calculation will improve the modeling of the free surface flow in the interfacial region.

References

[1] Hirt, C. W.; Nichols B.D.: Journal Comp. Phys. 39 (1981), 201-225.

[2] Idelsohn, S.R; Onate, E.: Finite Volumes and Finite Elements "Two good Friends", Int. Journal Numer. Meth. Eng. 37 (1994), 3324-3341.

[3a] Dhatt, G.; Gao, D. M.; Ben Cheikh, A.: Finite Element Simulation of Metal Flow in Moulds, Int. Journal of Num. Meth. Eng. 30 (1990), 821-831.

[3b] Gaston, L.; Glut, B.; Bellet, M.; Chenot, J.-L.: An Arbitrary Lagrangian Eulerian Finite Element Approach to Non-Steady State Fluid Flows. Application to Mould Filling., Proc. 7th Int. Conf. on Modeling of Casting, Welding and Advanced Solidification Processes, TMS (1995), 163-170.

[4a] Kothe, D.B.; Rider, W.J.: Comments on Modeling Interfacial Flows with Volume-of-Fluid Methods, submitted to Journal Comp. Phys., http://lune.mst.lanl.gov/Telluride/Text/publications.html.

[4b] Kothe, D.B.; Rider, W.J.; Mosso, S.J.; Brock, J. S.; Hochstein, J. I.: Volume Tracking of Interfaces Having Surface Tension in Two and Three Dimensions, AIAA-96-0659, http://lune.mst.lanl.gov/Telluride/Text/publications.html.

[5] Smoller, J.: Shock Waves and Reaction-Diffusion Equations, Springer, New York-Heidelberg-Berlin (1988).

[6] Collela, P.; Woodward, P.: The Piecewise Parabolic Method (PPM) for Gas-Dynamical Problems, Journal Comp. Phys. 54 (1984), 174-201.

[7] Reddy, A. K.; Kothe, D.B.; Beckermann, C.; Ferrell, R.C.; Lam, K. L.: High Resolution Finite Volume Parallel Simulations of Mould Filling and Binary Alloy Solidification on Unstructured 3D Meshes, LANL Report LA-UR-97-136 (1997), http://lune.mst.lanl.gov/Telluride/Text/publications.html.

[8a] Jacqmin, D.: An Energy Approach to the Continuum Surface Tension Method, 34th Aerospace Sciences Meeting and Exhibit, January 15-18, 1996, Reno, AIAA-96-0858.

[8b] Landau, L. D.; Lifschitz, E. M.: Lehrbuch der theoretischen Physik, Bd. 5: Statistische Physik Teil I, Akademieverlag Berlin (1987).

[9] Delanaye M.; Esser J.A.: Finite volume scheme with quadratic reconstruction on unstructured adaptive meshes applied to turbomachinery flows, Journal of Turbomachinery 119, 263-269, (1997).

[10a] Hediger, F.; Hofmann, N.: Process Simulation for Directionally Solidified Turbine Blades of Complex Shape, Proc. 5th Int. Conf. on Modelling of Casting, Welding and Advanced Solidification Processes, TMS (1991), 611.

[10b] Steinbach, I.; Franke, D.: A Virtual Crystallization Furnace of Solar Silicon, Proc. of the First World Conference on Photovoltaic Energy Conversion (1994).

[10c] Hofmann, N.; Neises, J.; Hediger, F.; Proff, U.: Numerical Simulation of Radiation Heated Mobile Casting Moulds, Computing Systems in Engineering 6 (1995), 73-81.

[11] Young, L.C.: A Finite-Element Method for Reservoir Simulation, Soc. Petrol Eng. S. 21 (1981), 115-128.

[12] Neises, J. ; Steinbach, I.: Finite Element Integration for the Control Volume Method, Comm. Num. Meth. Eng. 12 (1996), 543-555.

[13] Steinbach, I.; Neises, J.: A control volume treatment of finite elements and its application to a solidification problem, in: Numerical Methods in Thermal Problems, eds.: Lewis, R.W.; Durbetaki, P., Vol IX (1995), 466-473.

[14] Hahn, C.: Untersuchungen der Volume-of-Fluid Methode zur Beschreibung freier Randflächen mit diffusen Grenzschichten, Diploma Thesis RWTH Aachen (1996).

[15] Sirrell, B.; Holliday, M.; Campbell, J.: The Benchmarktest 1995, Proc. 7th Int. Conf. on Modeling of Casting, Welding and Advanced Solidification Processes, TMS (1995), 915-932.

Turbulent convection driven by an imposed temperature gradient in the presence of a constant vertical magnetic field

Oleg Zikanov and Andre Thess
Center for Physical Fluid Dynamics, Department of Mechanical Engineering
Dresden University of Technology, D-01062 Dresden, Germany
Joel Sommeria
Ecole Normale Supérieure de Lyon, Laboratoire de Physique, 46
Allee d'Italie, 69364 Lyon Cedex 07, France

Summary

The influence of a constant uniform magnetic field on a turbulent thermal convection is investigated using direct numerical simulation. The problem is simplified by utilization of the homogeneity assumption. Namely, we study the flow in a box with periodic boundary conditions driven by a constant imposed mean temperature gradient. The case of small Prandtl number and both the magnetic field and temperature gradient in vertical direction is considered. The main feature of the flow is the development of two antiparallel vertical jets providing an effective mechanism for the vertical heat transfer. The magnetic field is shown to stabilize the jets and, thus, to increase heat transfer and enhance the anisotropy of the flow.

1 Introduction

The ability of a constant uniform magnetic field to suppress the turbulent motions of electrically conducting fluids is a subject of growing interest. This phenomenon is particularly important for the problems including heat transfer such as construction of liquid metal cooling blankets for fusion reactors or optimization of semiconductor crystal growth because the magnetic field can lead to substantial changes of the flow structure and transfer properties. It was shown analytically [1], [2], [3], experimentally [4], [5], and numerically [6], [7], [8] that responsible for this suppression is the anisotropic Joule dissipation. For many laboratory and technical flows of liquid metals the assumptions of low magnetic Reynolds and Prandtl numbers

$$Re_m \equiv \frac{uL}{\eta} \ll 1, \qquad P_m \equiv \frac{\nu}{\eta} \ll 1 \tag{1.1}$$

are valid. In (1.1) ν is the kinematic viscosity, u is the mean velocity, L is the typical length scale, and $\eta = (\sigma\mu_0)^{-1}$ is the magnetic diffusivity, σ and μ_0 being the electric conductivity and magnetic permeability of the liquid. Under such assumptions, the quasi-static approximation can be applied for the perturbations of the magnetic field due to fluid motions [9]. The rotational part of the Lorentz force reduces to a linear functional of the velocity

$$\boldsymbol{F}[\boldsymbol{u}] = -\frac{\sigma B_0^2}{\rho}\Delta^{-1}\frac{\partial^2 \boldsymbol{u}}{\partial z^2} \tag{1.2}$$

with the Fourier transform

$$\hat{\boldsymbol{F}}[\hat{\boldsymbol{u}}] = -\frac{\sigma}{\rho}\frac{(\boldsymbol{B_0}\cdot\boldsymbol{k})^2}{k^2}\hat{\boldsymbol{u}}(\boldsymbol{k},t) = -\frac{\sigma B_0^2}{\rho}\cos^2\phi\,\hat{\boldsymbol{u}}(\boldsymbol{k},t), \tag{1.3}$$

where $\boldsymbol{B_0}$ is a constant magnetic field directed along the z-axis, $\boldsymbol{k}$ is a wavenumber vector, and ϕ is an angle between $\boldsymbol{B_0}$ and $\boldsymbol{k}$.

The additional Joule dissipation represented by (1.2) or (1.3) is anisotropic. It is maximum for modes with $\boldsymbol{B_0}\|\boldsymbol{k}$ and zero for modes independent of z-coordinate. The dissipation tends to eliminate velocity gradients in the direction of $\boldsymbol{B_0}$ and elongate the velocity structures in this direction. The magnetic field tends to transform the flow into a two-dimensional state where the velocity field depends only on the coordinates in the plane perpendicular to $\boldsymbol{B_0}$. An opposite tendency is due to the nonlinear energy transfer which tends to restore isotropy of the flow.

The estimate of the ratio of Joule to nonlinear term is given by the magnetic interaction parameter

$$N \equiv \frac{\sigma B_0^2 L}{\rho u} \tag{1.4}$$

that can also be defined as the ratio of the large-eddy turnover time $\tau_{tu} = L/u$ to the typical time of Joule dissipation $\tau_J = \rho/\sigma B_0^2$.

An extensive numerical study of the transformation of an initially isotropic liquid metal flow under the influence of imposed constant magnetic field was performed in [8]. The case of a homogeneous flow in a cubic box with periodic boundary conditions was considered. To force the flow, an artificial energy supply into large scale modes was employed. It was shown that there are three types of the flow transformation dependent of the interaction parameter N. If N is small, the flow remains three-dimensional, turbulent, and approximately isotropic. In the case of large N (strong magnetic field) the rapid irreversible transformation into purely two-dimensional steady state was observed. An intermediate value $N \sim 1$ was found to lead to the intermittent solution with the periods of quasi-two-dimensional, laminar and three-dimensional, turbulent behavior interchanging.

From the point of the scalar transfer problem, one interesting result found in [8] is that even in the case $N \gg 1$, when the flow becomes independent of the coordinate z in the direction of the magnetic field, the velocity component u_3 in this direction does not vanish (for a review of experimental indications see [10]). This implies the possibility of non-zero scalar transfer in the direction of $\boldsymbol{B_0}$. Being very interesting for the industrial applications this possibility has not received yet proper consideration.

In the present paper we study the turbulent Rayleigh–Benard convection under the influence of a constant magnetic field. The case of low Prandtl number is considered. Aside from the possibility of studying the effect of magnetic field on the heat transfer, this formulation allows us to replace an artificial forcing used in [8] by the natural mechanism of buoyancy. On the other hand, the problem remains highly idealized (and numerically treatable) because we consider the case without rigid boundaries. The flow in a three-dimensional box with periodic boundary conditions is driven by an imposed constant temperature gradient parallel to the magnetic field. Such a flow without magnetic field was first considered in [11] where the term "homogeneous thermal convection" was proposed.

2 Basic equations and method of solutions

We consider the three-dimensional convective motion of an electrically conducting viscous fluid (e. g. liquid metal). The flow is assumed to be homogeneous and contained in a box of a square horizontal cross-section with the horizontal side length 2π and vertical side length 8π. There can be a uniform magnetic field imposed in vertical direction. The energy supply into the system is provided by the buoyancy force created through the imposed mean temperature gradient $\nabla T = -\boldsymbol{e}_z$.

Assuming the Boussinesq approximation for the fluid and the quasi-static approximations for the perturbations of the magnetic field the basic equations can be given as

$$\partial_t \boldsymbol{u} + (\boldsymbol{u} \cdot \boldsymbol{\nabla})\boldsymbol{u} = -\boldsymbol{\nabla} p + \nu \Delta \boldsymbol{u} + \theta \boldsymbol{e_z} - \sigma B_0^2 \Delta^{-1} \partial^2 \boldsymbol{u} / \partial z^2 \qquad (2.5)$$

$$\partial_t \theta + \boldsymbol{u} \cdot \boldsymbol{\nabla} \theta = \kappa \Delta \theta + u_3 \qquad (2.6)$$

$$\boldsymbol{\nabla} \cdot \boldsymbol{u} = 0. \qquad (2.7)$$

We suppose here that the temperature field can be decomposed into the fluctuating part $\theta(\boldsymbol{x},t)$ subject to periodic boundary conditions and the constant mean part $T(z) = T_0 - z$. The boundary conditions for the velocity perturbations $\boldsymbol{u} = (u_1,u_2,u_3)$ are also periodic. We adopt for this study the mean density $\rho_0 = 1$ and the Boussinesq coefficient $\alpha g = 1$. The Prandtl number $Pr = \nu/\kappa$ is 0.025 (liquid mercury) and the Rayleigh number

$$Ra = \frac{\alpha g d^3 \Delta T}{\nu\kappa} = \frac{(8\pi)^4 Pr}{\nu^2} \tag{2.8}$$

can be varied by changing the viscosity, the other physical parameters being fixed.

After application of the operator rot^2 to (2.5) and Fourier transform to (2.5) and (2.6) the system is solved using the standard pseudo-spectral technique based on the fast Fourier transform. The aliasing errors are not removed, which allows one to reduce the cost of calculations by a factor of about 2. The time-stepping technique includes a second-order leap-frog scheme for nonlinear term and exponential solution for the linear terms. To suppress the oscillatory instability inherent in leap-frog methods, the solutions at two subsequent time layers are averaged every 20th time step.

The system considered here seems to be a reasonable model for the turbulent convection in the regions with approximately constant mean temperature gradient far from the walls, e. g. in the middle of the long vertical cavity heated from below. An advantage of this system is the possibility of effective numerical simulation. There is also a specific disadvantage stemming from our highly idealized formulation. One can easily see that the most unstable modes of the linear stability problem are those with the zero horizontal velocity components and the only nonzero Fourier modes being

$$\hat{u_3}, \hat{\theta} \text{ with } \boldsymbol{k} = (1,0,0) \text{ or } \boldsymbol{k} = (0,1,0). \tag{2.9}$$

Spatially, this solution have a form of two (one ascending and another descending) vertical jets. Such a solution is usually forbidden by rigid or stress-free boundary conditions applied in z-direction.

The degeneration of our formulation manifests itself in the fact that the exponentially growing modes (2.9) are exact solutions of the nonlinear problem. One consequence is that the statistically steady turbulence state is difficult to achieve starting from arbitrary initial conditions if one uses a square computational box and low resolution 32^3 (high viscosity). The modes (2.9) grow and the other modes are dissipated before the nonlinear interaction becomes able to establish an effective energy transfer. Even in the case of an elongated box we had to start the calculations using the full resolution.

Another consequence of the degeneration described above is that the calculations with strong magnetic field do not provide physically reasonable solutions. If the magnetic interaction parameter N is large and the flow approaches a kinematically two-dimensional state in accordance with [8], the modes (2.9) become dominating since they are not affected by the Joule dissipation and the efficiency of the nonlinear energy transfer is poor in a two-dimensional flow at moderate Reynolds number. Therefore, the case $N \gg 1$ was not considered.

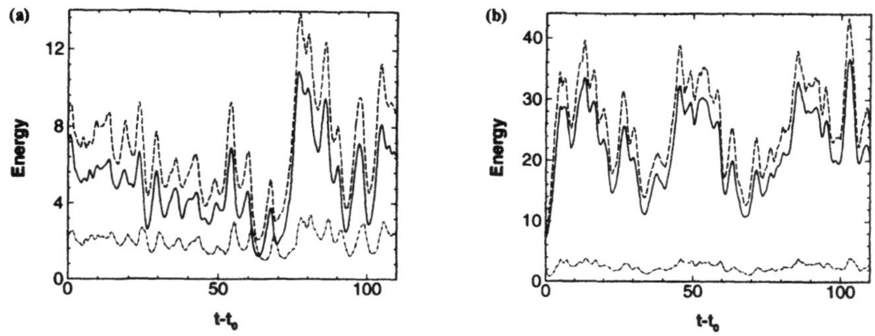

Figure 1 Mean energies of velocity and temperature perturbations as functions of time for the case without *(a)* and with *(b)* magnetic field. — · — · —, horizontal velocity $(E_1 + E_2)/2$; - - -, vertical velocity E_3; ——, temperature fluctuations E_T

3 Numerical experiments

Two numerical runs were performed. Numerical resolution was 64^2 functions in horizontal plane and 256 functions in vertical direction. The kinematic viscosity was $\nu = 0.027$ in both cases. This corresponds to the Rayleigh number (2.8) $Ra \approx 1.37 \cdot 10^7$. The initial conditions at $t = t_0$ were common for both runs and presented the developed turbulent flow obtained in the result of calculations without magnetic field starting from arbitrary initial conditions and lasting several turnover times τ_{tu}. In the first run the magnetic field was switched on at $t = t_0$ and remained constant till the end of calculations, the initial value of the magnetic interaction parameter (1.4) being $N(t_0) = 2.0$. In the second run, which was performed for comparison, the magnetic field was zero.

Figures 1–4 show the temporal behavior of different integral characteristics of the flow. Corresponding time averages are given in table 1. The averaging was performed using 2000 time points taken with the step 0.05 One can see in figure 1 that the mean energies of the velocity components and temperature perturbations

$$E_i = \frac{1}{2}\langle u_i^2 \rangle, \quad i = 1,2,3, \qquad E_T = \langle \theta^2 \rangle \tag{3.10}$$

(symbol $\langle \ldots \rangle$ stands hereafter for the space averaging) are strongly fluctuating with time. These fluctuations described early in [11] are relatively slow. Their typical time scale is of the order of several turnover times τ_{tu} (it can be seen in table 1 that our time unit is about $2\tau_{tu}$ at $N = 0$ and $3\tau_{tu}$ at $N > 0$). It was also shown in [11] and is discussed below in the present paper that the fluctuations appear in the result of the instabilities of large scale vertical convective jets developing in the flow.

Table 1 Time averages and variances of the integral characteristics of the flow

f	$N=0$ $\bar{f}$	$N=0$ σ_f	$N(t_0)=2.0$ $\bar{f}$	$N(t_0)=2.0$ σ_f
u	1.98	0.25	2.30	0.26
v	1.97	0.27	2.28	0.25
w	3.58	0.65	7.47	0.88
θ	2.16	0.42	4.86	0.61
Nu	6.21	2.13	27.79	7.07
G_1	0.71	0.08	0.40	0.05
G_2	0.66	0.09	0.36	0.04
$2w^2/(u^2+v^2)$	3.34	0.87	10.68	1.49
L_1	1.25	0.26	0.73	0.14
L_2	1.20	0.20	0.79	0.12
L_3	1.56	0.35	2.35	0.12
ϵ_1	1.75	0.72	4.42	1.62
ϵ_2	1.76	0.72	4.26	1.53
ϵ_3	2.34	0.93	9.72	2.81
μ_1			3.74	1.02
μ_2			3.81	1.05
μ_3			4.45	0.92

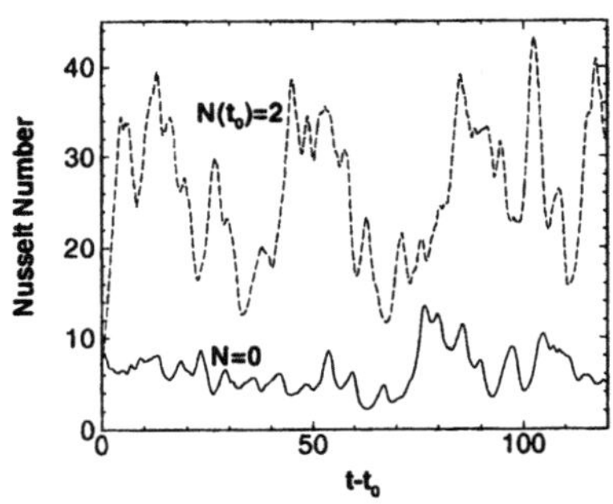

Figure 2 Nusselt number for the flow without (——) and with (- - -) magnetic field

The amplitude of the vertical velocity component is much larger than the amplitudes of horizontal components. This difference becomes more pronounced when the magnetic field is applied because, as can be seen in figure 1 and table 1, the magnetic field increases considerably vertical velocity as well as the amplitude of the temperature perturbations. Obviously, this leads to the growth

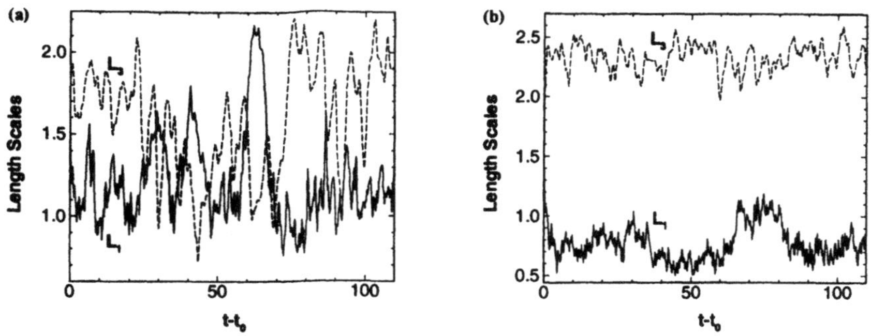

Figure 3 Integral length scales (3.12) for the case without *(a)* and with *(b)* magnetic field. ——, horizontal scale L_1; – – –, vertical scale L_3

of the heat transfer in vertical direction. As an illustration figure 2 and table 1 show the Nusselt number calculated as

$$Nu = 1 + \frac{H}{\kappa}, \tag{3.11}$$

where $H = \langle u_3\theta \rangle$ is the mean vertical heat flux. In the case with magnetic field the averaged Nusselt number is about 4.5 times larger than without magnetic field.

The imposed mean temperature gradient as well as the magnetic field leads inevitably to the flow which is anisotropic in a vertical cross-section. The coefficients quantifying the anisotropy are shown in figures 3-4 and table 1. Figure 3 presents the integral length scales in different directions

$$L_i = \frac{\pi}{2E_i} E_i(0), \quad i = 1{,}2{,}3, \tag{3.12}$$

where $E_i(k_i)$ is an one-dimensional longitudinal spectrum of the velocity component u_i and E_i in the denominator is defined by (3.10). In an isotropic case all L_i must be equal to the integral lenth scale $L = \pi/2u^2 \int_0^\infty k^{-1}E(k)dk$ used usually for homogeneous turbulence [12]. One can see in figure 3a and table 1 that in the case without magnetic field the vertical length scale L_3 is only slightly larger than L_1 and L_2. The magnetic field elongates the flow structures in vertical direction decreasing L_1 and L_2 and increasing L_3.

Another way to quantify anisotropy of the flow is to calculate different anisotropy coefficients such as

$$A = \frac{2E_3}{E_1 + E_2}, \tag{3.13}$$

where E_i are the energies of the velocity components (3.10), and

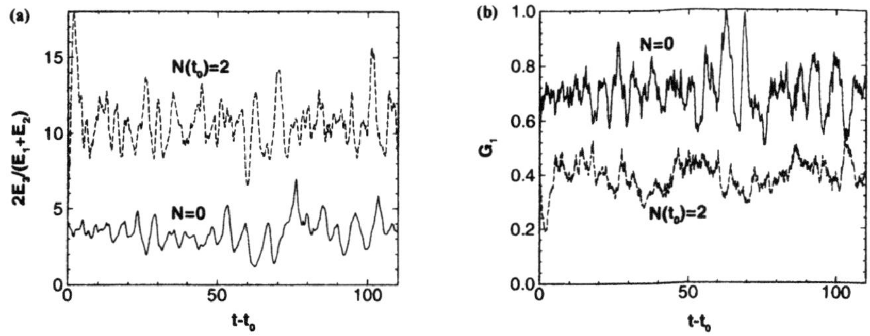

Figure 4 Anisotropy coefficients for the flow without (——) and with (- - -) magnetic field. *(a)* Large scale anisotropy coefficient (3.13). *(b)* Moderate scale anisotropy coefficient (3.14)

$$G_1 = \frac{\langle(\partial u_2/\partial z)^2\rangle}{2\langle(\partial u_2/\partial y)^2\rangle}, \quad G_2 = \frac{2\langle(\partial u_3/\partial z)^2\rangle}{\langle(\partial u_3/\partial y)^2\rangle}. \tag{3.14}$$

All three coefficients must be equal to unity in an isotropic flow. The coefficient A, calculated for homogeneous thermal convection in [11], is an estimate of the large-scale anisotropy of the flow. The coefficients G_1 and G_2 were used in [6], [8] to follow the damping of vertical velocity gradients by the Joule dissipation and can be considered as characteristics of the anisotropy at smaller scales. G_1 and G_2 must be zero in a purely two-dimensional flow independent of z-coordinate.

One can see in figure 4a that even the flow without magnetic field possesses considerable anisotropy at large scales. The time-averaged value $\bar{A} = 3.34$ is in agreement with [11] where $\bar{A}$ varied with numerical resolution between 2.7 and 3.9. The coefficients G_1 and G_2 demonstrate for this flow a moderate degree of anisotropy at smaller scales. In the presence of the magnetic field both A and G_1, G_2 change considerably implying the substantial growth of anisotropy at large and moderate scales. On the other hand, the magnetic field is clearly not strong enough to dissipate all the vertical gradients and the flow remains three-dimensional.

During the numerical runs the time-averaged two-dimensional energy distributions $E(k,\phi)$, $E_T(k,\phi)$ were calculated. Here $E(k,\phi)$ and $E_T(k,\phi)$ correspond to the energies of velocity and temperature perturbations contained in the Fourier modes with $|\boldsymbol{k}| = k$ and the angle between $\boldsymbol{k}$ and the z-axis equal to ϕ. The results are partly shown in figure 5.

Figure 5a presents the wavenumber spectra obtained by integration of $E(k,\phi)$, $E_T(k,\phi)$ over $\phi \in [0,\pi/2]$. One can see that the magnetic field does not change greatly the velocity and temperature spectra. Our resolution is too poor to make any reliable conclusions about the inertial range scaling. We can only

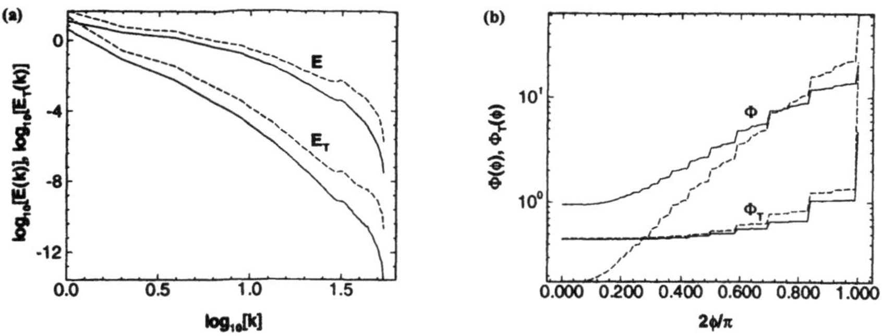

Figure 5 *(a)* Wavenumber spectra and *(b)* angular distributions (3.15) of the velocity (E, Φ) and temperature (E_T, Φ_T) perturbations. ——, flow without magnetic field; - - -, flow with magnetic field.

state that in a region near $4 < k < 10$ the velocity spectra are approximated to some degree by $k^{-5/3}$ which is in agreement with the retention of three-dimensionality of the flow [1]. The spectra $E_T(k)$ of the temperature fluctuations are much steeper than $E(k)$. What is more, they are steeper than k^{-4} in whole range of k. This seems to be in agreement with the theoretical predictions and experimental data relevant for our case of small Prandtl number and moderate Rayleigh number. The experiments [13] with sodium ($Pr \approx 5 \times 10^{-3}$) provided scaling $E_T \sim k^{-3.96}$. For a theoretical explanation we can use a classical Kolmogorov-Oboukhov phenomenology and estimate the so-called conductive (diffusive) wavenumber k_c for the temperature spectrum (see e. g. [14]). At $Pr < 1$ the estimation is $k_c \sim Pr^{3/4} k_d$, where k_d is Kolmogorov dissipative wavenumber. In our calculations the estimation gives k_c between 1 and 2 and we can conclude that the temperature spectrum consists of inertial-conductive (at small k) and viscous-conductive (at larger k) ranges. For the former, the classical scaling is $E_T(k) \sim k^{-4} E(k)$.

The angular energy spectra are shown in figure 5b. Instead of the angular distributions

$$E(\phi) = \int_0^\infty E(k,\phi)dk$$

we use the integrals

$$\Phi(\phi) = \int_0^\phi E(\lambda)d\lambda \qquad (3.15)$$

[1] It was shown in [8] that in the two-dimensional flow developing under the influence of a strong magnetic field at moderate Reynolds number the velocity spectrum is even steeper than k^{-3}.

corresponding to the total energy of the modes contained in the cone of vertical axis and semiangle ϕ. In a purely two-dimensional flow independent of the z-coordinate this spectrum becomes

$$\Phi(\phi) = \begin{cases} E & \text{at} \quad \phi = \pi/2 \\ 0 & \text{at} \quad \phi < \pi/2 \end{cases} .$$

On the contrary, in an isotropic flow, $\Phi(\phi)$ is a linear function of ϕ. The angular velocity spectra for the flow without magnetic field shown in figure 5b confirm the conclusion drawn above using the anisotropy coefficients. The flow is anisotropic, with the energy concentrating in the modes with wavenumber vectors in horizontal plane. This concentration is visibly enhanced by the applied magnetic field. Very interesting are the angular spectra of the temperature perturbations. One can see that the deviation from the isotropic form is much more pronounced for the temperature than for the velocity field. This can be related to the fact that, according to the wavenumber spectra, the energy of temperature perturbations is more concentrated in the large scale modes and the anisotropy is stronger at these scales.

Additional information can be drawn from the mean viscous (ϵ_i) and magnetic (μ_i) dissipations calculated for each velocity component. The time averages of ϵ_i and μ_i are given in table 1. One can see that the relative dissipations ϵ_i/E_i and μ_i/E_i of the vertical velocity component are much smaller than the corresponding values for the horizontal velocity components. This can be explained using the analysis of two-dimensional spectra. The difference between the value of E_3 on one hand and values of E_1, E_2 on the other hand is mostly due to the presence of modes (2.9) in the vertical velocity component. Clearly, the magnetic dissipation is zero and viscous dissipation is very small for these modes.

It is known (see e. g. [15]) that there is an exact relation between the mean heat flux and mean viscous dissipation expressing the energy balance in the Boussinesq equation. In our case this relation can be rewritten as

$$(Nu - 1)\kappa = \sum_i \epsilon_i + \mu_i . \tag{3.16}$$

Simple calculations demonstrate that (3.16) is satisfied by the time averages in table 1 with the accuracy within 5%. Taking into account the variances shown in table 1 this seems to be a good accuracy.

The spatial structure of the flow was analyzed using the snapshots of the vorticity and temperature fields. The main conclusion is in full agreement with [11]. The flow dynamics is determined by the evolution of two vertical jets. One jet is ascending and hot and another is descending and cold. An illustration is presented in figure 6 where the snapshots of the temperature field are shown using the isosurfaces of the temperature perturbations θ.

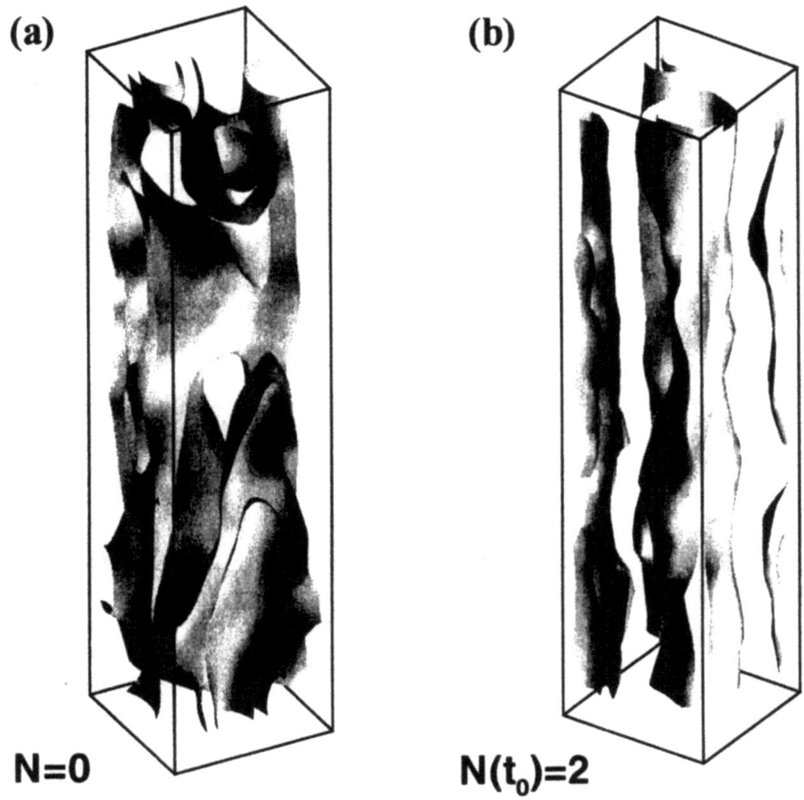

Figure 6 Snapshots of the field of temperature perturbations θ in the flow without *(a)* and with *(b)* magnetic field. Two isosurfaces are shown for both positive and negative θ.

As was mentioned above the modes (2.9), which are the main participants of the formation of the jets are exponentially growing exact solutions of the basic equations (2.5–2.7). The viscous dissipation alone is not able to damp this growth created by the buoyancy force. The only mechanism that can prevent the full solution from the exponential growth is the nonlinear energy transfer manifesting itself as the inherent instability of the jets. The instability leads to their bending (see figure 6a) and, from time to time, to disintegration. This process is responsible for the fluctuations of flow energy, heat flux and other integral characteristics shown in figures 1–4.

It can be seen when comparing figures 6a and 6b that the applied vertical magnetic field stabilizes the jets. This is clearly due to the additional Joule dissipation acting primarily on the modes with $\boldsymbol{k}\|\boldsymbol{B}_0$, that is on the modes responsible for the instability.

4 Concluding remarks

We have studied the influence of a uniform vertical magnetic field $\boldsymbol{B}_0$ on the homogeneous turbulent convection driven by a mean temperature gradient directed parallel to $\boldsymbol{B}_0$. The utilization of the quasi-static approximation and the assumption of spatial homogeneity allowed to reduce the problem to an extremely simple model that can be easily treated numerically. On the other hand, the use of periodic boundary conditions introduces a degree of ambiguity into the formulation. We can only assume that the model is appropriate for real flows such as, for example, the convective flow in a long vertical cavity heated from below.

The part of our results concerning the flow without magnetic field is in full agreement with [11]. The flow consists primarily of the two vertical jets, one ascending and another descending, accelerated by the buoyancy force and subject to the jet instability. The jets provide an effective mechanism for the generation of the mean heat flux in vertical direction. The flow is anisotropic, especially at large scales. Its integral characteristics fluctuate strongly with time as the jets disintegrate and reappear.

The main conclusions concern the influence of a constant vertical magnetic field on the flow. Additional Joule dissipation tends to eliminate vertical velocity gradients and, thus, to stabilize the jets. This leads to increase of the amplitudes of vertical velocity component and temperature perturbations. As a result, mean heat flux grows substantially. The jets are elongated in vertical direction, their typical horizontal scale decreasing and vertical scale increasing. The flow becomes more anisotropic but remains turbulent and three-dimensional.

The last conclusion is especially interesting when being compared with the results of [8]. Considering the homogeneous turbulence with an artificial forcing it was found that at moderate Reynolds number there are three types of the evolution of an initially isotropic flow under the influence of a constant magnetic field. The flow remains three-dimensional, turbulent and approximately isotropic if the magnetic interaction parameter $N \ll 1$, transforms into purely two-dimensional steady state if $N \gg 1$, and has temporarily intermittent two-dimensional, laminar — three-dimensional, turbulent behavior at an intermediate value of N. The results of the present work demonstrate the realization of another scenario proposed early in [5]. The flow becomes highly anisotropic but remains three-dimensional and turbulent, and retains a statistically steady level of the anisotropy.

Acknowledgment This work was supported by the Deutsche Forschungsgemeinschaft under Grant INK 18/A1-1 (Innovationskolleg Magnetofluiddynamik) and by the French-German collaboration program under Grant Th497/11-1. The computations have been performed at the CRAY-T90 at the

Höchstleistungsrechenzentrum Jülich, Germany.

References

[1] H. K. Moffatt, "On the suppression of turbulence by a uniform magnetic field," J. Fluid Mech. **23**, N3, 571–592 (1967).

[2] J. Sommeria and R. Moreau, "Why, how, and when, MHD turbulence becomes two-dimensional," J. Fluid Mech. **118**, 507–518 (1982).

[3] P. A. Davidson, " The role of angular momentum in the magnetic damping of turbulence," J. Fluid Mech. **336**, 123–150 (1997).

[4] A. D. Votsish and Yu. B. Kolesnikov, "Spatial correlations and vorticity in two-dimensional homogeneous turbulence," Magn. Gidrodin. **3**, 25–28 (1976); "Investigation of transition from three-dimensional turbulence to two-dimensional under a magnetic field," Magn. Gidrodin. **3**, 141–142 (1976).

[5] A. Alemany, R. Moreau, P. L. Sulem, and U. Frisch, "Influence of an external magnetic field on homogeneous MHD turbulence," J. de Méc., **18**, N2, 277–313 (1979).

[6] U. Schumann, "Numerical simulation of the transition from three- to two-dimensional turbulence under a uniform magnetic field," J. Fluid Mech. **74**, N1, 31–58 (1976).

[7] M. Hossain, " Inverse energy cascades in three-dimensional turbulence," Phys. Fluids B **3**, N3, 511-514 (1991).

[8] O. Zikanov and A. Thess, "Direct numerical simulation of forced MHD turbulence at low magnetic Reynolds number" accepted to J. Fluid Mech. (1998).

[9] R. Moreau, *Magnetohydrodynamics*, (Kluwer 1990).

[10] A. Tsinober, "MHD flow drag reduction." in *Viscous drag reduction in boundary layers*, edited by D. M. Bushnell and J. N. Hefner (AIAA Prog. Astron. Aeron. series, 1990).

[11] V. Borue and S. A. Orszag, "Turbulent convection driven by a constant temperature gradient," submitted to Phys. Rev. E (1996).

[12] J. O. Hinze, *Turbulence*, (McGraw-Hill, 1959).

[13] S. Cioni, S. Horanyi, L. Krebs, and U. Müller, "Temperature fluctuation properties in sodium convection," Phys. Rev. E **56**, N4, R3753-56 (1997).

[14] M. Lesieur *Turbulence in Fluids*, (Kluwer, 1997).

[15] S. Cioni, S. Ciliberto, and J. Sommeria, "Strongly turbulent Rayleigh-Benard convection in mercury: comparison with results at moderate Prandtl number," J. Fluid Mech. **335**, 111–140 (1996).

III. FLOWS OF REACTING GASES

Numerical Simulation of Combustion in Partially Premixed Turbulent Flows

V. Favier & L. Vervisch
Institut National des Sciences Appliquées de Rouen
LMFN - UMR CNRS - CORIA 6614, Avenue de l'Université - BP 8
76801 Saint Etienne du Rouvray Cedex, France

M. Herrmann, P. Terhoeven, B. Binninger, N. Peters
Institut für Technische Mechanik
Templergraben 64
52056 Aachen, Germany

Summary

To help to understand and to model lift-off in non-premixed turbulent combustion, triple flames and edge-flames are investigated in this work. In the first part of the paper the effects of unsteadiness at the edge of a diffusion flame are studied. Numbers representative of propagation and extinction properties of edge-flames are chosen, and direct numerical simulation of an edge-flame submitted to vorticity and unsteady micromixing are performed varying those key numbers. The outcome is a diagram delineating conditions for upstream and downstream propagation of an edge-flame in such environments. In addition to these direct numerical computations a modelling approach is presented to simulate partially premixed combustion in the flamelet regime which combines the flamelet models for non-premixed and premixed combustion. In the flamelet regime the calculation of the chemical processes can be separated from the flow calculation, so that the flow solver needs not resolve any chemical scales. To account for the combustion controlling mechanisms, mixing and flame propagation, transport equations for two scalars, the mixture fraction Z and the level set function G, are introduced. These quantities and the scalar dissipation rate χ serve as input parameters for pre-calculated triple flame libraries which provide the chemical composition of the flow and the burning velocity s_L. A hybrid tracking capturing scheme is used to solve the governing equations.

1 Direct numerical simulation of the effects of unsteadiness in edge-flames

1.1 Introduction

In non-premixed turbulent combustion the reaction zone is confined close to the highly convoluted stoichiometric surface of the mixture [1]. The reaction rate, however, is non uniform over that surface. For instance, there are zones where excessive thermal gradients will cause local extinctions of the flame. Such enhanced thermal gradients are expected to occur in a turbulent flow where velocity fluctuations are known to induce flame stretch. Direct numerical simulation (DNS) of non-premixed turbulent flames suggest that the characteristic flame structure that is observed at the edges of the stoichiometric surface bordering the extinction zone can be modeled locally by a triple flame [2]. This "triple flame" or "edge-flame" [3] consists of a curved flame front followed by a trailing diffusion flame (fig. 1).

One of the challenges of non-premixed turbulent combustion is the description and the modeling of lift-off phenomena observed in jets flames [4]. In this situation the combustion starts in a zone located at a distance h from the burning exit, and a partially premixed flame contributes to the stabilization of the flame.

To help to understand and to model lift-off in non-premixed turbulent flames, we have studied effects of unsteadiness at the edge of a diffusion flame. Unsteady large scales are known to play an important role in lift-off, and Large Eddy Simulation (LES) should emerge as an effective tool in this context. This work is motivated by the need of developing subgrid combustion models to perform these LES. First, numbers representative of propagation and extinction properties of edge-flames are chosen. Then, direct numerical simulation of an edge-flame submitted to vorticity and unsteady micromixing are performed varying those key numbers. The outcome is a diagram delineating conditions for upstream and downstream propagation of an edge-flame in such environment. It is found that edge-flames can take advantage of mixing to propagate faster than triple flames. This propagation was also observed for conditions at which a diffusion flame would be quenched. Moreover, local quenching of the trailing diffusion flame generates new ends of reaction zones, helping to sustain combustion. The properties of the relative velocity of the edge of the reaction zone are compared with the asymptotic solution obtained for a steady triple flame [5]. This helps to discriminate between edge-flames able to progress within a zone of intense unsteady mixing from those moving downstream with the flow ; in particular, it is observed that an edge-flame submitted to strong unsteady micromixing effects may lose its propagation properties, and evolves into a "hot-spot".

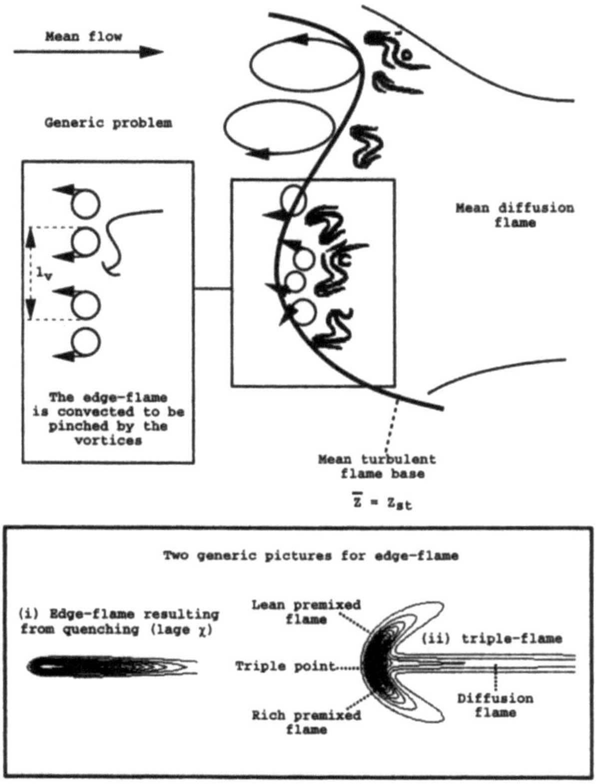

Figure 1 Sketch of the turbulent flame base. Quenching of the diffusion flame leads to edge-flame (i) of which generic model is the triple flame (ii). The edge-flames are convected by the large scales structures and undergo edge-flames/unsteady micromixing interaction.

1.2 Model problem

The control parameters of diffusion combustion are the mixture fraction Z introduced to measure the extent of mixing between fuel and oxidizer, ($Z = 0$ in pure air and $Z = 1$ in the fuel stream) and its dissipation rate $\chi = 2D|\nabla Z|^2$, where D is a diffusion coefficient [6].

In turbulent flows, an integral length scale $l_t \sim (k^{3/2}/\varepsilon)$ is usually introduced, where k is the turbulent kinetic energy and ε its dissipation rate. A mixing length l_d is defined from the mean distribution of mixture fraction $l_d \sim |\nabla \overline{Z}|^{-1}$ (an overline denotes an averaged quantity). Mixing by large eddies is controlled by the integral length scale of the flow and $l_d \sim l_t$. A micromixing length δ_d is obtained via the fluctuations in the mixing of the reactants, $\delta_d \sim (\overline{|\nabla Z'|^2})^{-1/2} = (D/\overline{\chi})^{1/2}$, where

$\overline{\chi} = D\overline{|\nabla Z'|^2}$ is the rate of dissipation of the fluctuations $\overline{Z'^2}$. We will now use these definitions of lengths based on $|\nabla Z|$ to parametrize a model problem.

As shown in the sketch of the turbulent flame base (fig. 1), some edge-flames can be convected by the flow so that they will avoid zones of intense micromixing. At the flame base when combustion is sustained, other edge-flames will be transported along a trajectory where they must be able to endure an increase in χ. Simulating the pinching of an edge-flame by a pair of vortices will mimic this situation (fig. 1). In this modeled configuration, the edge-flame is submitted to an unsteady distribution of Z, with χ increasing in front of the flame. In the simulations, the characteristic time of the modification of the Z field at the edge flame $\sim \chi/(\partial\chi/\partial t)$, is much smaller than the time required to strongly modify the system composed by the pair of vortices.

The simulations are performed using a fully compressible DNS code previously developed [7]. A triple flame is initially stabilized within a two-dimensional domain following a procedure proposed by Ruetsch et al [8]. The computations are realized with a Zel'dovitch number $\beta = 8$ and a heat release parameter $\alpha = (T_{burnt} - T_{fresh})/T_{burnt} = 0.8$, T_{fresh} and T_{burnt} denote the temperature at both sides of a stoichiometric unstrained premixed flame. The reaction rate of the one-step global chemistry is cast in the form of an Arrhenius law. The Lewis and Schmidt numbers are set to unity. The streamwise component of the velocity u is equal to the triple flame displacement velocity S^o_{TF}, and the pair of vortices is convected toward the edge-flame.

The strain rate $\varphi = (\partial u/\partial x_n)$ generated by the vortices defines a micromixing length $\delta_{m_{v'}} = (v'/\varphi)$. Here n denotes the direction normal to the stoichiometric surface and v' is the tangential velocity of the vortices. A characterization of the thickness of the mixing zone that a particular pair of vortices will impose to the triple flame is then obtained via δ_{m_v}. A scalar dissipation rate $\chi_v = (D/\delta^2_{m_v})$ corresponding to this length is also defined.

To fully parametrize the simulations, we compare χ_v with the scalar dissipation rate χ^o_{TP} calculated at the triple point of the initial steady edge-flame. The ratio $R_\chi = (\chi^o_{TP}/\chi_v)^{1/2}$ measures the intensity of the mixing that the vortices impose to the triple flame. An initial triple flame length $\delta^o_{m_{TP}} = |\nabla Z|^{-1}_{TP} = (D/\chi^o_{TP})^{1/2}$ is obtained, then $R_\chi = (\delta_{m_v}/\delta^o_{m_{TP}})$. Another important number is the ratio between the characteristic vortex velocity v' and the initial triple flame displacement velocity S^o_{TF}, we note $R_S = (v'/S^o_{TF})$. A Damköhler number is derived from those numbers, $Da = (\delta_{m_v}/v')/(\delta^o_{m_{TP}}/S^o_{TF}) = (R_\chi/R_S)$. Thus, the formulated problem includes at the same time parameters related to propagation (R_S) and to extinction (R_χ). Forty simulations have been carried out, some representative cases are given in table 1.

Table 1

R_χ	R_S	(l_v/l_r)	$D_a = (R_\chi/R_S)$	Propagation	
				Upstream	Downstream
0.095	2.5	4.93	0.038	•	
0.154	4.0	8.21	0.038	•	
0.174	3.9	9.36	0.044	•	
0.109	2.0	5.75	0.055	•	
0.165	5.5	9.36	0.030	o[a]	
0.066	6.0	3.03	0.011		•
0.080	6.5	4.10	0.012		•
0.073	3.5	3.26	0.021		•
0.139	5	7.39	0.028		•

[a] The trailing diffusion flame is quenched

We have also included in this table the ratio between l_v, the characteristic length of the pair of vortices (fig. 1) and l_r, the thickness of the reaction zone in the initial trailing diffusion flame.

2 Results

Different dynamics are obtained depending on the control parameters of the problems and results are conveniently summarized in a spectral stabilization diagram (fig. 2). It is seen in this diagram that the edge-flame/vortices interactions belong to one of the following two main groups :

1. Interactions that lead the edge-flame to propagate upstream of its initial position and will help flame stabilization ;
2. Interactions that push the edge-flame downstream of its initial position and will promote flame lift-off.

In case (1), high values of scalar dissipation rate are found between the vortices and the edge-flame maintains combustion by progressing through an hostile environment (fig. 3). A careful examination of the response of the triple flame displacement velocity with asymptotic solutions [5] shows that the main properties of laminar triple flames are preserved during this interaction. Since we observe an upstream movement of the edge-flame with a mean flow velocity equal to the initial triple flame propagation velocity, it is concluded that vorticity and unsteady micromixing can help an edge-flame to progress with an absolute velocity larger than the displacement velocity of a triple flame.

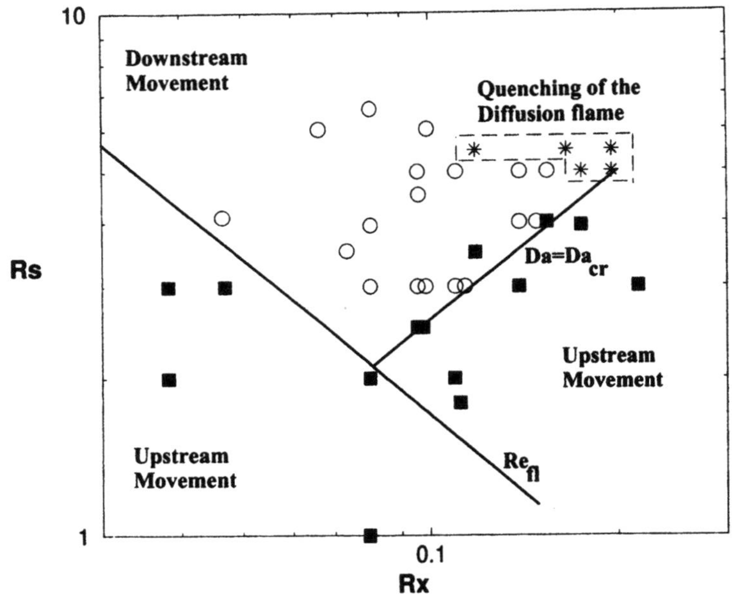

Figure 2 Diagram delineating various edge-flame behaviors. $R_\chi = (\chi^o_{TP}/\chi_v)^{1/2}$, $R_S = (v^{'}/S^o_{TF})$, $Re_{fl} = R_s R_\chi (S^o_{TF}/\chi^0_{TF}\delta^o_{TF})$ and $Da = (R_\chi/R_S)$. Square (case (1a)): edge-flames move upstream contributing to the stabilization. Star (case (1b)): upstream movement with quenching of the trailing diffusion flame. Circle (case (2)) : edge-flames move downstream (blow-out).

In case (2), unsteady micromixing effects are too strong, and the flame moves downstream of its initial position (see Da values in table 1). In this situation, a careful analysis of the imbalance between diffusion of heat and chemical source, shows that the leading-edge of the reaction zone does not feature the properties of a propagating partially premixed front, it has become a simple "hot-spot", with properties different from those of a laminar propagating triple flame.

2.1 Summary of edge-flame/micro-mixing interaction

When simulating triple flame/vorticity interaction, we observe that the edge of a diffusion flame can take benefit from vorticity and unsteady mixing to progress in an environment where a diffusion flame would be quenched, with an absolute velocity larger than its corresponding "triple flame burning velocity". Depending on positions in a diagram constructed from propagation and micromixing properties (fig. 2), edge-flames exhibit two different behaviors: (1) The edge-flame progresses within the flow,

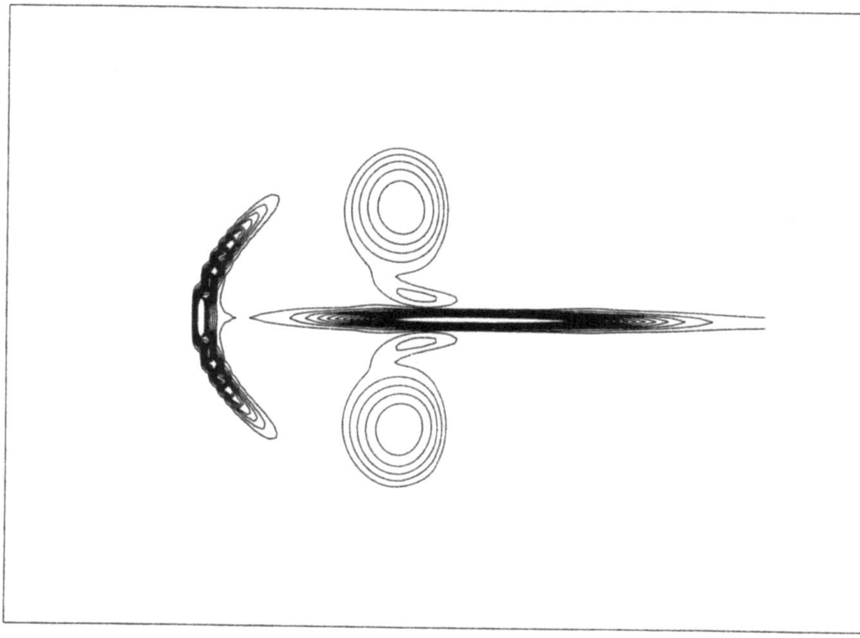

Figure 3 Isolines of heat release and vorticity in the case of an edge-flame moving upstream of its initial position.

then combustion is sustained within a stabilization zone ; (2) The edge of the reaction zones is pushed away downstream and blow-out occurs. A criterium exists to distinguish between (1) and (2), it is constructed from basic control parameters of diffusion (extinction) and partially-premixed (propagation) flames.

3 Modelling partially premixed combustion with the two-scalar level set approach

Since partially premixed combustion typically consists of both non-premixed and premixed combustion, the two-scalar level set approach aims to combine the flamelet model for non-premixed combustion [11] with the flamelet model for premixed combustion [10]. The two-scalar level set approach has first been used by Müller [9] to simulate partially premixed turbulent combustion. This model is valid in the flamelet regime [10],[11], where chemical time and length scales are much smaller than the corresponding flow scales, so that combustion takes place in thin flame zones. Outside these zones the flow is assumed to be in chemical equilibrium or inert. The thin flame zone and the surrounding inert mixing regions are called flamelets. In the flamelet regime the flame structure may be detached from the outer flow field and analyzed separately. The outer flow field acts upon the flamelet structure only through the scalar dissipation rate at stoichiometric mixture χ_{st} [11]. The structure of laminar flamelets may be determined either by experiments or DNS calculations using full or reduced chemical mechanisms. The results, namely the chemical composition, the mass fractions Y_i, of the flamelet, may be stored in flamelet libraries and made available to the outer flow calculation (fig. 4).

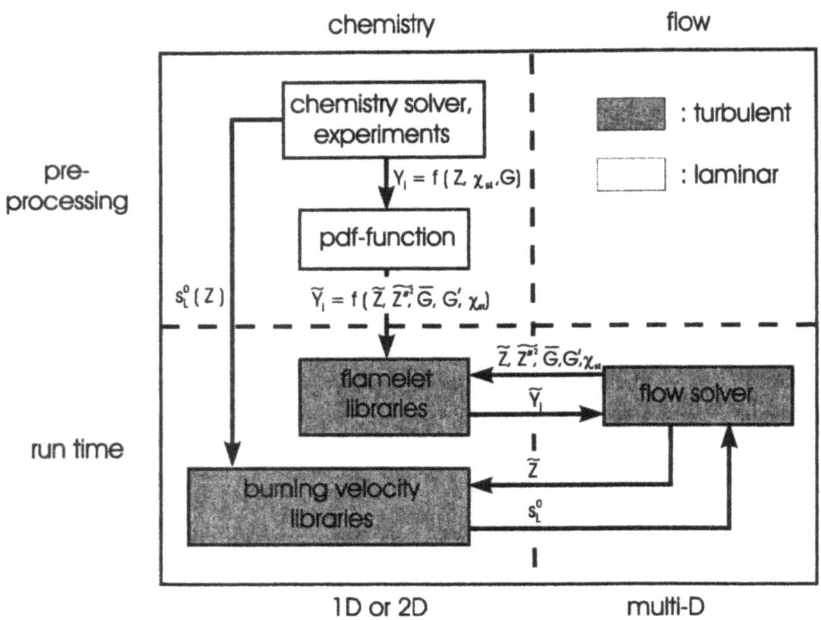

Figure 4 The flamelet concept for partially premixed combustion

3.1 The flamelet model for non-premixed combustion

In the flamelet regime non-premixed combustion can be described in terms of the mixture fraction Z, which is a measure for the ratio of mass from the fuel stream to the mass of the fuel stream and the oxidizer stream. Reactions take place in the vicinity of the iso-surface $Z = Z_{st}$, with the stoichiometric value of Z_{st} being fixed [11]. Assuming that the Lewis numbers of all species are equal to unity, the transport equation for the conserved scalar Z can be derived from the conservation equations of the mass fractions Y_i,

$$\frac{D\rho Z}{Dt} - \mathrm{div}(\rho D \mathrm{grad} Z) = 0 \,. \tag{3.1}$$

Since (3.1) does not contain a chemical source term, Z is independent of the chemical reactions. In a laminar flow the scalar dissipation rate χ is a function of the local mixture fraction gradient

$$\chi = 2D \left(\frac{\partial Z}{\partial x_\alpha} \right)^2 \tag{3.2}$$

and may be interpreted as a characteristic diffusion time.

For turbulent flows (3.1) is favre averaged yielding the conservation equations for the favre averaged mixture fraction $\tilde{Z}$

$$\frac{\partial \bar{\rho} \tilde{Z}}{\partial t} + \tilde{v}_\alpha \frac{\partial \bar{\rho} \tilde{Z}}{\partial x_\alpha} - \frac{\partial}{\partial x_\alpha} \left(\bar{\rho} D_t \frac{\partial \tilde{Z}}{\partial x_\alpha} \right) = 0 \tag{3.3}$$

and the mixture fraction variance $\widetilde{Z''^2}$

$$\frac{\bar{\rho} \widetilde{Z''^2}}{\partial t} + \tilde{v}_\alpha \frac{\partial \bar{\rho} \widetilde{Z''^2}}{\partial x_\alpha} - \frac{\partial}{\partial x_\alpha} \left(\bar{\rho} D_t \frac{\widetilde{Z''^2}}{\partial x_\alpha} \right) = 2 \bar{\rho} D_t \left(\frac{\partial \tilde{Z}}{\partial x_\alpha} \right)^2 - \bar{\rho} \tilde{\chi} \,, \tag{3.4}$$

where the non-closed turbulent terms have been modeled by a gradient flux ansatz [12]. The turbulent scalar dissipation rate $\tilde{\chi}$ is modeled by

$$\tilde{\chi} = C_\chi \frac{1}{t_{turb}} \widetilde{Z''^2} \tag{3.5}$$

with C_χ being a model constant ($C_\chi = 2.0$), and t_{turb} the turbulent time scale. If a k-ϵ turbulence model is used, t_{turb} may be expressed as

$$t_{turb} = \frac{\tilde{k}}{\tilde{\epsilon}} \,. \tag{3.6}$$

In the flamelet regime turbulent diffusion flames are interpreted as an ensemble of laminar, one dimensional flamelets [12]. The ensemble is defined by a partial density function, a general accepted form being the beta function [13]

$$\tilde{P} \left(Z, \tilde{Z}, \widetilde{Z''^2} \right) = Z^{\alpha - 1} \left(1 - Z \right)^{\beta - 1} \frac{\Gamma(\gamma)}{\Gamma(\alpha) \Gamma(\beta)}, \tag{3.7}$$

where Γ is the Gamma function and

$$\gamma = \frac{\tilde{Z} \left(1 - \tilde{Z} \right)}{\widetilde{Z''^2}} - 1 \,, \qquad \alpha = \tilde{Z} \gamma \,, \qquad \beta = \left(1 - \tilde{Z} \right) \gamma. \tag{3.8}$$

Using the pdf function $\tilde{P}$, the favre averaged mass fractions $\tilde{Y}_i$ can then be calculated from the mass fractions Y_i of the laminar flamelets, $\tilde{Z}$, $\widetilde{Z''^2}$ and χ_{st}

$$\tilde{Y}_i\left(\tilde{Z},\widetilde{Z''^2},\chi_{st}\right)=\int_0^1 Y_i\left(Z,\chi_{st}\right)\tilde{P}\left(Z,\tilde{Z},\widetilde{Z''^2}\right)dZ\,. \tag{3.9}$$

The scalar dissipation rate at stoichiometric mixture χ_{st} can be determined from [14]

$$\chi_{st}=\tilde{\chi}\frac{\exp\left(-2\left[\mathrm{erfc}^{-1}\left(2\tilde{Z}\right)\right]^2\right)}{\exp\left(-2\left[\mathrm{erfc}^{-1}\left(2\tilde{Z}_{st}\right)\right]^2\right)}\,. \tag{3.10}$$

Thus using (3.3) - (3.10), the local chemical composition of the turbulent flow can be calculated from the laminar flamelet libraries without having to resolve the chemical time and length scales in the flow solver (fig. 4). Equation (3.3) determines the position of the iso-surface of stoichiometric mixture and therefore the position of the turbulent diffusion flame in the flow field.

3.2 The flamelet model for premixed combustion

The central quantity within the flamelet model for premixed combustion is the scalar function G, which defines the position of the premixed flame front within the flow field as the iso-surface

$$G\left(\vec{x},t\right)=G_0\,. \tag{3.11}$$

The basic property of a premixed flame is its ability to propagate normal to itself. The transport equation for G may therefore be derived from kinematical considerations

$$\frac{\partial G}{\partial t}+\left(\vec{v}_u+s_L\vec{n}\right)\cdot\nabla G=0\,, \tag{3.12}$$

where $\vec{v}_u$ denotes the flow velocity in the unburnt mixture directly ahead of the flame front, s_L the laminar burning velocity and $\vec{n}$ the flame front normal

$$\vec{n}=-\frac{\nabla G}{|\nabla G|}\,. \tag{3.13}$$

The laminar burning velocity s_L is a function of the flame front curvature κ, the flame stretch and the laminar burning velocity of a plane premixed flame s_L^0 [12]

$$s_L=s_L^0-s_L^0\mathcal{L}\kappa+\mathcal{L}\vec{n}\nabla\vec{v}\vec{n}\,. \tag{3.14}$$

Here $\mathcal{L}$ is the Markstein length, which is a sensitivity parameter of the flow to flame curvature

$$\kappa=\nabla\vec{n} \tag{3.15}$$

and stretch. The laminar burning velocity of a plane flame s_L^0 may either be determined using approximation formulas derived by Göttgens [15], DNS calculations of the flamelets or experimental results. Since the approximation formulas of [15] are only valid for lean mixtures, in general the last two should be the methods of choice and their results should be stored in burning velocity libraries and made available to the flow calculation (fig. 4).

The definition of G, (3.11), is valid only at the location of the flame front itself. For all points away from the flame front it is convenient to define the level set function G as a distance function [16]

$$|\nabla G| = 1 \, . \tag{3.16}$$

Regions of burnt gas are then characterized by $G > G_0$, regions of unburnt gas by $G < G_0$.

In turbulent flows, the position of the averaged premixed flame front is defined by the averaged scalar $\bar{G}$. As in laminar flows, kinematic considerations for turbulent flows lead to the transport equations of the averaged level set $\bar{G}$

$$\frac{\partial \bar{G}}{\partial t} + \left(\bar{\vec{v}}_u + s_T \vec{n}\right) \cdot \nabla \bar{G} = 0 \, . \tag{3.17}$$

In (3.17) s_T is the turbulent burning velocity, which according to Wirth and Peters [17] is a function of s_L^0, the velocity fluctuations v', the averaged scalar dissipation rate $\bar{\chi}$ and the flame front curvature κ,

$$s_T = s_L^0 \left(1 + b_2 \sqrt{\frac{v'}{s_L^0}} + b_1 \frac{v'}{s_L^0}\right) \left(1 - \alpha_F \frac{\bar{\chi}}{\chi_q}\right) - \mathcal{L} \kappa v' \, , \tag{3.18}$$

with

$$\bar{\chi} = \frac{1}{t_{turb}} (\Delta Z)_F = 2 \frac{\tilde{\epsilon}_u}{\tilde{k}_u} Z_{st} \, . \tag{3.19}$$

To ensure a correct transport of the G_0 level set all quantities used to determine the burning velocity must be evaluated in the unburnt mixture directly in front of the flame front. Since the chemical length scales are small compared to the flow length scales and shall not be resolved by the flow solver in the flamelet concept, the premixed flame front represents a numerical discontinuity resulting in jumps of the conserved quantities. Standard capturing schemes, which are used in most flow solvers, however cannot treat the flame discontinuity correctly due to the high amount of numerical diffusion at the flame front. Godunov type schemes do not account for the flame discontinuity at all, since (3.17) is not in conservation form and cannot be solved with the Euler equations in a coupled manner. Tracking schemes could resolve this problem, but require an unacceptable high amount of numerical and logistic resources.

The hybrid tracking capturing scheme first proposed by Smiljanowski and Klein [19] is used here. It combines the advantages of a tracking scheme with the simplicity of a capturing scheme while ensuring a correct representation of the flame as a free boundary problem in the flow. The method is based on standard capturing methods, but tracks the flame front via G level sets and introduces the flame front discontinuity into the capturing methods using a so called reconstruction scheme.

The reconstruction scheme determines the flow quantities directly in front and behind the flame front from the cell averaged quantities of the capturing scheme and the jump conditions, which may be derived from the integral conservation equations across the flame front discontinuity. Assuming equal turbulent kinetic energy and dissipation in the burnt and unburnt gas directly in front and behind the flame front the jump conditions [20] are

$$\frac{\tilde{v}_{n,2}}{\bar{c}_1} = \frac{\gamma}{\gamma+1}\left(\frac{1}{\gamma\bar{M}_1} + \bar{M}_1 + \frac{2}{3}\frac{\tilde{k}_1}{\bar{M}_1\bar{c}_1^2}\right) + \left\{\left[\frac{\gamma}{\gamma+1}\left(\frac{1}{\gamma\bar{M}_1} + \bar{M}_1 + \frac{2}{3}\frac{\tilde{k}_1}{\bar{M}_1\bar{c}_1^2}\right)\right]^2 - \frac{2}{\gamma+1}\left[1 + \frac{\gamma-1}{2}\bar{M}_1^2 + \frac{2\gamma}{3}\frac{\tilde{k}_2}{\bar{c}_1^2} + (\gamma-1)\frac{\bar{\psi}_1-\bar{\psi}_2}{\bar{c}_1^2}\right]\right\}^{\frac{1}{2}} \tag{3.20}$$

$$\tilde{v}_{t_i,2} = \tilde{v}_{t_i,1} \tag{3.21}$$

$$\frac{\bar{\rho}_2}{\bar{\rho}_1} = \bar{M}_1\frac{\bar{c}_1}{\tilde{v}_{n,2}} \tag{3.22}$$

$$\frac{\bar{p}_2}{\bar{p}_1} = 1 + \gamma\bar{M}_1\left[\bar{M}_1 - \frac{\tilde{v}_{n,2}}{\bar{c}_1} + \frac{2}{3}\frac{1}{\bar{c}_1^2}\left(\frac{\tilde{k}_1}{\bar{M}_1} - \frac{\bar{c}_1}{\tilde{v}_{n,2}}\tilde{k}_2\right)\right] \tag{3.23}$$

$$\tilde{Z}_2 = \tilde{Z}_1 \tag{3.24}$$

$$\widetilde{Z_2''^2} = \widetilde{Z_1''^2} \tag{3.25}$$

$$\tilde{k}_2 = \tilde{k}_1 \tag{3.26}$$

$$\tilde{\epsilon}_2 = \tilde{\epsilon}_1\ . \tag{3.27}$$

Equations (3.20) - (3.27) are valid in a coordinate system that is moved with the flame front, so that

$$\tilde{v}_{n,1} = s_T \tag{3.28}$$

$$\tilde{v}_{n,2} = \tilde{v}_{n,b} - \tilde{v}_{n,u} + \tilde{v}_{n,1} \; . \tag{3.29}$$

Together with the definition of the cell averaged quantities

$$\bar{\vec{U}} = \alpha \vec{U}_b + (1-\alpha)\, \vec{U}_u \; , \tag{3.30}$$

where α is the burnt cell volume fraction, $\vec{U}$ the vector of conserved quantities $\vec{U} = \left(\bar{\rho}, \bar{\rho}\tilde{v}_n, \bar{\rho}\tilde{v}_{t_i}, \bar{\rho}\tilde{E}, \bar{\rho}\tilde{k}, \bar{\rho}\tilde{\epsilon}, \bar{\rho}\tilde{Z}, \bar{\rho}\widetilde{Z''^2}\right)$, and $\vec{U}_u$ and $\vec{U}_b$ are the vectors of conserved quantities in the unburnt and burnt gas directly in front and behind the flame front, equations (3.20) - (3.29) yield a system of equations which is closed by the equation for the turbulent burning velocity (3.18). This system of equations is solved by an iterative solution algorithm yielding the correct burning velocity s_T and the means to determine the correct fluxes in the unburnt $\vec{F}_u$ and burnt $\vec{F}_b$ regions of the flow field.

In addition, the flux calculation of the standard capturing scheme has to be modified, such that

$$\bar{\vec{F}} = \beta \vec{F}_b + (1-\beta)\, \vec{F}_u \; , \tag{3.31}$$

where β is the burnt cell face fraction. Also, gradients at the flame front must not be calculated using the cell averaged quantities, but rather have to be determined using the reconstructed quantities.

3.3 The flamelet model for partially premixed combustion

To simulate partially premixed combustion, the two-scalar level set approach combines the flamelet models for non-premixed and premixed combustion. The scalar $\bar{G}\,(\vec{x},t)$ is used to identify the premixed flame front locations in the flow field, $\bar{G} = G_0$, while the scalar $\tilde{Z}\,(\vec{x},t)$ fixes the location of the diffusion flame in the flow field, $\tilde{Z} = Z_{st}$. Because of the flamelet ansatz the flow calculation can be separated from the flamelet structure calculation. In contrast to [9], experiments and DNS of laminar triple flames have shown [18], that the chemical composition in the burnt gas is not only a function of Z and χ_{st}, but of G as well. The calculation of the flamelet structure can therefore no longer be performed using one dimensional laminar model flames but must be done using laminar model triple flames. DNS calculations of a plane laminar methane-air triple flame using a reduced 10 step mechanism with 13 species have been performed by Terhoeven. The results of the DNS calculations can now be used to construct triple flame libraries. Defining a temperature iso-surface of 1400K as the location of the premixed flame front ($G = G_0$), the massfractions Y_i of every species can be determined as a function of Z and G yielding the laminar triple flame library shown in fig. 5 for CH_4, O_2, CO and H.

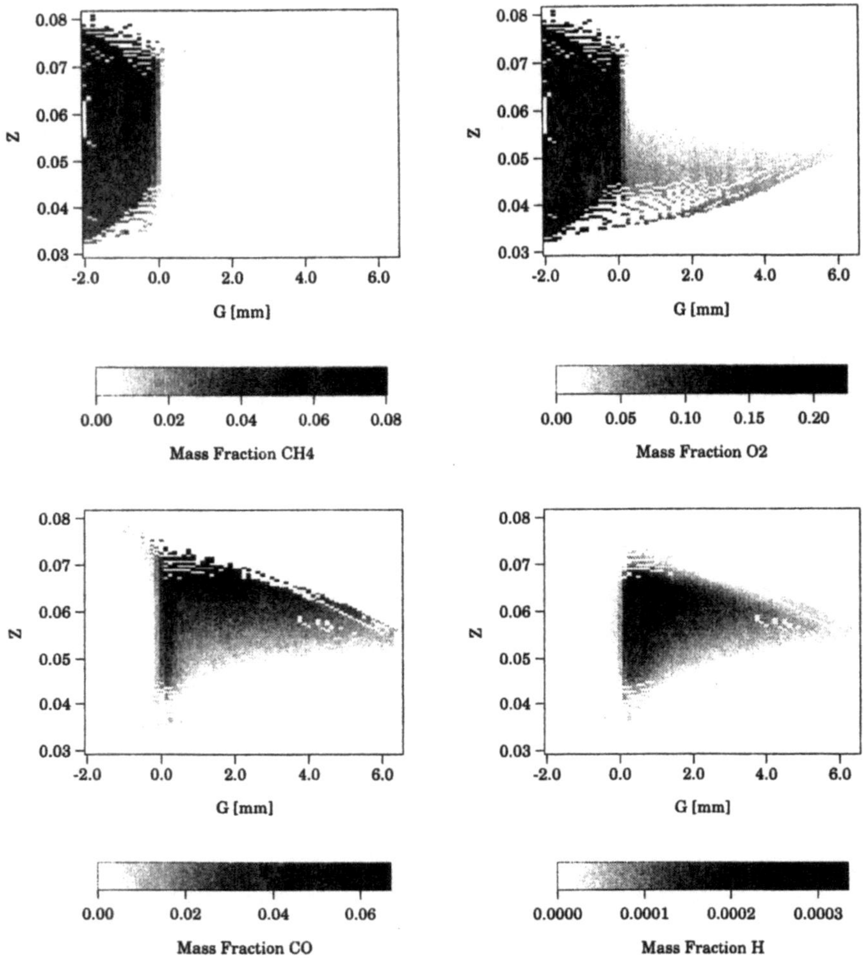

Figure 5 Triple flame library for CH_4, O_2, CO and H

One can clearly see that even on the rich side of the triple flame no fuel is left in the burnt gas. Instead stable species like CO and H_2 are consumed in the diffusion flame. In addition the burning velocity s_L may be determined from the DNS results as a function of the mixture fraction Z (fig. 6). In the following section the results for a laminar triple flame are presented using the mathematical modeling technique described above.

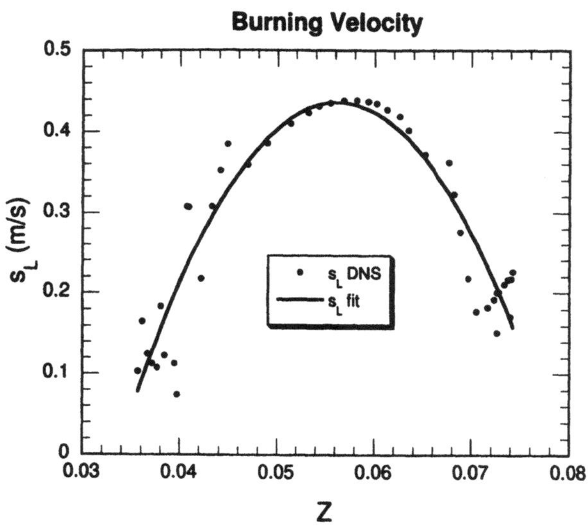

Figure 6 Burning velocity library for a triple flame

3.4 Calculation of a triple flame using the two-scalar level set approach

The two-scalar level set approach has been used to calculate a plane laminar methane-air triple flame using the triple flame libraries presented in the previous section. The flow solver AixCo solves the compressible Navier-Stokes equations using an operator splitting method by Strang [21]. The convection operator is solved using a Godunov type method with an approximate Riemann solver by Roe [22]. To achieve second order the upwind scheme is modified by a method of Yee, Harten and Warming [23]. The diffusion operator uses central differencing and a two step Runge-Kutta algorithm. Equation (3.12) is solved using a characteristic 2nd order method and a reinitialization by Sussman [16]. Inside the computational domain a mixture fraction distribution

$$Z(y) = \frac{1}{2}\left\{Z_{max} + Z_{min} - (Z_{max} - Z_{min})\cos\left[\frac{\pi}{2}\left(1 - \cos\frac{\pi y}{L_{ref}}\right)\right]\right\} \quad (3.32)$$

is initialized with a dimensionless mixture fraction thickness of

$$l_m = \frac{Z_{max} - Z_{min}}{l_F \left(\frac{\partial Z}{\partial y}\right)\Big|_{Z_{st}}} = 48\,, \quad (3.33)$$

$Z_{max} = 0.08$ and $Z_{min} = 0.03$. The pressure in the unburnt mixture is $p_u = 1\text{bar}$, the temperature is $T_u = 300\text{K}$. The Reynolds number evaluated using the burning velocity of a plane stoichiometric methane-air flame is $\text{Re}_{s_L} = 532$. Grid resolution is $\Delta x = \Delta y = 1\text{mm}$. A much finer resolution would disobey the flamelet assumption, that the premixed flame fronts (flame thickness $l_F \approx 0.2\text{mm}$) are to be discontinuities in the numerical scheme. Fig. 7 shows the starting conditions of the calculation and the choice of boundary conditions. In fig. 8 the temporal evolution of the premixed flame fronts beginning with the starting elliptic form are presented. One can see the

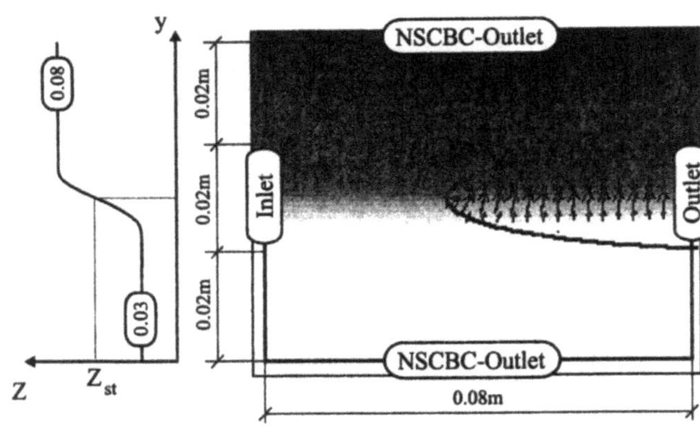

Figure 7 Starting conditions and boundary conditions for the triple flame calculation

typical characteristics of a triple flame with heat release, which have already been reported by Ruetsch [24]. The triple flame generates a divergence of the flow field in the unburnt mixture ahead of the triple point due to the expansion effect of the premixed flame fronts. This results in a reduction of the flow velocity just ahead of the triple point, so that the structural velocity of the whole triple flame is greater than the burning velocity of a plane flame at stoichiometric mixture. The structural velocity determined in this calculation is $v_{triple} = 1.1\text{m/s}$. The DNS calculations by Terhoeven have determined the structural velocity to be $v_{triple} = 0.85\text{m/s}$. This difference may be in part due to the fact that in AixCo only a simple equation of state was used.

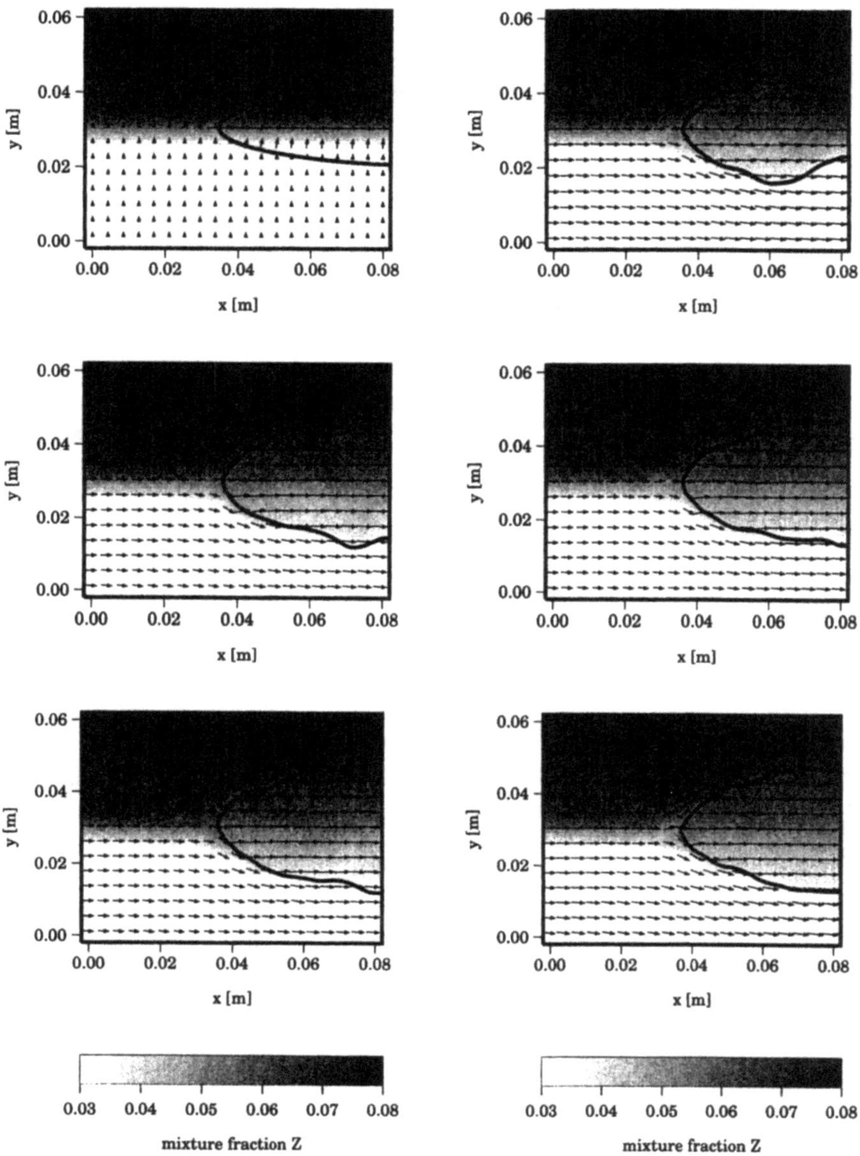

Figure 8 Mixture fraction, premixed flame fronts and velocity distribution at t=0ms, t=20ms, t=40ms, t=60ms, t=80ms and t=100ms of an methane-air triple flame

References

[1] A. Liñán. The asymptotic structure of counterflow diffusion flames for large activation energies. *Acta Astronautica*, 1007(1), 1974.

[2] L. Vervisch and T. Poinsot. Direct numerical simulation of non-premixed turbulent flame. *Annu. Rev. Fluid Mech.*, (30):655–692, 1998.

[3] J. Buckmaster and R. Weber. Edge-flame holding. In *Proceedings of the 26th Symp. (Int.) on Combustion*. The Combustion Institute, Pittsburgh, 1996.

[4] W.M. Pitts. Assesment of theories for the behavior and blow-out of lifted turbulent jet diffusion flame. In *Proceedings of the 22nd Symposium (Int) on Combustion*. The Combustion Institute Pittsburgh, 1988.

[5] S. Ghosal and L. Vervisch. Asymptotic theory of triple flame including effects of heat release. *Submitted to JFM*, 1997.

[6] N. Peters. Laminar flamelet concepts in turbulent combustion. In *Proceedings of the 21rst Symposium (international) on Combustion, Irvine*, pages 1231–1250. The Combustion Institute Pittsburgh, 1986.

[7] P. Domingo and L. Vervisch. Triple flames and partially premixed combustion in autoignition of non-premixed mixtures. In *Proceedings of the 26th Symp. (Int.) on Combustion, Naples*. The Combustion Institute, Pittsburgh, (1996).

[8] G.R. Ruetsch, L. Vervisch, and A. Liñán. Effects of heat release on triple flame. *Phys. Fluids*, 6(7):1447–1454, (1995).

[9] C.M. Müller, H. Breitbach, N. Peters, "Partially premixed turbulent flame propagation in jet flames", 25th Symposium (International) on Combustion, The Combustion Institute, 1986, pp. 1099-1106.

[10] N. Peters, "Propagating Thin Reaction Zones in Premixed Turbulent Combustion, Part I: Theory and Modelling Aspects", submitted to JFM, 1998.

[11] N. Peters, "Laminar Diffusion Flamelet Models in Non-Premixed Turbulent Combustion", Progr. Energy Combust. Sci., Vol. 10, 1984, p. 319-339.

[12] K.N.C. Bray, N. Peters, "Laminar Flamelets in Turbulent Flames", in: Turbulent Reacting Flows, Eds. P.A. Libby, F.A. Williams, Springer Verlag, 1994.

[13] W.P. Jones, J.H. Whitelaw, "Calculation Methods for Reacting Turbulent Flows: A Review", Combust. Flame, Vol. 48, 1982, p. 1.

[14] N. Peters, "ERCOFTAC-Lecture Series", Institut für Technische Mechanik, RWTH Aachen.

[15] J. Göttgens, F. Mauss, N. Peters, "Analytical approximations of burning velocities and flame thicknesses of lean hydrogen, methane, ethylene, ethane, acetylene and propane flames", 24th Symposium (International) on Combustion / The Combustion Institute, 1992, p. 129ff.

[16] M. Sussman, P. Smereka, S. Osher, "A level set approach for computing solutions to incompressible two-phase flow", preprint, Dept. Of Math., University of California, Los Angeles, 1994.

[17] M. Wirth, N. Peters, "Premixed Turbulent Combustion: A Flamelet Formulation and Spectral Analysis in Theory and IC-Engine Experiments", 24th Symposium (International) on Combustion / The Combustion Institute, 1992, p. 493ff.

[18] T. Plessing, P. Terhoeven, N. Peters, M. Mansour "An experimental and numerical study on a laminar triple flame", submitted to Combust. Flame, 1997.

[19] V. Smiljanowski, R. Klein, "Flame front tracking via in-cell reconstruction", Institut für Technische Mechanik, RWTH Aachen, 1994.

[20] M. Herrmann, "Implementierung des Flamelet-Modells zur Berechnung teilweise vorgemischter turbulenter Flammen", Diplomarbeit, Institut für Technische Mechanik, RWTH Aachen, 1995.

[21] G. Strang, "On the Construction and Comparison of Difference Schemes", SIAM, J. Num. Anal., Vol. 57, 1985, p. 506-517.

[22] P.L. Roe, "Approximate Riemann Solvers, Parameter Vectors, and Difference Schemes", J. Comp. Phys., Vol. 43, 1981, p. 357-372.

[23] H.C. Yee, R.F. Warming, A. Harten, "Implicit Total Variation Diminishing (TVD) Schemes for Steady-State Calculations", Journal of Computational Physics, Vol. 57, 1985, p. 327-360.

[24] D. Veynante, L. Vervisch, T. Poinsot, A. Linan, G. Ruetsch, " Triple flame structure and diffusion flame stabilization", Center for Turb. Res., Proc. of the Summer Program 1994, 1994, p. 55-73.

Numerical Modeling of Gas Flows in the Transition between Rarefied and Continuum Regimes

M. Günther*, P. Le Tallec**, J.P. Perlat**, J. Struckmeier*

*Universität Kaiserslautern
Fachbereich Mathematik
Postfach 3049
D 67653 Kaiserslautern

**INRIA
Domaine de Voluceau
Rocquencourt
Boîte Postale 105
F 78153 Le Chesnay Cedex

Summary

In the present paper we derive fluid dynamic equations by performing asymptotic analysis for the generalized Boltzmann equation for polyatomic gases. In particular, we consider the steady state, one-dimensional Boltzmann equation with one additional internal energy and different relaxation times. Moreover, we present a new approach to define coupling procedures for the Boltzmann equation and Navier-Stokes equations based on the 14-moments expansion of Levermore. These coupled models are validated by numerical simulations.

1 Introduction

The Boltzmann equation is an appropriate model to simulate rarefied gas flows as long as the mean free path of the gas is not too small. However, in the transition from rarefied to continuum flows, numerical simulations of the Boltzmann equation reach the limits of the existing hardware platforms, even when using massively–parallel systems.

Hence, it is worthwhile to investigate the transition from the Boltzmann equation to classical continuum mechanics, like Euler or Navier–Stokes equations and to combine both models in terms of a domain decomposition approach. Besides others, one of the challenging problems applying a domain decomposition procedure, is to detect (automatically) those regions, where it is possible to switch to the more simplified fluid dynamic models as well as to formulate appropriate coupling conditions between the different flow models. Recently, several authors investigated various domain decomposition methods based on

kinetic and continuum flow models, like the formulation of coupling conditions as well as the derivation of criteria to automatically detect kinetic and continuum regimes within the gas flow [1, 6, 9, 10, 11, 12, 17, 19].
The aim of the present investigation is twofolded: the first aim is to generalize the existing domain decomposition methods to the case of a rarefied gas with real gas effects, i.e. generalized Boltzmann equations, which include rotational and vibrational degrees of freedom as well as chemical reactions in the gas. Moreover, we like to present a new approach based on the fourteen moments expansion of Levermore [13], which combines a Galerkin approximation for the Boltzmann equation with a particular expansion of the solution of the Boltzmann equation, which is in contrast to the well–known Chapman–Enskog expansion strictly positive. In particular, we investigate how the resulting equations may degenerate to a more standard Navier–Stokes model.
The paper is organized as follows: in the next section we present briefly the generalized Boltzmann equation based on diatomic molecules, which include energy exchange between translational, rotational and vibrational degrees of freedom. At the end of the section we propose a simplified model problem, which is used in Section 3 to perform an asymptotic analysis to derive the limiting fluid dynamic equations for the Boltzmann equation with internal degrees of freedom. In Section 4 we show, how the Navier–Stokes equations may be obtained from the 14–moments expansion of Levermore using the well–known Hilbert expansion technique. Moreover, the somewhat original derivation clearly emphasize a kinetic interpretation of the moment expansion used by Levermore and we discuss the problem how to define a coupling process between the two models and give numerical simulations to validate the proposed coupling strategy.

2 The Boltzmann Equation for Diatomic Molecules

2.1 The Description of Internal Degrees of Freedom

In the following we consider a rarefied gas consisting of identical diatomic molecules, where each molecule carries besides the translational energy two additional degrees of freedom, namely a rotational and vibrational energy.
Due to the small level distance for rotational energy states, it is appropriate to simplify the resulting model considering the rotational energy as a continuous variable $e \in \mathbf{R}_0^+$, whereas the vibrational energy is assumed to take only discrete values $\nu_i \in \mathcal{D}$, where $\mathcal{D}$ denotes the finite set of different vibrational

levels.

Then, the generalized Boltzmann equation for diatomic molecules describes the time evolution of a kinetic density function $f(x,v,e,\nu_i)$ – the phase space density – in the form

$$\partial_t f + v\nabla f = \frac{1}{\varepsilon}Q(f,f), \tag{2.1}$$

where $Q(f,f)$ denotes the collision operator given by

$$Q(f,f) = \sum_{\nu_{i_1}} \int_{\mathbf{R}_0^+} \int_{\mathbf{R}^3} \underbrace{\sum\sum\int\int}_{0\leq\nu_{i'}+\nu_{i'_1}+e'+e'_1\leq E} \int_{S^2} W\,(f'f'_1 - ff_1)\,d\eta' de' de'_1 dv_1 de_1$$

and $f' = f(x,v',e',\nu_{i'})$, $f'_1 = f(x,v'_1,e'_1,\nu_{i'_1})$ etc. The collisions scattering kernel W (the transition probability) is assumed to take the form

$$W = |v - v_1|\,\sigma(E;\eta,\nu_i,\nu_{i_1},e,e_1 \to \eta',\nu_{i'},\nu_{i'_1},e',e'_1),$$

where the function σ denotes the differential cross section, E the total collision energy given by $E = \frac{1}{4}|v-v_1|^2 + e + e_1 + \nu_i + \nu_{i_1}$ and $\eta = (v-v_1)/|v-v_1|$. In (2.1) we already assumed a hydrodynamic scaling $x \to \varepsilon x$, $t \to \varepsilon t$ to relate the time evolution of the rarefied gas with the so–called Knudsen number $Kn = \lambda/L$, where λ denotes the mean free path of the gas, L a characteristic length of the problem, respectively.

An essential part in the generalized Boltzmann equation is to define appropriate, but in the same way simple models to specify the differential cross section σ. In the following, we use a differential cross section based on the generalized Borgnakke–Larsen model [18], which can be written in the form

$$\sigma_{gbl} = (1 - a(E) - b(E))\sigma_{el} + a(E)\sigma_{ve} + b(E)\sigma_{in}, \tag{2.2}$$

where $0 \leq a(E),b(E)$ and $a(E) + b(E) \leq 1$ for all $E \in \mathbf{R}_0^+$ and

$$\begin{aligned}
\sigma_{el} &= \varphi(\eta\cdot\eta')\sigma_{gbl}^{tot}(E)\delta(e-e')\delta(e_1-e'_1)\delta_{ii'}\delta_{i_1 i'_1}\\
\sigma_{ve} &= \frac{6}{(E-\nu_i-\nu_{i_1})^3}\varphi(\eta\cdot\eta')\sigma_{gbl}^{tot}(E)(E-e'-e'_1-\nu_{i'}-\nu_{i'_1})\delta_{ii'}\delta_{i_1 i'_1}\\
\sigma_{in} &= \frac{6}{\sum\limits_{\nu_{i'}+\nu_{i'_1}\leq E}(E-\nu_{i'}-\nu_{i'_1})^3}\varphi(\eta\cdot\eta')\sigma_{gbl}^{tot}(E)(E-e'-e'_1-\nu_{i'}-\nu_{i'_1}).
\end{aligned}$$

Moreover, we assume isotropic scattering for the angular dependence in the collisions, i.e.

$$\varphi(\eta\cdot\eta') = \frac{1}{4\pi}.$$

In the Borgnakke–Larsen–type models [8], one assumes a differential cross section, which is a combination of elastic, vibrational elastic and completely inelastic parts, i.e. in the first case we have no energy exchange between translation and internal energies, in the second case the vibrational energy is preserved and in the third case we have full energy exchange between all degrees of freedom. Hence, even the collision operator can be written as a sum of three different operators, $Q = Q_{el} + Q_{ve} + Q_{in}$. The functions $a(E), b(E)$ in (2.2) may be determined using macroscopic models [18], e.g. the Parker model for rotational relaxation determines the function $a(E)$, $b(E)$ may be fixed assuming a Millikan–White model for vibrational relaxation, where both models contain some parameters, which are fitted to experimental data.

2.2 Definition of a Model Problem

In particular at high temperatures, the relaxation times for translational, rotational and vibrational energies may differ strongly and this may influence the resulting fluid dynamic limits of the generalized Boltzmann equation. Here, one may expect to obtain modified Navier–Stokes equations, which include additional temperature equations for the rotational as well as vibrational states of molecules.

In the following section we will study the fluid dynamic limits of a simplified model problem, where the relaxation times between translational and rotational energies differ by some parameter δ, the vibrational states are kept frozen. Moreover, we restrict ourselve to a one–dimensional steady–state problem in a slab, with diffusive boundary conditions based on the two wall temperatures T_L and T_R on the left and right wall of the slab, respectively.

To validate the asymptotic expansions obtained from the generalized Boltzmann equation, the model problem has been treated numerically applying a particle method to simulate rarefied gas flows [16].

Macroscopic quantities like the density or the mean velocity are defined as moments of the density function $f(t,x,v,e)$, like

$$\begin{aligned}
\text{mass density} \quad &: \quad \rho(t,x) = \left\langle f \right\rangle_{v,e} \\
\text{mean velocity} \quad &: \quad u(t,x) = \frac{1}{\rho}\left\langle vf \right\rangle_{v,e} \\
\text{(kinetic) temperature} \quad &: \quad T(t,x) = \frac{2}{3}\frac{1}{\rho}\left\langle c^2 f \right\rangle_{v,e} \\
\text{rotational temperature} \quad &: \quad T_{rot}(t,x) = \frac{1}{\rho}\left\langle ef \right\rangle_{v,e}
\end{aligned}$$

$$\text{stress tensor} \quad : \quad \sigma(t,x) = -\Big\langle c \otimes c f \Big\rangle_{v,e}$$

$$\text{heat flux vector} \quad : \quad q(t,x) = \Big\langle \frac{1}{2} c^2 c f \Big\rangle_{v,e}$$

where $\langle \cdot \rangle_{v,e}$ denotes the integration over v and e and $c = v - u$.

3 Asymptotic Expansions

In the following we consider the steady–state, one–dimensional Boltzmann equation in a slab written in the form

$$v_x \partial_x f \quad = \quad \frac{1}{\varepsilon} Q_{el}(f,f) + \frac{\delta}{\varepsilon} Q_{ve}(f,f), \qquad x \in [0,1]. \tag{3.1}$$

where we assume, that the vibrational states of molecules are kept frozen and δ denotes a parameter to relate the translational and rotational relaxation times. In particular, we study the asymptotic limits as $\varepsilon, \delta \to 0$ to obtain macroscopic equations for the density ρ and the temperatures T and T_{rot}, the kinetic and rotational temperature of the gas. This is achieved using an asymptotic expansion for $f(t,x,v,e)$ and the most general Ansatz would be a double expansion in terms of ε and δ in the form

$$f = \sum_{k,l=0}^{\infty} \varepsilon^k \delta^l f_\delta^{(k,l)}.$$

A more simple analysis is to study the asymptotic limits only along particular traces $(\varepsilon, \delta) \to 0$, i.e. to assume a functional dependence between ε and δ in the form $\delta = \delta(\varepsilon)$. Then, the double expansion is reduced to a single expansion with respect to the small parameter ε given by

$$f(x,v,e) = \sum_{j=0}^{\infty} \varepsilon^j f^{(j)}(x,v,e). \tag{3.2}$$

For the single functions $f^{(j)}$, $j \geq 1$ we demand the normalization condition

$$\Big\langle \Psi(v,e) f^{(j)} \Big\rangle_{v,e} = 0, \tag{3.3}$$

where $\Psi(v,e) = (1,v,v^2,k(e))^t$, $k(e)$ denotes an arbitrary (non–constant) function in e. Condition (3.3) implies, that the macroscopic quantities ρ,u,T and T_{rot} are completely determined by $f^{(0)}$. Moreover, we use the (formal) decomposition $f^{(j)}(x,v,e) = h^{(j)}(x,v,e) g(x,e)$ for $j \geq 0$.

3.1 The Case $\delta = 0$

In this case the Boltzmann equation can be rewritten as

$$v_x \partial_x f = \frac{1}{\varepsilon} Q_{el}(f,f). \tag{3.4}$$

Inserting (3.2) into (3.4) yields

$$\begin{aligned} \varepsilon^{-1} &: \quad Q_{el}(f^{(0)}, f^{(0)}) = 0 \\ \varepsilon^{0} &: \quad v_x \partial_x f^{(0)} = Q_{el}(f^{(1)}, f^{(0)}) \\ \varepsilon^{1} &: \quad v_x \partial_x f^{(1)} = Q_{el}(f^{(2)}, f^{(0)}) + Q_{el}(f^{(1)}, f^{(1)}). \end{aligned}$$

From the equation at the ε^{-1}–level we can write

$$h^{(0)}(x,v) = M_{\rho,u,T}(v),$$

where $M_{\rho,u,T}(v) = \frac{\rho}{(\pi T)^{3/2}} \exp(-(v-u)^2/T)$ is a Maxwellian, $f^{(0)} = h^{(0)} g$ and the function $g(x,e)$ remains unknown at this level – up to the normalization condition

$$\left\langle g \right\rangle_e = 1. \tag{3.5}$$

Moreover, the velocity u is zero, because integrating (3.4) with respect to v and e yields

$$\partial_x(\rho u) = 0 \quad \Rightarrow \quad \rho u = c$$

and due to the diffusive boundary conditions, one has $(\rho u)(0) = (\rho u)(1) = 0$, i.e. $c = 0$. Since $\rho > 0$, it follows $u = 0$.
From the ε^0-level we obtain

$$v_x g \partial_x h^{(0)} + v_x h^{(0)} \partial_x g = Q_{el}(h^{(1)} g, h^{(0)} g).$$

Now we split the function $h^{(1)}$ into two parts, namely $h^{(1)}(x,v,e) = h_1^{(1)}(x,v) + h_2^{(1)}(x,v,e)$, to get a better representation of the first approximation and obtain from the ε^0–level the two equations

$$v_x g \partial_x h^{(0)} = Q_{el}(h_1^{(1)} g, h^{(0)} g) \tag{3.6}$$

$$v_x h^{(0)} \partial_x g = Q_{el}(h_2^{(1)} g, h^{(0)} g). \tag{3.7}$$

We demand, that $h_1^{(1)} g$ and $h_2^{(1)} g$ fulfill the condition (3.3).
Eqs. (3.6),(3.7) are conditions for the unknown functions $h_1^{(1)}$ and $h_2^{(2)}$. The right side can be regarded as a linear operator for $h_1^{(1)}, h_2^{(1)}$ and the kernel for $L(\cdot) = Q_{el}(\cdot h^{(0)} g, h^{(0)} g)$ is given by $\ker L =< 1, v, v^2, k(e) >$, where $k(e)$ is a non–constant function.
Then, (3.6) and (3.7) are solvable, if the conditions

$$\left\langle \Psi v_x g \partial_x h^{(0)} \right\rangle_{v,e} = 0, \tag{3.8}$$

$$\left\langle \Psi v_x h^{(0)} \partial_x g \right\rangle_{v,e} = 0 \tag{3.9}$$

with $\Psi(v,e) = (1,v,v^2,k(e))^t$ are fulfilled.
From (3.8) one gets

$$\partial_x(\rho T) = 0. \tag{3.10}$$

All other equations are identically equal to zero.
Then, the first order approximation is given by

$$h_1^{(1)}(x,v) = v_x \partial_x \ln T(x)(\frac{v^2}{T} - \frac{5}{2})\Theta_{\rho,T}(x)h^{(0)}(x,v) + (b_y v_y + b_z v_z)h^{(0)}(x,v).$$

If we use a collision model, such that the viscosity $\nu \sim T^p$, we obtain $\Theta_{\rho,T} \sim \frac{T^{p-1}}{\rho}$. In the case of Maxwellian molecules and rigid elastic spherical molecules we have $p = 1$ and $p = 0.5$, respectively.
The ε^1-order equation determines the function $f^{(2)}$: the solvability equations are

$$\left\langle \Psi \left(v_x \partial_x (h_1^{(1)} g + h_2^{(1)} g) - Q_{el}(h_1^{(1)} g + h_2^{(1)} g, h_1^{(1)} g + h_2^{(1)} g) \right) \right\rangle_{v,e} = 0 \tag{3.11}$$

From (3.11) we obtain

$$\partial_x(\lambda(\rho,T)\partial_x T) = 0 \tag{3.12}$$

$$\left\langle k(e) v_x \partial_x f^{(1)} \right\rangle_{v,e} = 0 \tag{3.13}$$

where

$$\lambda(\rho,T) = \left\langle v^2 \frac{v_x^2}{T}(\frac{v^2}{T} - \frac{5}{2})\Theta_{\rho,T} h_{\rho,T}^{(0)}(v) \right\rangle_v$$

is the heat conduction coefficient. Again, all other equations are vanishing trivially.
Eqs. (3.10) and (3.12) determine the parameters ρ and T. For example for the special model $\nu \sim T^p$, the solution is given by

$$\rho(x) = \frac{p_0}{((T_R^q - T_L^q)x + T_L^q)^{1/q}}, \quad p_0 \in \mathbf{R}_0^+ \qquad x \in [0,1] \tag{3.14}$$

$$T(x) = ((T_R^q - T_L^q)x + T_L^q)^{1/q}, \qquad x \in [0,1], \tag{3.15}$$

where $q = p + 1$, i.e. we have $q = 2$ for Maxwellian molecules and $q = 1.5$ for rigid spherical molecules.
Eqs. (3.10) and (3.12) represent the standard Navier-Stokes equations for the one-dimensional heat transfer, in particular we obtain a diffusion equation for the temperature T, which defines the translational temperature of the gas. Moreover, the heat conduction coefficient λ coincides with the expression obtained for a monoatomic gas.

However, we derived some more equations: for each $k(e)$ we obtain an additional equation (3.13); but, since we do not have an analytical expression for $f^{(1)}$, it is not possible to write the equation in terms of macroscopic quantities together with transport coefficients.
To verify our result, we performed numerical simulations using a particle method like discussed in Ref. [16]. Here, we used Maxwellian molecules at a

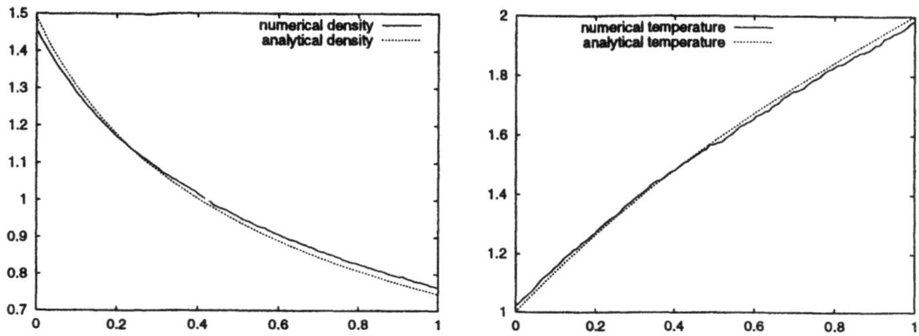

Figure 1 Density and Temperature Profile along the Slab

Knudsen number $Kn = 0.01$ and wall temperatures $T_L = 1$ and $T_R = 2$.
Fig. 1 shows the density and temperature profiles along the slab. The numerical results coincide nearly with the theoretical ones, which confirms the correctness of our derivation.

3.2 Representation of the Density g

It remains to find a representation for the internal energy density g. Since the internal energies do not influence the transport and collisions of the particles, the distribution should be a superposition of the internal boundary distributions. Therefore, we make the Ansatz

$$g(x,e) = \beta(x)g_L(e) + \alpha(x)g_R(e), \tag{3.16}$$

where g_L and g_R are the boundary conditions for the internal energy at $x = 0$ and $x = 1$, respectively. Condition (3.5) implies that $\beta(x) = 1 - \alpha(x)$ and the next step is to determine the function $\alpha(x)$: we separate the gas into two species, where the first one carries the information of the internal energy of the left boundary and the second one of the right boundary. The density of these two species are denoted by f^L and f^R, respectively. Then, the function α gives the mixing ratio, i.e. $\alpha = \rho^R/\rho$ and $1 - \alpha = \rho^L/\rho$, where ρ^L, ρ^R are the mass densities and $\rho = \rho^L + \rho^R$.

For the functions f^L, f^R we have the system

$$\begin{aligned} v_x \partial_x f^L &= \frac{1}{\varepsilon}(Q(f^L,f^L) + Q(f^L,f^R)) \\ v_x \partial_x f^R &= \frac{1}{\varepsilon}(Q(f^L,f^R) + Q(f^R,f^R)). \end{aligned}$$

where Q now denotes the standard monoatomic collision operator. For each density f^L, f^R we use an asymptotic expansion in the form

$$f^L = \sum_{j=0}^{\infty} \varepsilon^j f_L^{(j)}, \qquad f^R = \sum_{j=0}^{\infty} \varepsilon^j f_R^{(j)}.$$

Following [3, 4], we can write an approximation for f^L, f^R in the form

$$\begin{aligned} f^L &= f_L^{(0)} + f_L^{(1)} = f_L^{(0)}(1 + \Phi_L) \\ \Phi_L &= a_{-1}\,(\frac{v^2}{T} - \frac{5}{2})\,\frac{v_x}{\sqrt{T}}\partial_x \ln T + d_0\, a_0^{f^L}\,\frac{v_x}{\sqrt{T}}\, d_{12} + d_{-1}\,(\frac{v^2}{T} - \frac{5}{2})\,\frac{v_x}{\sqrt{T}} d_{12} \\ f^R &= f_R^{(0)} + f_R^{(1)} = f_R^{(0)}(1 + \Phi_R) \\ \Phi_R &= a_1\,(\frac{v^2}{T} - \frac{5}{2})\,\frac{v_x}{\sqrt{T}}\partial_x \ln T + d_0\, a_0^{f^R}\,\frac{v_x}{\sqrt{T}}\, d_{21} + d_1\,(\frac{v^2}{T} - \frac{5}{2})\,\frac{v_x}{\sqrt{T}} d_{21} \end{aligned} \tag{3.17}$$

where $f_L^{(0)} = (1-\alpha)M_{\rho,0,T}(v)$, $f_R^{(0)} = \alpha M_{\rho,0,T}(v)$ and all coefficients a_k, d_k, $k = -1,0,1$, d_{12}, d_{21} depend on ρ and T (for the exact form see [3]).
For the mass fluxes we have the conditions

$$q_{f^L} = \left\langle v f^L \right\rangle_v = const. \tag{3.18}$$

$$q_{f^R} = \left\langle v f^R \right\rangle_v = const. \tag{3.19}$$

With the approximation (3.17), we obtain for the mass flux q_{f^R}

$$q_{f^R} \sim C(T)\,\partial_x \alpha, \tag{3.20}$$

where $C(T)$ depends on the collision model. For the hard sphere model we have $C(T) \sim T^{1/2}$ and in the case of Maxwellian molecules it holds $C(T) \sim T$.
Condition (3.19) together with (3.20) yields $\alpha = c_1 T + c_2$ and applying the boundary conditions we obtain

$$\alpha(x) = \frac{T(x) - T_L}{T_R - T_L}.$$

If $T_L = T_R$, we can perform the limit $T_L \to T_R$ to get $\alpha(x) = x$. If α is determined, the resulting density g has the form

$$g = \frac{T_R - T}{T_R - T_L} g_L + \frac{T - T_L}{T_R - T_L} g_R. \tag{3.21}$$

To validate the expression for the density g as given in Eq. (3.21), we performed some numerical experiments assuming diffusive boundary condition for the internal energy, i.e.

$$g_L(e) = \frac{1}{T_{L,i}} \exp(-\frac{e}{T_{L,i}}), \qquad g_R(e) = \frac{1}{T_{R,i}} \exp(-\frac{e}{T_{R,i}}). \quad (3.22)$$

The first set of simulations have been realized with Maxwellian molecules at a Knudsen number $Kn = 0.002$ using about $1.2 \cdot 10^5$ particles and boundary temperatures $T_L = 1$, $T_R = 2$ and $T_{L,i} = 3$, $T_{R,i} = 4$. The computational results obtained for the first 4 moments of the density function showed a quite good agreement between the numerical and analytical result. A further simulation was done, which should verify the correctness of the computation of α under the assumption (3.16) for the density g. In this case we use boundary conditions of the type

$$g_L(e) = \delta(e - e_L), \quad g_R(e) = \delta(e - e_R), \quad \text{where} \quad e_L \neq e_R.$$

and the kinetic temperatures at the walls are $T_L = 1$, $T_R = 2$. The computations were done with Maxwellian molecules at Knudsen number $Kn = 0.01$ using 25.600 particles. Fig. 2 shows the fraction of particles, which have the internal energy e_R, where this number is equal to the function α.

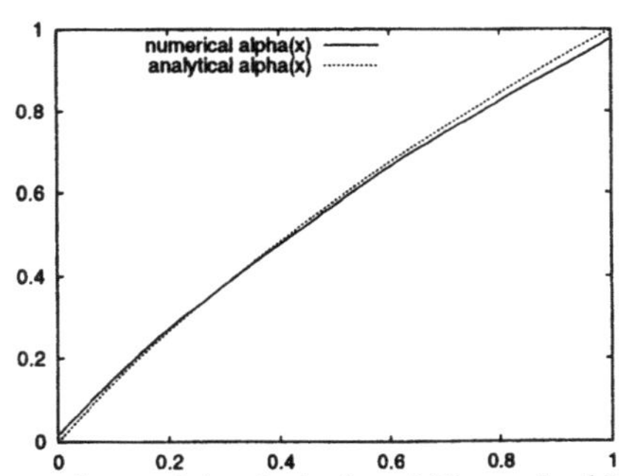

Figure 2 Comparison of Analytical and Numerical Results for $\alpha(x)$

All the simulations indicate, that the representation of α seems to be correct. The errors between the curves is explained by the numerical inaccuracy. The curves do not differ much, such that we can regard them as a confirmation of our assumption for the function g.

3.3 The Case $\delta = \varepsilon$

Taking $\delta = \varepsilon$, we have the additional collision term $Q_{ve}(f,f)$ on the right hand side of (3.4) and the first equations in ε now read

$$\begin{aligned}
\varepsilon^{-1} &: \quad Q_{el}(f^{(0)}, f^{(0)}) = 0 \\
\varepsilon^{0} &: \quad v_x \partial_x f^{(0)} = Q_{el}(f^{(1)}, f^{(0)}) + Q_{ve}(f^{(0)}, f^{(0)}) \\
\varepsilon^{1} &: \quad v_x \partial_x f^{(1)} = Q_{el}(f^{(2)}, f^{(0)}) + Q_{el}(f^{(1)}, f^{(1)}) + Q_{ve}(f^{(1)}, f^{(0)}).
\end{aligned}$$

The solvability conditions for $f^{(1)}$ yield

$$0 = \left\langle Q_{ve}(f^{(0)}, f^{(0)}) \right\rangle_{v,e} \qquad \partial_x(\rho T) = 0$$

$$0 = \left\langle v^2 Q_{ve}(f^{(0)}, f^{(0)}) \right\rangle_{v,e} \qquad 0 = \left\langle k(e) Q_{ve}(f^{(0)}, f^{(0)}) \right\rangle_{v,e}.$$

If we define $J(g)(e) = \left\langle Q_{ve}(h^{(0)}g, h^{(0)}g) \right\rangle_v$, we can write

$$0 = \int_{\mathbf{R}_0^+} \begin{pmatrix} 1 \\ k(e) \end{pmatrix} J(g)\, de.$$

which means, that the functions 1, $k(e)$ are orthogonal to $J(g)$. If we choose a basis b_i, s.t. $\left\langle b_i b_j \right\rangle_e = \delta_{ij}$, then $J(g) \perp b_i$ for all i. Moreover, if the basis b_i is complete, then $J(g) = 0$, i.e. $g \in \ker J$, but we obtain no more information on g.

One can easily show, that the Maxwellian is contained in $\ker J$ and numerical experiments indeed indicate, that the function g is a Maxwellian. The solvability conditions for $f^{(2)}$ yield

$$\begin{aligned}
\partial_x(\lambda(\rho,T)\partial_x T) &= \left\langle v^2\, Q_{ve}(h^{(1)}g, h^{(0)}g) \right\rangle_{v,e} \\
\left\langle k(e) v_x \partial_x f^{(1)} \right\rangle_{v,e} &= \left\langle k(e)\, Q_{ve}(h^{(1)}g, h^{(0)}g) \right\rangle_{v,e}.
\end{aligned}$$

As for $\delta = 0$, we obtain a constant pressure from the solvability conditions. The heat equations are similar as before except that the first order approximation of f appears in the source and loss term.

3.4 Numerical Results

Some additional simulations have been performed to test other scaling relations between the two parameters ε and δ numerically. All simulations used a fixed parameter (Knudsen number) $\varepsilon = 0.01$ and 25.600 particles. The boundary conditions are given by (3.22) with $T_L = T_R = 1.0$ and $T_{L,i} = T_{R,i} = 3.0$ and we used again the rigid–spherical molecule model. The different values for δ are $\delta = 0.0, 0.001(= \varepsilon^2), 0.01(= \varepsilon), 0.1(= \sqrt{\varepsilon}), 0.9(= 1 - \sqrt{\varepsilon}), 1.0$.

In Fig. 3 the curves are nearly constant, since there is no energy exchange between the kinetic and internal degrees of freedom. Small deviations appear in Fig. 4, which corresponds to the macroscopic equations. The case $\delta = 0.01$ (Figure 5) seems to be very interesting, since the curves just touch each other. Figs. 6–8 show that the kinetic and rotational temperature are equal in a domain between the walls. If we increase the δ-value, then this domain becomes larger.

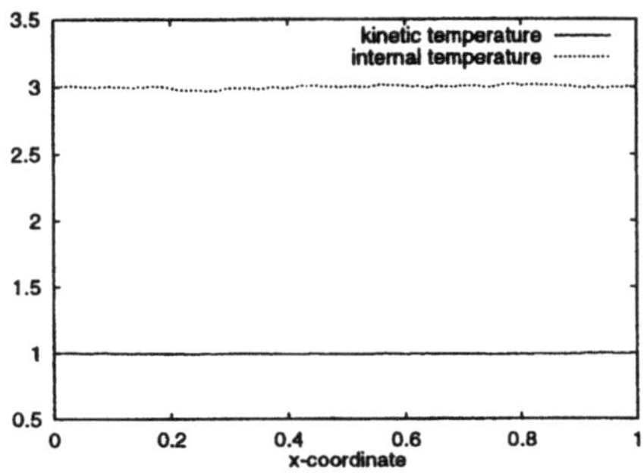

Figure 3 $\delta = 0.0, \varepsilon = 0.01$

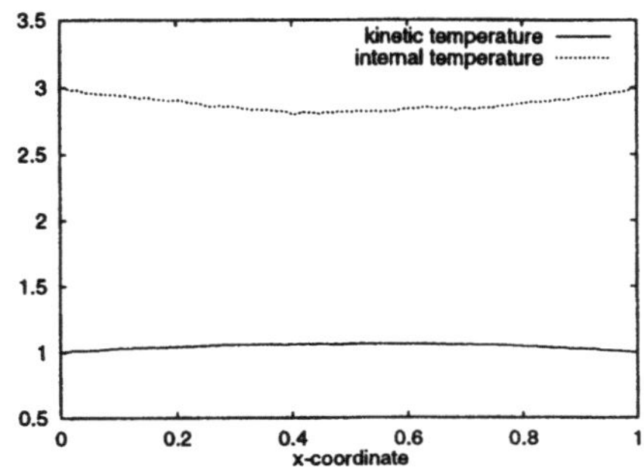

Figure 4 $\delta = 0.0001, \varepsilon = 0.01$

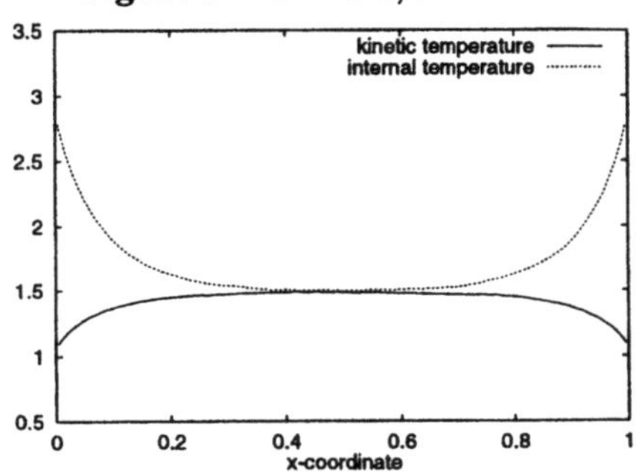

Figure 5 $\delta = 0.01, \varepsilon = 0.01$

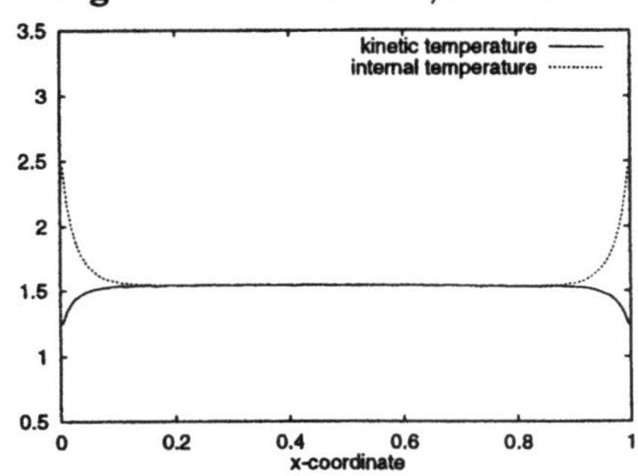

Figure 6 $\delta = 0.1, \varepsilon = 0.01$

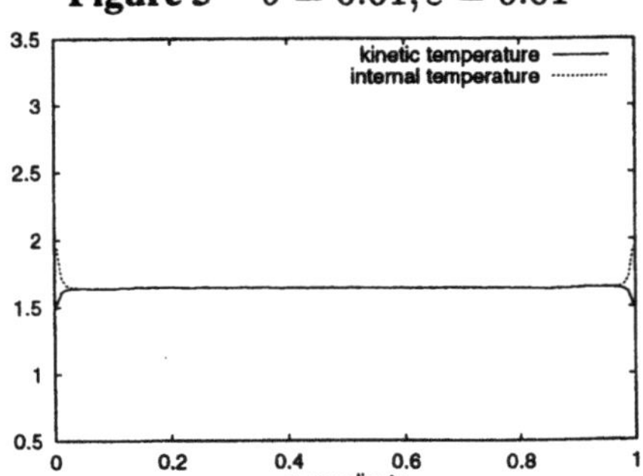

Figure 7 $\delta = 0.9, \varepsilon = 0.01$

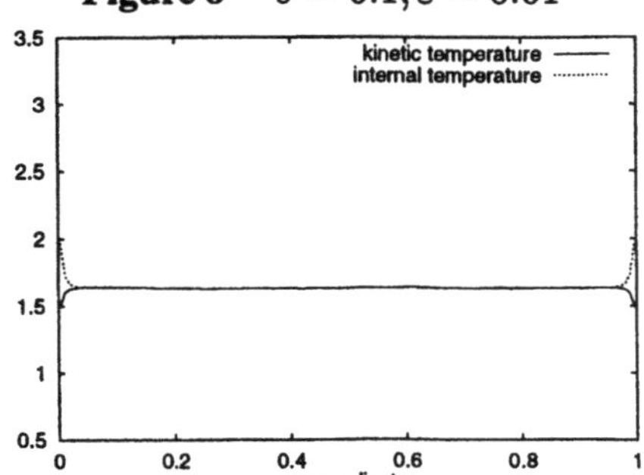

Figure 8 $\delta = 1.0, \varepsilon = 0.01$

4 Navier-Stokes Limit by Levermore Expansion

4.1 Approximation by Moment Expansions

The basic construction of an approximation in moments of the Boltzmann equation starts by choosing a linear space $\mathcal{E}_m$ of test functions in (v,e) and integrating the Boltzmann equation at each spatial point x against basis functions $m(v,e)$ of $\mathcal{E}_m$, which yields the system

$$\partial_t \left\langle mf \right\rangle_{v,e} + \operatorname{div} \left\langle v \otimes mf \right\rangle_{v,e} = \left\langle mQ(f,f) \right\rangle_{v,e}. \tag{4.1}$$

For physical reasons, the choice $\mathcal{E}_m$ of test functions cannot be arbitrary: first, because the resulting system must satisfy fundamental conservation laws in mass, momentum and energy, this space must contain the functions $m = \{1,v,v^2+e\}$, whose integral moments $\left\langle mf \right\rangle_{v,e}$ define the mass, velocity and total energy of the fluid. Second, the final system must be invariant under space translation and rotation, which means that the test space $\mathcal{E}_m$ itself must be stable under the action of any arbitrary translation and rotation.
The problem is, that the system (4.1) is not closed: equations above characterize the evolution of the averages $\left\langle mf \right\rangle_{v,e}$ by other averages $\left\langle v \otimes mf \right\rangle_{v,e}$ and the whole system can therefore only be solved, if one has additional information on the structure of the density function f. Different possibilities may be proposed at this level. Grad [5] builds the distribution function f using thirteen independent Hermite polynomials in v (for the monoatomic case). The resulting model predicts locally reasonable velocity jumps at solid boundaries, but it leads to a function f, which is not always positive. As a consequence, the notions of entropy and hyperbolicity are lost and the resulting boundary value problem is not well posed. A better choice, introduced in [13], which will be used in the following, is based on the Ansatz

$$F = F(\alpha) = \exp(\alpha \cdot m(v)),$$

with $m(v) = \{1,v,v \otimes v,v^2 v,v^4\}$ for the monoatomic case
and $m(v) = \{1,v,v \otimes v,I^2,(v^2+2I^2)v,(v^2+2I^2)^2\}$ for the diatomic case.
By construction, $F(\alpha) = \exp(\alpha \cdot m(v))$ is strictly positive. Moreover, it has nice entropy properties, since, as observed in [13], this distribution function is the formal solution of the entropy minimization problem

$$F(\alpha) = argmin_{f \in X_U} \left\langle f \ln f - f \right\rangle_{v,e}$$

over the set X_U of distributions with given moments U, i.e. the set $X_U = \{f : \left\langle mf \right\rangle_{v,e} = U\}$.

Substituting this ansatz in the moment equation (4.1), we obtain, that the kinetic moments $U_\alpha(\alpha)$, the kinetic fluxes $A_\alpha(\alpha)$ and the collision integrals $S(\alpha)$ with respect to the function $F(\alpha) = \exp(\alpha \cdot m)$ satisfy the system of partial differential equations with unknown α (or U_α) in $\mathbf{R}^{14}$

$$\partial_t U_\alpha(\alpha) + \operatorname{div} A_\alpha(\alpha) = S(\alpha). \tag{4.2}$$

The weak formulation of the above conservative system (4.2) is simply obtained by multiplying the system by a test function $\phi(x)$ and integrating by parts in x, which yields

$$-\int_\Omega \phi \partial_t U_\alpha \mathrm{d}x + \int_\Omega \partial_x \phi \cdot A_\alpha \mathrm{d}x - \int_{\partial\Omega} \phi \cdot \{A_\alpha \cdot n\} \mathrm{d}\gamma = -\int_\Omega \phi \cdot S_\alpha dx, \forall \phi(x). \tag{4.3}$$

The weak formulation (4.3) is certainly not well posed, because it lacks of boundary conditions specifying the boundary flux

$$A_\alpha \cdot n|_{\partial\Omega} := \int_{v \cdot n} v \cdot n\, m(v) F(v) dv de.$$

To do this while respecting the underlying kinetic boundary conditions imposed on the distribution f, a simple and consistent choice is to split this boundary flux into ingoing and outgoing fluxes

$$A_\alpha \cdot n = A_\alpha^+ + A_\alpha^-, \quad A_\alpha^+ := \int_{v \cdot n \geq 0} v.n\, m(v) F(v) dv de \tag{4.4}$$

with outgoing fluxes computed from the local value of the solution F and ingoing fluxes from the imposed reflected values at any wall

$$A_\alpha^- = \int_{v \cdot n \leq 0} v \cdot n\, m(v) \left[(1-k) \cdot M_{\rho, T_p, u} + k \int_{v' \cdot n > 0} \mathcal{R}(v, v') F(v') dv' \right] dv de \tag{4.5}$$

or imposed incoming values at infinity

$$A_\alpha^- = \int_{v \cdot n \leq 0} (v \cdot n)\, m(v) F_{imp}(v) dv de. \tag{4.6}$$

The above notion of ingoing and outgoing fluxes is in fact related to the microscopic notion of ingoing or outgoing particles. The corresponding boundary condition amounts in fact to average the microscopic kinetic boundary condition against the basis test functions $m(v)$.

With this choice, the problem to solve is the weak nonlinear variational problem (4.3) with boundary conditions (4.4)–(4.5)–(4.6).

4.2 Relation to Navier-Stokes

Easy algebraic manipulations transform the Levermore monoatomic system into five conservation laws (with $p = -Tr(\sigma)/3$)

$$\frac{\partial \rho}{\partial t} + u_k.\frac{\partial \rho}{\partial x_k} + \rho\frac{\partial u_k}{\partial x_k} = 0,$$

$$\frac{\partial u_i}{\partial t} + u_k.\frac{\partial u_i}{\partial x_k} - \frac{1}{\rho}\frac{\partial \sigma_{ik}}{\partial x_k} = 0,$$

$$\frac{\partial p}{\partial t} + \frac{\partial u_k p}{\partial x_k} - \frac{2}{3}\sigma_{jk}.\frac{\partial u_j}{\partial x_k} + \frac{2}{3}\frac{\partial q_k}{\partial x_k} = 0,$$

and eight plus one differential constitutive laws in stress, heat vector and an additional moment $\mathcal{R}_{jjik} = \left\langle |v|^4 f \right\rangle_v$.

$$\frac{\partial \sigma_{ij}}{\partial t} + \frac{\partial u_k \sigma_{ij}}{\partial x_k} + \sigma_{jk}.\frac{\partial u_i}{\partial x_k} + \sigma_{ik}.\frac{\partial u_j}{\partial x_k} - \frac{\partial Q_{ijk}}{\partial x_k} = \frac{1}{\epsilon}\left\langle v_i v_j Q(F,F)\right\rangle_v \quad (4.7)$$

$$\begin{aligned} &2\frac{\partial q_i}{\partial t} + 2\frac{\partial q_i u_k}{\partial x_k} + 2Q_{ijk}\frac{\partial u_j}{\partial x_k} + 2q_k\frac{\partial u_i}{\partial x_k} \\ &+3\frac{p}{\rho}\frac{\partial \sigma_{ik}}{\partial x_k} - 2\frac{\sigma_{ij}}{\rho}\frac{\partial \sigma_{jk}}{\partial x_k} + \frac{\partial \mathcal{R}_{jjik}}{\partial x_k} = \frac{1}{\epsilon}\left\langle v^2 v_j Q(F,F)\right\rangle_v \end{aligned} \quad (4.8)$$

$$\frac{\partial \mathcal{R}_{iijj}}{\partial t} + \frac{\partial \mathcal{R}_{iijj} u_k}{\partial x_k} + 4\mathcal{R}_{jjik}\frac{\partial u_i}{\partial x_k} + 8\frac{q_i}{\rho}\frac{\partial \sigma_{ik}}{\partial x_k} + \frac{\partial s_k}{\partial x_k} = \frac{1}{\epsilon}\left\langle v^4 Q(F,F)\right\rangle_v. \quad (4.9)$$

These constitutive laws are no longer explicit : they involve auxiliary moments Q and $\mathcal{R}_{jjik}$ to be obtained by first computing the underlying distribution function F^{14}. But nevertheless, this nonconservative writing gives a direct way to recover Navier–Stokes equations from the Levermore equations. Indeed, let us perform a standard asymptotic expansion of the unknowns in terms of the mean free path ε , i.e. $F = F^0 + \epsilon F^0 \alpha^1 \cdot m + O(\epsilon^2)$, where $F^0 = \exp(\alpha^0 \cdot m)$ and

$$\sigma_{ij} = \sigma^0_{ij} + \epsilon\sigma^1_{ij} + O(\epsilon^2),$$

$$Q_{ijk} = Q^0_{ijk} + \epsilon Q^1_{ijk} + O(\epsilon^2),$$

$$\mathcal{R}_{iijk} = \mathcal{R}^0_{iijk} + \epsilon\mathcal{R}^1_{iijk} + O(\epsilon^2),$$

with the normalization condition

$$\left\langle \Psi(v)\exp[\alpha^0 \cdot m(v)]\right\rangle_v = \left\langle \Psi(v)\exp[\alpha \cdot m(v)]\right\rangle_v \quad (4.10)$$

and $\Psi(v) = (1,v,v^2)^t$. At zero order, α^0 must cancel the collision term, and therefore F^0 correspond to a Maxwellian, yielding

$$\sigma^0_{ij} = -\rho RT\delta_{ij}, q^0_i = 0, \mathcal{R}^0_{iijj} = 15\rho R^2T^2.$$

At next order, because of (4.10), we have first $\left\langle \alpha^1 \cdot m(v)F^0 \right\rangle_v$ and after elimination of all time derivates by the conservation laws in a similar expansion of the constitutive laws (4.7,4.9,4.9), on gets the following equations in α^1,

$$\rho RT(\frac{\partial u_j}{\partial x_i} + \frac{\partial u_i}{\partial x_j} - \frac{2}{3}\delta_{ij}\frac{\partial u_k}{\partial x_k}) = \int_{v\in\mathbf{R}^3} 2v_iv_jQ(F^0\alpha^1 \cdot m,F^0)dv,$$

$$5\rho RT\frac{\partial RT}{\partial x_i} = \int_{v\in\mathbf{R}^3} 2v^2v_iQ(F^0\alpha^1 \cdot m,F^0)dv,$$

$$0 = \int_{v\in\mathbf{R}^3} 2v^4Q(F^0\alpha^1 \cdot m,F^0)dv.$$

By usual linearity and symmetry arguments (even-odd symmetry of integrals in v, invariance, linearity in $\alpha^1 \cdot m$ of the collision operator $Q(F^0\alpha^1 \cdot m,F^0)$), the solution in $\alpha^1 \cdot m$ is the standard Chapman-Enskog distribution

$$\epsilon\alpha^1 \cdot m(v) = -\frac{2}{5}\frac{\kappa}{\rho(RT)^2}(\frac{c^2}{2RT} - \frac{5}{2})c \cdot \frac{\partial T}{\partial x_i}$$
$$-\frac{\mu}{\rho(RT)^2}(c\otimes c - \frac{1}{3}c^2Id) : (\frac{1}{2}(\frac{\partial u}{\partial x} + \frac{\partial u}{\partial x}^T - \frac{2}{3}div(uId))$$

yielding at first order in ϵ the Navier-Stokes constitutive laws

$$\epsilon\sigma^1_{ij} = -\epsilon\left\langle c_ic_jF^0\alpha^1 \cdot m\right\rangle_v = \mu(\frac{\partial u_j}{\partial x_i} + \frac{\partial u_i}{\partial x_j} - \frac{2}{3}\delta_{ij}\frac{\partial u_k}{\partial x_k}),$$

$$\epsilon q^1_i = \epsilon\left\langle c^2c_iF^0\alpha^1 \cdot m\right\rangle_v = -\kappa\frac{\partial T}{\partial x_i},$$

and the additional law

$$\epsilon\mathcal{R}^1_{iijj} = \epsilon\left\langle c^4F^0\alpha^1 \cdot m\right\rangle_v = 0.$$

This means that any solution ρ, u, T, σ, q of the Navier-Stokes equations is a first order approximate solution

$$U^{NS}_\alpha = \begin{cases} \rho \\ \rho u \\ \rho u\otimes u - \sigma \\ \rho u^2u - tr(\sigma)u - 2\sigma\cdot u + 2q \\ \rho u^4 - 2tr(\sigma)u^2 - 4(\sigma\cdot u).u + 8q\cdot u + \frac{15p^2}{\rho} \end{cases}$$

of the fourteen moment Levermore's equation, and conversely.

In particular, we get by this technique a new kinetic interpretation of a Navier-Stokes solution: its corresponding moments U_α^{NS} are an approximate solution of a consistent entropic hyperbolic problem and are moments of a positive distribution function $F_{NS}(v) = \exp(\alpha \cdot m(v))$ with coefficients α given by

$$\left\langle m \exp(\alpha \cdot m(v)) \right\rangle_v = U_\alpha^{NS}.$$

The difference with more standard models is, that now the definition of F_{NS} is strictly positive, but no longer explicit and can only be obtained by solving a local entropy minimization problem.

4.3 Numerical Coupling Strategy

The additional advantage of the above kinetic interpretation is to lead to a simple multi–domain coupling strategy, which

1. solves the hyperbolic Levermore's model everywhere with semi–explicit schemes in time (collision implicit, transport terms explicit or linearly implicit) and finite volume in space,

$$U_{\alpha,j}^{n+1} - U_{\alpha,j}^{n} = -\frac{\Delta t}{\text{Vol } j} \sum_i (\mathcal{A}_{j,i}^n + \Delta t S(U_{\alpha,j}^{n+1}).$$

2. gegenerate locally the Levermore's constitutive laws to their asymptotic analytic Navier Stokes limit in all cells, except where analytic residuals of Levermore differential constitutive equations are large.

In this strategy, the interface fluxes between different cells are then given by using a standard flux splitting scheme for interfaces between two Navier-Stokes cells and a kinetic reconstruction F_{NS}, which preserves positivity and entropy, for all other interfaces

$$\mathcal{A}_{j,i} = \int_{t^n}^{t^{n+1}} \int_{\partial C_j \cup \partial C_i} v \cdot n F_{NS}^n(x - (t^{n+1} - t^n)v, v) d\gamma$$

with $\mathcal{A}_{ji}^n = A_{ji}^{n,+} + A_{ij}^{n,-}$, $A_{ji}^{n,+} = \int_{v.n_{ji} \geq 0} v \cdot n_{ji}\, m(v) F_{NS}^n(x_j, v) dv$ and $A_{ji}^{n,-}$ computed from outside Levermore's distribution. The practical implementation of the above coupled scheme still remains rather technical because

1. first, it involves nonlinear integrals in the velocity space, which can only be obtained by numerical quadrature [17], with a rule which must preserve local equilibria and must be used consistently everywhere (including at kinetic boundary conditions).

2. second, as observed in [7], for strong non–equilibrium states (inside strong shocks), there might be no coefficients α_k generating a given moment distribution U_k in step 1.

The simulation below considers a two dimensional flow of a monoatomic gas over a flat plate at an angle of attack of 10^0, Mach number 18.62, Reynolds number 30687 ($Kn = 1 \cdot e^{-3}$), temperature $194^o K$ at infinity and $1000^o K$ at wall. The viscosity is modeled by a hard sphere model, with $R = 287$. The gas is modeled locally by a Navier–Stokes model away from the boundary and a Levermore's model next to the wall. We present iso-density lines (Fig. 9) and temperature, Mach values at three cross–sections at $x = .25$, $x = .55$ and $x = .85$ (Fig. 10), where the results are compared with those of a full Boltzmann solution.

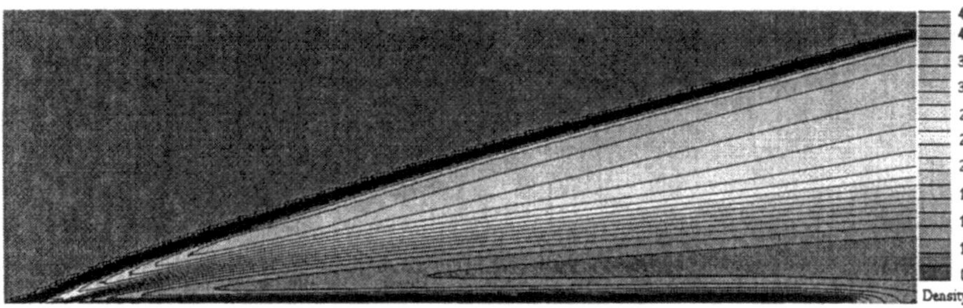

Figure 9 Iso–Density Values

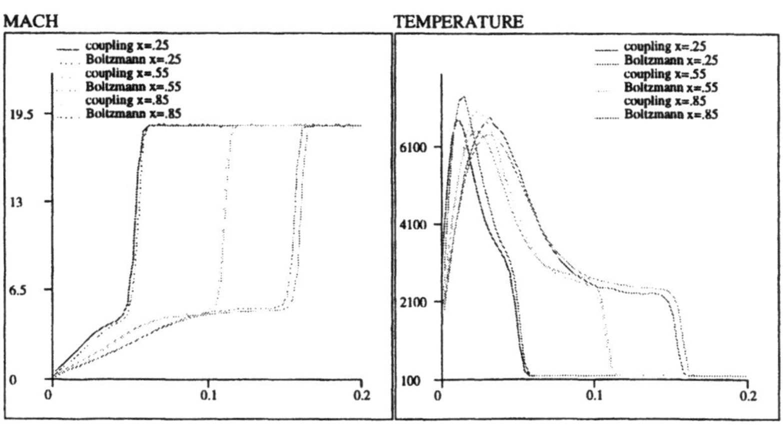

Figure 10 Mach and Temperature Values
Comparison with a Simulation for the Full Boltzmann Equation

5 Conclusion

We presented some preliminary results on the derivation of fluid dynamic limits for the generalized Boltzmann equation with real gas effects. In particular, we studied the steady–state, one–dimensional Boltzmann equation with one additional internal energy modelled as a continuous variable. The different relaxation times are related using a small parameter δ, which was later on related to the mean free path of the gas, assuming a fixed functional dependence between ε and δ. The resulting Navier–Stokes equations turned out to be enlarged by an additional temperature equation for the rotational temperature, with transport coefficients defined similar to the monoatomic case. Some numerical simulations were given to validate the asymptotic method to derive the fluid dynamic equations.
In the second part of the paper we present a new approach to define coupling procedures for the Boltzmann and Navier–Stokes equations, which are based on the 14–moments expansion of Levermore. In this expansion method, the density function remains strictly positive, which yields clear advantages with respect to the classical Grad's moment expansion. Moreover, we derived a new kinetic interpretation for the 14–moments method of Levermore and give numerical simulation results on the resulting domain decomposition approach.

References

[1] Bourgat, J.–F., Desvillettes, L., Le Tallec, P. and Perthame, B.: Microreversible collisions for polyatomic gases and Boltzmann's theorem, European Journal of Mechanics, B/Fluids, 13,no 2, 237-254 (1994).

[2] Charrier, P., Dubroca, B. and Feugeas, J.L.: Etude numérique de modèles aux moments de Levermore en 2 dimensions, personal communication, 1996.

[3] Chapman, S. and Cowling, T.G.: The Mathematical Theory of Non-uniform Gases, Cambridge University Press (3th edition, 1970).

[4] Ferziger, J.H. and Kaper, H.G.: Mathematical theory of transport processes in gases, North-Holland (1972).

[5] Grad, H.: On the Kinetic Theory of Rarefied Gazes, Comm. Pure Appl. Math 2, pp 331-407, 1949.

[6] Hash, D. and Hassan, A.: A Hybrid DSMC/Navier-Stokes Solver, AIAA 95-0410.

[7] Junck, M.: About the domain of definition of Levermore's five moment system, personal communication, June 1997.

[8] Kuščer, I.: Dissociation and Recombination in an Inhomogeneous Gas, Physica A, 176, 542-556 (1991).

[9] Le Tallec, P. and Mallinger, F.: Coupling Boltzmann and Navier-Stokes equations by half fluxes, Journal of Computational Physics, 136, 51-67, 1997.

[10] Le Tallec,P. and Perlat, J.-P. : Coupling Kinetic Models with Navier-Stokes equations, to appear in CFD Review, 1998.

[11] Le Tallec, P. and Perlat, J.-P., " Asymptotic Kinetic Models for Transitional Flows ", Proceedings of the " International Conference on Numerical Modelling in Continuum Mechanics ", M. Feistauer ed., Prague 1997.

[12] Le Tallec, P. and Perlat, J.-P. : Numerical Analysis of Levermore's Moment System, Rapport de recherche INRIA 3124 , Mars 1997.

[13] Levermore, D.: Moment Closure Hierarchies for Kinetic Theories, Department of Mathematics, University of Arizona,*submitted to the journal of statistical physics, May 1995.*

[14] Levermore, D. and Morokoff, W.J.: The Gaussian Moment Closure for Gas Dynamics, *SIAM J. on Applied Mathematics, submitted February 1996.*

[15] Levermore, D.: Entropy Based Moment Closures for Kinetic Equations. *Transport Theory and Statistical Physics, April 1996.*

[16] Neunzert, H. and Struckmeier, J.: Particle Methods for the Boltzmann Equation, ACTA NUMERICA 1995, Cambridge (1995).

[17] Perlat, J.P.: Modélisation et Calcul parallèle d'une couche limite cinétique, Universit Pierre et Marie Curie, Paris VI, Janvier 1998.

[18] Sack, W.: Modellierung und Numerik für reaktive Strömungen in verdünnten Gasen, PhD thesis, University of Kaiserslautern, October 1995.

[19] Tiwari, S. and Klar, A.: An Adaptive Domain Decomposition Procedure for Boltzmann and Euler Equations, to appear in Journal of Comp. & Appl. Math.

Numerical Techniques for Multi-Scale Weakly Compressible Reactive Flows

Rupert Klein
FB Mathematik & Informatik
Freie Universität Berlin
Arnimallee 2-6
14195 Berlin, Germany

Karl Josef Geratz, Peter Terhoeven
Institut für Technische Mechanik
RWTH Aachen
Templergraben 64
52068 Aachen, Germany

Claus-Dieter Munz, Sabine Roller
Institut für Aerodynamik und Gasdynamik
Universität Stuttgart
Pfaffenwaldring 55
70550 Stuttgart, Germany

Bruno Denet, Geoffrey Searby
Institute de Recherche des Problèmes Hors Equilibre, CNRS UMR 138
Universites d'Aix-Marseille I & II
Centre Universitaire de Saint Jèrôme -Service 252-
13397 Marseille Cedex 20, France

Summary

The numerical simulation of flame-acoustic interactions is very demanding due to the multiple length scales involved: There are chemical scales of about 1/10 mm and vortical flow structures of about 1-10 cm, while acoustic wave lengths are tens of centimeters to several meters.

Here we describe the current state of development of numerical techniques allowing us to overcome these scaling discrepancies. These include a multi-scale grid representation of long wave acoustics, its coupling to a quasi-incompressible flow solver and a capturing/tracking hybrid scheme for flame front discontinuities in the zero Mach number limit.

Experiments and theory reveal a stabilizing effect of acoustics at low forcing but a parametric flame instability at higher forcing. Using quasisteady flame jump conditions, as in the flame tracking method, one obtains these effects theoretically, but he fails to correctly assess the influence of Lewis number. Improved insight is gained by a more expensive numerical approach which resolves the flame structure. These computations correctly capture the influence of Lewis number.

We conclude that a flame front tracking approach must resort to modified jump conditions for highly unsteady flows in order to correctly operate under high acoustic forcing.

1 Introduction

Acoustic combustion instabilities induce unacceptable noise in technological combustion devices, generate turbulence and lead to flame acceleration including transition to detonation. Prominent examples are the knocking combustion in spark ignition engines and flow noise in the exhaust systems of large scale industrial burners. Of interest is purely gaseous combustion as well as flames in aerosol/air mixtures. The striking differences between these systems in terms of flame-acoustic interactions are as yet understood only qualitatively. Numerical simulation of these phenomena is very demanding due to the multiple length scales involved: There are, e.g., chemical scales on the order of 1/10 mm and vortical flow structures with a few centimeters in diameter, while acoustic wave lengths are tens of centimeters to several meters.

In this paper, we will describe numerical techniques to overcome the scaling discrepancies between the vortical flow patterns and typical reaction length scales on the one hand and between long wave acoustics and the small scale flow on the other. These techniques include a physics-induced multi-scale grid representation of long wave acoustics, its coupling to a quasi-incompressible reactive flow solver, and the implementation of a capturing/tracking hybrid scheme for flame front discontinuities in compressible flows that has been extended to cope with the zero Mach number limit. This approach produces very promising first results in that it exhibits a stabilization of the Darrieus-Landau instability under low acoustic forcing. This is in line with experimental evidence as well as with theoretical predictions.

However, a more detailed study of unsteady effects within the flame structure under higher forcing reveals inherent limitations: At higher acoustic forcing a parametric instability arises. It is shown through experiments and numerical simulations with resolved flame structures that the properties of this instability are essentially independent of Lewis number. While a numerical approach treating the flame as a standard Rankine-Hugoniot discontinuity *will* exhibit such a parametric instability, it will fail to correctly describe the fading influence of Lewis number for higher acoustic forcing. Following a suggestion by Joulin, we conclude that our flame front tracking/capturing hybrid scheme will need to be extended so as to account for modified jump conditions for highly unsteady flows.

2 Multiple Pressure Methods for low Mach number flows

Our Multiple Pressure Methods are based on an asymptotic analysis used to gain insight into the behavior of the solutions of the compressible Euler equations in the limit of vanishing Mach number, i.e., $M \to 0$. To cover small scale flow structures as well as the long-wave acoustic phenomena that are important for flame-acoustic interactions, a single time scale, multiple space scale analysis is performed. All flow variables f are expanded in terms of powers of the Mach number,

$$f(x,t;M) = f^{(0)}(x,\xi,t) + Mf^{(1)}(x,\xi,t) + M^2 f^{(2)}(x,\xi,t) + \dots \tag{2.1}$$

where

$$\xi = M\,x\,. \tag{2.2}$$

The resulting detailed expansion of the compressible Euler equations can be found in [2] for conservative variables and in [3] for primitive variables. Here we will merely provide a brief summary of the key results that directly affect the development of our numerical methods.

2.1 Results of the asymptotic analysis

The main result of the asymptotics concerns the role of "the pressure". It turns out that there is no longer one single pressure term to influence the leading order velocity, but that all the first three terms $p^{(0)}, p^{(1)}$, and $p^{(2)}$ within the pressure expansion

$$p = p^{(0)} + Mp^{(1)} + M^2 p^{(2)} + o(M^2) \tag{2.3}$$

have a leading order influence on the velocity field. The asymptotics allows a clean separation of the different physical effects associated with these pressure terms:

- The leading order term $p^{(0)} = P_0(t)$ is spacially homogeneous and acts as a thermodynamic variable satisfying the equation of state. It may retain time dependence if there is a global compression from the boundaries or if a reacting flow is enclosed in a constant volume chamber.

- The first order pressure $p^{(1)} = P^{(1)}(\xi, t)$ represents an acoustic amplitude. It is homogeneous in the small scale variable x just as $p^{(0)}$. Yet, it retains variation on the ξ-scale, thereby accounting for long-wavelength acoustic effects. Note that on bounded domains in terms of x, no acoustic wave can be accommodated for sufficiently small M and that $P^{(1)}$ may be absorbed in $P_0(t)$ in that case.

- The second order variable $p^{(2)}$is a balance-of-forces agent. It guarantees compliance with the elliptic divergence-constraint on the velocity field as $M \to 0$. In the limit, it decouples completely from the total pressure and, hence, from the equation of state.

We conclude that a numerical scheme for the low Mach number regime has to include at least two different pressure variables, the global term $P_0(t)$ and the second order term $p^{(2)}(x,\xi,t)$. If acoustic waves occur, the first order term $P^{(1)}(\xi,t)$ has to be included in addition. None of these terms can account for the effects of the other correctly for sufficiently small Mach number. It is conceivable to run a computation without such a pressure decomposition, but then the results will depend on the chosen numerical precision and generally on the machine the computation is run on. In order to avoid these uncertainties, we include a three-term pressure decomposition explicitly in our computational scheme for all subsonic Mach numbers.

2.2 Coarse-grid acoustic predictor

Since the asymptotic analysis is valid only for very small Mach numbers, we do not solve the asymptotic equations, but we use them as a predictor in a scheme that discretizes the full compressible flow equations. In fact, the corrector step will guarantee consistency with the complete compressible Euler (or Navier-Stokes) equations.

The predictor consists of three steps. First, the total pressure is decomposed into its contributions. Since the asymptotic analysis showed $p^{(0)}$ to be constant in space, and $p^{(1)}$ to vary only on the large acoustic scale, the decomposition is performed by standard averaging procedures over the computational domain and the acoustic scales, respectively. All global effects will then be captured by $p^{(0)}$, all large scale variations in $p^{(1)}$. The kinematic pressure $p^{(2)}$ is then defined as

$$p^{(2)} = \frac{1}{M^2}\left(p - p^{(0)} - Mp^{(1)}\right), \tag{2.4}$$

leading to a consistent decomposition of the total pressure.

Next, the time evolution of the global background pressure $p^{(0)}$ is computed by accounting for compression from the boundaries where either velocities are known or the global pressure is prescribed. Hence, this step essentially reduces to the integration of an ordinary differential equation.

The third step accounts for the evolution of the long wave effects by integrating the equations of linearized acoustics obtained from the asymptotic analysis,

$$\begin{array}{lcll} \overline{\vec{v}}_t & + & \frac{1}{\overline{\rho}}\nabla_\xi p^{(1)} & = 0, \\ p_t^{(1)} & + & \gamma p^{(0)}\nabla_\xi \cdot \overline{\vec{v}} & = 0. \end{array} \tag{2.5}$$

By construction, the data to which this operator is applied involve only the long acoustic wavelength. Short wave flow components are filtered out prior to solving the relevant equations. This is denoted by overbars, and ∇_ξ denotes the derivative with respect to the large scale variable. The evolution time scale matches that of the convective phenomena in the regime considered.

Despite this, a very restrictive CFL time step constraint would have to be observed for the fast acoustic signals if a standard explicit scheme were used on the given computational mesh. The more or less standard ansatz to overcome this problem involves solving the acoustic equations implicitly on the original mesh. Another approach uses an explicit acoustic solver, but avoids exceedingly small time steps by taking advantage of the long-wave nature of the data: Without loss of overall accuracy these data can be represented on a much coarser grid. Using an explicit solver on this coarse grid amounts to a CFL-based time step comparable to that for the small scale convection on the fine grid.

This approach reduces the computational costs in two ways. First, it allows us to stick to an explicit scheme and secondly it requires a complete updating procedure only on a much sparser grid matrix. On the other hand, it requires additional averaging and interpolation processes needed to manage the coarse-fine grid communication. The latter problems are related to those known from modern multi-grid techniques. In the regime considered here, these issues can be overcome by making explicit use of the knowledge gained through the asymptotic analysis. Due to the fact that the acoustic variables incorporate only large scale contributions, the only averaging necessary is that associated with the determination of the acoustic pressure and velocity fields. Straight injection is sufficient for the fine-to-coarse grid restriction, whereas for the coarse-to-fine grid prolongation straight-forward interpolation techniques with low numerical costs are sufficient. Fig. 1 shows an example that was calculated by this method. Errors introduced by the coarse-to-fine interpolation have small wavelength and are quickly captured by the $p^{(2)}$ terms.

Here, a right-running acoustic wave with wavelength of the length of the computational domain passes over a saw-tooth like transverse density layering which is thereby set into motion. The different density of two neighbouring points leads to a different acceleration under the force of the same acoustic pulse. This results in the occurence of a Kelvin-Helmholtz instability where one fluid is moving relative to the other. The initially horizontal interface starts rolling up into a hierarchy of vortical structures.

For one space dimension, a third approach has been used in [2], which is explicit and operates on the fine mesh. However, this "wave propagation method" does not straight-forwardly extend to more than one space dimension.

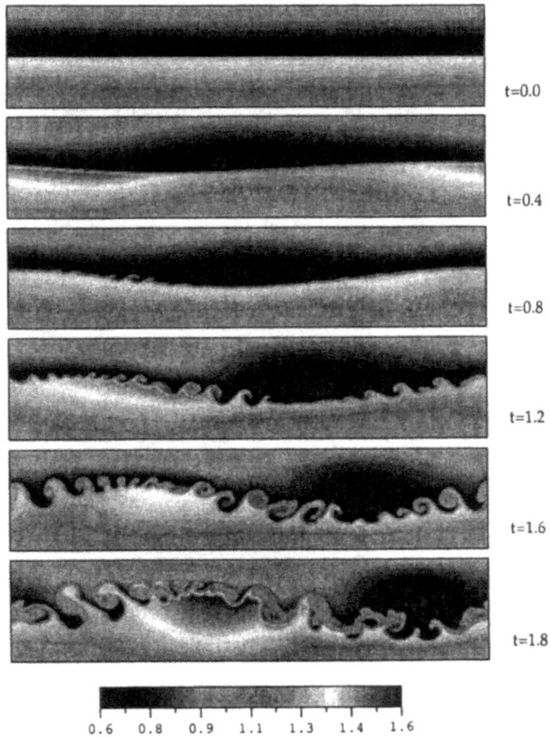

Figure 1 Temporal evolution of density layering under an acoustic pulse

2.3 Extension of incompressible flow solvers

In this section, we describe the corrector step as an extension of a standard incompressble flow solver to demonstrate clearly the connection between compressible and incompressible fluid flows in terms of their numerical representation. For this purpose, we use the compressible Euler equations in the primitive variable formulation, non-dimensionalized by choosing distinct reference values for fluid and sound velocity:

$$\begin{array}{lclcl} \rho_t & + & \nabla \cdot (\rho \vec{v}) & = & 0, \\ \vec{v}_t & + & (\vec{v} \cdot \nabla)\,\vec{v} + \frac{1}{M^2 \rho} \nabla p & = & 0, \\ p_t & + & \vec{v} \nabla p + \gamma p \nabla \cdot \vec{v} & = & 0. \end{array} \tag{2.6}$$

After insertion of the pressure expansion, terms involving $p^{(0)}$ and $p^{(1)}$ are collected on the rhs. The terms on the lhs will then survive the zero Mach number limit and lead to the known incompressible flow equations. A numerical approximation is applied to these terms that may be regarded as an extension of an incompressible method, as only the $p^{(2)}$-terms are left.

The right hand side contains the compressible effects, such as compression from the boundary and acoustic waves. The overall scheme might be considered as an incompressible flow solver with compressibility effects acting as source terms. In the zero Mach number limit these effects vanish by construction. Thus, the scheme in fact behaves uniformly for small but non-zero as well as for zero Mach number, [3, 4]. This is not the case for compressible algorithms acting on the total pressure. A detailed analysis of a SIMPLE-type compressible flow solver [1] revealed that, if only the overall pressure p is represented in a scheme, it will satisfy Laplace's equation with homogeneous boundary data in the limit. This would lead to a constant pressure which can not guarantee compliance with the divergence constraint for the incompressible velocity field. An appropriate discrete version of the pressure Poisson equation for incompressible flows is obtained only, if a separate equation for $p^{(2)}$ is derived and solved numerically. Thus, the pressure decomposition must necessarily be implemented numerically, if one intends to design a scheme that operates uniformly as $M \to 0$.

Several tests have been carried out to validate the inviscid as well as the viscous scheme. The example shown in Fig. 2 is an incompressible test case similiar to the acoustic example above. At Reynolds number $Re = 1500$, a constant density fluid is flowing from left to right at constant velocity. It is interrupted only by a narrow, centered jet flow in the opposite direction. Again, a Kelvin-Helmholtz instability occurs at the interface. The interface starts rolling up and a vortex street develops.

2.4 Zero Mach number flame front tracking

The capturing/tracking hybrid scheme from [5] for flame fronts in compressible flows has been extended to cope with zero Mach number flames as well. The key difficulty has been the discrete representation of the divergence constraint: A deflagration is a δ-singularity of flow divergence. The second order pressure $p^{(2)}$ (i.e., the "incompressible pressure") is discontinuous at the front. The associated problems have been overcome in [6] and a numerical scheme has been implemented successfully.

As in [5] the flame front evolution is followed explicitly by tracking the zero level set of a scalar function $G(x,t)$. Given that, the exact location of the pressure discontinuity is known and can be accounted for explicitly in the discretization of the relevant Poisson-type equation. In fact, whenever the Poisson stencil for one grid point includes the center of a cell lying across the flame front, a pressure jump is added to the neighboring pressure value. In this fashion, the Poisson stencil is never formed "across" the flame discontinuity.

The amplitude of the pressure jump is obtained, again as in [5], from an in-cell reconstruction procedure. Given the averages of mass, momentum and energy within a grid cell that is cut by the flame, it is possible to reconstruct the complete burnt and unburnt states using the following set of equations:

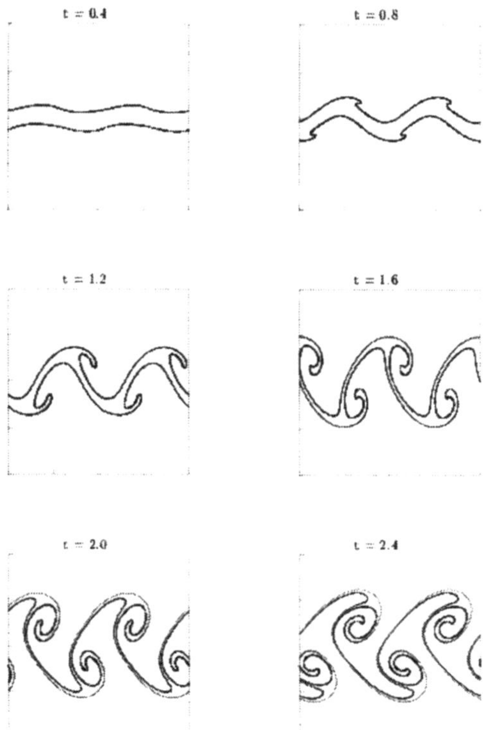

Figure 2 Time evolution a pair of vortex sheets bounding a 2D jet (on a square domain of unit size at Reynolds number 1500)

$$\bar{\mathbf{U}} = \alpha \mathbf{U}_u + (1-\alpha)\mathbf{U}_b, \tag{2.7}$$

$$D[\mathbf{U}] = \vec{n} \cdot [\vec{\mathbf{F}}(\mathbf{U})], \tag{2.8}$$

$$D = \vec{n} \cdot \vec{v}_u + s(\mathbf{U}_u)\,. \tag{2.9}$$

Here $\bar{\mathbf{U}}, \mathbf{U}_u, \mathbf{U}_b$ are the cell average, the unburnt and the burnt state vector of conservative variables, $\vec{\mathbf{F}}$ the flux tensor for these conservative variables, $\vec{v}_u$ the unburnt gas flow velocity immediately in front of the flame, $s(\mathbf{U}_u)$ the burning

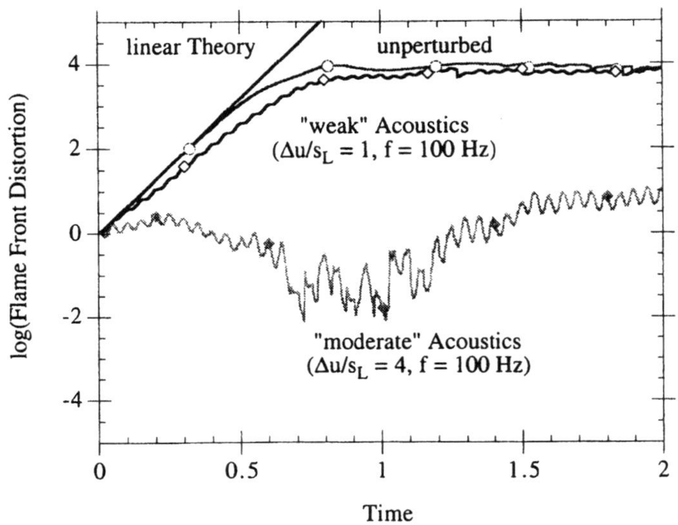

Figure 3 Temporal evolution of flame front perturbation amplitudes showing the suppression of the Darrieus-Landau instability by sufficiently strong acoustic acceleration (see Fig. 4 for graphs of the associated flame contours).

speed, D is the absolute front normal velocity. and α is the unburnt gas volume fraction within a "flame cell". Having solved this system of equations, the pressure jump across the flame is known as an implicit function of the unburnt gas state.

One of the test cases that is of primary importance for the present work concerns the influence of acoustic acceleration on flame stability. We have subjected an unstable flame in an open channel to oscillating accelerations at one end of the domain, thereby approximating the effects of long wave acoustics on the flow field next to the flame front. Figure 3 shows the amplitude histories of geometrical perturbations of an initially nearly plane flame for different levels of acoustic driving. The graph demonstrates that the numerical prediction is in agreement with the experimental observations by Searby et al. [11, 10]: Sufficiently strong (acoustic) oscillations suppress the standard Darrieus-Landau instability. At zero or sufficiently small acceleration amplitude the initial growth of perturbations closely follows the linear stability theory until nonlinear effects of front propagation lead to saturation. Figure 4 provides a collection of snapshots of the flame geometry for the three different oscillation amplitudes referred to in Fig. 3.

The capturing/tracking hybrid scheme for premixed flames in compressible flows from [5] has recently been extended to three space dimensions. As expected, the original concept from [5] could be applied without modifications. Merely technical details required some programming efforts. Results will be presented in a future report.

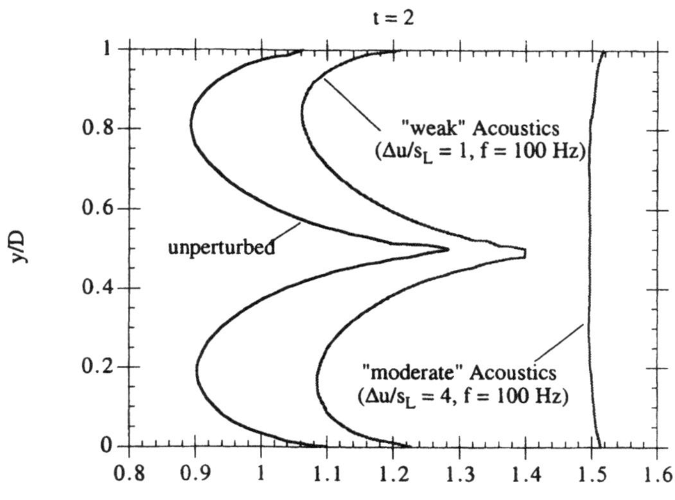

Figure 4 Snapshots of flame front geometries at time $t = 2.0$ for the cases considered in Fig. 3

3 High frequency response of a premixed flame

3.1 Markstein's model

The complete problem of flame acoustics interaction in a tube, involves a nonlinear feedback between geometrical flame evolution, unsteady development of the flame structure and weakly nonlinear acoustics. This complexity can be reduced in a simplified model if one follows the approach of Markstein. In this model, introduced in the fifties [7], one considers only the influence of the acoustic field on the flame, and not the retroaction of the flame on the acoustic amplitude. Although this is clearly an approximation not valid in the general case, an experiment has been designed by Searby [11] which is reasonably described by this model. In this experiment, the flame is stabilized in the tube close to a pressure node by tuning the mean flow, and furthermore submitted to an imposed acoustic field by using a loudspeaker located at the bottom of the tube. This configuration is thus rather different from the usual configuration of a propagating flame, where the flame itself amplifies the acoustic modes of the tube. In this section, we will describe a numerical study of Markstein's model, which extends the results of a first study of this model [14].

3.2 Equations and numerical method

We describe in this subsection the equations of premixed flames in the isobaric approximation, with a chemistry limited to a global one-step reaction. If we use as unit of length the flame thickness and as unit of time the transit time through the flame, we obtain the following equations:

$$\rho\frac{\partial T}{\partial t} + \rho(\vec{v}.\vec{\nabla})T = \Delta T + \Omega,$$

$$\rho\frac{\partial C}{\partial t} + \rho(\vec{v}.\vec{\nabla})C = \frac{1}{Le}\Delta C - \Omega,$$

with

$$\Omega = \frac{\beta^2}{2Le}C\frac{\rho}{1-\gamma}\exp\frac{\beta(T-1)}{1+\gamma(T-1)},$$

and

$$\frac{\partial\rho}{\partial t} + \vec{\nabla}.(\rho\vec{v}) = 0,$$

$$\frac{\partial(\rho\vec{v})}{\partial t} + \vec{\nabla}.(\rho\vec{v}\vec{v}) = -\vec{\nabla}P + \rho\vec{G} + \mu\Delta\vec{v} + (\lambda + \mu/3)\vec{\nabla}(\vec{\nabla}.\vec{v}),$$

$$\rho = (1 + \frac{\gamma}{1-\gamma}T)^{-1}.$$

T and C are the reduced temperature and mass fraction of the deficient reactant, made non-dimensional so as to vary from 0 to 1. Le, β and γ are respectively the Lewis number of the reactant (ratio of the thermal diffusivity to the molecular diffusivity of the deficient species), the reduced activation energy (or Zeldovich number) $\beta = E(T_b - T_u)/RT_b^2$ and the heat release parameter $\gamma = (\rho_b - \rho_u)/\rho_b$, the subscripts u and b referring to unburned and burned gases, ρ , v, P are respectively the density, velocity and pressure, G is a non-dimensional acceleration

$$G = gd/(u_l)^2 - \omega u\cos\omega t.$$

where u_l is the flame velocity, d the flame thickness, g the gravity acceleration, ω and u are the acoustic pulsation and amplitude. μ and λ are respectively the Prandtl numbers associated to shear and bulk viscosities.

In the isobaric approximation used in this section, valid for low Mach numbers, the pressure is to be interpreted as a dynamical pressure that serves to satisfy the continuity equation, and not as a thermodynamic pressure. Actually, this pressure is not written in the state equation, which is only a relation between ρ and T.

The idea of Markstein's model presented in previous subsection is to consider that the external acoustic field acts on the isobaric zone close to the flame (an internal zone) only through a modification of the acceleration seen by the flame (it could have been described in an equivalent way by a modification of the boundary conditions on the velocity field far from the flame). We recall that we consider the flame stabilized close to a pressure node, in such a way that the thermodynamic pressure is supposed constant in the state equation.

These isobaric equations, with a variable density, are to be handled by a specific solver. We use here the method presented in [13], with a finite differences/Fourier discretization. The real difficulty is to solve the pressure field so as to satisfy the continuity equation. In this method, we transform the discretized equations in order to obtain a Poisson equation for pressure, which can be solved with no inner iteration. We refer the reader to [13] for further details.

3.3 Qualitative results

In the case of a low forcing, the flame front is curved because of the Darrieus-Landau instability (see Fig. 5). Temperature contours are shown on this figure. It is found that, as suggested by Markstein, a moderate acoustic acceleration can suppress the hydrodynamic instability of an initially cellular flame, leading to a plane flame (see Fig. 5). A parametric instability of a flame in which the periodic acceleration of the acoustic field drives pulsating cellular structures of small wavelength whose frequency is one half that of the acoustic field, is observed for higher acoustic fields (see Fig. 5). These results are in qualitative agreement with experiments.

We describe now what happens in the parametric instability case. In Fig. 6 and Fig. 7 are shown the streamlines and temperature lines at two times differing by one acoustic period. It is seen that after one period, the solution is exactly reversed, so that the period of oscillation is twice the acoustic period. The cause of this rapid variation is to be searched in the velocity field. If one looks at Fig. 6, one sees regions where the streamlines converge (in the unburnt gases below the flame), causing an increase in velocity and slowing the flame advance. On the contrary, the velocity just before the cusps pointing towards burnt gases is negative, causing the flame to accelerate locally. As a consequence of this velocity, the flame has a tendency to reverse its shape.

3.4 Available theories on flame response to parametric forcing

We discuss in this subsection recent theoretical ideas to which the numerical results will be compared. A quasistationary theory has been obtained in [11], which improved a work of Markstein by using modern laminar flame theory. Specifically this theory predicts quantitatively the thresholds of the Darrieus-Landau and parametric instabilities by using classical quasistationary jump conditions across the flame. A comparison of experimental and theoretical results reveals however an important discrepancy. The effect of a high parametric forcing is predicted theoretically to depend on the diffusive properties of the mixture, i.e. on the Lewis number (ratio of the thermal to the molecular diffusivities). But experiments with very different gases, such as propane [11] and methane [9], do not show this effect.

G. Joulin [12] suggested that actually the boundary conditions used on the flame, in a modelization where the flame is considered as a discontinuity, should

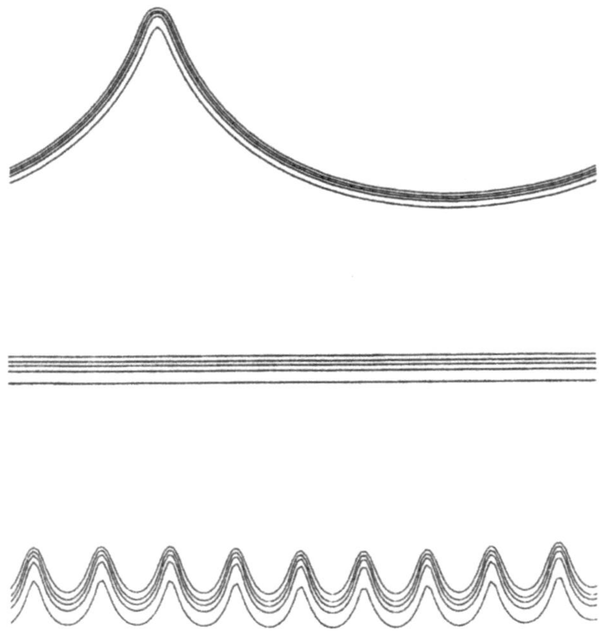

Figure 5 -Curved flame submitted to a low acoustic forcing -Plane flame submitted to a moderate acoustic forcing -Parametric instability of a flame submitted to a high forcing

be modified because of the high frequency of acoustics, compared to the usual quasistationary boundary conditions. These boundary conditions, obtained first in the thermal diffusive model where there is no retroaction of the flame on hydrodynamics, show that the flame structure retains a certain memory of its shape and of the velocity field at the previous times.

This work has been recently extended by Clavin and Joulin [15] to the complete hydrodynamic case. Compared to [12], the properties of the response functions (the frequency-dependent Markstein lengths describing the effect of strain and curvature on the flame velocity) are relatively close. Slight differences remain between these two analyses, but the essential effect is the same, i.e. the high frequency behaviour of these functions is independent of the Lewis number.

3.5 Quantitative Results

The quantitative properties of the thresholds of the various instabilities described above must be studied. In particular, do the memory effects suggested by Joulin imply that these thresholds are independent of the Lewis number, as suggested in the original paper? The numerical simulations answer to this question in a

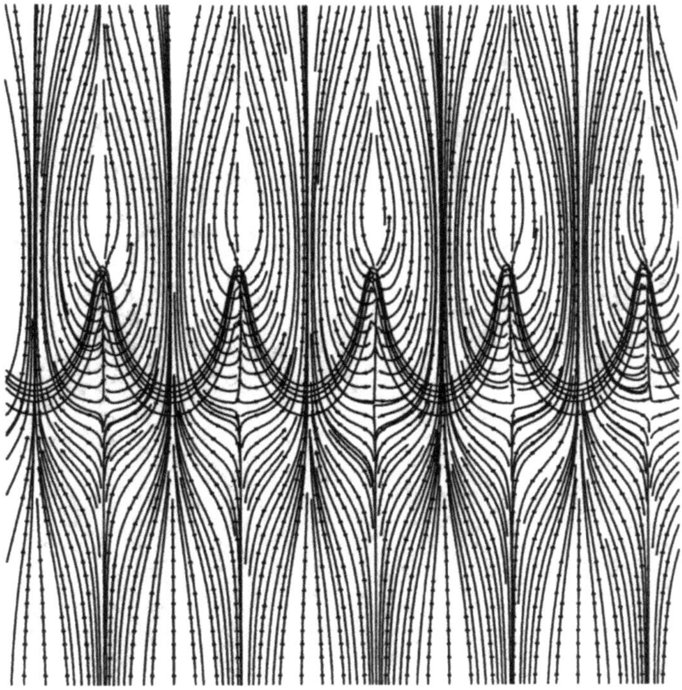

Figure 6 Streamlines and temperature lines of a flame in the parametric instability regime

surprising way (see Fig. 8).

On this figure can be seen the numerical and theoretical (quasistationary) curves for the thresholds in the plane (wave vector k, acoustic amplitude a) for two values of the Lewis number. The curves at the bottom correspond to thresholds of the Darrieus-Landau instability, the curves at the top to the thresholds of the parametric instability.

The first important thing that is seen in this figure is that, contrary to the quasistationary theoretical predictions, the parametric threshold does not depend on the Lewis number. This property is compatible with experimental results and the ideas of Joulin. If the flame shape fluctuates rapidly, then the memory effects are such that the response of the flame is independent of the Lewis number, in contrast with most other flame properties.

However, for a vanishing acoustic amplitude, the Darrieus-Landau zone collapses with the unstable zone of the Darrieus-Landau instability with no forcing, which is dependent on the Lewis number. How is it possible to reconcile this behaviour at zero amplitude, with what happens close to the parametric instability threshold? If one looks again at Fig. 8, it can be seen that actually the Darrieus-

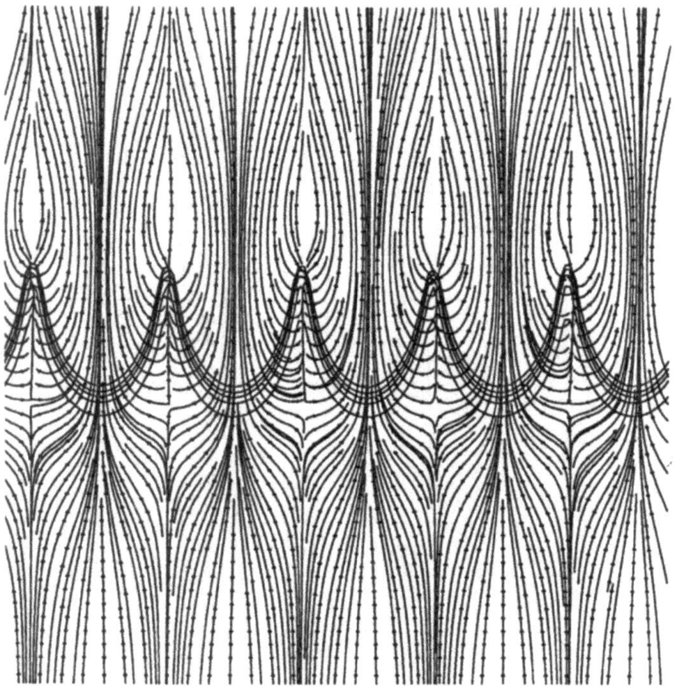

Figure 7 Streamlines and temperature lines of the solution of Fig. 6 one acoustic period later

Landau thresholds obtained numerically are dependent on the Lewis number, and that the agreement with the quasistationary theoretical predictions is reasonable. This agreement is better for higher Lewis numbers, where the unstable band of wavevectors is smaller, because the theory is actually valid for low k. So if the overall dynamics is slow as in this case, the acoustic perturbations which fluctuate rapidly, do not perturb the flame shape in a way sufficient for the Lewis number effect to disappear. A theoretical explanation of this effect has not been obtained for the moment, but it would increase our understanding of the behaviour of flames submitted to rapidly fluctuating acoustic fields.

4 Conclusions

The present paper summarizes recent joint efforts by IRPHE/CNRS Marseille, Universität Stuttgart and Freie Universität Berlin in the development and validation of numerical simulation techniques for flame-acoustic interactions. Experimental evidence generated at IRPHE as well as theoretical analyses have

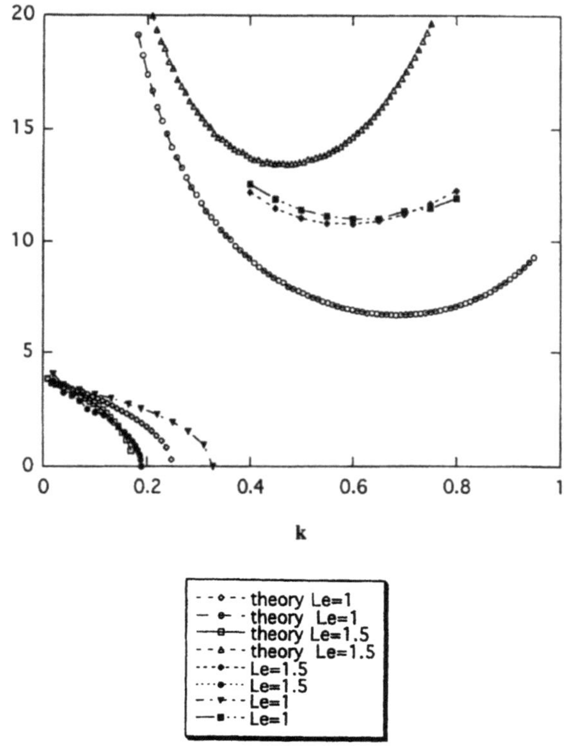

Figure 8 Numerical and theoretical thresholds of the hydrodynamic and parametric instabilities

shown a variety of stability/instability features which are not easily captured by standard numerical methods.

Two alternative numerical techniques are developed, one based on adaptive gridding and detailed flame structure resolution, the other using a flame front tracking/capturing hybrid approach. Both flame representations will be coupled to an outer long-wave acoustic solver in the near furure in order to allow for a true feedback from combustion to acoustics. Currently the influence of acoustics is merely modelled by an oscillating acceleration field imposed on the small-scale flow surrounding the flame(s).

While the resolved flame approach is naturally able to represent the observed effects of unsteady flame structure evolution, its application to realistic sized systems leads to both tremendous computational requirements and a considerable code complexity if the dynamic grid refinement ansatz is to be pursued consistently.

The alternative method of flame front tracking/capturing will not loose efficiency when applied to large scale systems, but requires a non-trivial modification when the abovementioned effects of unsteadiness are to be simulated.

Further developments will be geared towards a detailed numerical simulation of the experiments done at IRPHE in oder to quantify the capabilities of both numerical approaches. The insight gained during the first period of collaboration is the foundation for successful code validation and will allow us to optimize the schemes.

The group thanks CNRS and DFG for their generous support!

References

[1] Karki, K.C., Patankar, S.V., *Pressure Based Calculation Procedure for Viscous Flows at All Speeds in Arbitrary Configurations*, AIAA Journal, **27**(9), 1167–1174, (1989).

[2] Klein, R., *Semi-Implicit Extension of a Godunov-Type Scheme Based on Low Mach Number Asymptotics I: One-dimensional Flow*, J.Comput.Phys., **121**, 213–237, (1995).

[3] Munz, C. D., Klein, R., Roller, S., Geratz, K.J. *The Extension of Incompressible Flow Solvers to the Weakly Compressible Regime*, submitted to "Theoretical and Computational Fluid Dynamics", August 1997.

[4] Roller S., Dissertation, Institut für Aerodynamik & Gasdynamik, Univ. Stuttgart, (1998), in preparation.

[5] Smiljanovski V., Moser V., Klein R., *A Tracking/Capturing Hybrid Scheme for Deflagration Discontinuities*, Combustion Theory & Modelling, **1**, 183–215, (1997).

[6] Terhoeven P., *Ein numerisches Verfahren zur Berechnung von Flammenfronten bei kleiner Mach-Zahl*, Dissertation, Institut für Technische Mechanik, RWTH Aachen, (1997), in preparation.

[7] Markstein G.H., *Instability Phenomena in Combustion Waves*, Fourth Symposium in Combustion , 44 , Baltimore, Williams and Wilkins (1953).

[8] Clavin P., Pelce P. and He L., *One-dimensional Vibratory Instability of Planar Flames Propagating in Tubes*, J. Fluid Mech. **216**, 299-322, (1990).

[9] Searby G., *Experimental Investigations of Acoustic Instabilities in Laminar Premixed Flames*, Proceedings of the NATO ASI Summer School "Non linear phenomena related to growth and form", Edited by Ben Amar M., Pelcé P. and Tabeling P., Plenum Press (1991).

[10] Searby G., *Acoustic Instability in Premixed Flames*, Combust. Sci. & Tech. **81**, 221-231, (1992).

[11] Searby G., Rochwerger D., *A parametric Acoustic Instability in Premixed Flames*, J. Fluid Mech. **231**, 529-543, (1991).

[12] Joulin G., *On the response of premixed Flames to Time-dependent stretch and curvature*, Combust. Sci. Tech. **97**, 219, (1994).

[13] Denet B., Haldenwang P., *A numerical study of premixed Flames Darrieus-Landau instability*, Combust. Sci. Tech. **104(1-3)**, 143, (1995).

[14] Denet B., Toma A., *Numerical study of premixed Flames parametric acoustic instability*, Combust. Sci. Tech. **109**, 23-33, (1995).

[15] Clavin P., Joulin G., *High frequency response of premixed Flames to weak stretch and curvature: a variable density analysis* , to appear in Combust. Theor. Mod. , (1997).

[16] Pelce P., Rochwerger D., *Vibratory instability of cellular Flames propagating in tubes*, J. Fluid Mech. **239**, 293, (1992).

Computation of Vortex-Shedding Flows Past a Square Cylinder Employing LES and RANS

D. Lakehal, F. Thiele
Hermann Föttinger Institut für Strömungsmechanik, TU-Berlin,
Müller-Breslau-Strasse 8, D-10623 Berlin, Germany

L. Duchamp de Lageneste, M. Buffat
LMFA, Ecole Centrale de Lyon, 69313 Ecully Cedex, France

SUMMARY

The paper addresses the predictive capabilities of different computational modelling practices which employ various eddy-viscosity models for turbulence, that are based on a linear, a quadratic and two cubic representations of the Reynolds stress tensor in terms of strain and vorticity rates respectively, in the context of a comparative study between Reynolds–Averaged Navier–Stokes Equations and Large Eddy Simulation of vortex-shedding flows. The test case considered here refers to the familiar vortex–shedding flow past a square cylinder at $R_e = 22.000$ studied experimentally by Lyn *et al.*. All the models were found to reproduce fairly well the shedding dynamics with the exception of eddy–viscosity models employed with a high rate of oncoming turbulence. The total kinetic energy was fairly well predicted by all models, whereas the turbulent part was significantly underestimated. Comparison of this latter quantity has emphasized the effective role of turbulence anisotropy in rendering the effects of shedding mechanisms. The integral parameters were determined with much better agreement with measurements than the averaged quantities. The essential finding is that the explicit algebraic stress models predict much better the global parameters with increasing the order of anisotropy. The confusing results of Large Eddy Simulation, i.e. satisfactory agreements of the time–averaged velocity and energy profiles to the expense of a less accurate prediction of the global coefficients, did not support the superiority of the concept. Still, the practice has revealed a particular sentivity to the near–wall treatment.

INTRODUCTION

A variety of turbulent flows of industrial relevance are the subject of meticulous investigations, more precisely those involving transient reactive processes. The most recurrently cited example is the mixing in combustion chambers, with, eventually, flame stabilization using a bluff body. The potentialities of modern computational tools together with the recent developments in CFD practice, are likely to substantially enhance the understanding of the main mechanisms, and contribute to establish efficient solutions in conception. Turbulence plays a major role in vortex-shedding flows, and its accurate representation is the key element for a successful investigation devoted to this purpose. Theoretically, this can be achieved via the simulation of the whole spectrum of turbulent scales, a task which is exclusively within reach of direct numerical simulation (DNS) or large eddy simulation (LES). The LES concept seems to be a promising way of solving such flow problems. In LES the large eddies that depend strongly on the special flow configuration are resolved numerically whereas only the fine–scale turbulence has to be modelled by a subgrid-scale model. While the possibility of employing DNS as a prediction tool is still confined to low Reynolds number flows, and will not be applicable in the foreseeable future, the method based on the solution of the *Reynolds Averaged Navier-Stokes Equations*, combined with statistical turbulence models, is a viable alternative for its robustness and its potential of reduced computational expenses. Still, the method is justifiable provided that it accurately renders the basic mechanisms related to turbulence.

One route of scrutinizing efficacy of these models can be achieved through the computation of flows with organized-wave motion. Here, the comparison of integral parameters, namely the lift and drag coefficients (c_l and c_d), and Strouhal number ($St = f\,D/U$, f is the frequency of shedding, D the diameter and U the free-stream velocity) is a reliable indicator of the predictive performance of the model. The selected test-case represents the well-documented flow past a slender square-cylinder at $R_e = 22.000$ which is basically two-dimensional in the mean. The experiment reports that the flow has a transitional behaviour which is essentially set at flow regions near the cylinder side walls.

The conclusions drawn up to date from the different contributions to the computation of this test case can be summarized as follows: Linear eddy-viscosity models (EVM) with wall functions and without any *ad-hoc* modification are unlikely to yield shedding due to the excessive damping that is introduced through the spurious turbulence production generated by these models on the one hand, and to the near wall treatment which avoids to solve the semi-viscous sublayer on the other. As an exception is the work of Bosch [1] who employed non-reflective boundary conditions instead of zero-gradient conditions at outflow, together with a second-order differencing scheme. Still, the mean drag coefficient was drastically underpredicted. The vortex-shedding motion was found to really persist only by solving the semi-viscous sublayer, using either a one-equation near-wall model (Franke & Rodi [2] and Bosch [1]) or a low-Re model (Kawamura & Kawashima [3]). While these latter models have shown better performance in predicting global parameters, they have still underpredicted the time-mean turbulent kinetic energy and overpredicted the recirculation length along the centre-line of the cylinder. However, there is a significant difference to emphasize between the computation methods of these contributions, and of many others, concerning the turbulent inflow conditions and in particular of ε, which are a non-negligible source of deviation between results. The discussion of this point is given in the last section before the conclusion. Reducing the excessive turbulence production at flow stagnation regions by means of *ad-hoc* measures, either by suppressing explicitly the turbulence production or by introducing the Kato & Launder [4] modification was in all cases partially successful. In that case, $\overline{c_d}$ was well predicted to the expense of a tremendously overpredicted recirculation zone. The full Reynolds Stress Equations transport models (RSM) which are arguably preferred, yielded indeed the best agreement concerning the prediction of St and $\overline{c_d}$, but provided an overpredicted rate of momentum exchange in the wake by exaggerating periodic motions vis-à-vis stochastic fluctuations (Franke & Rodi [2]). The flow was extensively simulated with LES in the *Workshop on LES of Flows past Bluff Bodies* (Rodi *et al.* [5]). On the basis of the different results, which were of a very mixed quality, the conclusion reached was that LES does not automatically provide superior results and many numerical details have to be controlled carefully.

The present contribution tends to shed the light on the efficiency and behaviour of various non-linear eddy-viscosity models that are derived from different strategies for this particular flow. The calculations performed by the computation program (ELAN2D) of TU-Berlin were entirely achieved by solving the near-wall viscosity-affected layer. The results obtained by employing various models based respectively on a linear, a quadratic and two cubic functional formulations of the Reynolds stress are therefore compared to those obtained by LES using the code (NaturLES) of EC-Lyon and measurements of Lyn *et al.* [6].

INSTANTANEOUS, MEAN AND FILTERED FLOW EQUATIONS

Considering an incompressible turbulent motion of a Newtonian, neutral, viscous fluid, the velocity field u_i and kinematic pressure P are obtained by solving the mass conservation and Navier-Stokes equations which read

$$\frac{\partial u_i}{\partial x_i} = 0; \quad \frac{Du_i}{Dt} = -\frac{\partial P}{\partial x_i} + \nu \nabla^2 u_i , \tag{1}$$

where $D/Dt = \partial/\partial t + u_j \partial/\partial x_j$ denotes the mean convective derivative. In the decomposition procedure of the fluctuating signal representing a turbulent flow with periodic unsteadiness proposed by Reynolds & Hussein [8], a flow variable (f) is decomposed under the form

$$f = \overline{f} + \tilde{f} + f' \quad \text{and} \quad \langle f \rangle = \overline{f} + \tilde{f} , \tag{2}$$

where $\overline{f}$ is the time-averaged part, $\tilde{f}$ is the alternate wave and f' denotes the turbulent fluctuations. The phase- or *ensemble-average* is represented by the entity $\langle f \rangle$, which turns to $\overline{f}$ in steady state. Note that this decomposition is applicable to high Reynolds number flows in which the stochastic three-dimensional fluctuations are superimposed on the two-dimensional periodic motion. The phase-averaged continuity and Navier-Stokes equations representing the mean flow are obtained when applying the decomposition

(2) to the system of equations (1). The resulting system of Phase or *Reynolds-Averaged Navier-Stokes* equations (RANS) which takes the form

$$\frac{\partial \langle u_i \rangle}{\partial x_i} = 0; \quad \frac{D\langle u_i \rangle}{Dt} = -\frac{\partial \langle P \rangle}{\partial x_i} + \nu \nabla^2 \langle u_i \rangle - \frac{\partial \tau^*_{ij}}{\partial x_j} \tag{3}$$

is said being mathematically closed, provided that the Reynolds stress tensor $\tau^*_{ij} = \langle \overline{u'_i u'_j} \rangle$ is modelled in a physically consistent way.

In the LES approach a similar decomposition procedure is applied to the system of equations (1). However, in contrast to the RANS approach the averaging or rather filtering is now accomplished with respect to space and not to time, such that the different length scales in the turbulent flow field can be separated,

$$\underbrace{f}_{total} = \underbrace{\overline{f}}_{resolved} + \underbrace{f'}_{subgrid} . \tag{4}$$

Energy or large-scale structures which can be resolved directly are separated from dissipation or small-scale structures (subgrid–scales) which cannot be captured on a given grid. The governing equations for LES have the same form as for the RANS approach, however with a different meaning of some variables: for instance the stress tensor τ_{ij} which results from the filtering of the convective terms,

$$\tau_{ij} = \overline{u_i u_j} - \overline{u_i}\,\overline{u_j} \tag{5}$$

now refers to the **subgrid–scale** stress tensor. Modelling of the Reynolds stress tensor in the RANS strategy signifies that the time– or ensemble– averaged mean flow is subject to the influence of the whole spectrum of turbulent scales. In LES, however, a large part of the spectrum of turbulent motions is directly computed by the numerical scheme and only the subgrid scales, which are arguably simpler to model (more homogeneous and isotropic) have to be modelled (Breuer *et al.* [7]). By analogy with the eddy viscosity concept or the mixing length model of Prandtl, the subgrid stress tensor is modelled so far using a similar approach which links τ_{ij} linearly to the mean strain rate tensor $S_{ij} = (u_{i,j} + u_{j,i})$.

TURBULENCE TRANSPORT EQUATIONS AND SGS MODELLING

The non-linear representation of $\overline{u'_i u'_j}$ in the $K - \varepsilon$ modelling framework

The basic concept : Recently, a variety of sophisticated explicit algebraic stress models (EASM) have arised from the developments in RSM practice, and being alternatively routed just above EVM preserving the simple architecture of the concept, to overcome the deficits afflicted to the Boussinesq linear representation of τ^*_{ij}. Theoretically, it is argued that the platform of these models which are essentially derived from Reynolds stress transport equations, is constructed strictly in the limit of homogeneous turbulence in equilibrium (Gatski & Speziale [9]). In spite of this idealization, the method still retains a *precious* task of physical relevance, namely the distinction between the Reynolds stress deriving straightforward to a logical representation of the mechanism of turbulence production. On the basis of series-expansion analysis, the general formulation retaining terms up to third order which links the Reynolds stresses anisotropy tensor ($b_{ij} = \tau^*_{ij}/\langle K \rangle - 2\delta_{ij}/3$) to the mean rate of strain ($S^*_{ij} = \langle u_{i,j} \rangle + \langle u_{j,i} \rangle$) and rotation ($\Omega^*_{ij} = \langle u_{i,j} \rangle - \langle u_{j,i} \rangle$) in an inertial frame of reference, renders the familiar form

$$\begin{aligned} b_{ij} = & - \frac{\langle \nu_t \rangle}{\langle K \rangle} S^*_{ij} + C^*_1 \left(S^*_{ik} S^*_{kj} - \frac{\delta_{ij}}{3} S^*_{lk} S^*_{kl} \right) + C^*_2 \left(\Omega^*_{ik} S^*_{kj} + \Omega^*_{jk} S^*_{ki} \right) \\ & + C^*_3 \left(\Omega^*_{ik} \Omega^*_{kj} - \frac{\delta_{ij}}{3} \Omega^*_{lk} \Omega^*_{kl} \right) + C^{**}_4 \left(S^*_{ki} \Omega^*_{lj} + S^*_{kj} \Omega^*_{li} - \frac{2}{3} S^*_{km} \Omega^*_{lm} \delta_{ij} \right) S^*_{kl} \\ & + C^{**}_6 \left(S^*_{ij} S^*_{kl} S^*_{kl} \right) - C^{**}_7 \left(S^*_{ij} \Omega^*_{kl} \Omega^*_{kl} \right) , \end{aligned} \tag{6}$$

where $C^*_i = C_i \frac{\langle \nu_t \rangle}{\langle \varepsilon \rangle}$, and $C^{**}_i = C_i \frac{\langle \nu_t \rangle \langle K \rangle}{\langle \varepsilon \rangle^2}$; the model coefficients C_i with $i = 1 - 7$ are given in Table 1. In the context of two-equations turbulence modelling, in which the local state of turbulence is characterised through the turbulent kinetic energy ($K = 1/2\, \overline{u'_i u'_i}$) and its rate of dissipation ($\varepsilon = \nu\, \overline{u'_{i,j} u'_{i,j}}$), the generalized relation which determines the turbulent viscosity $\langle \nu_t \rangle$ takes the form

$$\langle\nu_t\rangle = C_\mu f_\mu \langle K\rangle^2/\langle\tilde{\varepsilon}\rangle \qquad \text{with: } \langle\tilde{\varepsilon}\rangle = \langle\varepsilon\rangle - \mathcal{D}\,, \tag{7}$$

Note that $\langle\tilde{\varepsilon}\rangle$ denotes the so-called *isotropic* dissipation rate which vanishes at the wall. The distributions of $\langle K\rangle$ and $\langle\tilde{\varepsilon}\rangle$ are determined from the following model-transport equations, solved together with those describing the mean flow :

$$\frac{D\langle K\rangle}{Dt} = \frac{\partial}{\partial x_i}\left(\Gamma_k \frac{\partial\langle K\rangle}{\partial x_i}\right) + \mathcal{P} - \langle\tilde{\varepsilon}\rangle - \mathcal{D} \tag{8}$$

$$\frac{D\langle\tilde{\varepsilon}\rangle}{Dt} = \frac{\partial}{\partial x_i}\left(\Gamma_\varepsilon \frac{\partial\langle\tilde{\varepsilon}\rangle}{\partial x_i}\right) + C_{\varepsilon 1} f_{\varepsilon 1}\frac{\langle\tilde{\varepsilon}\rangle}{\langle K\rangle}\mathcal{P} - C_{\varepsilon 2} f_{\varepsilon 2}\frac{\langle\tilde{\varepsilon}\rangle^2}{\langle K\rangle} + E\,, \tag{9}$$

where $\Gamma_k = (\nu + \langle\nu_t\rangle/\sigma_k)$ and $\Gamma_\varepsilon = (\nu + \langle\nu_t\rangle/\sigma_\varepsilon)$. The model functions f_μ, $f_{\varepsilon 1}$ and $f_{\varepsilon 2}$ are given in Table 1. The turbulence production ($\mathcal{P}$) resulting from the interaction of turbulent structures and velocity gradients reads

$$\mathcal{P} = -\langle\overline{u_i' u_j'}\rangle \frac{\partial\langle u_i\rangle}{\partial x_j}\,. \tag{10}$$

The EASM of Gatski & Speziale [9] *-GS-* : The present explicit ASM is constructed on tensorially quadratic stress–strain relations that rely on the use of integrity basis from linear algebra. The model employs the linear pressure–strain correlation model of Speziale, Sarkar & Gatski [10] (SSG) for homogeneous turbulent flows that are in the limit of equilibrium. The model consists of an anisotropic eddy viscosity with strain–dependent coefficients. Based on the SSG second–order closure, the calibration of the coefficients α_1 - α_4 and g was made on the basis of data for homogeneous shear flow, and their approximations to constants (see Table. 1) was justified due to the linear dependence on b_{ij}. However and in regard to the test case considered here where the flow is far from equilibrium, the following points were modified to cope with the features of an impinging flow (Rung [11]): the parameter α_5 which appears in g–expression (denoted hereafter by g^*) is no longer set to the equilibrium value $\mathcal{P}/\varepsilon = 1.89$, but is made strain–dependent (in Table 1, the modified GS model is denotd by GS*)

$$g^* = f_3\,(\alpha_1^* - 1) + \frac{S^2}{6. + 1.5S} \qquad \text{with: } f_3 = 1. + 0.95\,\left[1 - Tanh(0.22\,S^2)\right]\,. \tag{11}$$

Furthermore, the C_μ–regularization procedure proposed by Gatski & Speziale [9] which consists of employing a first order Padé approximation for η^2 is abandoned here, due to the fact that in highly strained stagnating flows, η can be > 1. Finally, the application of the model is extended to low-Re conditions by having recourse to Lien & Leschziner's proposal [12] which will be discussed later. It should be noted that neither the regularized nor the unregularized original variants (GS) were successful for this flow.

The EASM of Lien, Chen & Leschziner [13] *-LCL-* : The model is conceptually more elaborate for including the complete cubic contribution in b_{ij} proposed by Suga [14] to cope with streamline curvature effects in his proposal of a second–order closure. In addition, the model's form adopts ideas of Shih *et al.* [15] in respect of sensitizing the eddy-viscosity to strain and vorticity invariants; a concept which was originally routed on the basis of realizability analysis. Note however that due to the cubic extension, the model is no longer realizable. The coefficients C_1 - C_7 appearing in b_{ij} are similarly strain-dependent. In order to account for the semi-viscous near-wall effect, the eddy viscosity is damped using a similar function to the Norris & Reynolds' [16] one–equation model (Table 1). The closure constants are assigned the standard values.

The EASM of Craft, Launder & Suga [17] *-CLS-* : Here, the constitutive expressions which relate the Reynolds stresses to velocity gradients integrate the cubic terms by reference to Suga's second-order model [14] as well. Originally, the model was proposed to remedy the spurious normal-strain-induced production of turbulence resulting from conventional eddy-viscosity representation of τ_{ij}^*. As it is shown in Table. 1, all the coefficients that appear in b_{ij} are set to constants, with the exception of C_μ which is sensitized to strain and vorticity invariants. The approach is routed in conjunction with the Launder & Sharma [18] model in regard to the model functions of ε–equation, while it uses a different C_μ wall–damping relation, as it is reported in Table. 1. Note that the third fragment of the first cubic group in the CLS model; it does not intervene in the recent variant of Craft et al. [19] neither. It should be noted that in shear flow regions, all of the non-linear models were found to yield values of C_μ which fall beyond 0.09.

The EASM of Shih, Zhu & Lumley [20] *-SZL-* : The general constitutive relation applied to yield this model was obtained using the invariance theory in continuum mechanics. Both of the constraints based on rapid-distorsion theory and turbulent realizability principle were imposed. In addition, the model was deliberately truncated by the authors to its tensorial quadratic form. In the present situation however, this model was used in its linear form ($C_2 = 0$) as an alternative route to reduce the excessive $\mathcal{P}$ that is generated by conventional EVM. In the particular situation of flow acceleration, the isotropic energy component $\overline{u'^2} = 2K/3 - C_\mu S_{11} K^2/\varepsilon$ becomes negative for large S_{11}, and consequently the first realizability principle is violated. Thus, in case of linear EVM the realizability principle imposes a constraint exclusively on C_μ. Now, under the other extreme condition of flow impingement with abrupt separation, where $S >> \Omega$, the above model modification established by reference to realizability concept, is likely to remedy the shortcoming of the overpredicted ν_t by simple EVM. The model was applied together with the low-Re model of Abe *et al.* [21] which is to be introduced later. Still, the calibration of the model parameter A_0, which depends on the nature of the flow considered, is to be determined (originally $4.0 < A_0 < 6.5$). In order to ensure that C_μ does not return values which widely exceed 0.1 in simple shear flows, a new strain-based relation is proposed for A_0. It reads

$$A_0 = 10 - 4.1\, tanh\left(0.55\frac{K}{\varepsilon}\sqrt{2S_{lk}S_{kl}}\right). \tag{12}$$

The Near Wall Treatments in RANS

The low-Re $K-\varepsilon$ model of Abe *et al.* [21] *-AKN-* : The low-Re model developed by Abe et al. was extensively tested for separated flows, and showed excellent behaviour under various conditions. The main improvement is achieved by the introduction of the Kolmogorov length scale $u_\varepsilon = (\nu\varepsilon)^{1/4}$, instead of U_τ or $K^{1/2}$. It is argued that using u_ε yields no singularity at a separating or a reattaching point. Furthermore, the above length scale is composed of the characteristics quantities of a turbulent flow, as it denotes the intensity of the fluctuation of the instantaneous wall shear-stress $\tau_w(t)$. The model functions f_μ and $f_{\varepsilon 2}$, the near-wall non-dimensional scales R_t and y^*, and the near-wall boundary conditions employed are reported in Table 1. The closure constants have been re–evaluated as well, to take the values $C_{\varepsilon 1} = 1.5$, $C_{\varepsilon 2} = 1.9$, and $\sigma_k = \sigma_\varepsilon = 1.4$.

The low-Re $K-\varepsilon$ variants of Lien & Leschziner [12] *-LL-* : The approach results from the combination of the following ideas: The derivation of ν_t from both frameworks of one- and two-equation turbulence models is equivalent, and the correct near-wall level of ε is returned. The model functions f_μ and $f_{\varepsilon 2}$ are introduced by reference to Wolfshtein's [22] one-equation model and Jones & Launder's [23] prescription respectively (see Table 1). The correct distribution of near-wall dissipation rate is ensured by substituting $(1 + \mathcal{P}'/\mathcal{P})$ to $\mathcal{P}$ in $\langle\tilde{\varepsilon}\rangle$–equation, where $\mathcal{P}'$ stands for the apparent production term. The model keeps using the standard set of closure constants. At the wall, the variables are treated in a similar way to the former model, with the exception of ε_w to which are prescribed the algebraic relations reported in Table 1. In fact there is a slight difference between the LL model used with the GS EASM and its recent variant applied with the LCL model (see Table 1).

The Subgrid Scale Modelling (SGS) in LES

In principle the LES concept leads to a similar closure problem as the RANS approach, which justifies a similar classification of turbulence models. But as the SGS turbulence in a LES is much less flow–dependent than the large–scale one, this latter one can be represented by relatively simple models, e.g. zero-equation models. In this investigation, the SGS dynamic modelling procedure of Germano [24] was employed. In this, the Smagorinski constant (C_s) which appears in the mostly used model of the same name (Smagorinsky [25])

$$\tau_{ij} = -(C_s\,\overline{\Delta})^2\,\tilde{S}\,\tilde{S}_{ij} \qquad \text{with: } \tilde{S} = \sqrt{1/2\,\tilde{S}_{ij}\,\tilde{S}_{ij}} \qquad \overline{\Delta} = (\Delta x\,\Delta y\,\Delta z)^{1/3} \tag{13}$$

Table 1: **Constitutive relations and coefficients of Eq. 6 and model functions in Eqs. 7–9.**

Model	*Code*	C_μ	C_1	C_2	C_3	C_4	$C_6,\ C_7$
Gatski & Speziale [9]	*GS*	$\frac{2-1.5\alpha_2}{g(3-2\eta^2+6\xi^2)}$	$\frac{2-\alpha_3}{2g}$	$\frac{1-\alpha_4/2}{2g}$	0.0	0.0	0.0
Gatski & Speziale (modified)	*GS**	$\frac{2-1.5\alpha*_2}{g*(3-2\eta^2+6\xi^2)}$	$\frac{2-\alpha*_3}{2g*}$	$\frac{1-\alpha*_4/2}{2g*}$	0.0	0.0	0.0
Shih, Zhu & Lumley [20]	*SZL*	$\frac{1}{A_0+A_s\sqrt{S^2+\Omega^2}}$	0.0	$\frac{\sqrt{1-9C_\mu^2 S^2}}{2C_\mu(1.+6S\Omega)}$	0.0	0.0	0.0
Craft, Launder & Suga [17]	*CLS*	$\frac{0.3\left[1-exp\left(\frac{-0.36}{exp(-0.75S^*)}\right)\right]}{1.+0.35(S^*)^{1.5}}$	-0.1	0.1	0.26	$-10C_\mu^2$	$-5C_\mu^2$
Lien, Chen & Leschziner [13]	*LCL*	$\frac{0.667}{1.25+S+0.9\Omega}$	$\frac{0.75}{1000+S^3}$	$\frac{3.75}{1000+S^3}$	$\frac{4.75}{1000+S^3}$	$-10C_\mu^2$	$-2C_\mu^2$
Model	***Code***	f_μ	$f_{\epsilon 1} = 1. + \mathcal{P}'/\mathcal{P}$	$f_{\epsilon 2}$	$\mathcal{D}$	$\mathcal{E}$	$\tilde{\varepsilon}_w;\ \varepsilon_w$
Abe, Kondoh & Nagano [21]	*AKN*	$\frac{\left[1-exp\left(-\frac{y^*}{14}\right)\right]^2}{\left[1+5Re_t^{-3/4}exp\left(-\frac{2.5Re_t^2}{10^5}\right)\right]^{-1}}$	$f_{\epsilon 1} = 1.0$	$\frac{\left[1-exp\left(-\frac{y^*}{3.1}\right)\right]^2}{\left[1-0.3exp\left(-\frac{Re_t^2}{42.2}\right)\right]^{-1}}$	0.0	0.0	$\varepsilon_w = \frac{2\nu K}{y^2}$
Lien & Leschziner [12]	*LL*	$\frac{1-exp\left(-\frac{R_y}{62.5}\right)}{1-exp\left(-\frac{R_y}{3.8}\right)}$	$\mathcal{P}' = \frac{C_{\epsilon 2} f_{\epsilon 2} K^{3/2} exp\left(-\frac{R_y^2}{10^3}\right)}{3.53y\left[1-exp\left(-\frac{R_y}{3.8}\right)\right]}$	$1 - 0.3\,exp\,(-Re_t^2)$	0.0	0.0	$\varepsilon_w = \frac{2\nu K}{y^2}$
Craft, Launder & Suga [17]	*CLS*	$Min\ [1.,\ 0.2 + \frac{0.8Re_t}{50}]$	$f_{\epsilon 1} = 1.0$	$1 - 0.3\,exp(-Re_t^2)$	$2\nu\left(\frac{\partial\sqrt{K}}{\partial x_j}\right)^2$	$\frac{S\nu_t K^2}{16.66\varepsilon}$	$\tilde{\varepsilon}_w = 0.$
Lien, Chen & Leschziner [13]	*LCL*	$\frac{\left[1-exp\left(-\frac{R_y}{50.5}\right)\right]}{\left[1+\frac{5.29}{R_y}\right]}$	$\mathcal{P}' = \frac{1.33 f_{\epsilon 2}\left(\mathcal{P}+\frac{2\nu K}{y^2}\right)}{\left[exp\left(-\frac{R_y}{266.8}\right)\right]^{-1}}$	$1 - 0.3\,exp(-Re_t^2)$	0.0	0.0	$\varepsilon_w = \frac{2\nu K}{y^2}$

$g = \alpha_1 + \alpha_5 - 1$; $\alpha_1 = 3.4$; $\alpha_2 = 0.36$; $\alpha_3 = 1.25$; $\alpha_4 = 0.4$; $\alpha_5 = 1.9$; $g*$ (see Eq. 11); $\alpha*_1 = 2.6$; $\alpha*_2 = 0.4$; $\alpha*_3 = 0.45$; $\alpha*_5 = \frac{\mathcal{P}}{\varepsilon}$
$S^* = max(S, \Omega)$; $\eta^2 = \frac{(C_1 S)^2}{8}$; $\xi^2 = \frac{(C_2\Omega)^2}{8}$; $S = \frac{K}{\varepsilon}\sqrt{1/2\, S_{lk} S_{kl}}$; $\Omega = \frac{K}{\varepsilon}\sqrt{-1/2\,\Omega_{lk}\Omega_{kl}}$; $S_{ij} = (u_{i,j} + u_{j,i})$; $\Omega_{ij} = (u_{i,j} - u_{j,i})$
$Re_t = K^2/\nu\varepsilon$; $y^* = u_\varepsilon y/\nu$; $u_\varepsilon = (\nu\epsilon)^{1/4}$; $R_y = \sqrt{K}\, y/\nu$; $A_s = \sqrt{6}\cos\left[\frac{1}{3} arccos\left(\sqrt{6}W\right)\right]$; $W = \frac{S_{ij} S_{jk} S_{ki}}{(S_{ij} S_{ij})^{3/2}}$

is calculated as a function of time and position from the smallest scales of the resolved motion. Thus, in contrast to the Smagorinsky model C_s is no longer a constant but a local, time-dependent variable. The method allows therefore the determination of C_s from the information provided by the resolved velocity field. The fluctuations of the resulting Smagorinsky parameter are clipped to avoid too high (e.g. superior to 0.2) or too low (e.g. inferior to -0.1) values, as well as negative eddy viscosities. Still, the implementation of this particular SGS model has required to overcome an additional difficulty which is not familiar to finite-volume structured codes, namely the choice of an adequate geometry-based test filter related to unstructured grids. The reader is referred to Duchamp *et al.* [26] for further details.

COMPUTATIONAL METHODOLOGIES

Numerical procedures

The numerical program *ELAN2D* is based on a multi-block, multi-grid finite-volume approach for solving implicitly the incompressible Navier-Stokes equations. The algorithm employs the strongly conservative form of arbitrary non-orthogonal coordinates, employing a fully co-located grid arrangement for the independent variables. The momentum-interpolation technique of Rhie & Chow [27] is used to prevent the odd-even decoupling problem. The pressure correction procedure is achieved by means of the SIMPLE algorithm. Different high-order bounded and unbounded schemes can be used to approximate the advective volume-face fluxes, namely the TVD/MUSCL and QUICK schemes. The present computations were performed employing the QUICK scheme to all variables. The central difference approximation scheme was used for diffusion terms. The time marching process was conducted using a second-order backward scheme. The resulting system of equations is solved in an un-coupled manner using the Strongly Implicit Procedure of Stone [28].

The LES code *NaturLES* developed at LMFA uses a finite element discretization for the viscous gradients and a finite volume discretization for the convective terms of the compressible Navier-Stokes equations. This method was initially developed by Dervieux *et al.* [29] for unstructured grids in two dimensions and extended to three dimensions by Steve [30]. The main interest of this mixed approach is to allow the use of high order upwind scheme (Riemann solver: *Roe/MUSCL* scheme) to treat the hyperbolic part of the equations while preserving the precision of the finite element method to discretize the viscous and source terms. The utilization of upwind schemes in LES is particularly non-suited because they generate an excessive amount of numerical diffusion. A new method which consists of introducing two additional parameters (known also as the β–γ scheme) into such schemes in order to control the amount of numerical dissipation, was implemented into the program (Duchamp et al. [26]). The time-marching discretization is performed by a four-step Runge-Kutta algorithm of fourth-order accuracy, which allows to perform long-range simulations without excessive dissipation by preserving the TVD properties of the spatial differencing. The code is parallelized following a domain partitioning approach.

Grids and boundary conditions

In the RANS computations, the dimensions of the computational domain were set as follows: each lateral boundary is located at $6D$ from the centre line, the inflow and outflow planes being $4.5D$ upstream and $30D$ downstream the cylinder, respectively. Grid sensitivity studies indicated that grid-independent results can be obtained with a mesh consisting of 226 × 156 grid points in the $x-$ and $y-$ directions. Still, it was found that grid concentration at near-wall zones can affect the results, and a second grid-sensitivity test was necessary to clearly define how important is the distance to which the wall-adjacent control volume is set. For the same grid with 226 × 156 nodes, the time-averaged variables were almost unaltered, while the global parameters deviate with 0.4% error, for two grid refinements, namely when $y^+ < 2$ (*Grid1*) and $y^+ < 1$ (*Grid2*), respectively (see Table 2). Further, it was found that using *Grid1* for the computations with the GS model leads to numerical instabilities in near-wall regions, due to the very steep gradient of the rotational and irrotational strains η^2 and ξ^2 which intervene in model coefficients expression. At the upper boundary of the domain, symmetry condition was employed. At the inflow plane, a constant streamwise velocity was prescribed, while $K-$ and ε-profiles were specified using uniform distributions corresponding to a free-stream turbulence intensity of $T_u = 2\%$ and $\nu_t = 90\nu$. The non-reflective conditions were employed at the outflow only to the velocity components, whereas zero-gradient condition was held for the other variables. The computational domain employed in LES with the various boundaries is shown in Fig. 1. The location of the inflow and outflow boundaries and of the lateral boundaries were set with regard to the recommendations of a part of the LES–Workshop contributors to this test case (Rodi *et al.* [5]). Therefore, it was assumed that fixing the domain boundaries at the following locations: $-4.5L < X < 20L$, $-7L < Y < 7L$ and $-1.6L < Z < 1.6L$ in the $x-$, $y-$ and $z-$ directions respectively, would avoid any influence on the calculations.

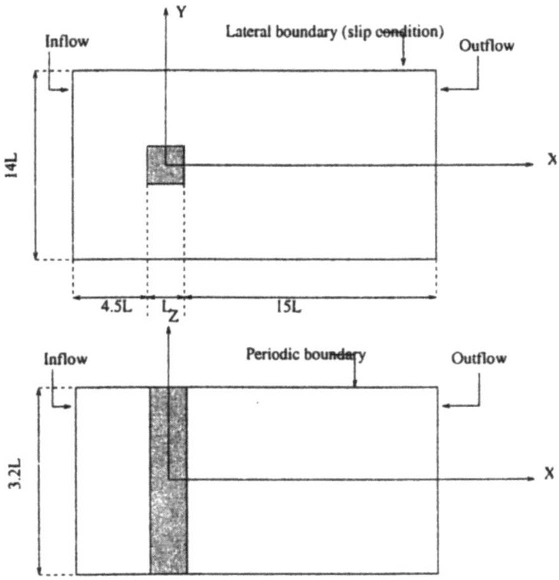

Fig. 1 : Geometry of the test case indicating the boundary conditions.

The grid is unstructured and consists of 106900 nodes. Non-reflective boundary conditions were employed at the outflow boundary, while slip conditions were imposed at the lateral planes and periodicity condition in the $z-$ direction at $Z = -1.6L$ and $Z = 1.6L$. Out of the first few thousands time steps, it was found that no-slip conditions prescribed preliminary on the cylinder walls were not successful, probably because grid resolution near the wall was not sufficient for this Re number. This condition was then replaced using the wall functions following Werner & Wengle [31] for the rest of the computations. The RANS computations were performed on the Cray T90 vector machine of the University of Kiel. 40 Hours CPU time were required for the calculation of the 30 periods (400 time-steps for 1 period, with $\Delta t = 2.0 * 10^{-2}$) selected for treating the signal. In the LES, the simulation of the 130000 time steps ($\Delta t = 1.0 * 10^{-3}$) took approximatively 10000 hours mono-processor on 64 processors of the CNRS' Cray T3E located at the *Institut du Developpement et des Ressources en Informatique Scientifique - IDRIS* ($\approx$ 160 Hours of calculation). The statistics were achieved over the last 70000 iterations (70 Seconds) which is relatively short for phase averaged quantities, as each phase needs approximatively 7 seconds.

RESULTS AND DISCUSSION

The first step of this section is devoted to a few particular points related to the pure numerical aspect of the problem, including grid sensitivity, boundary and inflow conditions and differencing and time-marching schemes. These elements were found very important and need therefore clarification before further discussions of turbulence or SGS modelling.

Effects of the computational details: Most of these observations were revealed from the RANS computation of *reference*, namely the one which uses the linear truncated SZL model combined with the AKN low-Re model for turbulence. It is important to note that the correction brought by the model to C_μ was in this context employed as an *ad-hoc* measure to suppress the excessive turbulence production, and therefore its application was restricted to the front of the body. As noted before, the attempt to establish grid dependency only on the basis of comparison of time-averaged velocity components might be insufficient. The results obtained using *Grid1* and *Grid2* which provide $y^+ < 2$ and $y^+ < 1$ at the wall respectively, showed that though these time-averaged quantities are almost identical, the integral parameters are subject to a slight deviation (see Table. 2). The results indicate that RMS values of the fluctuations of drag and lift coefficients $(c_{d,rms})$ and $(c_{l,rms})$ are the most sensitive to near-wall grid resolution, $\overline{c_d}$; $\overline{c_l}$ and St being almost the same. Concerning the capability of the spatial discretization schemes, it was found that calculations with the QUICK and TVD/MUSCL schemes exhibit no difference. The same can be said about employing a first–order or a second–order time–marching scheme, with a difference in convergence rate, i.e. the computation time required for the first vortex–shedding to appear. In the same manner, choosing non-reflective boundary condition rather than the conventional zero–gradient one at the outflow has shown a difference only in triggering the periodic motion. Uncertainty with regard to turbulence inflow conditions has posed problems: computations employing the AKN model alone, based on a value of ε in the oncoming flow corresponding to a ratio of eddy-viscosity to laminar viscosity equal to 50 and 90 were in vain. This fact might explain the relative success of Bosch [1] who used $\nu_t/\nu = 10$ with regard to Franke & Rodi's findings [2] who chose ν_t/ν ten times higher. This observation is valid for low-Re model calculations as well: for instance Kawamura & Kawashima [3] have had success with $\nu_t/\nu \approx 20$, while Deng *et al.* [32] failed with the Nagano & Tagawa's low-Re model, for having set a higher value to ν_t/ν. It seems now evident that a higher level of turbulence intensity in the oncoming flow constitutes another source of damping the vortex–shedding motion. Therefore, any further attempt to ignore this fact is now difficult to conceive. Finally, we believe that the choice of the low-Re model is not the overwhelming task for this problem. The AKN model was preferred to the LL one on the only basis of its good behaviour in the case of a vortex-shedding flow past a circular cylinder (Lakehal [33]).

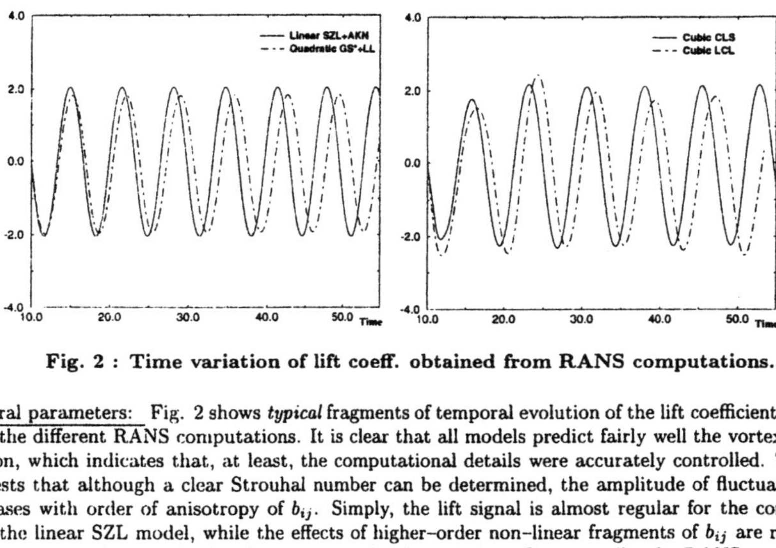

Fig. 2 : Time variation of lift coeff. obtained from RANS computations.

Integral parameters: Fig. 2 shows *typical* fragments of temporal evolution of the lift coefficients obtained with the different RANS computations. It is clear that all models predict fairly well the vortex shedding motion, which indicates that, at least, the computational details were accurately controlled. The figure suggests that although a clear Strouhal number can be determined, the amplitude of fluctuations of c_l increases with order of anisotropy of b_{ij}. Simply, the lift signal is almost regular for the computation with the linear SZL model, while the effects of higher–order non–linear fragments of b_{ij} are responsible of an important lower to higher frequency amplitude variation. Conceptually, the RANS computations which impose two-dimensionality of the flow yield rather a regular shedding behaviour, while the present result is somehow a denial. Still, this has to be supported by more evidence. In fact, there are a few elements which support this latter argument: the fluctuations of c_l signal are not a result of numerical instability, because the residuals were found to drop regularly in all calculations for the same time step (Δt) and number of iterations per Δt. Further, this is probably not a consequence of model coefficients of the quadratic and cubic terms being linked to strain invariant as those of the CLS model are constants. Finally, the influence of the damping function f_μ is naturally of less significance in the quadratic and cubic representation of τ_{ij} than in the linear one, because of C_μ functional forms being in all models related to strain invariant. This is also supported by the fact that calculations with the recent variant of the CLS model proposed by Craft et al. [19], which employs another form of the Re_t-related function f_μ, gave virtually the same result (results not included). Detailed analysis of integral parameters obtained with RANS and LES calculations and compared to experiment are reported in Table 2. This includes the dimensionless shedding frequency St, the time-averaged drag coefficient $\bar{c_d}$, the phase-averaged drag

Table 2: **Integral parameters from experiment, LES and RANS**

Contribution	$\bar{c_d}$	$\tilde{c_d}$	$c_{d,rms}$	$\tilde{c_l}$	$c_{l,rms}$	St	l_r
Experiment (Lyn et al., 1995)	2.1	–	–	–	–	0.132	1.38
Large Eddy Simulation							
Present (NaturLES, SGS2)	2.57	–	0.32	–	1.58	0.11	1.22
UKAHY2, SGS1 (Rodi et al. [5])	2.30	–	0.14	–	1.15	0.13	1.46
TAMU2, SGS2 (Rodi et al. [5])	2.77	–	0.19	–	1.79	0.14	0.94
Eddy Viscosity Models							
Present - *Grid1* (SZL+AKN)	2.265	2.422	0.080	2.07	1.320	0.148	1.09
Present - *Grid2* (SZL+AKN)	2.268	2.428	0.095	2.04	1.380	0.149	1.09
$k-\varepsilon$ & WF (Bosch [1])	1.64	–	≈ 0	–	0.305	0.134	2.84
$k-\varepsilon$ & TL (Bosch [1])	1.72	–	≈ 0	–	0.426	0.137	1.25
Algebraic Stress Models							
Present - *Grid2* (GS + LL)	2.24	2.424	0.0724	1.90	1.368	0.144	1.230
Present - *Grid2* (CLS)	2.27	2.405	0.107	2.12	1.534	0.135	1.225
Present - *Grid2* (LCL)	2.18	2.490	0.187	2.08	1.538	0.128	1.223
Reynolds Stress Models							
RSM & WF (Franke & Rodi [2])	2.15	0.383	0.27	2.11	1.49	0.136	0.98
RSM & TL (Franke & Rodi [2])	2.43	2.079	0.06	1.84	1.3	0.159	1.00

WF = wall functions; TL = 2-layer; SGS1 = Smagorinski model; SGS2 = Dynamic model; UKAHY2 = University of Karlsruhe; TAMU2 = Tokyo Inst. of Technology; **St based on peaks** in c_l-spectrum. Keys for codes SZL, ... as in Table 1; Grid 1 = $y^+ < 2$; Grid 2 = $y^+ < 1$.

and lift coefficients $\tilde{c_d}$ and $\tilde{c_l}$, the RMS values of the temporal variation of drag and lift coefficients $c_{d,rms}$ and $c_{l,rms}$, and the length of the time-averaged separation in the wake l_r. In addition, the table includes on the one hand two LES results selected from the *LES Workshop* of Rodi *et al.* [5], namely UKAHY2 and TAMU2, and on the other EVM results of Bosch [1] and RSM results of Franke & Rodi [2]. The Strouhal number obtained by the linear model is overpredicted and the same can be said about the mean drag coefficient. The shorter reattachment length in the wake tells that the model predicts a higher momentum exchange in the wake, which can result from a pronounced suppression of $\mathcal{P}$ in front of the cylinder. Note also that the global parameters provided by the *zonal* SZL model are comparable to those obtained with the TL RSM of Franke & Rodi [2]. When comparison is restricted to SZL and Bosch's TL computations, the results reveal that on the light of experiment and LES the former method predicts more realistic $\bar{c_d}$, $c_{d,rms}$ and $c_{l,rms}$ than the latter one. The predictive capability of the EASM is already demonstrated from the resulting integral parameters. Even if St shows no sensitivity to numerical details, it exhibits a significant reaction to turbulence modelling. The length of the separation in the wake seems to remain unaltered by any of the quadratic and cubic models and is predicted very close to experiment. The GS approach overpredicts St while both of the cubic models yield excellent agreement with experiment. Phase-averaged global quantities $\tilde{c_d}$ and $\tilde{c_l}$ are within the range of what the TL RSM provided, but the mean drag coefficient is comparatively well determined. Still, the result of the cubic LCL model with regard to $\bar{c_d}$ is distinguishable from all the others: the agreement with experiment is very clear. Note in addition that $\tilde{c_d}$ is much less sensitive to the order of anisotropy of b_{ij} than $\tilde{c_l}$. The LES calculations obtained with NaturLES yield results which stands globally in the bracket of UKAHY2 and TAMU2 LES. The Strouhal number is slightly underpredicted. It is important to say that for pure technical reasons, the statistics were achieved only on one half of the entire lift signal, which provides probably a partial explanation to the underestimation of St by LES. Also this is probably because the calculations have switched meanwhile from no-slip condition to wall functions. It is to underline that the magnitude of both l_r and $\bar{c_d}$ are better predicted with NaturLES than in TAMU2 LES which employs the same SGS model. Further, it was observed from LES calculations reported in [5] that l_r decreases with the utilization of no-slip boundary conditions at the wall.

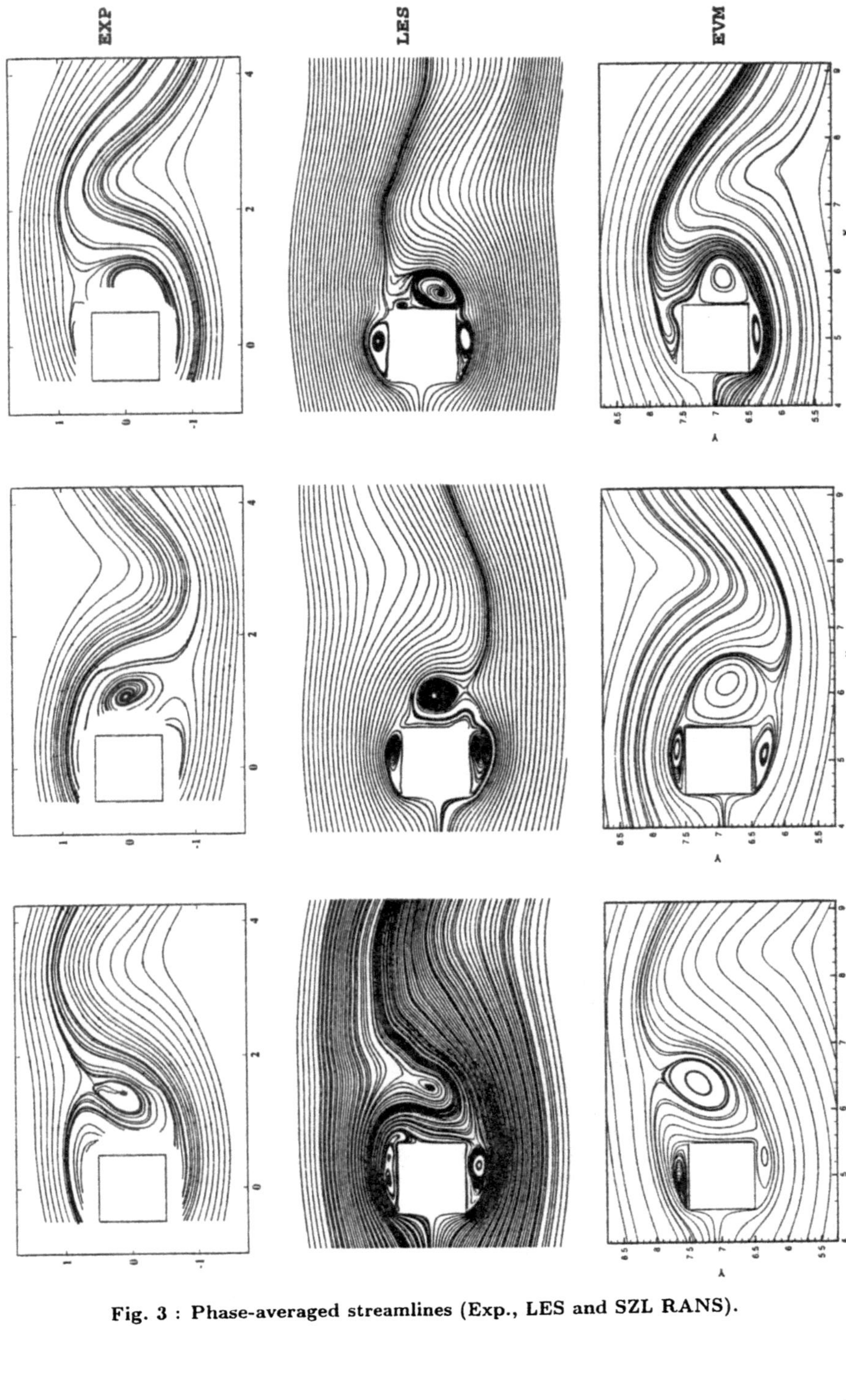

Fig. 3 : Phase-averaged streamlines (Exp., LES and SZL RANS).

Phase-averaged quantities: Comparison of the phase-averaged quantities is limited here to flow streamlines at three phases (Fig. 3) and $\langle u \rangle$ profiles at leeward plane of cylinder at two phases (Figs. 4 and 5), obtained by the SZL RANS and LES respectively. The streamlines are compared to experimental data of Lyn *et al.* [6], while former experimental data of Lyn (private communications) are used to compare the distributions of $\langle u \rangle$. The non-linear models yield in fact a similar picture of streamlines to that of the linear one. Clearly, there is a reasonable agreement between both computations (RANS and LES) and measurements, although some information on the structure of attached/separated shear layers on side walls of cylinder is missing by experimental data. The LES calculations reproduce fairly well the shedding dynamics in the body wake with a clear consistence to the selected LES contributions reported by Rodi *et al.* [5]. A close examination of Fig. 3 suggests yet that the shedding motion is more sustained in the SZL RANS calculations than in the LES and experiment, which is conform with the abovementioned discussion on integral parameters. Similar behaviour of the SZL RANS calculations to RSM computations of Franke & Rodi is observed from Figs. 4 and 5: RANS computations are not able to predict a reverse flow on both sides of the body at once, while LES does with the predicted results at phase 1 being in closer agreement to measurements than at phase 9.

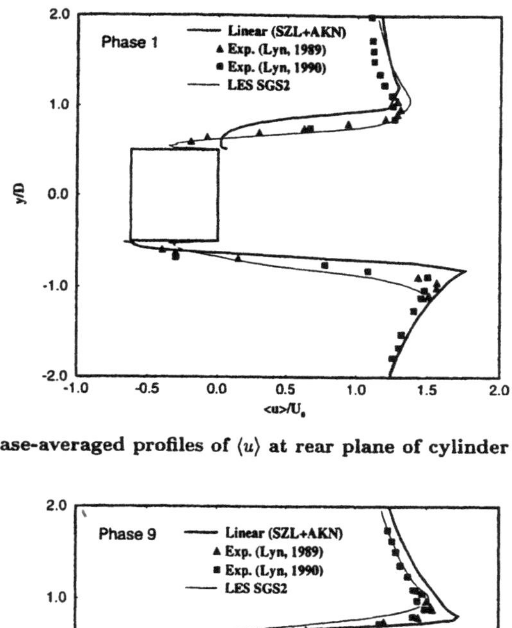

Fig. 4 : Phase-averaged profiles of $\langle u \rangle$ at rear plane of cylinder at phase 1.

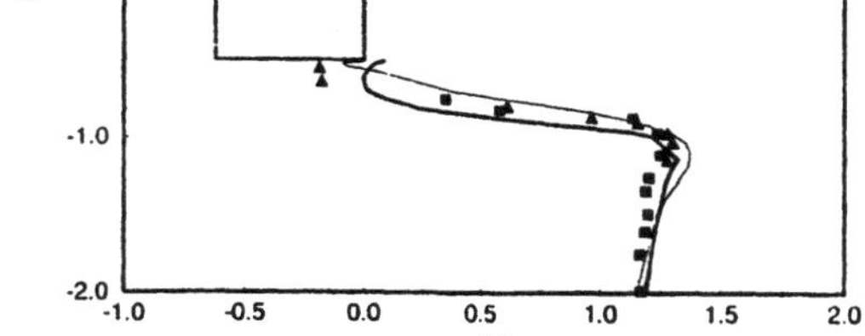

Fig. 5 : Phase-averaged profiles of $\langle u \rangle$ at rear plane of cylinder at phase 9.

Time-averaged quantities: In Fig. 6, results for the time-averaged streamwise velocity on the centre plane of the cylinder are compared. The information concerning the length of recirculation zone already discussed is confirmed: l_r is underestimated by the SZL RANS computation, while the non-linear models clearly improve it. Still, none of the models is able to correctly predict the minimum level of $\overline{U}$. The superiority of LES calculation in this context is evident. However, apart from the linear model calculation which, as it was to be expected, provides a very fast approach to the free-stream velocity due to an exaggerated rate of momentum exchange in body wake, there is a spread difference among the other models regarding this point. In all situations, flow recovery to the non-perturbated state is predicted too fast. Consistency between flow recovery and $\bar{c}_d$ prediction is noticeable in all cases. The closest agreement to experiment is brought by LCL model calculation. What is not entirely clear is the behaviour of the CLS model. The approach to free-stream velocity in the LES is predicted somewhat faster. Note that the *plateau* provided by experiment ($\approx 0.6\,U_0$) is $\approx$ 20% lower than that one of Durao.

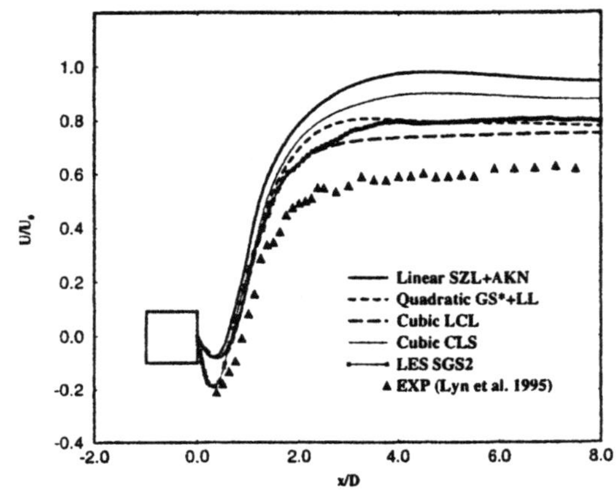

Fig. 6 : Mean $\overline{U}$-velocity distribution on centre-line of cylinder.

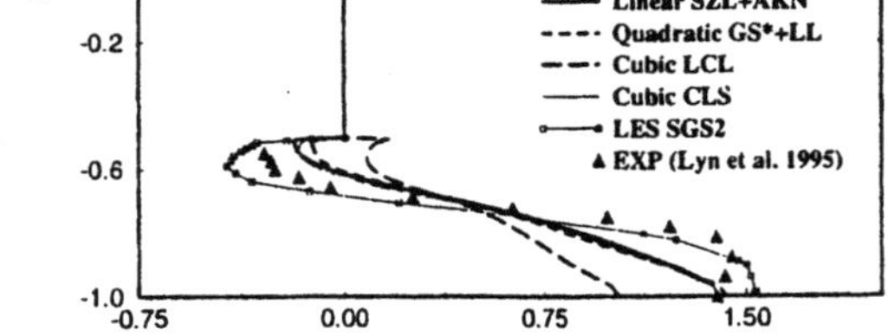

Fig. 7 : Mean $\overline{U}$-velocity distribution on leeward plane of cylinder.

Fig. 7 presents a close-up distribution of mean velocity $\overline{U}$ at leeward section of cylinder. With the exception of LCL RANS calculation, all other models reproduce a non-reattached separated shear layer. The closest agreement from RANS calculations to experiment is not evidently established: it is suggested here that grid resolution well above the region of negative velocity is probably not sufficiently high. In this case, LES performs much better than RANS, but it astonishingly provides an overpredicted magnitude of $|\overline{U}_{max} - \overline{U}_{min}|$. The comparison of mean streamwise and vertical velocity profiles at a vertical section of cylinder (middle) are displayed in Figs. 8 and 9. The mean flow field in this region is particularly well described by all models. Total agreement between RANS and experiment is almost achieved. In addition, an intriguing superior behaviour of the SZL model regarding prediction of the reverse flow very close to the wall is to be mentioned (Fig. 8). The desired depression of eddy-viscosity (and shear-stress then) proceeded in front of the body is probably the explanation. Such behaviour was already observed when the Kato & Launder $\mathcal{P}$–suppression measure was employed for the flow past a mounted cube by Lakehal & Rodi [34]. Relative deviations between EASM calculations occur solely in the tiny flow portion very close to the wall. It should be mentioned yet that this particular flow region is widely dominated by shear, which indicates that the impact of the quadratic and cubic parts of b_{ij} play a secondary role, provided that no spurious turbulent viscosity is swept from the upstream region.

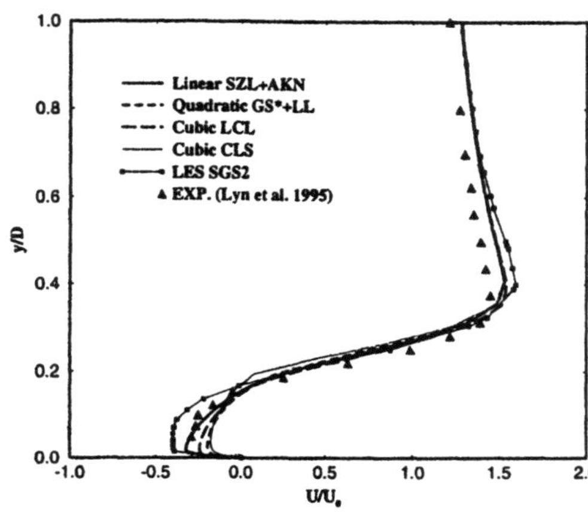

Fig. 8 : Mean $\overline{U}$ profiles on $x = -0.5D$-location of cylinder side-wall.

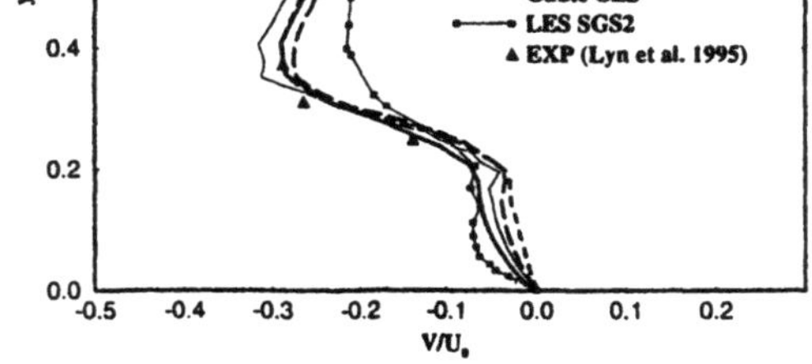

Fig. 9 : Mean $\overline{V}$ profiles on $x = -0.5D$-location of cylinder side-wall.

In contrast with the results reported in Fig. 7, there is a rather weak performance of LES in this region indicating that grid resolution is crucial regardless to which kind of near-wall treatment is applied. In Fig. 10, results of the mean turbulent kinetic-energy contribution of the fluctuations obtained from all models are compared with the LES and experimental data. It should be noted that for LES both the resolved and SGS parts are included. Overall the level of turbulent kinetic energy is much too small in the RANS calculations than in LES. The results of LES are in closer agreement with the experimental data even though they exhibit a slight overprediction of K. However, despite the wide deviation between LES and RANS results, a substantial amount of energy is brought by the two cubic EASM (LCL and CLS) vis-à-vis the linear SZL; the peak in K is predicted with a rate of 50% above. The scatter in the computed profiles of K increases with degree of anisotropy of b_{ij}: the quadratic GS model provides more energy in the wake than the SZL one, but less than the cubic models. Further, the cubic formulations provide a better behaviour than other models which have been so far employed, including high- and low-Re conventional EVM (Bosch [1], Kawamura & Kawashima [3]) and (surprisingly) RSM (Franke & Rodi[2]).

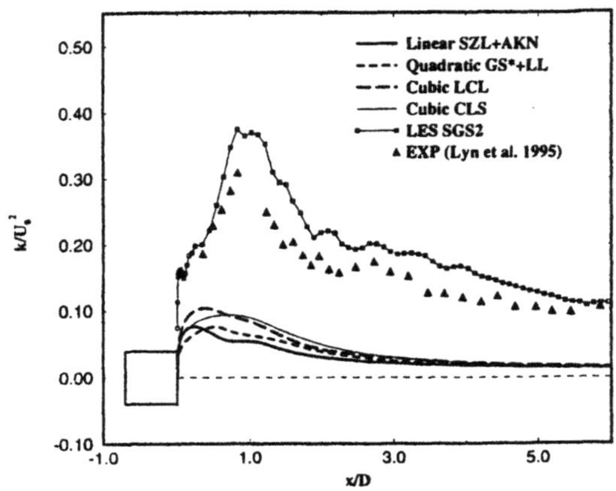

Fig. 10 : Mean turbulent kinetic energy ($1/2\,\overline{u'_k u'_k}$) on centre-line of cylinder.

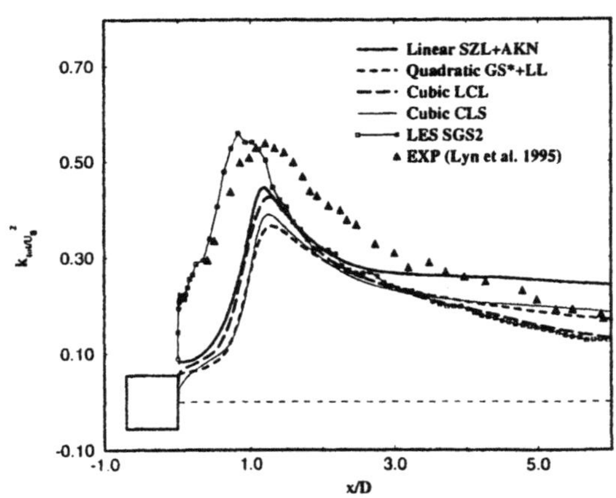

Fig. 11 : Total kinetic energy ($1/2[\overline{u'_k u'_k} + \bar{u}_k \bar{u}_k]$) on centre-line of cylinder.

Obviously, the non-linear models which seem to return a higher level of $\overline{u'^2}$ in the body wake than the linear model, are capable of reproducing anisotropy via the inclusion of non-linear terms in b_{ij}–expression. It is unlikely that the scatter of turbulence energy distributions obtained from model calculations in the wake is solely a matter of the functional dependence of $C_\mu(S,\Omega)$. This was demonstrated through the incapacity of the SZL model to simply perform, when such a functional dependence of C_μ was employed across the whole flow domain. However, there is really no clear indication on how important the sensitivity to streamline curvature in wake flow portion is brought through the inclusion of cubic elements. One understands that this kind of flows are extreme CPU-time consumers, and a systematic analysis of each group of b_{ij} is for the time being beyond our capacity. Fig. 11 compares the total kinetic energy (K_{tot}) of fluctuations (periodic and turbulent) obtained from all models with LES and measurements. The figure shows that LES results are roughly in agreement with the experimental data, although they exhibit a rapid deficit far downstream the region where the peak value is reached. All RANS calculations give a good tendency in comparison with measurements, and as it was to be expected, the linear SZL model yields rather a plateau than the smooth slope of K_{tot}. Still, it can be concluded that this closer agreement obtained from all RANS computations, along with the drastic lower level of turbulent energy discussed above is an indicator of an exaggerated periodic dynamic occurring in the wake.

CONCLUSION

The *challenging* vortex-shedding flow past a square cylinder at Re = 22.000 has been extensively investigated by four different Reynolds stress algebraic equation models that include up to cubic representations of $\overline{u_i'u_j'}$, as well as LES with the Dynamic SGS modelling procedure. The application of all RANS was achieved by solving the semi-viscous sublayer, each using its particular low-Re scheme. The considered flow poses more than one difficulty. The flow is first characterized by an organized periodic motion in its wake, and the time-dependent RANS approach used here forces the resolved dynamics to two-dimensionality and assumes that the 3D turbulent fluctuations are superimposed on the mean flow. The flow has a transitional behaviour: it is quasi-laminar upstream and transition takes place in the separated free shear layers on both side walls of the body. In addition, the flow impinges massively and therefore is highly strained; a fact that strongly questions the predictive capability of all models because the resulting rate of strain tensors deviate seriously from those of a simple shear flow for which most of the employed models have been calibrated.

The conventional $K-\varepsilon$ model used together with the AKN low-Re approach was found not successful. Too much turbulence production produced at impingement has resulted in damping of the vortex-shedding motion. The combination of the SZL and AKN models, which was used only as a measure to bring the desired effect, returned a noticeable success, despite of an exaggerated effect of periodic fluctuations. The comparison of both phase- and time-averaged velocity components showed a very good behaviour of the non-linear models. All RANS calculations provided a common picture concerning the prediction of turbulent kinetic energy: the turbulence fluctuations were severely underpredicted, despite a visible improvement brought through the inclusion of cubic terms in the LCL and CLS explicit algebraic stress models. The main reason is that such 2D RANS calculations, simply do not consider any 3D effects. Out of the analysis of the lift signal obtained from calculations employing the cubic models, it appeared that a low frequency variation of the shedding dynamic was captured, yielding therefore the highest magnitude of turbulence fluctuations. The total kinetic energy was fairly well predicted by all models. The global parameters were determined with much better agreement with measurements, including the recirculation length behind the body. It is important to note that the modification brought to the GS model to account for decelerating flows purely invoking irrotational condition ($\Omega \approx 0$) was very positive, in that, the potential of C_μ to become zero was overcome without taking recourse to any sort of regularization. Also, all RANS calculations were found to respect the realizability constraint. Further developments, calibration and improvements are of course needed to cope with more complicated than a simple homogeneous shear flow.

LES calculations seem to be clearly superior, even if better results were in reach of the present investigation if necessary computation time was used. Finally, the LES results revealed that the practice can be drastically sensitive to numerical details including the near-wall modelling method.

ACKNOWLEDGEMENTS

Financial support was provided by DFG and CNRS within the French-German Research program *Numerical Flow Simulation*. The computations were performed on Cray T90 of the University of Kiel and on Cray T3E of IDRIS (CNRS).

References

[1] Bosch, G. : *Experimentelle und theoretische Untersuchung der instationären Strömung um zylindrische Strukturen*, Ph.D Thesis, University of Karlsruhe, (1995).

[2] Franke R., Rodi W. : *Calculation of vortex shedding past a square cylinder with various turbulence models*, in Turb. Shear Flows **8**, Durst et al. (Eds.), Springer Verlag, (1993).

[3] Kawamura, H., Kawashima, N. : *An Application of a near Wall $k-\tilde{\varepsilon}$ model to the Turbulent Channel Flow with Transpiration and to the Oscillatory Flow around a Square Cylinder*, Proc. 2nd Symp. Turbulent Heat Mass Transfer, Delft (1997).

[4] Kato, M., Launder, B.E. : *The Modeling of Turbulent Flow around Stationary and Vibrating Square Cylinders*, Proc. Turb. Shear Flows **9**, Kyoto, 10-4-1, (1993).

[5] Rodi W., Ferziger J., Breuer M., Pourquier M. : *Status of Large Eddy Simulation. Results of a Workshop*, J. Fluid Mech., **119**, pp. 248–262, (1997).

[6] Lyn, D.A., Einav, S., Rodi, W., Park, J.H. : *A Laser-Doppler Velocimetry Study of Ensemble-averaged Characteristics of the Turbulent Near Wake of a Square Cylinder*, J. Fluid Mech., **304**, pp. 285–319, (1995).

[7] Breuer, M., Lakehal, D., Rodi, W. : *Flow Around a Surface Mounted Cubical Obstacle: Comparison of LES and RANS-Results*, in Notes of Numerical Fluid Mech., Vieweg Verlag, **53**, pp. 22–30, (1996).

[8] Reynolds, W.C., Hussein, A.K.M.F. : *The mechanisms of an Organized wave in Turbulent Shear Flow. Part 3. Theoritical Models and Comparisons with Experiments*, J. Fluid Mech., **54**, pp. 263-288, (1972)

[9] Gatski, T. B., Speziale, C. G. : *On explicit algebraic stress models for complex turbulent flows*, J. Fluid Mech., **254**, pp. 59–78, (1993).

[10] Speziale, C. G., Sarkar, S., Gatski, T. B. : *Modelling the presure–strain correlation of turbulence: an invariant dynamical systems approch*, J. Fluid Mech., **227**, pp. 245–272, (1991).

[11] Rung, T. : *Private communication*, Hermann Föttinger Institute, TU-Berlin, Germany (1998).

[12] Lien, F.S., Leschziner, M.A. : *Computational Modelling of 3D Turbulent Flow in S-Diffuser and Transition Ducts*, Eng. Turbulence Modelling and Experiments 2, Elsevier Sce. Pub. (1993).

[13] Lien, F.S., Chen, W.L., Leschziner, M.A. : *Low-Reynolds-Number Eddy-viscosity modelling based on non-linear stress-strain/vorticity relations*, Proc. 3rd Int. Symposium of Engineering Turbulence Modeling and Measurements, Crete, Greece (1996).

[14] Suga, K. : *Eddy-viscosity modelling with deformation invariants and non-linear elements*, First year Ph.D report, Dept. Mech. Eng., UMIST, (1993).

[15] Shih, T.H., Zhu, J., Lumley, J.L. : *A realisable Reynolds stress algebraic equation model*, NASA TM-105993, (1993)

[16] Norris, L.H., Reynolds,W.C. : *Turbulent channel flow with a mooving wavy boundary*, Rept. No. FM-10, Stanford University, Dept. Mech. Eng., (1975).

[17] Craft, T.J., Launder, B.E., Suga, K. : *Extending the applicability of eddy-viscosity models through tie use of deformation invariants and non-linear elements*, Proc. 5th Int. Symp. Refined Flow Modelling and Turbulence Measurements, Paris (1993).

[18] Launder, B.E., Sharma, B. : *Application of the energy dissipation model of turbulence for the calculation of flow near a spinning disc*, Lett. Heat Mass Transfer, **1**, pp. 131–138, (1974).

[19] Craft, T.J., Launder, B.E., Suga, K. : *Development and application of a cubic eddy-viscosity model of turbulence*, Int. J. Heat and Fluid Flow, **17**, pp. 108–115, (1996).

[20] Shih, T.H., Zhu, J., Lumley, J.L. : *A New Reynolds stress algebraic equation model*, Comput. Methods Appl. Mech. Engrg., **125**, pp. 287-302, (1995).

[21] Abe K., Kondoh T., Nagano, Y. : *A new turbulence model for predicting fluid flow and heat transfer in separating and reattaching flows–I. Flow field calculations*, int. J. Heat Mass Transfer, **37**, 1, 139–151, (1994).

[22] Wolfshtein, M. : *The velocity and temperature distribution in one–dimensional low with turbulence augmentation and pressure gradient*, Int. J. Heat Mass Transfer, **12**, pp. 301–318, (1969).

[23] Jones, W.P., Launder, B. E. : *Prediction of relaminarization with a two-equation turbulence model*, Int. J. Heat Mass Transfer, **15**, pp. 301–314, (1972).

[24] Germano, M.; Piomelli, U.; Moin, P.; Cabot, W. H. : *A dynamic subgrid-scale eddy viscosity model*, Phys. Fluids A. **3** (7), pp. 1760–1765, (1991).

[25] Smagorinsky, J. : *General circulation experiments with the primitive equations, I, The basic experiment*, Mon. Weather Rev. **91**, pp. 99–165, (1963).

[26] Duchamp de Lageneste, L., Buffat, M., Leribault, C. : *Development of large eddy simulation methods on unstructured grids*, Internal Report 22 pp., LMFA, Ecole Centrale de Lyon, France (1997).

[27] Rhie, C.M., Chow, W.L. : *A numerical study of the turbulent flow past an isolated airfoil with trailing edge separation*, AIAA-J., **21**, pp. 1225–1532, (1983).

[28] Stone, H.L. : *Iterative solution of implicit approximations of multidimensional partial differential equations*, SIAM J. on Num. Anal., **5**, pp. 530–558, (1968).

[29] Dervieux, A. : *Steady Euler flow simulation using unstructured meshes*, VKI L. Series 1884-04, (1985).

[30] Steve, H. : *Schémas implicites linéaires décentrés pour la resolution des équations d'Euler en plusieurs dimensions*, Thèse Université de Provence Aix-Marseille 1, (1988).

[31] Werner, H., Wengle, H. : *Large-Eddy Simulation of Turbulent Flow over and around a Cube in a plate Channel*, Proc. 8th Symp. Turb. Shear Flows, Stuttgart, Springer Verlag, (1993).

[32] Deng, D.W., Piquet, J., Queutey, P., Visonneau, M. : *Vortex-shedding flow predictions with eddy-viscosity models*, Engng. Turbulence Modelling & Measurements 2, Elsevier, pp. 143–152, (1993).

[33] Lakehal, D. : *Computation of vortex shedding past bluff bodies employing statistical turbulence models*, Institutesbericht 02-1998, Hermann Föttinger Institute, TU-Berlin, Germany (1998).

[34] Lakehal, D., Rodi, W. : *Calculation of the Flow Past a Surface-Mounted Cube with Two-Layer Turbulence Models*, J. Wind Eng. Ind. Aerod., **67-68**, pp. 65–78, (1997).

Numerical simulation of turbulent high speed flows

PART I : Supersonic Combustion.
U.Wepler*, W. Koschel*, S.Melen†, S.Sasse*
PART II : Atomization and mixing.
A.Stoukov†, D.Vandromme†, X.Silvani†, H.Ha Minh‡
* Institute for Jet Propulsion and Turbomachinery
Aachen University of Technology, D-52062 Aachen, Germany
† LMFN,CORIA,UMR 6614
CNRS - Universite/INSA de Rouen, Campus du Madrillet BP8,
76801 St Etienne du Rouvray, France
‡ IMFT CNRS URA 005, Toulouse, France

Summary

The solution of the compressible Navier-Stokes equations have been extended to different problems of turbulent flows. A first part describes the application of a Probabilistic Eulerian-Lagrangian Model (PEuL) to the non-premixed gaseous hydrogen/oxygen flame in a supersonic flow. A second part is devoted to the subsonic mixing and atomization of a liquid oxygen stream into a hydrogen one. The common part of these works is the development of physical models for complex flows with applications to high speed propulsion devices.

Part I: Supersonic Combustion

An unstructured finite element method for the solution of the compressible Navier–Stokes equations including the chemistry for supersonic turbulent non-premixed combustion is presented. Turbulent mixing is modeled by a high Reynolds number $k-\varepsilon$ model using a modification to take compressibility effects into account. A Probabilistic Euler Lagrangian (PEuL) model is applied to describe the non-premixed flame. The numerical integration of the transport equations is based on a Runge–Kutta type time marching scheme using unstructured adaptive meshes. Finally results of the numerical simulation of a reacting mixing-layer are presented.

I-1 Introduction

In this paper the progress of the research group at the RWTH Aachen within the French–German collaboration program on numerical flow simulation will be described. The renewed interest in hypersonic vehicles causes a substantial need in developing efficient numerical codes to model combustion chamber flows under supersonic conditions. For the application of supersonic combustion in airbreathing propulsion engines the mixing and the reaction process of the fuel are the major problems. The incoming fuel has to mix with the surrounding air before reaction takes place. The turbulent fluctuations have a great influence on mixing which in terms controls the chemical reaction rates. Therefore they could not be neglected in supersonic reactive flows. It is not possible to express the mean chemical reaction rates in terms of the mean temperature and chemical concentrations. Here a statistical description of the flow field is needed. This description may be provided by the use of a probability density function (PDF) approach, which has been developed in a common approach between CORIA Rouen and RWTH Aachen. Here the PEuL model [1] including such a statistical description of the reaction rate is used.

I-2 Governing equations

I-2.1 Navier–Stokes equations

For a compressible flow of a chemically reacting gas the Navier–Stokes equations are considered in their Favre–averaged (mass weighted) conservative form. These equations are the continuity equation:

$$\frac{\partial}{\partial t}(\overline{\rho}) + \frac{\partial}{\partial x_j}(\overline{\rho}\widetilde{u}_j) = 0 \tag{I-1}$$

the species continuity equation:

$$\frac{\partial}{\partial t}\left(\overline{\rho}\widetilde{Y}_n\right) + \frac{\partial}{\partial x_j}\left(\overline{\rho}\widetilde{Y}_n\,\widetilde{u}_j\right) = \frac{\partial}{\partial x_j}\left(\overline{J_{j_n}} - \overline{\rho\, u_j'' Y_n''}\right) + \overline{\rho}\widetilde{\omega}_n \tag{I-2}$$

the momentum equation:

$$\frac{\partial}{\partial t}(\overline{\rho}\widetilde{u}_i) + \frac{\partial}{\partial x_j}(\overline{\rho}\widetilde{u}_i\widetilde{u}_j) = -\frac{\partial \overline{p}}{\partial x_i} + \frac{\partial}{\partial x_j}\left(\widetilde{\sigma}_{ij} - \overline{\rho\, u_i'' u_j''}\right) \tag{I-3}$$

and finally the energy equation:

$$\frac{\partial}{\partial t}\left(\overline{\rho}\widetilde{E}\right) + \frac{\partial}{\partial x_j}\left(\left(\overline{\rho}\widetilde{E} + \overline{p}\right)\widetilde{u}_j\right) = \\ \frac{\partial}{\partial x_j}\left(\left(\widetilde{\sigma}_{ij} - \overline{\rho\, u_i'' u_j''}\right)\widetilde{u}_i + \widetilde{q}_j + \overline{\rho\, u_j'' H_T''}\right) \tag{I-4}$$

where $\widetilde{E} = \frac{\widetilde{u}_i\widetilde{u}_i}{2} + \widetilde{e} + k$ is the specific total energy with the mass averaged turbulent kinetic energy $k = \frac{\widetilde{u_i'' u_i''}}{2}$. The mass flux $\overline{J_{j_n}}$ and the turbulent diffusion flux $\overline{\rho\, u_j'' Y_n''}$ is given by:

$$\overline{J_{j_n}} = \overline{\rho} D_n \frac{\partial \widetilde{Y}_n}{\partial x_j} = \frac{\mu}{Sc_n} \frac{\partial \widetilde{Y}_n}{\partial x_j} \tag{I-5}$$

$$\overline{\rho u_j'' Y_n''} = -\frac{\mu_t}{Sc_{t_n}} \frac{\partial \widetilde{Y}_n}{\partial x_j} . \tag{I-6}$$

The Newton stress tensor $\widetilde{\sigma}_{ij}$ and the turbulent stress tensor $\overline{\rho u_i'' u_j''}$ are calculated from:

$$\widetilde{\sigma}_{ij} = \mu \left(\frac{\partial \widetilde{u}_i}{\partial x_j} + \frac{\partial \widetilde{u}_j}{\partial x_i} \right) - \frac{2}{3} \mu \frac{\partial \widetilde{u}_k}{\partial x_k} \delta_{ij} \tag{I-7}$$

$$-\overline{\rho u_i'' u_j''} = \mu_t \left(\frac{\partial \widetilde{u}_i}{\partial x_j} + \frac{\partial \widetilde{u}_j}{\partial x_i} - \frac{2}{3} \frac{\partial \widetilde{u}_k}{\partial x_k} \delta_{ij} \right) - \frac{2}{3} \rho k \delta_{ij} . \tag{I-8}$$

The internal energy is obtained by:

$$\widetilde{e} = \sum_{n=1}^{N_{sp}} \left(\widetilde{Y}_n \int_0^T c_{v n} dT + Y_n h_n^0 \right) . \tag{I-9}$$

The temperature dependency of the molecular viscosity μ is considered by the Sutherland equation. The turbulent viscosity is given by:

$$\mu_t = C_\mu \frac{\overline{\rho} k^2}{\varepsilon} . \tag{I-10}$$

The heat flux $\widetilde{q}_j$ and its turbulent transport $\overline{\rho u_j'' H_T''}$ is denoted as:

$$\widetilde{q}_j = -c_p \frac{\mu}{Pr} \frac{\partial \overline{T}}{\partial x_j} - \sum_{n=1}^{N_{sp}} \frac{\mu}{Sc_n} h_n(\overline{T}) \frac{\partial \widetilde{Y}_n}{\partial x_j} \tag{I-11}$$

$$\overline{\rho u_j'' H_T''} = \frac{\mu_t}{Sc_{t_n}} \frac{\partial H_T}{\partial x_j} . \tag{I-12}$$

The equation set described above is completed by addition of the thermal equation of state:

$$\overline{p} = \rho R \overline{T} = \rho \mathcal{R} \sum_{n=1}^{N_{sp}} \frac{\widetilde{Y}_n}{\mathcal{M}_n} \overline{T} . \tag{I-13}$$

There are the turbulent energy k and the dissipation rate ε, representing the influence of turbulence, and the chemical source term $\widetilde{\omega}$, representing the influence of combustion, left to close the system. They will be considered in the following sections.

I-2.2 Turbulence modeling

The turbulence is modeled by the standard $k-\varepsilon$ model for high Reynolds number suggested by Launder and Spalding [2] with the addition of the Sarkar model [3] to take compressibility effects into account. The model consists of the transport equation for the turbulent kinetic energy k:

$$\frac{\partial}{\partial t}(\overline{\rho} k) + \frac{\partial}{\partial x_i}(\overline{\rho}\,\widetilde{u}_i\, k) = P_t + \frac{\partial}{\partial x_i}\left(\left(\mu + \frac{\mu_t}{\sigma_k}\right)\frac{\partial k}{\partial x_i}\right) - \overline{\rho}\,(\varepsilon_s + \varepsilon_d) \quad \text{(I-14)}$$

and for the solenoidal dissipation rate ε_s:

$$\frac{\partial}{\partial t}(\overline{\rho}\varepsilon_s) + \frac{\partial}{\partial x_i}(\overline{\rho}\widetilde{u}_i\varepsilon_s) = C_{\varepsilon 1} P_t \frac{\varepsilon_s}{k} + \frac{\partial}{\partial x_i}\left(\left(\mu + \frac{\mu_t}{\sigma_\varepsilon}\right)\frac{\partial \varepsilon_s}{\partial x_i}\right) - C_{\varepsilon 2}\overline{\rho}\frac{\varepsilon_s^2}{k} \,. \quad \text{(I-15)}$$

Sarkar postulated that the solenoidal dissipation ε_s is unaffected by compressibility. Sarkar supposed further on that the dilatation dissipation ε_d should be a function of turbulence Mach number M_t defined by

$$M_t^2 = \frac{2\,k}{a^2} \quad \text{(I-16)}$$

with a equal to the speed of sound. The dilatation dissipation ε_d is assumed to be proportional to ε_s:

$$\varepsilon_d = \alpha_1 M_t^2 \varepsilon_s \quad \text{with} \quad \alpha_1 = 1\,. \quad \text{(I-17)}$$

The turbulent production term P_t is derived from:

$$P_t = \mu_t\left(\frac{\partial \widetilde{u}_i}{\partial x_j} + \frac{\partial \widetilde{u}_j}{\partial x_i} - \frac{2}{3}\frac{\partial \widetilde{u}_k}{\partial x_k}\delta_{ij}\right) - \frac{2}{3}\rho\, k\,\frac{\partial \widetilde{u}_k}{\partial x_k}\delta_{ij}\,. \quad \text{(I-18)}$$

I-2.3 Combustion modeling

The last unknown term is the chemical source term $\widetilde{\omega}_n$. Due to the strong non-linear dependence of this term on temperature and the species concentration and due to their strong fluctuation in turbulent flow, it's not possible to compute $\widetilde{\omega}_n$ from the mean values of those quantities. It's necessary to apply a statistical description by introducing a probability density function (PDF).

In order to take the mutual influence between turbulence and chemistry into account Borghi [1] proposed a concept which is based on a Lagrangian description of a fluid particle. This PEuL (Probabilistic Euler Lagrangian) model was further developed by Vervisch [4] and Melen [5].

For the chemistry, the first hypothesis is the very fast chemical reaction with a presumed PDF approach for a conserved scalar. For study effects of finite-rate chemistry in turbulent non-premixed flames, a presumed PDF approach is used where the chemistry is described through one global reaction step:

$$\text{fuel} + r_f\ \text{oxidizer} \longrightarrow (1 - r_f)\ \text{product} \quad \text{(I-19)}$$

and r_f is the stoichiometric coefficient. The aim of the method proposed is to built an approximated PDF of the thermo-chemical variable. In this model only two variables, the mixture fraction Z and the oxidizer mass fraction Y_{ox}, are required for the description of the reactive flow field.

$$\widetilde{\omega}_n = \int_0^1 \int_0^1 \dot{\omega}_n \left(Y_{ox},\, Z\right) \widetilde{P}\left(Y_{ox},\, Z\right) dY_{ox}\, dZ\,. \tag{I-20}$$

The joint PDF is reduced to a two-dimensional PDF of the mixture fraction and the reactive species. Each computational volume is represented by the mean of the mixture fraction and the reactive species, and by a spectrum of the turbulent mixing time scales. A fluid particle may chemically react and at the same time mix with the surrounding fluid. These processes are described by a Lagrangian equation where the mixing process is modeled by the IEM (Interaction with the mean) model [6]. In mixture fraction space the Lagrangian equation for species n is written as

$$\frac{\partial Y_n}{\partial Z} = \frac{\left(\widetilde{Y}_n - Y_n\right) + \tau_{mix}\, \dot{\omega}_n}{\widetilde{Z} - Z}\,. \tag{I-21}$$

Equation (I-21) describes the evolution of a fluid particle and defines a trajectory in mixture fraction space. Fig. 1 shows schematically such trajectories for the species oxygen. The three lines represent the following physical conditions. If the chemical time is very small compared to a mixing time scale, chemical equilibrium is reached, which is represented by the line $C^{(2)}$. In contrast, if the chemical time scale is very large compared to a mixing time scale, the chemical reaction is to small and only mixing occurs, line $C^{(1)}$.

For intermediate values a competition between chemistry and mixing is observed which gives line $C^{(3)}$. This Line is computed with a turbulent mixing time:

$$\tau_{mix} = C_t\, \frac{k}{\varepsilon} \tag{I-22}$$

where C_t is a constant of order one. The PEuL model approximates the joint PDF with these three lines. Each line contributes to the final solution depending on the weighting coefficient α_s, which are found from the following conditions:

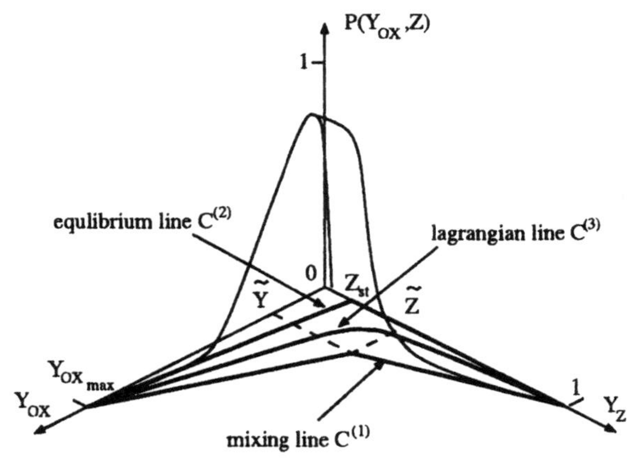

Figure 1 Trajectories of a fluid particle

$$\widetilde{Y}_n = \int_0^1 P(Z) \sum_{s=1}^{3} \alpha_s\, Y_n^s(Z)\, dZ\,; \quad \sum_{s=1}^{3} \alpha_s = 1 \text{ with } \alpha_1 = 0 \text{ or } \alpha_2 = 0\,. \tag{I-23}$$

Finally, the mean chemical source term is obtained by:

$$\widetilde{\omega}_n = \int_0^1 P(Z) \sum_{s=1}^{3} \alpha_s \widetilde{\omega}_n^s(Z)\, dZ. \tag{I-24}$$

The PDF $P(Z)$ is presumed as a β function with the two parameters mixture fraction $\widetilde{Z}$ and its variance $\widetilde{Z''^2}$. For these two variables two transport equations are solved:

$$\frac{\partial}{\partial t}\left(\overline{\rho}\widetilde{Z}\right) + \frac{\partial}{\partial x_i}\left(\overline{\rho}\widetilde{u}_i\widetilde{Z}\right) = \frac{\partial}{\partial x_i}\left(\frac{\mu}{Sc_z}\frac{\partial \widetilde{Z}}{\partial x_i} + \frac{\mu_T}{Sc_t}\frac{\partial \widetilde{Z}}{\partial x_i}\right) \tag{I-25}$$

$$\frac{\partial}{\partial t}\left(\overline{\rho}\widetilde{Z''^2}\right) + \frac{\partial}{\partial x_i}\left(\overline{\rho}\widetilde{u}_i\widetilde{Z''^2}\right) = \frac{\partial}{\partial x_i}\left(\left(\frac{\mu}{Sc_z} + \frac{\mu_T}{Sc_t}\right)\frac{\partial \widetilde{Z}}{\partial x_i}\right) + 2\frac{\mu_T}{Sc_t}\left(\frac{\partial \widetilde{Z}}{\partial x_i}\right)^2 - C_Z\,\widetilde{Z''^2}\,\frac{\widetilde{\varepsilon}}{\widetilde{k}}. \tag{I-26}$$

I-3 Numerical scheme and mesh generation

I-3.1 Runge–Kutta scheme

The transport equations are integrated by the application of the Galerkin weighted residual method in combination with a Rung–Kutta type time marching scheme [7, 8]. The five step time integration algorithm involves the weak Galerkin formulation applied to each sub step ν:

$$\int N_K \left(\frac{\partial U}{\partial t}\right)^{\nu+1} d\Omega = \int \frac{\partial N_K}{\partial x_j} F_j^\nu\, d\Omega - \int N_K F_j^\nu n_j\, d\Gamma \tag{I-27}$$

where the convective fluxes F_j^ν are represented by a C^0 linear approximation:

$$F_J^\nu = N_I\, F_J^{I\nu}. \tag{I-28}$$

The discretization above can be regarded as a node–centered scheme with an arithmetic averaging of the fluxes normal to the edges of each element. For more detailed information about the the numerical scheme and solution method see the work of Rick [8].

I-3.2 Mesh generation

The triangulation of arbitrary shaped computational domains with embedded interior boundaries is accomplished by an automatic mesh generation scheme following the generalized advancing front method [9]. Points and elements are introduced simultaneously and therefore significant changes in the local mesh structure are possible. The shape of the triangles is controlled by the parameters: element size, element stretching and stretching direction. A spatial distribution of these mesh parameters is provided by elements when generating an initial mesh.

Within the automatic mesh generation scheme, allowing to change all mesh parameters significantly, the initial mesh is acting as a background grid providing the spatial distribution of the mesh parameters. In order to determine the nodal values of these parameters in an optimal manner, it is necessary to apply some method of error estimation. The second derivatives of a scalar key variable, e.g. density, mixture fraction or Mach number, are used to give some indication of error magnitude and direction. The condition of a uniformly distributed error indicator within the domain leads to the optimal nodal values of the mesh parameters.

I-4 Computational Results

As a first test case computations were performed for compressible shear layers of two air flows ranging from subsonic to supersonic conditions.

For these cases, a lot of different experimental results [10, 11, 12] are available showing the dependence of the relative shear layer growth (compressible to incompressible) on the convective Mach number (Fig. 2). The experimental data obtained at NASA Langley indicate the well-known behavior of the reduced shear layer growth at supersonic conditions. The results of the numerical simulation fit if the compressibility correction for the $k-\varepsilon$ model is applied. They also correspond to the computational results obtained by Melen [5] for the French group. For code validation of turbulent reactive flow fields a supersonic mixing layer is calculated. Figure 3 displays the flow conditions for this flow simulation. The flow field in the channel with constant area is splitted into an

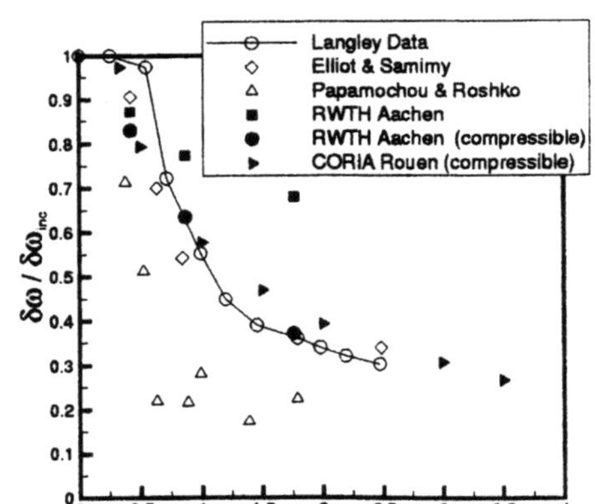

Figure 2 Relative growth of the shear layer

upper air stream and into a gaseous hydrogen flow. Together they generate a shear layer in the middle of the channel. The computational mesh is presented in Figure 4 showing a fine grid in the shear layer zone which leads to a good resolution of the gradients. For this case two different computations were performed. First the shear layer was simulated without reaction in order to get a close view of the turbulent mixing process. Subsequent by the reaction of hydrogen was initiated by increasing the incoming air temperature to a self-ignition level.

The contour plots of constant mixture fraction $\widetilde{Z}$ shown in Figure 5 reveal the growth rate of the mixing layer. It depends on the convective Mach number. In Figure 6 the contour plots of the variance $\widetilde{Z''^2}$ of the mixture fraction are shown. The maximum can be seen in the center of the mixing zone slightly moved to the hydrogen side because of differences between air and hydrogen with respect to their densities.

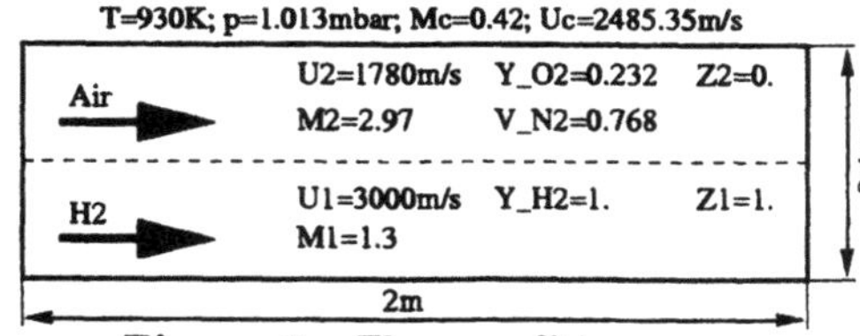

Figure 3 Flow conditions

Figure 7 shows the contour plots of the resulting temperature isolines from the calculation of a non-reacting mixing layer. The temperature raises from $930K$ in the free stream value to $960K$ in the center of the mixing layer due to the shear stresses. This temperature rise is finally responsible for the ignition of the shear layer in the reactive case.

In Figure 8 the contour plot of the constant temperature lines of a reacting mixing layer is presented.

For this calculations the PEuL-model was used. The results show that the ignition delay can be simulated by the PEuL-model. The temperature contours correspond to the contour of the reaction product water. The results agree very well with computations performed by the French research group [5] applying a different numerical approach for solving the Navier-Stokes equations.

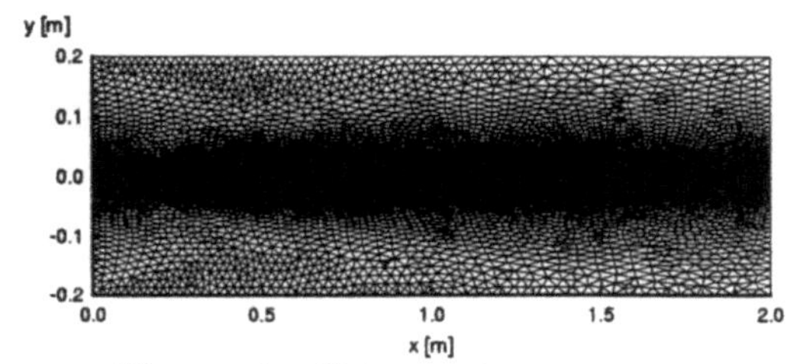

Figure 4 Netgeometry

I-5 Conclusion

A FEM flow solver for supersonic turbulent and reactive flow fields was presented. The code was developed by the research group at the RWTH Aachen with support of the French Group concerning the implementation of the PEuL-model and has been successfully validated for the case of compressible shear layer growth. Furtheron it has been applied to the simplified case of a reactive mixing layer of gaseous hydrogen and air at supersonic conditions. The results obtained by the calculations are reasonable in regard to physics and mixing behaviour. The PEuL–model predicts the ignition delay well.

As further work within the French–German collaboration the code will be applied to other reactive test cases in order to improve the numerical code. The solver will also be used to simulate an experiment carried out by Nuding [13] at the DLR, Germany to investigate the shock wave – shear layer interaction under supersonic non-reactive and reactive flow conditions [14].

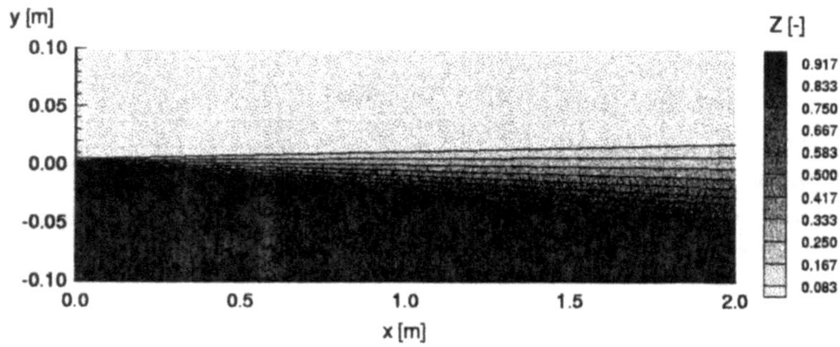

Figure 5 Mixture fraction contours

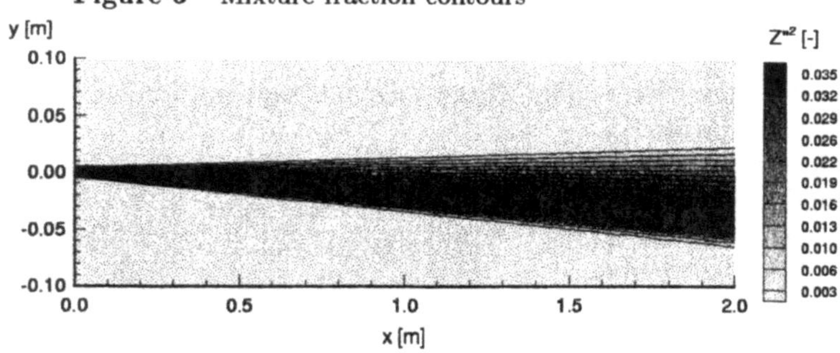

Figure 6 Variance of the mixture fraction contours

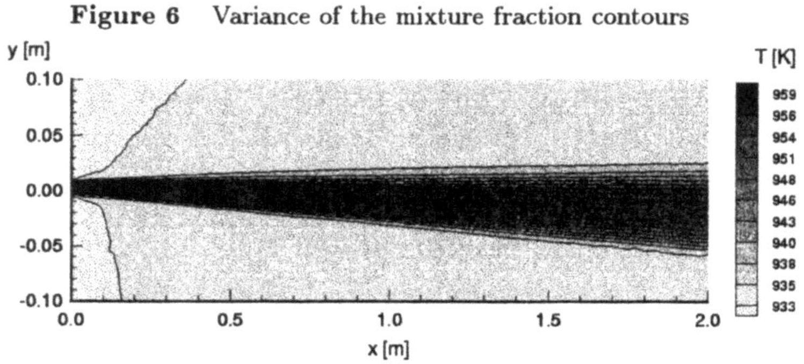

Figure 7 Temperature contour of a non reacting shear layer

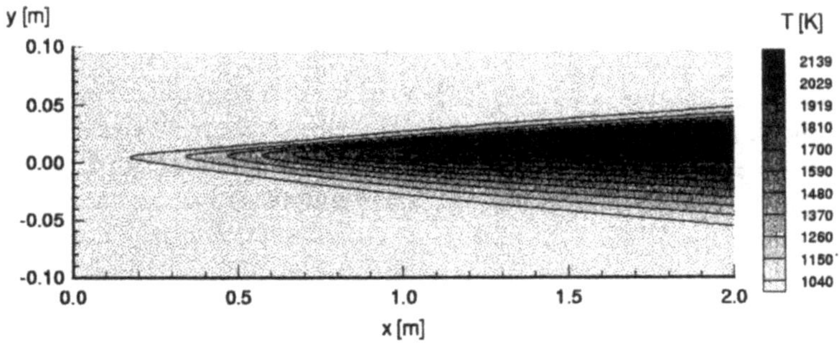

Figure 8 Temperature contours of the reacting shear layer

Part II: Atomization and mixing

In this work, we suppose that the primary phase of atomization of a low velocity liquid oxygen jet into high speed hydrogen flow can be modelled without accounting for the surface tension. This assumption is valid if primary instabilities are supposed to be controlled by effects of high Reynolds number, i.e the developed turbulence for which one can describe macroscopic effects statistically as in the case of a turbulent mixture. Thus, statistical models should represent the phenomena correctly . The obvious limit is that such a model cannot provide deterministic information about droplet formation, which depends on the surface tension. With such a statistical description of the behavior of the interface, it is possible to evaluate the properties of the gas/liquid mixture (or rather heavy/light gas fluid mixture) with a model of equivalent droplet.

This model is based on the knowledge of the volume fraction of the heavy/light gas and the density of interface per volume (3D) or area (2D) unit. From these quantities, it is possible to calculate the size of equivalent droplet created at the primary phase of atomization.

Current work gathers two complementary contributions:

- direct numerical simulation of a heavy and slow jet in the fast flow of a light gas.
- Eulerian statistical modeling of primary atomization.

The first part is devoted to the description of the behavior of the heavy/light gas interface, defined by a given value of a scalar in order to supply the statistical modeling of the second part. This last one consists on implementation of the transport equations for the volume fraction and the interface density. These last equations allow the evolution of equivalent drop size.

II-1 Introduction

Considering the various interactions between phases in the injection flow is a major difficulty for the development of the rocket motors with liquid propellants. Various approaches have already been studied... These works deal primarily with the approaches of interface capturing method recommended by Zaleski [15], [16], [17] [18] and with the statistical representation of the liquid phase dispersed within the gas one from a Lagrangian point of view [19] or Eulerian [20], [21]. Between them, no single method allows to go from a dense phase regime (interface capturing approach) to a completely atomized and dispersed regime.

Pulverization was primarily studied with the point of view of the instability of the interface. But we can consider also that as turbulent mixture with strong density differences and that capillary forces act only at small scales. With the traditional approach of single phase turbulence, one can think that the effect of the capillary forces is similar to the one of viscosity and molecular diffusivity; Then, the consequence is that the properties of atomization (dispersion of the liquid) would not be dependent on the surface tension and viscosity, whereas the properties on small scales would. This approach is chosen in the following paper.

II-2 Numerical simulation

II-2.1 Governing equations

Taking into account the introductory assumptions, in the two-dimensional case, the flow is described by the following equations:
Continuity

$$\frac{\partial}{\partial t}\rho^L + \frac{\partial}{\partial x_j}\rho^L u_j = D\frac{\partial}{\partial x_j}Y^L$$

$$\frac{\partial}{\partial t}\rho^G + \frac{\partial}{\partial x_j}\rho^G u_j = D\frac{\partial}{\partial x_j}Y^G$$

where ρ^L and ρ^G stands for the heavy and light gas density.
Momentum

$$\frac{\partial}{\partial t}\rho u_i + \frac{\partial}{\partial x_j}(\rho u_i u_j + p\delta_{ij}) = \frac{\partial}{\partial x_j}\sigma_{ij}$$

Energy

$$\frac{\partial}{\partial t}\rho E_t + \frac{\partial}{\partial x_j}[u_j(\rho E_t + p)] = \frac{\partial}{\partial x_j}\sigma_{ij}u_i - \frac{\partial}{\partial x_j}q_j.$$

The mixture density is $\rho = \rho^L + \rho^G$, the pressure is given by the equation of perfect gas state $P = \rho rT$. We recall that, surface tension is neglected.

II-2.2 Numerical method

The numerical method employed for the resolution of the system above is based on the Harten-Yee TVD scheme for the convective terms [22], [14] and the centered scheme for the viscous ones. It is explicit and second order accurate in space and time.

Simulation requires a great number of calculation points as well as a significant number of time steps. In order to preserve reasonable memory size and calculation time, the code was parallelized using multi-block technique, implemented with PVM Library [23]. As an example, calculation on a domain containing 2048*1024 points, distributed on 64 processors, requires 30 CPU hours on CRAY T3E for 60000 time steps.

II-2.3 Flow configuration

Boundary conditions

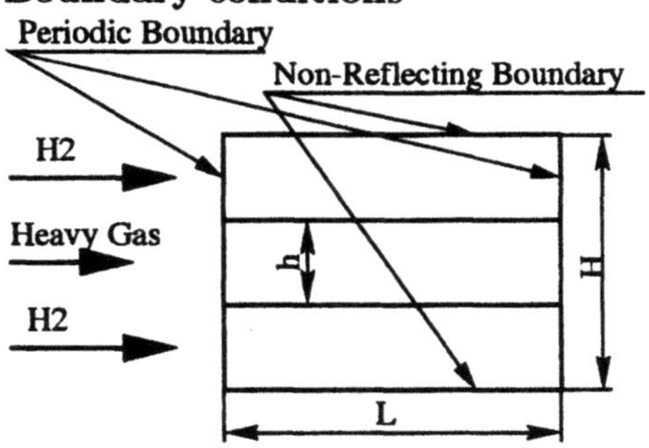

The simulation is temporal and two-dimensional, so that boundary conditions are periodic in longitudinal direction and non-reflective at top and bottom (*cf* figure 9).

Figure 9 Flow configuration

Simulation protocol

To get closer to real injector conditions, the height of the jet $h = 5mm$ is retained. The calculation domain dimensions are $L*H = 20h*10h$ and there are 2048*1024 grid points. The main parameters are presented in the following table:

Gas	ρ, kg/m^3	U, m/s	Re	η_k	P, Pa	T, K	$\frac{\rho_1}{\rho_2}$	$\frac{\rho_2 U_2^2}{\rho_1 U_1^2}$
Heavy(1)	3.2	30	2639	$1.36 \cdot 10^{-6}$	$1.013 \cdot 10^5$	300	40	13.6
H_2(2)	0.081	700	1539	$2.03 \cdot 10^{-6}$				

Here the Reynolds number is $Re = \frac{\rho U \delta_{CL}}{\mu}$, on which δ_{cl} is the initial thickness of the boundary layer. The Kolmogorov scale η_k is calculated using this Reynolds number. In this simulation, the size of a mesh is of $0.5 \cdot 10^{-6} m$. Thus, concerning the spatial resolution, our simulations use the cell size close to the Kolmogorov scale.

Initial conditions

At initial time $t = 0$ two mixing zones are prescribed at the interface between the two streams. The longitudinal velocity profile is similar to a "boundary layer" one with a thickness $0.1h$ and the density profile is "mixing layer" type with a thickness $0.025h$, (*cf.* figure 10). At the time $t = 0$, the mixture zone is disturbed using a transverse velocity with a white noise spectrum. The maximum amplitude of the disturbance is equal to 1% of the local longitudinal velocity.

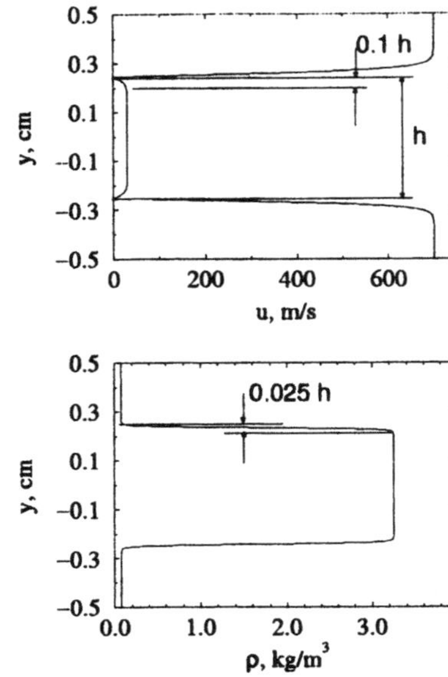

Figure 10 Velocity and density initial profiles

II-2.4 Simulation results

The temporal evolution of the heavy gas mass fraction is presented on the figures 11-14. Dimensionless time is obtained with the heavy jet velocity and height:

$$\bar{T} = t\frac{u_1}{h}.$$

One observes on figure 11 that the initial disturbances are amplified and the large scale structures with given size appear clearly. By the time, these structures grow producing ligaments. Further on, the potential core ondulates, figure13 and leads to the jet break-up, figure14.

Figure 11 Heavy gas mass fraction at time $\bar{t}=4$

Figure 12 Heavy gas mass fraction at time $\bar{t}=5$

Figure 13 Heavy gas mass fraction at time $\bar{t}=7$

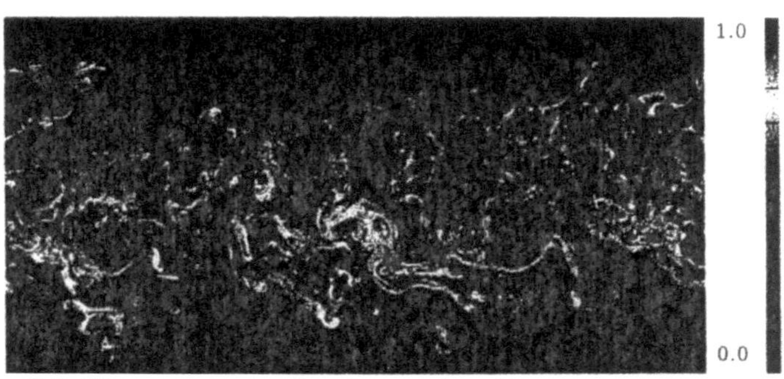

Figure 14 Heavy gas mass fraction at time $\bar{t}=10$

To find the wavelength selected by the flow, we calculated the Lomb periodograms [24] in the homogeneous direction. Thus It is possible to evaluate the significant level of the spectral power associated with the wavelength of the most amplified mode. From this calculation, the wavelength histogram on the whole line ($y = cste$) is built for each time. One finds at times $\bar{t} = 4$ and $\bar{t} = 5$ (figure 15) the maxima corresponding to the wavelength $\lambda \approx 1.1cm$. This is in good agreement with results of Villermaux [25], who proposes the following relation to estimate the most amplified mode:

$$\lambda = \frac{2\pi}{1.5}\left(\frac{\rho_1}{\rho_2}\right)^{1/2}\delta.$$

In our case, this relation gives $\lambda = 1.32cm$.

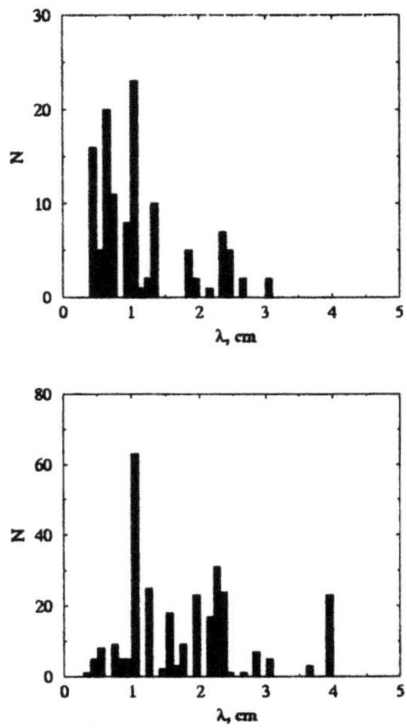

Figure 15 Wavelenght histogramme, $\bar{t}=4$, $\bar{t}=5$

The first efforts to supply modeling starting from simulation relate to the estimation of the density of the interface light gas/heavy gas. In our case, this interface can be seen like a given isovalue line. According to the work of Vervisch and *al.* [26] the average density of interface is defined as it follows:

$$\bar{\Sigma}(\psi,\mathbf{x},t) = \overline{(|\nabla\varphi(\mathbf{x},t)||\varphi(\mathbf{x},t) = \psi)}\bar{P}(\psi,\mathbf{x},t),$$

relative on the isovalue $\varphi(\mathbf{x},t) = \psi$.

Within the framework of temporal simulation, the averages are calculated in the homogeneous space direction. Initially, the probability density function $\bar{P}$ was calculated directly. From the distribution of the interface average density , it is possible to consider the " equivalent droplet size" (Borghi [27]):

$$\bar{D} = \frac{6\bar{\alpha}_l}{\bar{\Sigma}},$$

where α_l is the volume fraction of the heavy gas. The computational results of the interface average density and the "equivalent drop size" for various times are presented on the figure 16. The interface have been defined at mass fraction level represented by the isovalue line $Y^L = 0.5$.

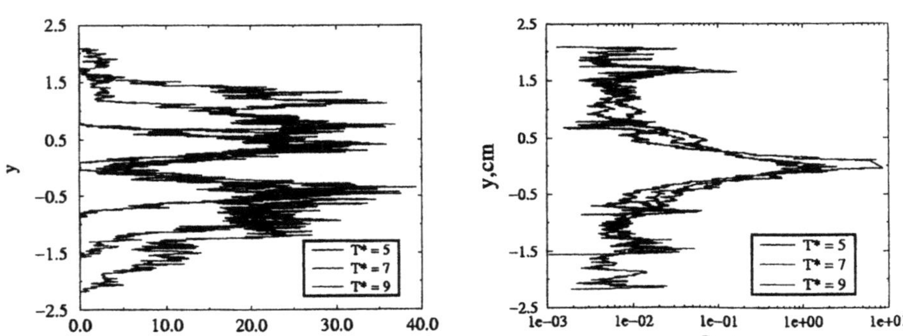

Figure 16 Mean interface density and equivalent drop size

On this figure, the widening of the developed interface zone and the tendency of the interface average density to evolve to a constant value in the mixing zone can be observed. The curve represented for $\bar{t} = 9$ shows this convergence rather clearly. This is translated into terms of " equivalent size " by a value of about $100\mu m$, which is coherent with reality. Nevertheless, the curves have strong oscillations, which makes them currently not exploitable for model development. Work is going on to refine the calculation of $\bar{\Sigma}$.

II-3 Modeling

The main assumption is to consider a single mixture velocity of both phases. This assumption is not too restrictive because it does not exclude that the distribution can be governed by a probability density function with two peaks. The consequence of this single velocity is that , because of the dispersion around the mixture velocity, there is a term of turbulent diffusion in the conservation equations of the components. Moreover the use of a single mixture velocity allows not to duplicate the model of turbulence for each phase.

The approach of modeling of traditional turbulence remains valid even in this case, considering that large scales of the fluctuations control the mixture and that the influence of the capillary forces remain restricted on the small scales. Thus, a two-phase flow is described by the following system of Favre averaged equations:

Mixture:

$$\frac{\partial \bar{\rho}^L}{\partial t} + \frac{\partial \bar{\rho}^L \widetilde{u}_i}{\partial x_i} = \frac{\partial}{\partial x_i}\left(\frac{\mu_t}{Sc_t}\frac{\partial \widetilde{Y}^L}{\partial x_i}\right)$$

$$\frac{\partial \bar{\rho}^G}{\partial t} + \frac{\partial \bar{\rho}^G \widetilde{u}_i}{\partial x_i} = \frac{\partial}{\partial x_i}\left(\frac{\mu_t}{Sc_t}\frac{\partial \widetilde{Y}^G}{\partial x_i}\right)$$

Momentum:

$$\frac{\partial}{\partial t}\overline{\rho}\widetilde{u}_i + \frac{\partial}{\partial x_j}[\overline{\rho}\widetilde{u}_i\widetilde{u}_j + p^*\delta_{ij} - (\widetilde{\sigma_{ij}} + \widetilde{r_{ij}})] = 0 \quad p^* = \bar{p} + 2/3\bar{\rho}k$$

Energy:

$$\frac{\partial}{\partial t}\bar{\rho}\widetilde{E}_t + \frac{\partial}{\partial x_j}\left[(\bar{\rho}\widetilde{E}_t + p^*)\widetilde{u}_j - (\widetilde{\sigma_{ij}} + \widetilde{r_{ij}})\widetilde{u}_i - \gamma(\frac{\overline{\mu}}{Pr} + \frac{\mu_t}{Pr_t})\frac{\partial \widetilde{e}}{\partial x_j}\right] = 0$$

Turbulent mixing:

$$\frac{\partial \bar{\rho} k}{\partial t} + \frac{\partial \bar{\rho}\widetilde{u}_i k}{\partial x_i} = \frac{\partial}{\partial x_i}\left(\mu_k \frac{\partial k}{\partial x_i}\right) + P_k - \bar{\rho}\epsilon$$

$$\frac{\partial \bar{\rho}\epsilon}{\partial t} + \frac{\partial \bar{\rho}\widetilde{u}_i \epsilon}{\partial x_i} = \frac{\partial}{\partial x_i}\left(\mu_\epsilon \frac{\partial \epsilon}{\partial x_i}\right) + C_{\epsilon 1}\frac{\epsilon}{k}P_k - C_{\epsilon 2}\bar{\rho}\frac{\epsilon^2}{k}$$

where

$$\mu_t = \bar{\rho} C_\mu \frac{k^2}{\epsilon} \quad \mu_k = \mu + \frac{\mu_t}{Pr_k} \quad \mu_\epsilon = \mu + \frac{\mu_t}{Pr_\epsilon}$$

and

$$P_k = \tau_{ij} \frac{\partial \widetilde{u}_i}{\partial x_j}$$

Liquid/gas interface density equation:

$$\frac{\partial \bar{\rho} \Sigma}{\partial t} + \frac{\partial \tilde{u}_i \bar{\rho} \Sigma}{\partial x_i} = \frac{\partial}{\partial x_i} \left(\frac{\mu_t}{Sc_t} \frac{\partial \Sigma}{\partial x_i} \right) + C_4 \frac{1}{k} P_k \Sigma + \bar{\rho} D_a \Sigma - Va \Sigma^2.$$

In this equation, the second term on the right of the equality models the effect of the stretching of the liquid by the mean velocity gradients. The third term relates to the stretching of the interface by the fluctuations velocity, with

$$D_a = C_5 Re_t^{1/2} \frac{\epsilon}{k}.$$

The last term models the influence of the surface tension on the interface stretching:

$$V_a = C_6 \frac{\sigma \epsilon^{1/2}}{\alpha^L \nu^{1/2} \rho^G k} \quad and \quad \alpha^L = \frac{\bar{\rho} \widetilde{Y}^L}{\rho^L}.$$

Initially, the constants $C_3 - C_6$ are equal to the unit.

II-3.1 Modeling results

The equations presented above are solved using the same solver as that used for simulation. Space calculation is carried out in a two-dimensional field of the size $6h * 4h$, comprising $80 * 200$ points, where h corresponds to its value taken for simulation (5 mm). The main flow parameters are presented in the section II-2.3. The mean longitudinal velocity profile at the inlet is into 1/7 and that of the density is discontinuous. The value of Σ at the entry is equal to $\frac{4\alpha^l}{h}$.

The figure 17 presents the mean density. The length of the potential core is easily identified, which is about $3h$, and that is in good agreement with the length given by Villermaux [25]:

$$\frac{L}{D} = \frac{6}{M^{1/2}} = 3.16,$$

The axial distribution of equivalent drop size is presented on the figure 18. In the potential core, it is quasi-constant and corresponds to the height of the jet. After the rupture, the size decreases strongly up to values close to zero. It is noted that one does not find a step on this curve, specific to the dispersed phase. Indeed, the set of constants must be adjusted and the coupling between σ and the coefficients of transport must be carried out.

Figure 17 Mean density

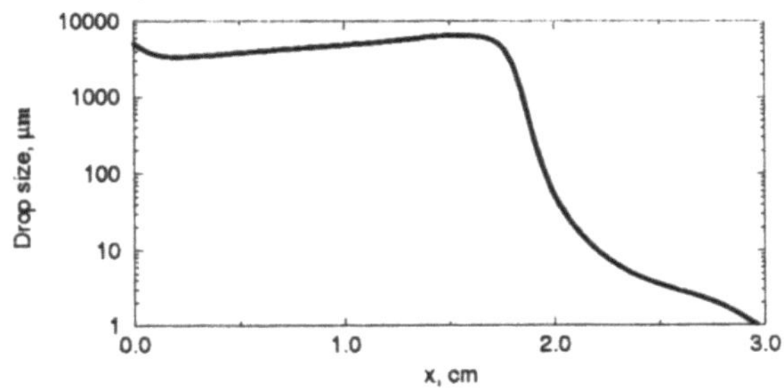

Figure 18 Equivalent drop size

II-4 Conclusion

Temporal simulation for the density ratio 40 was carried out. The wavelength of the most amplified mode, estimated from the simulation corresponds to experimental and theoretical results. The development of the tools for post-processing is going on. The space evolving simulation of a real geometry case and temporal simulation 3D will be considered.

With regard to modeling, the length of the potential core is correctly estimated. On the other hand, the drop size distribution from the jet does not present the expected behavior. This problem requires the calibration of the constants in the model as well as a finer representation of transport coefficients. This will be done using the experimental results of DLR (DLR Cryogenic Injector facility).

Acknowledgments

The first part was funded by the Deutsche Forschungsgemeinschaft (DFG) within the French–German collaboration program "Numerische Strömungssimulation - Simulation Numérique d'Écoulements". The authors thank Prof. Luc Vervisch.

The second is financed by the SEP and CNRS. The authors are grateful to Prof. Roland Borghi for his permanent implication in their work and to Emmanuel Villermaux for fruitful discussions. The data-processing resources are provided by the CRIHAN (Centre de Ressources Informatiques de HAute Normandie) and the IDRIS (Institut du Developpement et des Ressources en Informatique Scientifique).

References

[1] R. Borghi. Turbulent combustion modelling. *Progress in Energy and Combustion Science*, 14:245–292, 1988.

[2] B.E. Launder. *Mathematical Models of Turbulence*. Academic Press, 1972.

[3] Sarkar, S., Erlebacher, G., Hussaini, M. Y., Kreiss, H. O. The analysis and modeling of dilatation terms in compressible turbulence. *ICASE Report No. 89-79*, December 1989.

[4] L. Vervisch. *Prise en compte d'effets de cinétique chimique dans les flammes de diffusion turbulentes par l'approche fonction densité de probabilité*. PhD thesis, Universite de Rouen, 1991.

[5] S. Melen. *Modelisation et etude numerique de la combustion supersonique turbulente non-premelangee, approche probaliste*. PhD thesis, Universite de Rouen, 1995.

[6] S.B. Pope. The relationship between the probabiltity approach and particle models for reaction in homogenes turbulence. *Combustion and Flames*, 35:41–45, 1979.

[7] D.J. Mavriples. Accurate multigrid solution of the euler equations on unstructured and adaptive meshes. *AIAA-Journal*, 28, 1990.

[8] W. Rick. *Adaptive Galerkin Finite Elemente Verfahren zur numerischen Strömungssimlation auf unstrukturierten Netzen*. PhD thesis, Institut für Strahlantriebe, RWTH Aachen, 1994.

[9] Bikker, S., Greza, H., Koschel, W. Parallel computing and multigrid solution on adaptive unstructured meshes. *Notes on Numerical Fluid Mechanics*, 47:9–16, 1994.

[10] Spencer, B.W., Jones, B.G. Statistical investigation of pressure and velocity fields in the turbulent two-stream mixing layer. *AIAA-71-613*, 1971.

[11] D. Papamoschou and Roschko A. The compressible turbulent shear layer: an experimental study. *Journal of Fluid Mechanics*, 197:453–477, 1988.

[12] Elliott, G.S., Samimiy, M. and Arnette, S.A. Study of compressible mixing layers using filtered rayleigh scattering based visualizations. *AIAA-Journal*, 30:2567–2569, 1992.

[13] J.-R. Nuding. Interaction of compressible shear layers with shock waves: an experimental study, part i. *AIAA-96-4515*, 1996.

[14] A. Stoukov. *Simulation numérique de l'auto-allumage et de la combustion dans une couche de mélange réactive supersonique*. PhD thesis, Université de Rouen, 1996.

[15] S. Zaleski. Simulation par capture d'interface de la formation d'intrusions liquides dans les jets cisaillés. Rapport au PRC CNRS/SEP/CNES., 1992.

[16] F.X. Keller, A. Vallet, D. Vandromme, and S. Zaleski. Etude numérique de la pulvérisation des jets liquide–gaz. Rapport au PRC CNRS/SEP/CNES, 1992.

[17] F.X. Keller, J. Li, A. Vallet, D. Vandromme, and S. Zaleski. Direct numerical simulation of interface breakup and atomisation. ICLASS 94, 1994.

[18] B. Lafaurie, R. Nardone, S. Scardovelli, S. Zaleski, and G. Zanetti. Modelling merging and fragmentation in multiphase flows with surfer. *Journal of Computational Physics*, 113:134–147, 1994.

[19] A. Berlemont. *Modélisation eulerienne et lagrangienne de la dispersion particulaire en écoulement turbulent*. PhD thesis, Université de Rouen, 1987.

[20] W.A. Sirignano. Fluids dynamics of sprays - 1992, freeman scholar lecture. *Journal of Fluids Engineering*, 115:345–377, 1993.

[21] R. Borghi. Droplets and spray combustion. Modelling of Combustion and Turbulence, CNRS-CSAMI-CRIHAN, Aussois., 1993.

[22] H.C. Yee. A class of high-resolution explicit and implicit shock-capturing methods. Computational Fluid Dynamics, March 1989, Rhode-St-Genèse, Belgium, Von Karman Institute for fluid dynamics, Lecture Series 1989-04, 1989.

[23] E. Kessy, A. Stoukov, and D. Vandromme. Numerical simulation of reacting mixing layer with a parallel implementation. In *Parallel Computing Technologies*. Verlag, 1995.

[24] H.W. Press, S.A. Teukolsky, W.T. Vetterling, and B.P. Flannery. *Numerical Recipes in FORTRAN*. Cambridge University Press, 1992.

[25] E. Villermaux. Mixing and spray formation in coaxial jets. Presented at the 3rd International Symposium on Space Propulsion, Beijing, China, August 11-13, 1997.

[26] L. Vervisch, E. Bidaux, K.N.C. Bray, and W. Kollman. Surface density function in premixed turbulent combustion modelling, simularities between probability density function and flame surface approaches. *Phys. Fluids*, 7(10):2496–2503, October 1995.

[27] R. Borghi. Commentaires au sujet de la pulvérisation de jets. Note GDR, septembre 1994.

IV. TURBULENT FLOWS

Simulation of Spatially Developing Plane and Round Jets

P. Comte[1], Y. Dubief[1], C. Brun[1]
and
M. Meinke[2], C. Schulz[2], Th. Rister[2]
[1]Institut National Polytechnique de Grenoble, LEGI/IMG
BP 53, 38041 Grenoble
France
[2]Aerodynamisches Institut
Rheinisch-Westfälische Technische Hochschule (RWTH) Aachen
Wüllnerstraße zw. 5 u. 7
D-52062 Aachen
Germany

Summary

Large-Eddy Simulations of jets and wall-bounded flows obtained from different numerical techniques are presented. In the jet case, it is found that a sensitive parameter is the nature of the upstream perturbations, stochastic forcing yield an unphysical drop in turbulence intensity near the inflow boundary, whatever the method, resolution and subgrid-scale model used. With deterministic perturbations smooth statistics are obtained. A second investigation is concerned with the performance of two different solution schemes (a second-order scheme of AUSM type and a compact finite difference scheme), which are applied to a plane turbulent plane jet. Both methods produce similar results, disregarding whether a subgrid scale model is used or not. Finally, the recycling technique proposed by Lund, Wu and Squires (CTR Annual Research Briefs, 1996, pp. 287–295) for turbulent boundary layers at zero-pressure gradient has been proved to resist the pressure fluctuations caused by D-type roughness element, which is very promising for forthcoming simulations of wakes behind bluff bodies.

1 Introduction

After impressive success in academic situations (see in particular the review paper [16]), Large-Eddy Simulations have been identified as a promising tool for industrial applications. The straightforward application of classical LES techniques to more applied situations, however raises difficulties, that will have to be overcome shortly. These come essentially from the large Reynolds numbers and the inherent numerical dissipation of the schemes capable of coping with strong grid distortions and the large gradients deliberately left under-resolved in regions considered as non-critical.

The motivation of this paper is to identify the most crucial difficulties, where effort should be put in priority. The strategy is to confront two different approaches:

- the classical approach mentioned above, with subgrid-scale models of the spectral or structure-function families, as described in [16], associated to spectral/compact schemes in the incompressible regime, and up to fourth-order weakly dissipative schemes in compressible situations, without tuning parameters.
- an approach, which uses reconstructions in an AUSM context, or 4th order explicit filtering associated with 6-4-3 compact finite differences. These solution schemes seem to perform without visible degradation of accuracy, with respect to the former approach. This even holds when used without subgrid-scale model at least for the relatively low Reynolds number flows considered here.

2 Mathematical Model

The formalism of compressible LES is shortly recalled below: written in tensor notation and in terms of dimensionless conservative variables for a Cartesian coordinate system, the Navier-Stokes equations for an ideal gas read:

$$\frac{\partial \boldsymbol{Q}}{\partial t} + (\boldsymbol{F}^C_\beta - \boldsymbol{F}^D_\beta)_{,\beta} = 0 \quad , \qquad \boldsymbol{Q} = [\varrho, \varrho u_\alpha, \varrho E]^T \quad , \tag{2.1}$$

where $\boldsymbol{Q}$ is the vector of the conservative variables, and $\boldsymbol{F}^C_\beta$ denotes the vector of the convective and $\boldsymbol{F}^D_\beta$ of the diffusive fluxes:

$$\boldsymbol{F}^C_\beta - \boldsymbol{F}^D_\beta = \begin{pmatrix} \varrho u_\beta \\ \varrho u_\alpha u_\beta + p\delta_{\alpha\beta} \\ u_\beta(\varrho E + p) \end{pmatrix} + \frac{1}{Re} \begin{pmatrix} 0 \\ \sigma_{\alpha\beta} \\ u_\alpha \sigma_{\alpha\beta} + q_\beta \end{pmatrix} \quad . \tag{2.2}$$

The stress tensor $\sigma_{\alpha\beta}$ is written as a function of the strain rate tensor $S_{\alpha\beta}$

$$\sigma_{\alpha\beta} = -2\,\nu(S_{\alpha\beta} - \tfrac{1}{3} S_{\gamma\gamma}\delta_{\alpha\beta}) \qquad \text{with} \qquad S_{\alpha\beta} = \tfrac{1}{2}(u_{\alpha,\beta} + u_{\beta,\alpha}) \quad . \tag{2.3}$$

The dynamic viscosity μ is assumed to depend only on temperature, and is given by empirical laws such as Sutherland's. The system is closed with Fourier's law of heat conduction applied to the heat flux q_β, and the equation of state for ideal gases

$$q_\beta = -\frac{k}{Pr(\gamma - 1)} T_{,\beta} \quad , \quad p = \frac{1}{\gamma}\varrho T \quad , \quad p = (\gamma - 1)\left(\varrho e - \frac{\varrho}{2} u_\beta u_\beta\right) \quad , \tag{2.4}$$

where γ is the ratio of the specific heat capacities, T the temperature and Pr the Prandtl number.

The compressible LES equations are obtained by convolution of Eqs. (2.1) by a low-pass filter of width Δ which corresponds here to a local average in each grid cell (box filter). The resulting subgrid scale (SGS) stresses are modelled with two alternate SGS-models, the Smagorinsky, the filtered structure function, [4], see also [16], and the dynamic model, [9]. The compressible generalization of the Smagorinsky model for the anisotropic part of the SGS-stresses τ is given by, [8]

$$\tau_{\alpha\beta} - \tfrac{1}{3}\tau_{\gamma\gamma}\delta_{\alpha\beta} = -\varrho(C_S\tilde{\Delta})^2\,|2S|\left(S_{\alpha\beta} - \tfrac{1}{3}S_{\gamma\gamma}\delta_{\alpha\beta}\right) \quad , \tag{2.5}$$

where the filter width is set to the average of the spatial steps in all space directions $\tilde{\Delta}=(\Delta_1\Delta_2\Delta_3)^{\frac{1}{3}}$. For compressible flow at low Mach numbers, several terms from the SGS energy flux can be neglected without introducing appreciable errors [8], [23]. Only the heat flux q_t due to the fluctuating internal energy is modelled with an eddy diffusivity SGS-model, [7]:

$$q_t = -\frac{\varrho}{\gamma-1}\frac{\nu_t}{Pr_t}T_{,\beta} \quad . \tag{2.6}$$

The turbulent Prandtl number Pr_t in (2.6) is chosen to be 0.5.

The main advantage of the Smagorinsky model is its simplicity. The main disadvantages are a non vanishing eddy viscosity in laminar flow regions, a problem-dependent constant C_S, and that it does not allow for backscatter of energy from small to large scales. The dynamic model proposed by Germano tries to overcome these deficiencies. Information contained in the large-scale field is used for a dynamic determination of the Smagorinsky constant C_S. More details about this model can be found in [9], more details of the present implementation can be found in [31]. Instead of adjusting locally its coefficient, the filtered structure-function mode proposed in [4] keeps it constant, to the value derived from the same inertial arguments as for $C_S = 0.18$. Instead of $\tilde{\Delta}|2S(\boldsymbol{u})|$, its velocity scale is $\sqrt{\tilde{F}_{2_\Delta}(\boldsymbol{x},t)}$, in which $\tilde{F}_{2_\Delta}(\tilde{\boldsymbol{u}})$ denotes the discrete second-order structure-function of $\tilde{\boldsymbol{u}}$, obtained from $\boldsymbol{u}$ after discrete Laplacian high-pass filtering iterated three times. This makes the model highly insensitive to large-scale fluctuations, and renders possible low-cost LES of the complete transition of spatially-growing boundary layers, and the observation of pairings between the streamwise vortices of (spatially-growing) mixing layers, as conjectured from experimental hot-wire data [24]. Taking into account the transfer function of the triply-iterated Laplacian filter, the constant of the model is $0.0014\,C_K^{-3/2}$ with the classical value of the Kolmogorov constant $C_K = 1.4$.

Until recently, numerical considerations have generally been excluded from discussions around LES. For industrial applications, numerical diffusion cannot be zero. The practical implementation of the subgrid-scale models designed from theoretical arguments should thus take into account the truncation error of the scheme. In case of an implicit grid filter and a finite-volume scheme, the subgrid scale model and the truncation error are coupled to the spatial steps of the computational grid. Both terms lead to damping of the smaller resolved scales which can be shown to be of the same order of magnitude for second-order schemes, at least in the time averaged sense, [10]. Finite-difference or finite-volume methods often exhibit a large amount of numerical dissipation, so that it has to be carefully checked whether they are suited for the LES of turbulent flows. In principle an explicit filtering with a filter width larger than the spatial steps should alleviate the poor resolution properties of finite difference schemes in the high wavenumber range, at the expense of an increased computational effort, [18]. Until now, no improvement could be obtained with this approach, if compared to simulations with an implicit grid filter and the same computational resources.

In the following section, a low-dissipation second-order scheme which produced results for the LES of pipe and channel flows in good agreement with DNS and experimental data is briefly described. Additionally, a higher-order compact finite difference scheme following Lele [15], is formulated by for grids with general curvilinear coordinates.

3 Method of Solution

The turbulent flows were simulated with different solution methods, partially described in detail already previously, see [16]. Two other methods, which are applied for the simulation of a plane turbulent jet are described briefly in the following. In that case results of a second-order finite volume method based on a AUSM method and a compact finite difference scheme of 6th-order are compared.

3.1 Second-order AUSM-Scheme

The governing equations, (2.1), are discretized on a block-structured grid in general curvilinear coordinates. A modified AUSM-scheme is used for the discretization of the convective fluxes on a collocated grid with a second-order accuracy.

$$\boldsymbol{F}_\beta^C = \begin{pmatrix} F_1^C \equiv \frac{M_\beta}{2}((\rho a)^L + (\rho a)^R) + \frac{|M_\beta|}{2}((\rho a)^L - (\rho a)^R) \\ \frac{F_1^C}{2}(u_\alpha^L + u_\alpha^R) + \frac{|F_1^C|}{2}(u_\alpha^L - u_\alpha^R) + (\frac{p^L+p^R}{2} + \chi(p^L u_\alpha^L - p^R u_\alpha^R))\delta_{\alpha\beta} \\ \frac{F_1^C}{2}(\frac{(\rho e)^L+p^L}{\rho^L} + \frac{(\rho e)^R+p^R}{\rho^R}) + \frac{|F_1^C|}{2}(\frac{(\rho e)^L+p^L}{\rho^L}) - \frac{(\rho e)^R+p^R}{\rho^R}) \end{pmatrix}. \quad (3.7)$$

Superscript ^{L} and ^{R} denote right and left interpolated variables which are obtained with a quadratic MUSCL interpolation of the primitive flow variables. $M_\beta = \frac{u_\beta}{a}$ is the average of the left and right interpolated Mach number. The speed of sound a is also computed as an average of the left and right interpolated values. The parameter χ is used to control the amount of numerical dissipation within this scheme. Its influence on the solution is demonstrated in [20, 31]. In case χ is set to zero, a central discretization of the pressure derivative is obtained, which exhibits only a very little numerical dissipation and was therefore found to be best suited for the LES of the different flow problems investigated. For strongly curvilinear or stretched grids a value of χ=0 may result in an unstable scheme, so that slightly higher values must be used. The discretization of the viscous stresses was carried out with central differences of second order accuracy.

3.2 Higher-Order Compact Finite Difference Scheme

The higher-order compact difference scheme proposed by Lele is implemented for the skew-symmetric formulation of the governing equations. The inner scheme of sixth-order accuracy is given by

$$F'_{p+1} + 3F'_p + F'_{p-1} = \frac{1}{12}(F_{p-2} - F_{p+2}) + \frac{7}{3}(F_{p-1} - F_{p+1}) \quad . \tag{3.8}$$

F'_p denotes the first derivative of the convective fluxes F^C_β at the grid point p defined by the subscript. The formulation given here is only valid for an application in the equidistant grids. For the application in general curvilinear coordinates the metric terms of the coordinate transformation are computed with the same numerical scheme. Near boundaries with non-periodic conditions the order of the scheme is reduced. The accuracy is reduced from 6th to 4th and then to 3rd order approaching the boundary. A detailed investigation concerning the boundary treatment is presented in [2].

The reduction of accuracy near boundaries leads locally to larger errors and to the generation of high-frequency oscillations. These oscillations could be avoided with a higher grid resolution in those regions. A reduction of the spatial steps may be acceptable for direct numerical simulations, but not in a context of a large-eddy simulation. Therefore an explicit filtering procedure is applied every 5th time step with an one-parameter family of fourth-order accuracy:

$$\alpha\overline{Q}_{p+1} + \overline{Q}_p + \alpha\overline{Q}_{p-1} = \frac{5+6\alpha}{8}Q_p + \frac{1+2\alpha}{4}(Q_{p+1} + Q_{p-1}) + \frac{2\alpha-1}{8}(Q_{p+2} + Q_{p-2}) \quad . \tag{3.9}$$

$\overline{Q}_p$ denotes the filtered variable at the node p. The parameter α allows a control of the filtering. Values only slightly smaller than $\alpha = 0.5$ result in a cutoff filtering with a filter width equal to the grid spacing. Values of α in the range of [0.49-0.499] were used to damp high-frequency oscillations. Near the boundary nodes the filtering scheme has to be modified, since values for Q_p outside the domain are not available. Details of the present implementation are presented in [29].

The discretization of the viscous fluxes is performed with central differences of second-order accuracy. Test simulations with a sixth-order scheme for the viscous fluxes did not provide more accurate results compared to those obtained with the second-order scheme.

3.3 Integration in Time

The integration in time of the differential equations is carried out with an explicit Runge-Kutta method for both discretization schemes. For the second-order scheme a low storage method, [11], with second-order accuracy is used. The coefficients in the Runge-Kutta steps are chosen as α_l=(0.25, 0.1667, 0.375, 0.5, 1), which are optimized for maximum stability for a central scheme. For the higher-order method a classical Runge-Kutta method of 4th order and, for comparison, also the low storage version is applied. A comparison of results showed that the same accuracy is obtained with the two different integration schemes for the cases considered here, since the time step is relatively small due to stability restrictions. The computational effort for the integration of the same physical time is about four times higher for the higher-order scheme than for the second-order scheme. This is due to the fact, that the computation of the space derivatives is about two times more expensive for the compact scheme. The other factor of two comes from the smaller time step in the 4th order Runge-Kutta method, which can only be chosen half the size of that for the low storage version. Therefore, the low storage version was used in most of all cases with the compact scheme.

4 Results

4.1 Turbulent Jets with Random Upstream Forcing

In Fig. 1 pressure and vorticity magnitude isosurfaces obtained from a LES of an incompressible turbulent jet performed with the cartesian spectral/compact code mentioned in the introduction are shown. Since the lateral boundary conditions are periodic by construction, there is some doubt that entrainment, which is a key feature of free jets, can be properly accounted for with such methods. Nevertheless their application is attractive because of their minimal numerical dissipation.

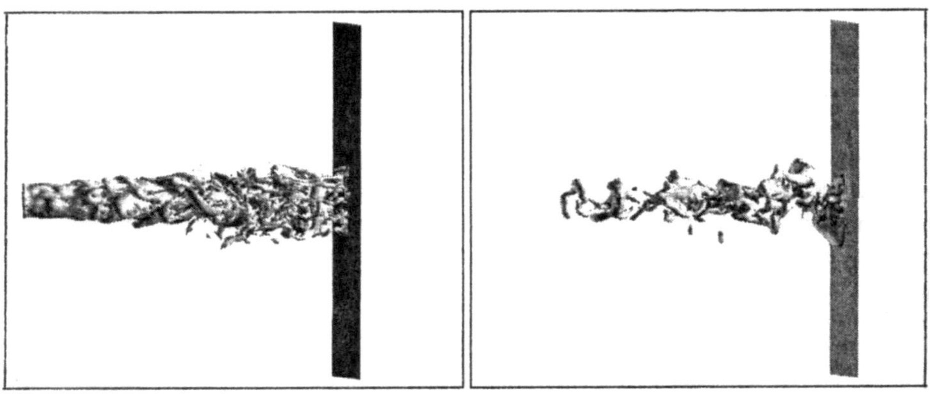

Figure 1 Incompressible jet with the same upstream mean and fluctuation velocity profiles as in experiment [3]. **Left:** isosurface of vorticity magnitude $||\vec{\omega}|| = \omega_i/3$ (where ω_i is the maximum value of the mean profile prescribed at the upstream boundary. **Right:** low-pressure isosurface (threshold $\langle p \rangle - p_{rms}$) at the same time, showing a combination of axisymmetric and oblique modes, as expected from [21].

The flow conditions of this simulation correspond to an experiment setup [3], in which a turbulent jet enclosed within a duct (of square cross-section of side $L = 285mm$), with a low-speed constant co-flow $U_e = 0.1\,U_c$ (where $U_c = 12ms^{-1}$ stands for the jet's centerline velocity at its origin, a sharp-trailing-edge fully-developed pipe flow of diameter $D_j = 26mm$) is investigated. The motivation of this experiment was to extract high-order statistics relevant to free jets with the aid of PIV techniques. It has therefore been carefully checked that the duct, necessary to solve seeding problems, is large enough so that the jet spreading is independent of the duct size L. The departures with respect to a free jet, fully documented in Ref. [3], have thus been checked to be only due to the co-flow. Free-jet statistics can thus be inferred from the ducted statistics by means of Abramovitch-Sabin relations, which makes this experiment an outstanding candidate test case for LES validations (in particular, the momentum of the co-flow is such that the Craya-Curtet criterion is satisfied, which eliminates recirculations near the origin).

With resolution 101×288^2, the dynamics of the large vortical structures is correctly reproduced, at a qualitative level at least. In particular, Fig. 2 shows that the Widnall instability which yields the streamwise vortices does not suffer from the cartesian arrangement of the collocation points.

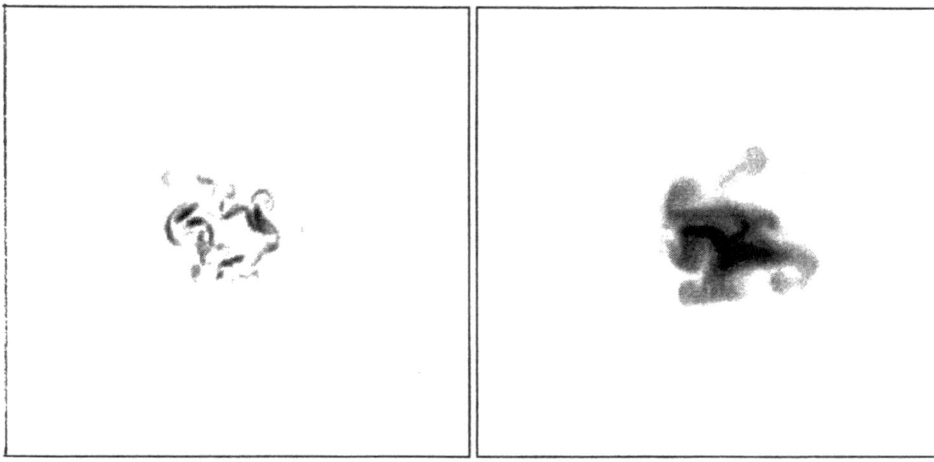

Figure 2 Cross-cut of the vorticity magnitude (left) and the axial velocity (right) at $x = 4.5D_j$

Figure 3 Streamwise evolution of the time-averaged half-velocity jet width $2L_x$ (left) and energy $\sqrt{\langle u'u' + v'v' + w'w'\rangle_t/3}$ (right), for different LES.

Despite the fact that the mean and fluctuation velocity profiles match the experimental measurements (at the most upstream streamwise location, i.e. $x = 0.2\ D_j$), the length of the potential core departs from the experimental value by about 25%, and this for different subgrid-scale models and resolutions, independently of the profile and the nature of the stochastic velocity fluctuations prescribed at the inflow boundary, as shown in Fig. 3. In this figure, the run mentioned above is denoted TAU3 and appears in thick solid lines. Run COMP corresponds to a compressible simulation at convective Mach number $M_c = 0.25$, with analytic mean and fluctuation upstream profiles fitted on the experimental data, on a cartesian grid. The counterpart of this simulation on a cylindrical grid yields quasi-identical statistics in this respect. Fig. 4 shows an example of the vortical structures obtained at a slightly higher Mach number.

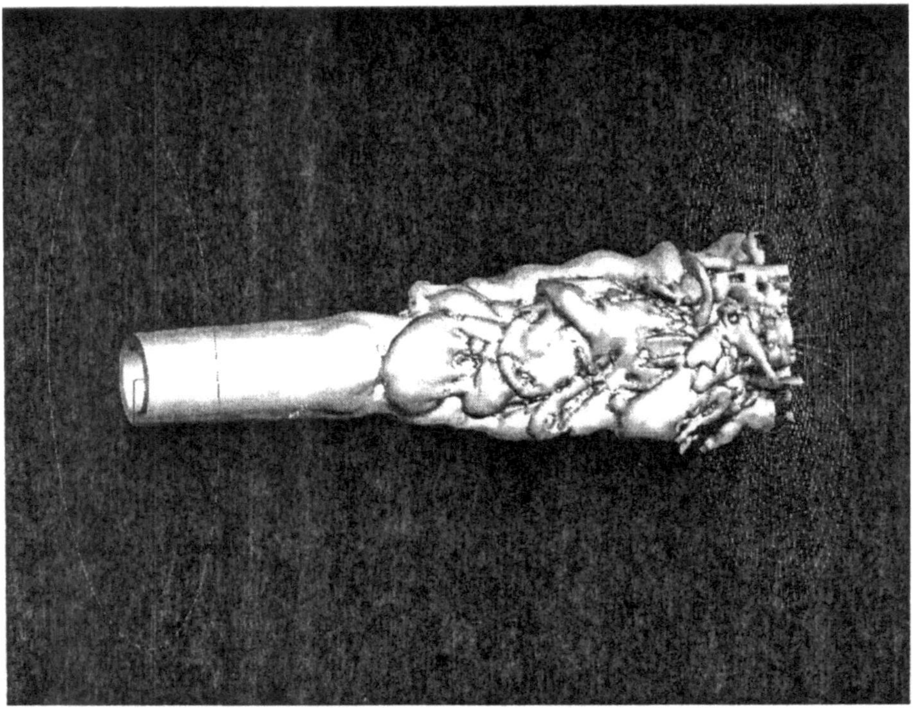

Figure 4 Isosurface $||\vec{\omega}|| = \omega_i/3$ in a compressible jet at upstream convective Mach number $M_c = 0.4$.

This set of results shows that the main source of errors comes from the stochastic upstream perturbations, which do not satisfy the Navier-Stokes equation. This is visible in particular in the right-hand plot in Fig. 3: in contrast with the other curves, the solid line starts with the same initial energy as the experimental value, but it decays almost immediately to the same level as the other curves, which correspond to analytic upstream profiles. For information, the mean part of these is given by

$$\frac{U(r) - U_e}{U_j - U_e} = \frac{1}{2}\left[1 - \tanh\left(\frac{1}{2}\frac{R}{\theta}\left(\frac{r}{R} - \frac{R}{r}\right)\right)\right] \tag{4.10}$$

as in the stability calculations of Michalke and Hermann [21], with $R/\theta = 6$ (the value which best fits the experimental profile). For the fluctuations, the same Gaussian profile with amplitude $2\ 10^{-3}(U_j - U_e)$ is prescribed for all components, which is much smaller than the experimental value, which is $0.15\ (U_j - U_e)$ for the streamwise component for example.

The conclusion of this first part is that deterministic upstream conditions have to be used to get correct statistics, which motivates the further sections.

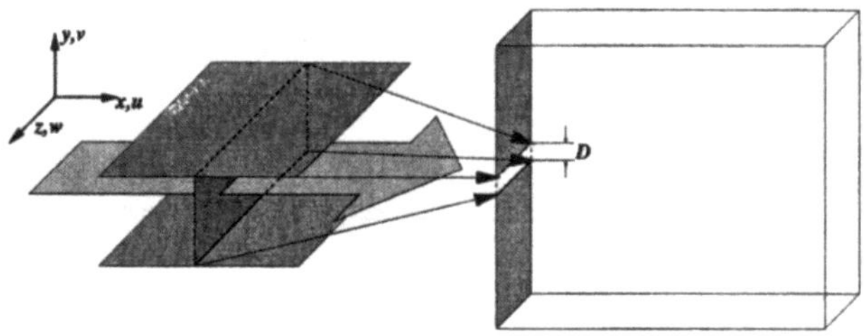

Figure 5 Inflow boundary condition and coordinate system for the LES of the turbulent plane jet flow.

4.2 Turbulent Jets with Deterministic Inflow Conditions

Inflow Boundary Condition

It is known that jet inflow conditions have a large influence on the downstream development of the jet, [30]. In order to provide meaningful inflow conditions for the mass flow, a turbulent inflow condition is used, which is generated by the LES of a fully developed turbulent channel or pipe flow. Both, the jet and pipe or channel simulation, are conducted simultaneously, so that at each time level an instantaneous solution from a cross section of the channel and pipe flow can be inserted in the inflow plane of the round and plane turbulent jet, respectively. This is shown schematically in Fig. 5 together with the coordinate system used further below.

Table 1 Grid parameters for the LES of the fully developed turbulent channel flow

Re_τ	Re	Sim.	Δx^+	Δy^+_{min}	Δz^+	L_x^+	L_z^+	N_x	N_y	N_z	Case
200	3800	LES	20.4	2	15.0	2000	600	99	49	41	1
			31.2	2	31.2		1000	65	33	33	2
200	3800	LES	20.4	2	15.0	1000	600	51	49	41	3
			31.2	2	31.2		1000	33	33	33	4
200	3800	LES, [26]	53	0.24	13	2500	630	48	65	64	-
180	3300	DNS, [12]	12	0.05	7	2300	1150	192	129	160	-

The fully developed turbulent channel or pipe flow is simulated for a Mach number of Ma=0.1. Quasi periodic boundary conditions are used in streamwise direction, in which pressure and temperature fluctuations and the mass flow rate are considered to be periodic. Periodic boundary conditions are used in spanwise direction for the channel flow. The channel flow has been investigated both numerically by DNS [12], [22], LES

e. g. [27], [26] and experimentally [6], [35] so that a large data base is available for comparison. For the pipe flow experimental and DNS data is available in [34, 5]. A detailed description of the results obtained for the fully developed channel and pipe flows can be found in [20]. Here, only few results are presented, which were used for the plane jet simulations, their characteristic grid and flow parameters are summarized in table 1.

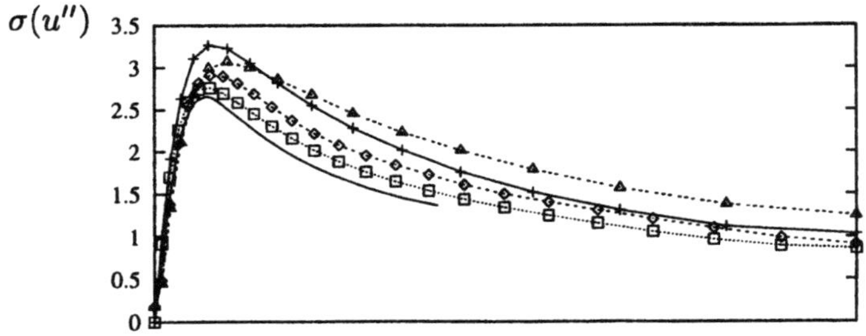

Figure 6 Turbulence intensities in streamwise direction for the channel flow. Calculation with the AUSM scheme without SGS-model. —: *DNS* [12] $(192 \times 129 \times 160)$, □: $(99 \times 49 \times 41$, case 1), +: $(65 \times 33 \times 33$, case 2), ◇: $(51 \times 49 \times 41$, case 3), △: $(33 \times 33 \times 33$, case 4)

In Fig. 6 the turbulence intensities in streamwise direction obtained for case 1-4 are shown. Case 1 is a fine, case 2 is a coarse grid calculation. These results are compared to simulations conducted on grids of the same spatial resolution but with only one half of the extent in streamwise direction, case 3 and case 4 respectively. First, an acceptable agreement with the reference data [12] can be observed for the fine grid calculations case 1 and case 3. The deviation of results for case 2 and case 4 from the reference data is much larger, which shows that the difference in the resolution has a much larger effect on the accuracy than the difference in the extent of the grid in streamwise direction. Therefore, the shorter computational grids are used for the generation of the inflow boundary condition for the jet flow to keep the computational effort within reasonable limits.

A result, which demonstrates that the level of turbulence energy is maintained after the inflow boundary is shown for a round jet at a Reynolds number of 7600 and a Mach number of 0.1 based on the mean velocity and its diameter in Fig. 8. More detailed results of this case can be found in [20].

AUSM and Compact Scheme Simulations of a Plane Turbulent Jet

The plane turbulent jet flow is computed with both the second-order AUSM and the compact finite difference scheme. The corresponding grid and flow parameters are summarized in table 2. The grid points of the channel flow mesh (case 3 and case 4) are continued identically in the mesh for the jet flow.

The other boundary conditions correspond to the case of a turbulent jet exhausting

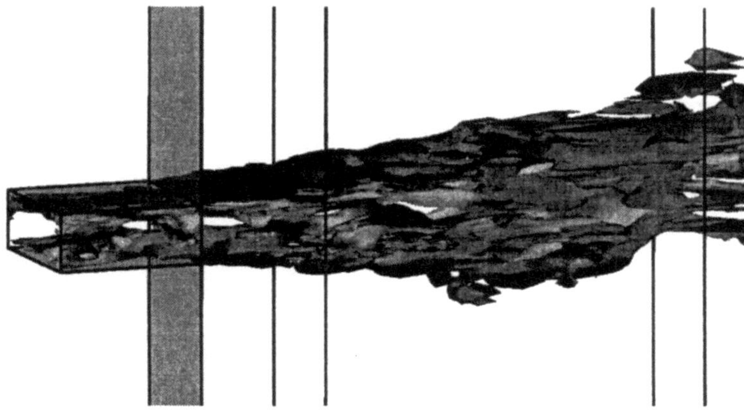

Figure 7 Instantaneous surface of constant vorticity for a plane turbulent jet at a Reynolds number of 7600 and a Mach number of 0.1.

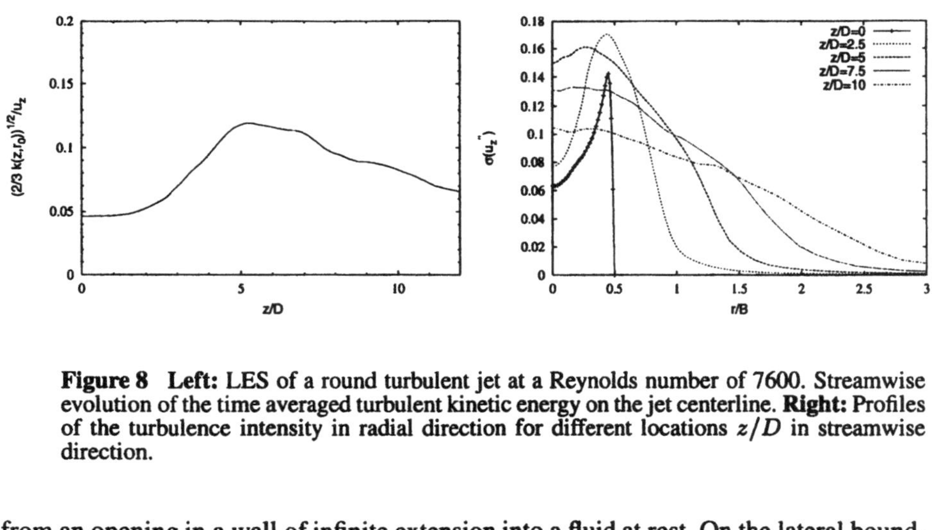

Figure 8 **Left:** LES of a round turbulent jet at a Reynolds number of 7600. Streamwise evolution of the time averaged turbulent kinetic energy on the jet centerline. **Right:** Profiles of the turbulence intensity in radial direction for different locations z/D in streamwise direction.

from an opening in a wall of infinite extension into a fluid at rest. On the lateral boundaries the mass flux on the boundary was assumed to be the same as for the first inner surface and the pressure was held constant. In the exit plane non-reflecting boundary conditions, proposed by Thompson, [33], and Poinsot & Lele, [28] are applied, formulated for general curvilinear coordinates, [29]. This boundary condition allows to use a small domain of integration and thus leads to a reduced computational effort.

The Reynolds number is 7600 based on the width D of the jet which corresponds to Re_τ=200 of the channel flow calculation. An instantaneous surface of constant vorticity is shown in Fig. 7. In Fig. 9 the instantaneous velocity distribution in a plane of the jet are shown for case 1 and case 2. A comparison clearly shows smaller turbulent structures on the finer grid. For both cases the non-reflecting outflow boundary condition

Table 2 Grid parameters for the LES of the turbulent plane jet at a Reynolds number of 7600 and a Mach number of 0.1. Δx_1 and Δy_1 are the smallest spatial steps in x- and y-direction, respectively.

jet grid	grid points	$L_x \times L_y \times L_z$	Δx_1	Δy_1	channel grid
case 1	$81 \times 149 \times 65$	$10D \times 8D \times 2.5D$	0.05D	0.005D	case 3
case 2	$65 \times 129 \times 33$	$12.5D \times 8D \times 2.5D$	0.15D	0.005D	case 4

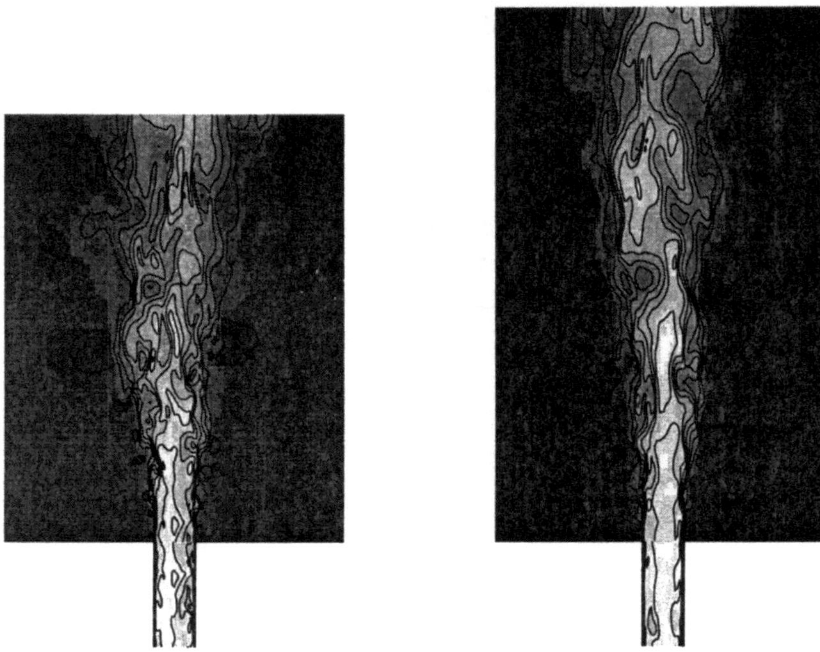

Figure 9 Instantaneous distribution of the streamwise velocity u for the plane turbulent jet at a Reynolds number of 7600. Calculation performed with the AUSM scheme without SGS-model. **Left:** case 1, **Right:** case 2.

does not introduce artificial disturbances into the interior domain.

In Fig. 10 solutions of the second-order scheme are compared with those from the higher-order Padé–3/4/6 scheme. For the compact finite difference scheme an explicit filtering with α=0.499 is used to remove high frequency oscillations in the flow field, which are generated by the reduced accuracy near boundaries. It can be seen that the same length of the potential core is predicted with case 1 and case 2 grids and also with the different discretization schemes. The decay of the centerline velocity is predicted with the same accuracy until the end of the domain of integration. The spreading rate is also in very good agreement with the theoretical value of 0.097 (Fig. 10).

In Fig. 11 the turbulence intensities $\sigma(u'')$ at different locations near the inlet sections

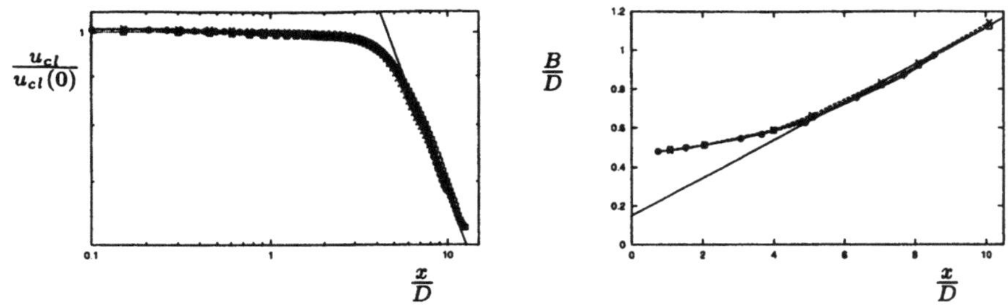

Figure 10 Simulation of a plane turbulent jet at a Reynolds number of 7600. **Left:** decay of centerline velocity $u_{cl}/u_{cl}(x=0)$, **Right:** Half width B of the jet. o: case 1, without model, PADÉ-3/4/6 with $\alpha = 0.499$; □: case 2, dynamic model, AUSM; +: case 1, without model, AUSM; ×: case 2, without model, AUSM

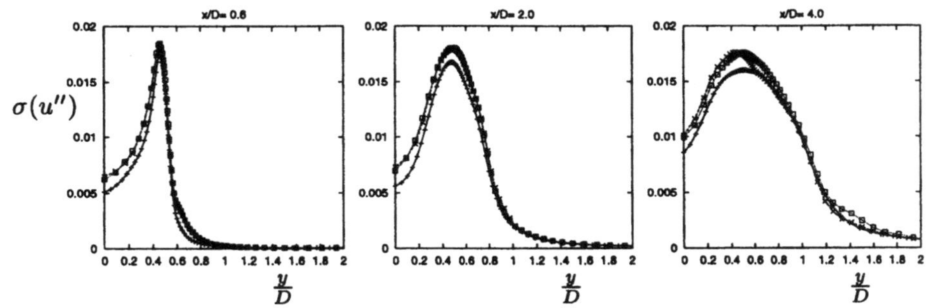

Figure 11 Comparison of the turbulence intensity in streamwise direction $\sigma(u'')$. Results of the AUSM scheme with different SGS-models. +: case 1, without model, ×: case 2, without model, □: case 2, dynamic model.

obtained with the AUSM scheme are presented. Only negligible differences between the calculation with and without SGS-model on the coarse grid (case 2) are visible. The turbulence intensity computed on the finer grid (case 1) differs from that on the coarser grid (case 2) especially in the core region. This is partly due to the different solution for the channel flow, i. e. the inflow condition and partly due to the different resolution in the jet domain.

The comparison of the second moments in Fig. 12 of the Padé-3/4/6 and the AUSM scheme for case 1 and case 2 calculations again shows only small differences for the cases with and without SGS-model. The turbulence intensities for the compact scheme on the coarse grid (case 2) are higher in the vicinity of the inlet section, because of small high-frequency oscillations generated by the decreased order of accuracy near the physical boundaries. The usual way to remove such oscillations is by grid refinement near these boundaries. On the finer grid, which has a three times smaller spatial step in streamwise direction, these oscillations vanish totally.

Further variations of numerical parameters were carried out, but are not presented

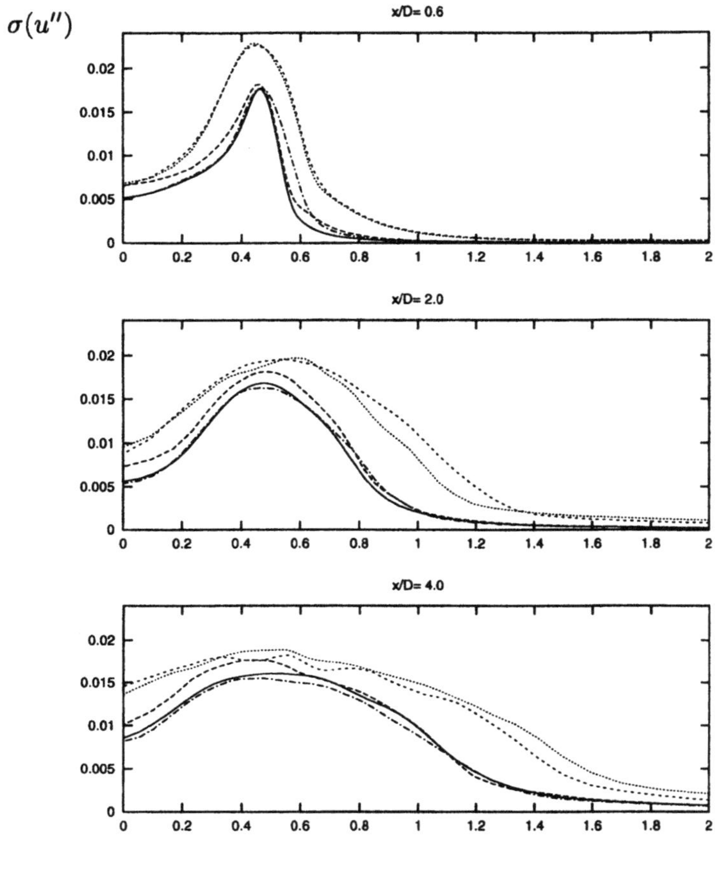

Figure 12 Turbulence intensity in streamwise direction $\sigma(u'')$ at different locations near the inlet section: —·—·— : case 1, without model, PADÉ-3/4/6 with $\alpha = 0.499$; – – –: case 2, dynamic model, PADÉ-3/4/6 with $\alpha = 0.499$; ....: case 2, without model, PADÉ-3/4/6 mit $\alpha = 0.499$; ————: case 1, without model, AUSM; – – – –: case 2, without model, AUSM

here. Even for a higher Reynolds number of 22.000, the same tendency is observed. A detailed description of the numerical methods and all results can be found in [29].

4.3 Turbulent boundary layers perturbed by a small spanwise groove

For forthcoming LES of the wake behind bluff bodies, it is necessary to demonstrate that it is possible to provide reliable simulations of the incoming boundary layers. One step beyond the simulation of a plane channel flow is that of a spatially-growing simulation of a zero-pressure gradient flat-plate boundary layer, as in Ref. [4]. Another step has been made since, in the form of a local change in the adherence condition, in order to respond to a problem risen more than 15 years ago by Prof. Townsend, Bradshaw, Morkovin and Antonia. In this paper, the motivation is just to show that the procedure of generation of deterministic upstream conditions used (derived from Ref. [19]) is capable of resisting unsteady pressure gradients. This procedure consists in re-injecting at the inlet the fluctuations taken further downstream (i.e. about 1500 wall units), through proper (i.e. both in outer and wall units) rescaling.

Fig. 13 below shows the effect of a spanwise groove (D-type roughness) whose depth d is identical to the visual thickness δ_0 of the incoming boundary layer. The Reynolds number of the flow based on the free stream velocity U_0 and δ_0 is 5100. It corresponds to the numerical simulation of a turbulent boundary layer performed by Spalart [32] at $R_\theta = 670$ (where θ is the momentum thickness of the flow). The Reynolds number is also roughly the same as that of Pearson ***et al.*** [25]'s experiment ($R_\theta = 700$ for the flow visualization and 1320 for the measurements). The resolution for groove block is $41 \times 101 \times 40$, that of the two adjacent flat-plate blocks being $121 \times 51 \times 40$. The minimal grid spacing at the wall in the vertical direction corresponds to $\Delta y^+ = 1.2$. The streamwise grid spacing goes from $\Delta x^+ = 3.2$ near the groove edges to 20 at the outlet. The spanwise resolution is $\Delta z^+ = 16$. The grid is stretched by an hyperbolic tangent type transformation similar to that used by Le & Moin [14] for a backward-facing step flow.

From Fig. 13, it is evident that the main effect of the groove is to isotropize partly the vortical structures, which is confirmed by statistical results such as maps of the invariants of the Reynolds-stress anisotropy tensor introduced by Lumley & Newman [17].

Fig. 14 shows spanwise correlation distances before and after the groove. The spanwise wavelength of the streaks is estimated from the location of the negative maximum of the correlation. For u, this maximum occurs at 50 wall units and the spanwise wavelength is therefore $\lambda_z^+ = 100$ (see Kim ***et al.*** [13]). Downstream of the groove, λ_z^+ drops to approximately 60, which is evidence that the groove strongly affect the near-wall dynamics. In contrast, the outer region is much less influenced.

5 Conclusion

The feasibility of two different LES approaches have been demonstrated in the case of turbulent jets and simulations of wall bounded flows. In the case of a round jet at $Re \simeq 20000$ forced upstream by stochastic perturbations, it has been found that the turbulent kinetic energy is reduced to an unphysical low level after the inflow boundary. This type of forcing was the main source of error, the code accuracy or the type of subgrid-scale model appearing as of secondary influence, which meets the conclusions of Akselvoll and Moin [1] in the case a backward-facing step flow. In order to improve upon the upstream boundary conditions, results of a temporally-growing plane channel flow have been injected at the inlet of a plane-jet simulation (at Re=7600 and also forRe=7600), yielding better statistics. Different numerical methods have been used

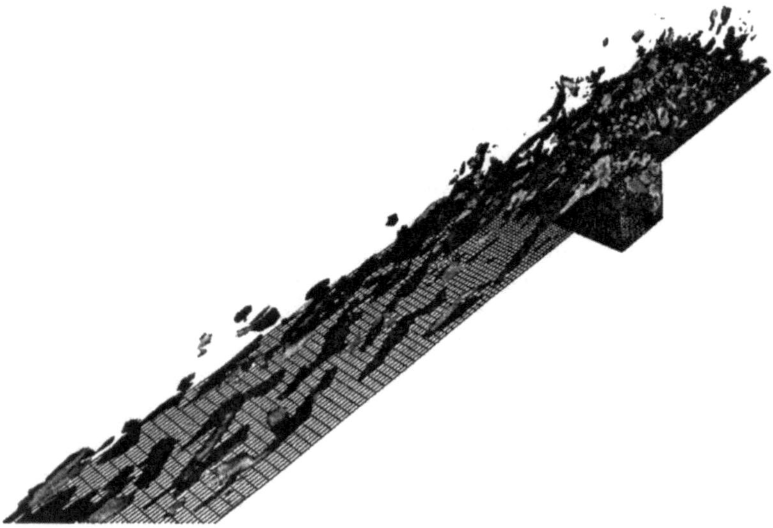

Figure 13 Isosurfaces of vorticity fluctuations filtered by positive $Q = (\Omega_{ij}\Omega_{ij} - S_{ij}S_{ij})/2, \omega = 0.3\omega_i$ and shaded/colored by the sign of longitudinal vorticity: black/blue $\omega_x \leq -0.1$, grey/red $\omega_x \geq 0.1$.

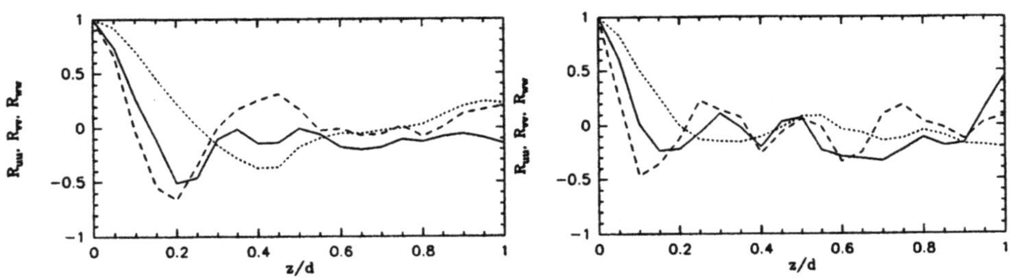

Figure 14 Spanwise correlation of the velocity components, upstream and downstream of the groove (left-hand and right-hand plots, respectively): ——, R_{uu} ; ----, R_{vv}; ·····, R_{ww};

for this purpose: a second-order (weakly dissipative) and a sixth-order compact scheme associated with a low-pass filter. Only small difference in the turbulence statistics were found in the results of the two methods regardless of the subgrid scale model used. Finally, another technique of generation of deterministic upstream conditions, proposed by Lund ***et al.*** [19], has been tested in the case of a spatially-growing boundary layer perturbed by a small spanwise groove, yielding highly positive conclusions with respect to this approach.

References

[1] K. AKSELVOLL and P. MOIN. Large eddy simulation of turbulent confined coannular jets and turbulent flow over a backward facing step. Technical Report 63, Stanford University, 1995.

[2] M. H. Carpenter, D. Gottlieb, and S. Abarbanel. The stability of numerical boundary treatments for compact high–order finite–difference schemes. ***J. Comput. Phys.***, 108:272–295, 1993.

[3] T. DJÉRIDANE. ***Contribution à l'étude expérimentale de jets turbulents axisymétriques à densité variable***. PhD thesis, Université d'Aix-Marseille II, 1994.

[4] F. DUCROS, P. COMTE, and M. LESIEUR. Large eddy simulation of transition to turbulence in a boundary layer developing over a flat plate. ***J. Fluid Mech.***, 326:1–36, 1996.

[5] F. Durst, J. Jovanović, and J. Sender. LDA measurements in the near-wall region of a turbulent pipe flow. ***J. Fluid Mech.***, 295:305–335, 1995.

[6] H. Eckelmann. The structure of the viscous sublayer and the adjacent wall region in a turbulent channel flow. ***J. Fluid Mech.***, 65:439–460, Sept. 1974.

[7] T. M. Eidson. Numerical simulation of the turbulent Rayleigh-Benard problem using subgrid modelling. ***J. Fluid Mech.***, 158:245–268, June 1985.

[8] G. Erlebacher, M. Hussaini, C. Speziale, and T. Zang. Toward the Large-Eddy Simulation of compressible turbulent flows. ***J. Fluid Mech.***, 238:155–185, 1992.

[9] M. Germano, U. Piomelli, P. Moin, and W. Cabot. A dynamic subgrid-scale eddy viscosity model. ***Phys. Fluids***, 3(7):1760–1765, July 1991.

[10] S. Ghosal and P. Moin. The basic equations for the large eddy simulation of turbulent flows in complex geometry. ***J. Comput. Phys.***, 118:24–37, 1995.

[11] A. Jameson. Solution of the Euler equations for two-dimensional transonic flow by a multigrid method. ***Applied Math. and Comp.***, 13:327–355, 1983.

[12] J. Kim, P. Moin, and R. Moser. Turbulence statistics in fully developed channel flow at low Reynolds number. ***J. Fluid Mech.***, 177:133–166, 1987.

[13] J. KIM, P. MOIN, and R. MOSER. Turbulent statistics in fully developed channel flow at low reynolds number. ***J. Fluid Mech.***, 177:133–166, 1987.

[14] H. LE and P. MOIN. Direct simulation of turbulent flow over a backward-facing step. Technical Report 58, NASA, 1994.

[15] S. K. Lele. Compact finite difference schemes with spectral-like resolution. ***J. Comput. Phys.***, 103:16–42, 1992.

[16] M. LESIEUR and O. MÉTAIS. New trends in large eddy simulations of turbulence. ***Ann. Rev. Fluid Mech.***, 28:45–82, 1996.

[17] J. L. LUMLEY and G. R. NEWMAN. The return to isotropy of homogeneous turbulence. ***J. Fluid Mech.***, 82:161–178, 1977.

[18] T. Lund. Large-eddy simulation of a boundary layer with concave stream-wise curvature. In ***Ann. Res. Briefs***, pages 185–196. Center Turb. Res., 1994.

[19] T. S. LUND, X. WU, and K. D. SQUIRES. ***On the Generation of Turbulent Inflow Conditions for Boundary Layer Simulations***, pages 287–295. Center For Turbulence Research, 1996.

[20] M. Meinke, C. Schulz, and T. Rister. LES of Spatially Developing Jets. In R. Friedrich and P. Bontoux, editors, ***Computation and visualization of three-dimensional vortical and turbulent flows. Proceedings of the Fifth CNRS/DFG Workshop on Numerical Flow Simulation***, volume NNFM 64. Vieweg Verlag, 1998.

[21] A. MICHALKE and G. HERMANN. On the inviscid instability of the circular jet with external flow. ***J. Fluid Mech.***, 114:343–359, 1982.

[22] P. Moin and J. Kim. Numerical investigation of turbulent channel flow. ***J. Fluid Mech.***, 118:341–377, 1982.

[23] E. D. P. COMTE and M. LESIEUR. Un formalisme pour la simulation des grandes échelles d'écoulements compressibles. ***preprint LEGI, for submission to C.R. Acad. Sci. Paris***, 1998.

[24] J. S. P. COMTE and P. BÉGOU. Streamwise vortices in large-eddy simulations of mixing layers. ***Eur. J. Mech. (to appear)***, 1998.

[25] B. R. PEARSON, R. ELAVARASAN, and R. A. ANTONIA. The response of a turbulent boundary layer to a square groove. ***J. Fluids Eng.***, 1996. To be published.

[26] U. Piomelli. High Reynolds number calculations using the dynamic sgs model. ***Phys. Fluids***, A 5(6):1484–1490, June 1993.

[27] U. Piomelli, P. Moin, and J. Ferziger. Model consistency in LES of turbulent channel flows. ***Phys. Fluids***, 31(7):1884–1886, July 1988.

[28] T. Poinsot and S. Lele. Boundary Conditions for Direct Simulations of Compressible Viscous Flows. ***J. Comput. Phys.***, 101:104–129, 1992.

[29] T. Rister. ***Grobstruktursimulation schwach kompressibler turbulenter Freistrahlen - ein Vergleich zweier Lösungsansätze***. Dissertation, Aerodyn. Inst. RWTH-Aachen, 1998.

[30] S. Russ and P. Stykowski. Turbulent structure and entrainment in heated jets: The effect of initial conditions. ***Phys. Fluids***, A 5(12):3216–3225, Dec. 1993.

[31] C. Schulz. ***Grobstruktursimulation turbulenter Freistrahlen***. Dissertation, Aerodyn. Inst. RWTH-Aachen, 1997.

[32] P. R. Spalart. Direct simulation of a turbulent boundary layer up to $R_\theta = 1410$. ***J. Fluid Mech.***, 187:61–98, 1988.

[33] K. Thompson. Time dependent boundary conditions for hyperbolic systems. ***J. Comput. Phys.***, 68(1):1–24, 1987.

[34] F. Unger. ***Numerische Simulation turbulenter Rohrströmungen***. Dissertation, Technische Universität München, Lehrstuhl für Fluidmechanik, Feb. 1994.

[35] T. Wei and W. Willmarth. Reynolds-number effects on the structure of a turbulent channel flow. ***J. Fluid Mech.***, 204:57–95, 1989.

Large Eddy Simulation of Flow around Circular Cylinders on Structured and Unstructured Grids

J. Fröhlich[1], W. Rodi[1], Ph. Kessler[2], S. Parpais[2], J.P. Bertoglio[2], D. Laurence[3]
[1] Institut für Hydromechanik, Universität Karlsruhe, 76128 Karlsruhe, Germany
[2] LMFA UMR 5509, Ecole Centrale de Lyon, 69131 Ecully, France
[3] EDF DER-LNH, 6 quai Watier, 78041 Chatou, France

Summary

The paper presents LES computations of subcritical flow around circular cylinders at $Re = 3900$ and $Re = 140000$. For the former the results obtained with the Finite–Volume code LESOCC employing a structured grid and the Finite–Element code N3S using unstructured tetrahedra are reported. The results obtained with the two codes are compared with each other and with results reported in the literature in order to assess the performance and the potential of the two approaches.

1 Introduction

The flow around bluff bodies is characterized by complex interactions between different phenomena such as boundary layers, separation and reattachment, shear layers, large two– and/or three–dimensional vortical structures, etc. Such flows are relevant for many applications in mechanical, chemical and civil engineering, aerodynamics and other areas. In particular when fluid–structure aerodynamic coupling or aerodynamic noise production is of concern, the unsteady flow has to be determined with high accuracy. Although models based on the Reynolds Averaged Navier–Stokes equations (RANS) have been applied successfully in many practical computations, these tend to fail for flows involving large unsteady vortical structures. Here, the Large Eddy Simulation (LES) technique clearly has a better potential since no model is required for the large flow structures that depend on the individual geometry of the problem. Only the fine–scale turbulence, which has more universal character, needs to be modelled. A comparison of LES and RANS modelling for a certain test problems has been given recently in a series of papers [7], [29], [30], [31].

The circular cylinder in uniform cross flow is a classical example for bluff body flow. Although the geometry is fairly simple the physics are extremely rich and have been investigated in a large number of papers [42]. Since the flow is very sensitive to experimental conditions such as aspect ratio, blockage, end conditions, turbulence level etc., such a compilation also reveals that the experimental information available is often of limited accuracy (see also the compilation of [11]).

Applying LES to bluff body flows requires calculation of the flow close to the wall. In cases where the geometry exhibits corners, such as for a square cylinder or a cube, the separation is fixed and the solution is less sensitive to modelling in the near wall region. For a round cylinder, in contrast, the separation depends on the details of the attached boundary layer, be it laminar or turbulent. Another aspect is the use of wall functions. These have generally been developed for situations with flat boundaries so that their application to bodies with entirely different shape is a delicate task, in particular when the separation point is not fixed by the geometry. Attempts are currently being made to develop better strategies for near wall regions [1], [10]; the issue is however far from being settled. For the present computations we therefore decided to avoid the need for wall–function modelling. At the Reynolds numbers considered the boundary layer is laminar and resolved with the grid employed. The simulations therefore are of mixed type with high resolution and direct or almost direct simulation close to the wall and coarse resolution and hence LES in the wake. The results of these computations will be used in future work as a basis for the development of improved wall functions.

In the present configuration, as for any complex geometry, the mesh is non–uniform in order to represent the geometry and to achieve sufficient resolution in critical regions. Since in an LES the inherent filter is generally related to the grid used for the computation, the effect of an inhomogeneous discretization is of concern. We have therefore undertaken a set of separate model computations to elucidate this point.

The main part of the paper deals with the calculation of flow around a circular cylinder in the subcritical regime. We have employed two entirely different codes for this task: a structured Finite–Volume code and an unstructured Finite–Element code. The computational results obtained with each method are interesting on their own but it is their detailed comparison which is main novelty of the present work.

2 Computational methods

2.1 Structured Finite Volume method

The code LESOCC (Large Eddy Simulation On Curvilinear Coordinates) has been developed in Karlsruhe by Breuer and Rodi [8], [9]. It employs a Finite Volume (FV) discretization with a non–staggered arrangement of variables using momentum interpolation to avoid velocity–pressure decoupling. Any structured orthogonal or non–orthogonal grid can be used. Both, convective and diffusive fluxes are discretized by 2nd order central differences. The time scheme consists of an explicit predictor step by a 3–step Runge–Kutta scheme for the velocities and an implicit corrector step where a Poisson equation is solved for the pressure correction (SIMPLE). The overall time scheme is of 2nd order as the pressure is not updated in each of the Runge–Kutta steps for computational efficiency. Different subgrid–scale models are implemented. One is the Smagorinsky model with van Driest damping near solid walls. Furthermore, several variants of the dynamic model with the least–squares approach of Lilly are available. They differ in the way of averaging in space and/or time and in the way of clipping negative viscosities. Different wall function models are implemented but have not been employed in the present computations. The code is highly vectorized and has been validated extensively as described in the references cited.

2.2 Unstructured Finite Element method

Unstructured grids have the advantage that complex geometries can be meshed easily and that local, solution–dependent mesh refinement is possible. Very few attempts of such a discretization in LES have been made [18], [2]. The code N3S was initially developed for the RANS equations, using an upwind scheme and a classical Finite Element (FE) discretization on tetrahedra (linear for the pressure, quadratic for the velocity [27]). While applying this code to LES, Rollet–Miet [32] has shown that a centered scheme is preferable to minimize numerical diffusion since the small time-steps, required for physical reasons in the LES context, are more stringent than the stability criteria of the centered scheme. Next, the standard textbook choice of finite elements, that allows far less degrees of freedom for the pressure than for the velocity, was shown unsuitable for LES. Indeed in this context, a collocated arrangement is preferable to capture the small scales of turbulence which exhibit high frequencies in both pressure and velocity. This N3S-LES code [32] has been used in the present study with the following features: centered Adams-Bashford discretization for the non-linear term, and Crank-Nicolson scheme for the diffusion term. Like the velocity, the pressure is discretized by linear trial functions on the smallest elements. Continuity is obtained by a projection method, yielding a Poisson equation for the pressure. Pressure oscillations are eliminated by the Arakawa method. These oscillations, which can be observed when the tetrahedron elements are obtained from a structured rectangular grid, actually vanish completely when the mesh is fully unstructured. The N3S-LES code was developed and tested by P. Rollet-Miet for grid turbulence, channel flow and the flow through a tube bundle (experiment of [36]). This configuration has been found specifically challenging for RANS models in several ERCOFTAC Workshops [35]. In particular the flow from the axis of the wake to the impingement on the next downstream cylinder was badly predicted, the mean flow and especially the Reynolds stresses. Figure 1 shows the good agreement obtained with N3S-LES and the fairly small influence of the subgrid-scale model.

2.3 Investigation of the influence of the filter size

Most subgrid–scale models rely on ideas and concepts that were developed for isotropic homogeneous turbulence. Theoretical works supporting the existing subgrid models are essentially related to analyses performed in terms of energy flux across a spectral cut-off supposed to be constant [19], [13] and [4].

In the case of an LES on a non–uniform grid, turbulence advected by the mean flow experiences a change in the filter width. The response of the subgrid model to this effect has received little attention in the past. It is here addressed using a two-point

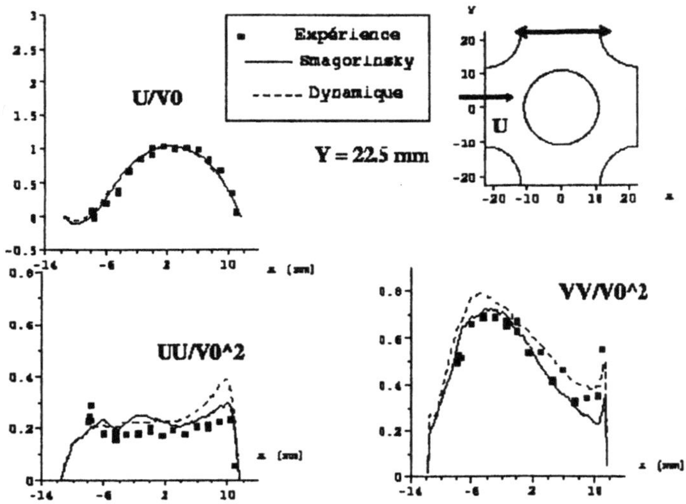

Figure 1 Results of tube bundle computations with N3S obtained with the Smagorinsky model (continuous lines), and the dynamic model (dotted lines). Upper right: geometrical configuration, x and y coordinate in mm. Others: mean and fluctuating velocities versus x at $y = 22.5\ mm$. Symbols indicate experimental values of [36].

closure model (Eddy Damped Quasi-Normal Markovian) extended to inhomogeneous turbulence. The closure model is described in detail in [25]. The main feature of the model is that, at each point of the Reynolds–averaged flow field, a transport equation for the turbulent energy spectrum, $E(K, x)$, is solved. By wave–number integration, the usual turbulent quantities, such as the kinetic energy, can be deduced. For applications to the problem of subgrid models, a wave–number cut–off is introduced, and the effect of small eddies is modeled via a subgrid eddy viscosity. A simple situation is investigated, namely the flow between two flat plates. The wave–number cut–off K_c is supposed to vary with the streamwise direction $K_c = K_c(x_1)$. The channel is divided into three parts. In the first region, ($x_1 < 2.5h$, in Figure 2, where h is half the channel width), the cut–off is uniform and situated at small scales. In the second section ($2.5h < x_1 < 7h$), a much smaller value for K_c is used (5 times smaller), whereas in the last region, K_c is set back to its initial value. Consequently, an abrupt increase in the filter width, followed by an abrupt decrease, are experienced by a fluid particle. The results in Figure 2a show that the mean flow remains nearly unaffected. In Figure 2b, it can be observed that the turbulent kinetic energy decreases in the large filter width region, which is simply related to the fact that there is less energy captured in the resolved field. A more refined analysis, in terms of spectra, shows that the large scales are left nearly unaltered in this case.

When the flow re–enters the fine–resolution domain, the turbulent kinetic energy is overestimated. The explanation is that, the cut–off being suddenly shifted to a higher wave–number, it takes some time before the spectrum recovers an equilibrium state. During this time, the dissipation is underestimated, inducing the overestimation of the turbulent kinetic energy. Further investigations are necessary before definite conclusions can be reached: a more realistic geometry and more progressive variations in the filter width should be introduced. However, the present results already provide an optimistic information concerning LES of wakes, the worse situation being encountered when the eddies are advected from a coarse mesh region to a more refined zone, which is not the case in the flow around a cylinder where turbulence is essentially generated on a fine mesh and then advected to a coarser mesh region.

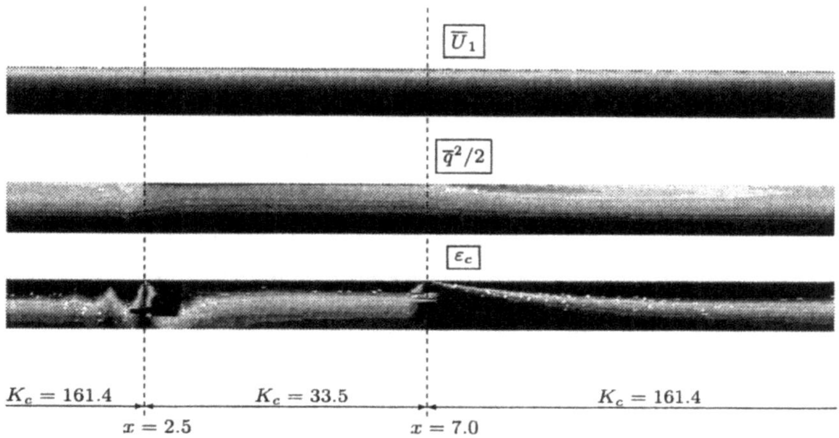

Figure 2 Channel flow with spatially varying filter (sharp spectral cut-off $K_c(x_1)$, only the upper half of the domain is shown.

3 Results for the circular cylinder

3.1 Various flow regimes and choice of Reynolds number

The large amount of literature on the flow around circular cylinders is reviewed in detail in the recent book of Zdravkovich [42]. Different regimes are observed depending on the Reynolds number $Re = Du_\infty/\nu$ where D is the cylinder diameter, u_∞ the free stream velocity and ν the kinematic viscosity. The two Reynolds numbers selected for the computations below both belong to the subcritical regime ($Re = 350 - 200000$). In this regime the boundary layer on the cylinder wall extending from stagnation point to separation point is laminar while the transition to turbulence occurs after a transition length L_t in the shear layer which forms behind the separation point. Further downstream these vortices roll up to form the larger, alternating von Karman vortices. This alternate shedding causes the boundary layer and the stagnation point to oscillate accordingly [26].

The subcritical regime can further be divided into a lower (up to $Re = 1500$), an intermediate ($Re = 1500 - 30000$), and an upper subcritical regime (above $Re = 30000$). In the intermediate regime the shear layer persists for a while after separation before it becomes unstable due to Kelvin–Helmholtz type vortices and spanwise instability. With increasing Reynolds number the transition length L_t decreases from $1.5D$ to $0.2D$, i.e. it takes place immediately after separation, while the formation length changes from $2.5D$ to about $1D$ [42]. The three–dimensional structure of the near wake has been visualized in e.g. [39] and investigated in [14], [15] by measurement of the streamwise vorticity component. Computations of the correlation lengths of this quantity for $Re = 5000$ in spanwise and cross–stream direction yield minimum values of $\lambda_z = 0.55D$ and $\lambda_y = 0.1D$ close to the separation point, respectively, both increasing in streamwise direction [14]. The ratio λ_z/λ_y corresponds to the one observed for mixing layers [15] suggesting that the mechanisms involved are similar.

In the upper subcritical regime the time–averaged flow changes only little with Re. As in the intermediate regime the flow exhibits considerable three–dimensionality . Variations of the flow also occur from cycle to cycle, so that e.g. the Strouhal number may temporarily differ from the mean by as much as 10% with drag and lift varying accordingly. Detailed phase–averaged measurements of the near wake at $Re = 140000$ have been published in [11].

The Reynolds numbers $Re = 3900$ and $Re = 140000$ have been selected for the LES computations below for the following reasons. The lower one was studied in some detail by LES performed at the CTR, Stanford, [3], [21], [22]. It allows a comparison with these results and also with the experiments in [24]. Breuer [6] also investigated this flow in independent computations with LESOCC focussing particularly on the influence of the convection scheme and the subgrid–scale model. The higher Reynolds number was chosen because only for this value detailed phase–resolved experimental data are available for comparisons [11]. Up to now no successfull LES has been published for this case. Such higher Reynolds numbers are more relevant for applications since they are frequently encountered in chemical, mechanical, and nuclear engineering [42].

3.2 Boundary conditions and mesh

The coordinate system is oriented such that x designates the streamwise, y the normal, and z the spanwise coordinate, respectively. All lengths are normalized with the cylinder diameter D, whereas velocities are nondimensionalized with the freestream velocity. Angles are counted from the upstream average stagnation point.

For the structured grid used by LESOCC we employed a circular computational domain with radius 15. It allows to use an O-mesh, depicted in Figure 3, which is advantageous due to its constant quality all along the cylinder wall. (Previous RANS computations in [5] have shown that with an H-mesh inadequate resolution close to the stagnation point considerably affected the result.) The points are clustered in the wake and near the wall using geometric stretching [8], [6]. A laminar inflow condition was employed for $x < 0$, i.e. $u = 1$, $v = w = 0$. The outflow boundary condition was a convective one. In spanwise direction we used periodic conditions with period length

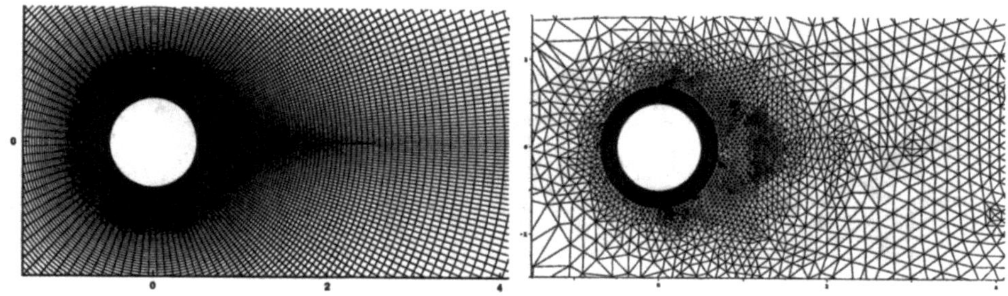

Figure 3 Grid for computations at $Re = 3900$ in the $x - y$-plane, identical zoom near the cylinder. Left: structured FV grid for LESOCC, right: unstructrued FE grid for N3S.

π, the same as in [3]. A no–slip condition was applied at the cylinder wall. The mesh consists of $N_i = 166$ points in radial, $N_j = 166$ in circumferential, and $N_k = 32$ or $N_k = 48$ in spanwise direction, respectively, yielding a total of 881792 and 1322688 points. (The effective number of unknowns is smaller as in directions of periodicity 6 points are employed to implement this condition.) Three computations employing the Smagorinsky model were performed at $Re = 3900$: LRUN1: $N_k = 32, C_s = 0$, i.e. no SGS–model, LRUN2: $N_k = 32, C_s = 0.1$, LRUN3: $N_k = 48, C_s = 0.1$.

The grid for the N3S computations was constructed after the first results with LESOCC had been obtained which allowed to benefit from the available information. The computational domain was chosen to be a rectangular box with $x \in [-5, 15], y \in [-5, 5], z \in [0, \pi]$. This implies a blockage ratio of 0.1 which is higher than for the circular domain but still comparable to most of the experiments [42]. The unstructured grid employs thetraedral elements and was set up in a zonal way. A two–dimensional grid comprising an annulus of thickness 0.1 (20×120 equidistant points) and a V–shaped domain of unstructured triangles covering the wake was constructed. This grid was repeated in 39 parallel planes in spanwise direction and completed to yield a tetrahedral mesh. The remaining space was filled with a fully 3d Voronoï triangulation so that the final grid contains 176286 points. As for the structured code, the inflow condition was laminar, the outflow condition was a convective one, and periodicity was imposed in the spanwise direction. A slip condition was used on the lateral boundaries. For the wall boundary condition the wall function approach of Werner and Wengle [40] was used. However, the mesh is sufficiently fine so that this condition is actually applied in the viscous sublayer. Since in this region a linear velocity distribution is assumed the wall function becomes effectively a no–slip condition. The same subgrid–scale model was employed as with the structured code, namely the Smagorinsky model with $C_s = 0.1$.

A zoom of both grids in the vicinity of the cylinder is shown in Figure 3.

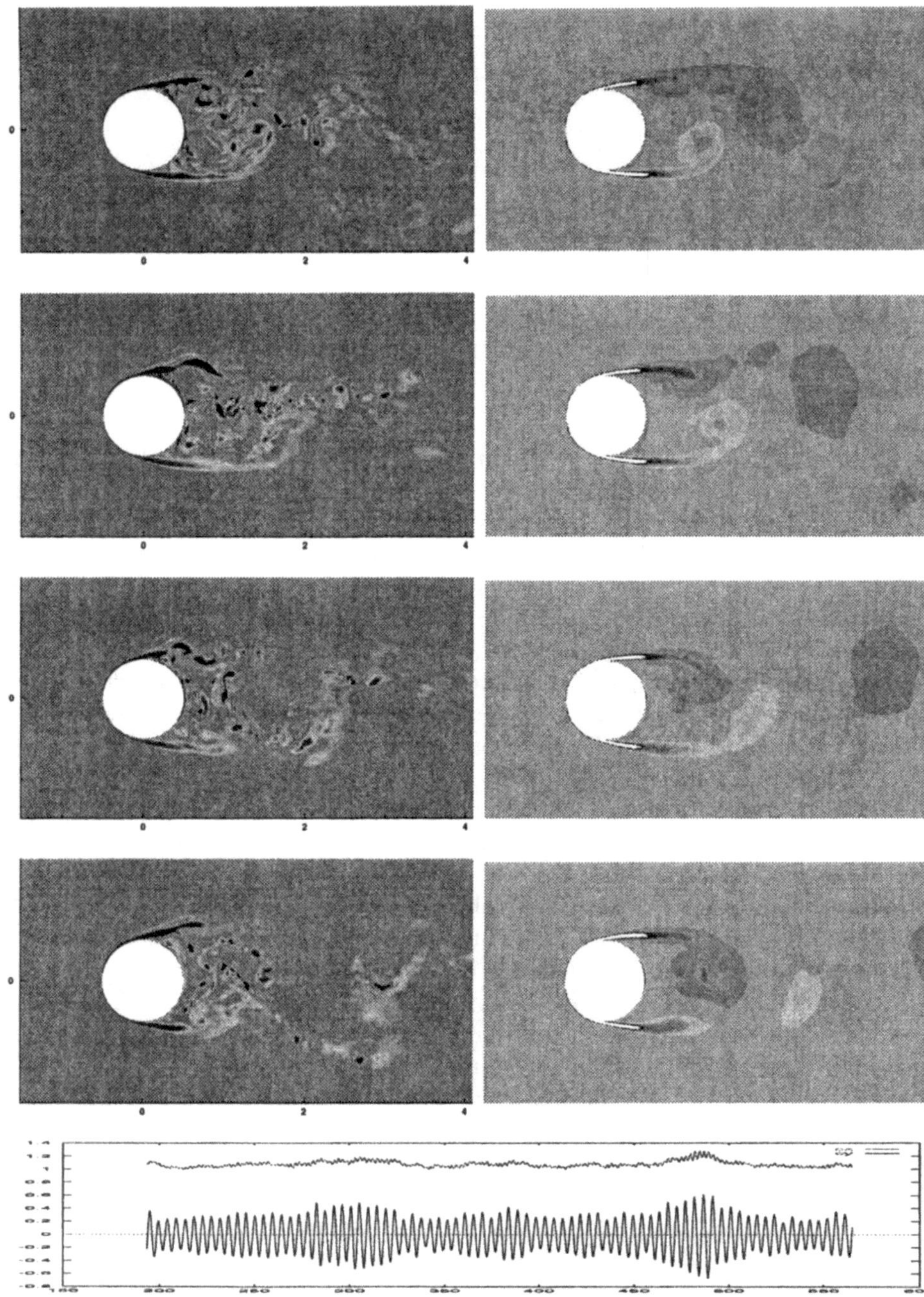

Figure 4 Axial component of the vorticity vector for $Re = 3900$. Left: LRUN2, $\Delta t_{frame} = 0.96$, right: NRUN1, $\Delta t_{frame} = 1$. Bottom window: Time record of drag coefficient C_D (top) and lift coefficient C_L (bottom), in LRUN2 over 80 shedding cycles.

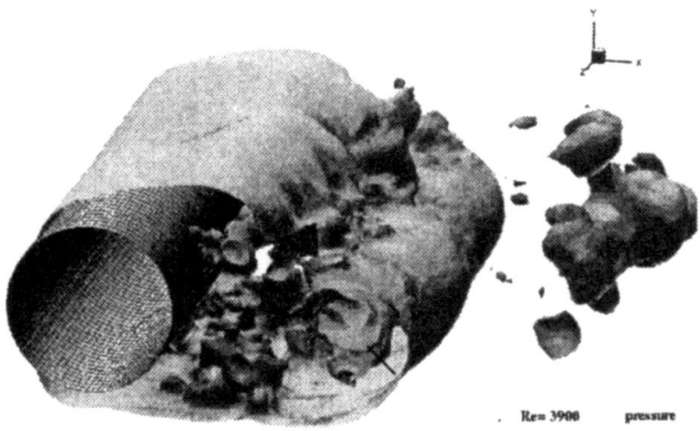

Figure 5 Instantaneous pressure surfaces at $p = -0.34$ and $p = -0.67$, respectively. The latter represents two rolls inside the former (arrows), the remaining plot corresponds to the higher value.

3.3 Instantaneous flow field

This section and 3.4–3.6 are concerned with the results for $Re = 3900$. We compare the solutions obtained with the structured and unstructured method and results reported in the literature.

In order to give an impression of the computed time–dependent solution of the LES with both methods, Figure 4 displays the axial component of the instantaneous vorticity vector in a cut normal to the cylinder axis at similar time intervals. Due to the highly three–dimensional and irregular flow which will be further highlighted below it is of course impossible to obtain virtually corresponding plots. Even from one period to another the solution is different in the same computation.We note that the computed solution corresponds very well to the physical description of the flow given above. In particular, the shear layers forming after separation can be discerned as well as their breaking up into smaller vortices. The larger von Karman vortices are also visible. In the figure, one of them travels to the lower right corner while a second one forms on the upper side. It is also obvious that, due to different grid scales in the two computations, the granularity of the solution is higher in the FV computation than for the FE solution. The large vortices, however, are similar.

The bottom frame of Figure 4 contains a plot of drag and lift coefficient obtained with the structured code. The oscillations are quite regular with respect to their period. The magnitude of the oscillations reflects the spanwise correlation and depends on the length of the cylinder (see [41] for a study at somewhat higher Re.) The modulation of the amplitude, on the other hand, is no deficiency of the computation but is observed similarly in experiments [38] [42]. This irregularity is due to in the irregular three–dimensional break up of the vortices. This is highlighted in Figure 5 displaying two instantaneous iso–pressure surfaces in the FV computation. The figure shows particularly well the instability of the shear layer in spanwise direction resulting from the mechanism of secondary vortices proposed in [39].

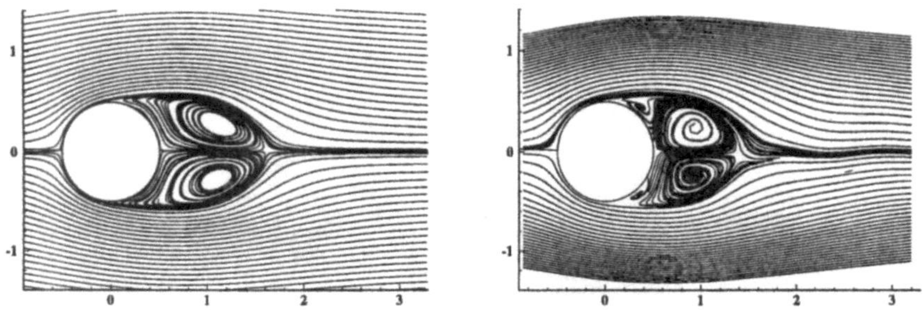

Figure 6 Average stream lines of LRUN3 (left) and NRUN1 (right).

The considerable three–dimensionality of the wake flow is further reflected by the spanwise velocity component (not shown here). In the center plane ($y = 0$) its extrema exceed 70% of the free stream velocity.

3.4 Mean flow and global quantities

After a suitable start–up phase, the flow field has been averaged with respect to the spanwise coordinate and in time (20 cycles for NRUN1 and 86 cycles for LRUN3). The resulting quantities are denoted by brackets. Streamlines of the two–dimensional average flow are presented in Figure 6. Apart from the large recirculation zone behind the cylinder, two small secondary recirculation zones are obtained with LRUN3. These have been obtained in [3] as well and are reported in [37] to exist at $Re = 5000$. In the results of NRUN1 the secondary recirculation zones appear to be more pronounced.

Figure 7a displays the mean streamwise velocity along the centerline. Comparing the curves for NRUN1 and LRUN3 reveals that the former is not very smooth in the far wake which is a consequence of the coarse grid employed in this region as well as the shorter averaging time. The recirculation length is underpredicted in NRUN1. The curve of LRUN3 agrees fairly well with the experiment [20] in the recirculation zone and with the data of [24] in the wake (the data of [20] seem to be questionable around $x = 3$). The other parts of Figure 7 contain lateral profiles at $x = 1.54$, i.e. close to the end of the mean flow separation point in Figure 6. The result of LRUN3 matches the streamwise velocity of the experiment while NRUN3 yields a positive value of the mean velocity, a consequence of the shorter recirculation length. It should be noted that at $y = \pm 5$, the border of the computational domain in NRUN1, $< \bar{u} >= 1.06$ while in LRUN3 $< \bar{u} >= 1.03$ and is decreasing further in the outward direction. This amount of acceleration due to blockage seems acceptable. Figure 7c also displays profiles of $< \bar{v} >$ at $x = 1.54$. The extrema appear at the edges of the mean recirculation bubble and are larger in the present computations than measured in the experiment. Note however that the experimental data exhibit uncertainty monitored by the unsymmetry of the measurements [3].

Figure 7 also compares the present results to those of [3] obtained with a staggered discretization of $144 \times 136 \times 48$ cells. We have included the curves obtained with the

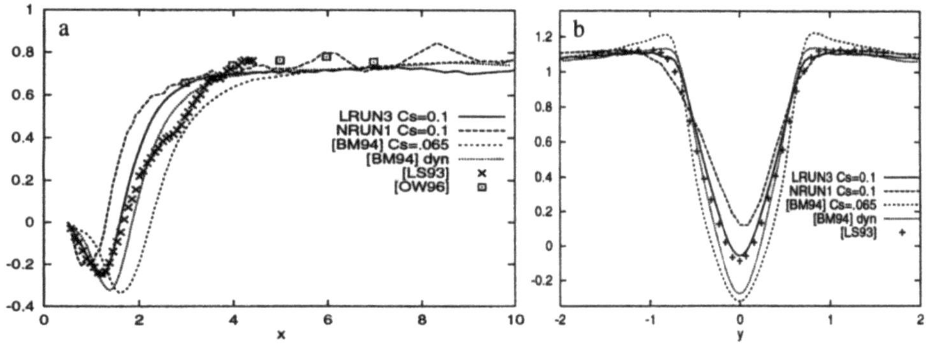

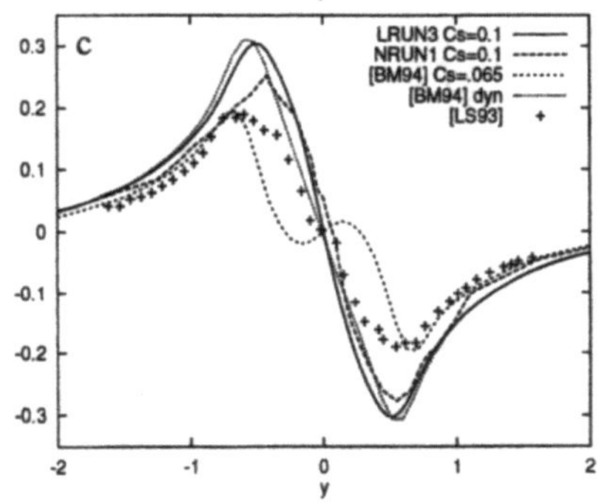

Figure 7 Averaged velocity components: comparison of the results of both methods, comparison to the LES of [3] and experiments [20], [24].
a) Centerline streamwise velocity,
b) streamwise velocity at $x = 1.54$,
c) normal velocity at $x = 1.54$.

Smargorinsky model ($C_s = 0.065$) as well as those computed with the dynamic model since the latter were the most satisfactory in that reference. The mean recirculation length is predicted considerably higher in these computations than in the present ones which agree fairly well with the experimental data. The more reliable of the two $< \bar{v} >$ profiles is very close to the present results.

Table 1 supplements the above comparisons with more quantitative data: the Strouhal number agrees very well with the experimental results. However, in the regime considered here this quantity is relatively insensitive to the details of the calculation. A similar observation has been made for the square cylinder [31]. The drag coefficient is slightly overpredicted in the results of LESOCC. This is related to a somewhat higher separation angle (but still $\Theta < 90°$) and a lower base pressure. The former also yields a shorter recirculation length. These observations illustrate the interaction between the upstream boundary layer and its separation with the downstream recirculation zone. With the N3S code, the drag coefficient is overpredicted significantly more.

3.5 Influence of subgrid–scale model and spanwise discretization

Figure 9 compares the three runs with LESOCC listed in Table 1. With respect to the case LRUN2 the influence of the subgrid–scale model can be assessed by comparing to LRUN1, whereas the influence of the spanwise discretization is revealed by comparison to LRUN3. These results have been obtained in parallel to [6] to which the reader

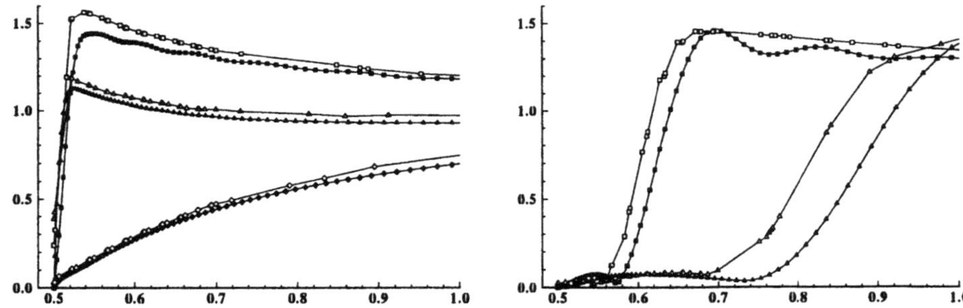

Figure 8 Radial profiles of the absolute value of the mean velocity vector. Solid symbols: LRUN3, cell centers in the structured FV mesh, hollow symbols: NRUN1, intersections of the cut with element faces in the FE mesh. Left: profiles through the attatched boundary layer at 2° (◇), 41° (△), 80° (□), respectively, Right: profiles at 114° (□), and 138° (△), respectively, where the boundary layer has separated.

Table 1 Global quantities for cylinder flow comptations at $Re = 3900$, comparision of the present results to LES of Beaudan, Moin [3] as well as to the experiments: St= Strouhal number, C_D= drag coefficient, C_{bp}= back pressure coefficient, Θ= separation angle in degrees, L_r= length of recirculation zone.

	N3S	LESOCC			BM [3]		Experiment
	NRUN1	LRUN1	LRUN2	LRUN3	n.m.	Smag.	
St	0.216	0.210	0.210	0.216	0.216	0.209	0.215 ± 0.005 [12]
C_D	1.45	1.17	1.08	1.08	0.96	0.92	0.98 ± 0.05 [23]
C_{bp}	−1.59	−1.15	−1.06	−1.03	−0.89	−0.81	-0.90 ± 0.05 [23]
Θ	90.0	89.0	88.0	88.1	85.3	84.8	85 ± 2 [37]
L_r	0.80	0.96	1.09	1.09	1.56	1.74	1.33 ± 0.2[12], 1.19[20]

is refered for further discussion, in particular concerning the effect of different convection schemes. Figure 9a shows the turbulent kinetic energy. It rises steeply behind the cylinder and attains its maximum near the end of the main recirculation zone. The small difference between the computations with and without subgrid–scale model reflects the amount of energy contained in the subgrid–scales. Due to the typical decay of the spectral energy this portion is naturally limited since the large, energy–containing vortices are resolved. On the other hand, inspection of Table 1 indeed shows that the solution changes only little between LRUN1 and LRUN2.

The same holds for the Reynolds stress components shown in Figure 9b and 9c. Note that the streamwise gradient of k is fairly high. Since its principal contribution is the Reynolds stress $< \bar{v}'\bar{v}' >$, the profiles of this quantity at $x = 1.54$ are very sensitive to streamwise dislocations in the flow field. An increase in spanwise resolution improves the result concerning the global quantities in Table 1. It yields slightly increased

Reynolds stresses which is simply due to a shift of the cut–off scale to higher wave numbers in spanwise direction. It is interesting to note, however, that closely behind the cylinder the turbulent kinetic energy decreases further.

We conclude that the results in this section show that the subgrid–scale model and the spanwise discretization are of similar importance in for the present computation. This is in accordance with [3] [6] and [22]. It led to the implementation of a Fourier discretization for the z-direction in the last reference.

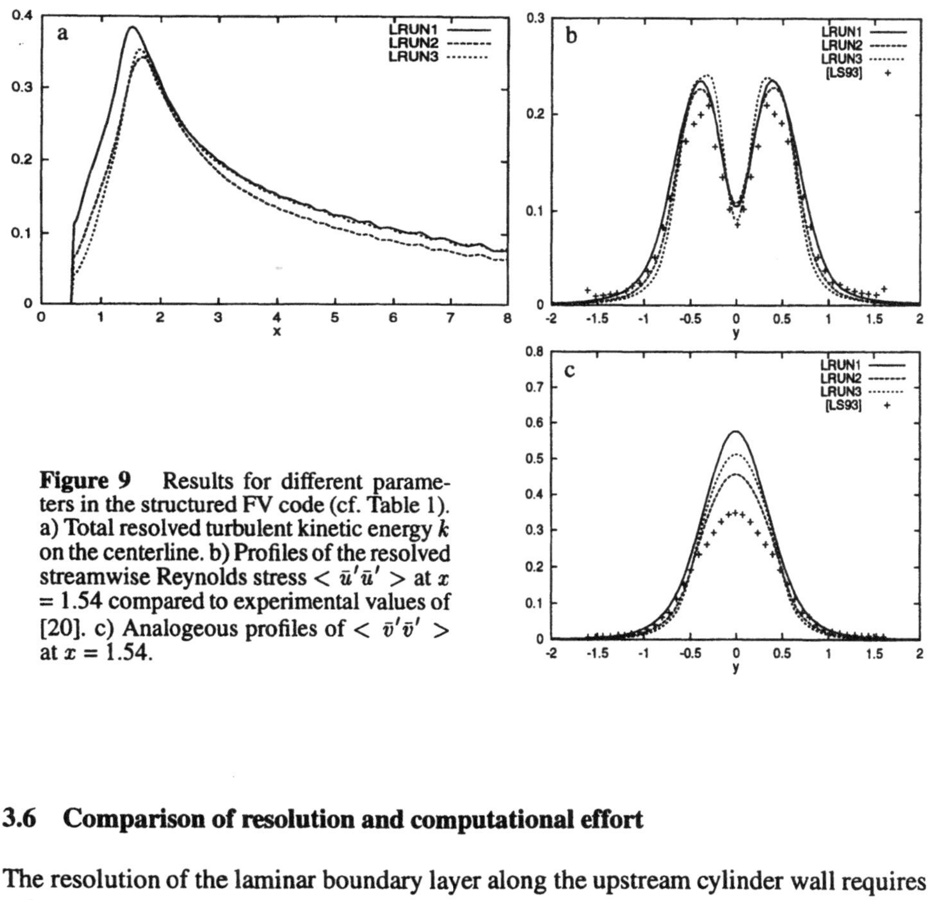

Figure 9 Results for different parameters in the structured FV code (cf. Table 1). a) Total resolved turbulent kinetic energy k on the centerline. b) Profiles of the resolved streamwise Reynolds stress $< \bar{u}'\bar{u}' >$ at $x = 1.54$ compared to experimental values of [20]. c) Analogeous profiles of $< \bar{v}'\bar{v}' >$ at $x = 1.54$.

3.6 Comparison of resolution and computational effort

The resolution of the laminar boundary layer along the upstream cylinder wall requires a fine mesh in radial direction. This also limits the azimuthal mesh size for numerical reasons. On the rear side a small radial mesh size is less important. The wake can be discretized with an increasingly coarser mesh towards the outflow boundary. The employed structured grid employed in LESOCC meets the above resolution requirements. The radial cell size near the cylinder wall is $\Delta r_1 = 0.0025$ yielding $max\{y_1^+\} \approx 1$. Radial profiles of the absolute value of the velocity vector in Figure 8 illustrate the boundary layer resolution.

With an unstructured grid the discretization can be adapted better to the flow considered. In the present case a grid with about 7.5 times fewer points than the structured grid was employed. In particular the upstream region can be resolved more efficiently. Note that a coarsening of the grid in spanwise direction can easily be achieved in regions where the flow is known to be almost two–dimensional. The elements adjacent to the wall are of size $\Delta r_1 = 0.005$, similar to [3]. Radial profiles of the average solution are displayed in Figure 8. Comparing the curves for the two computations we notice the slight acceleration due to the different domain size mentioned above. The curves at larger angles are quite sensitive to the actual form of the recirculation bubble. Those of NRUN1 are to the left of the companion ones due to the slightly larger separation angle. The symbols in Figure 8 reflect the respective discretization. It is somewhat finer in the structured grid with, e.g., 28 instead of 20 points within a distance of 0.1 from the cylinder wall. However, the difference is not as drastic as it might appear. The symbols representing the intersection with the element faces are clustered and, e.g., the shear layer at $\theta = 114°$ is discretized with 19 and 16 points, respectively. An interesting feature is the increased smoothness of the mean velocity with N3S particularly visible at $\theta = 114°$. Note that this is a location where the flow is still laminar. The oscillations persisting in the average are presumably due to the existence of a prefered direction in the structured mesh, whereas this radial anisotropy is not present in the bulk of the FE mesh (cf. Figure 3). Finally, the discussion in Section 3.5 and the results of [21], [22] and [6] show that sufficient resolution in spanwise direction is important. The higher the number of points on the cylinder wall in this direction, the larger is the potential saving with an unstructured discretization.

Comparing the performance of two entirely different codes run on different machines is of course a delicate task and can only result in gross estimates. Note for example that N3S is a complex general purpose code whereas LESOCC is "streamlined" for the present task. The price to be paid for the saving of grid points with an unstructured discretization is a higher complexity of the code and a larger CPU time per point. The advantage of this approach highly depends on the flow considered and the particular discretization. The results presented were obtained with LESOCC at 620 MFLOPS on a VPP300[1] and with N3S at 440 MFLOPS on a CRAY98[2]. The CFL number was $0.7-1$ for LRUN3 and 0.3 for NRUN1. One time step of LRUN3 took $9.3sec$ and $18sec$ in NRUN1. Taking into account the different FLOP–rate the performance of both codes is similar for the present case. For higher spanwise resolution or problems with more complex geometry the unstructured method is likely to be more efficient.

3.7 Results for Re = 140000

The Reynolds number considered is in the upper subcritical regime: the boundary layer on the cylinder is still laminar, but the transition to turbulence takes place very shortly after separation. For slightly higher values, i.e. around $Re = 2 - 3 \times 10^5$ the boundary layer becomes turbulent, the Strouhal number increases and the well–known drag crisis is observed. Hence, at the border of the upper subcritical regime the flow is fairly sensitive to disturbances in experimental or numerical conditions [11] [28].

[1] Rechenzentrum Universität Karlsruhe

[2] IDRIS (Institut du Dévelopement et des Ressources en Informatique Scientifique) and EDF, Paris.

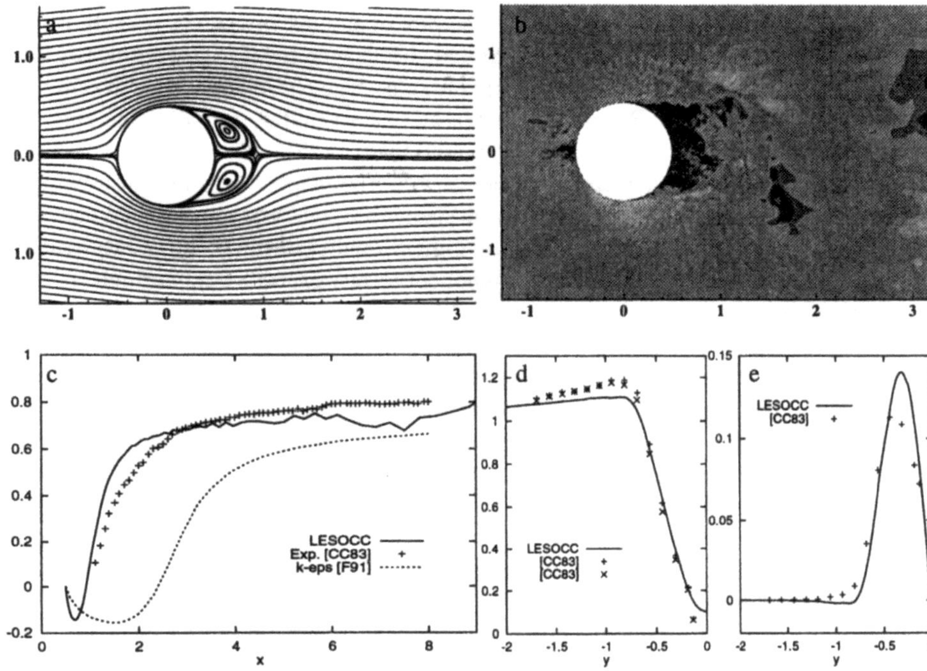

Figure 10 Results for the cylinder flow at $Re = 140000$: a) streamlines of the mean flow, b) magnitude of the instantaneous velocity in a plane normal to the cylinder axis, the maximum is about 2.25, c) mean streamwise velocity at $y = 0$, d) mean streamwise velocity at $x = 1$, e) resolved shear stress $< \bar{u}'\bar{v}' >$. Symbols indicate measured total shear stress from [11] obtained by adding phase–averaged and random contribution.

The flow was computed with LESOCC and a no–slip condition on the wall. Since the boundary layer thickness behaves like $1/\sqrt{Re}$ [34] the grid spacing should be smaller by a factor of 6 than in the computations above to achieve the same quality of resolution. It is clear, however, that this cannot be obtained with the available means. Reducing the minimal mesh size in radial direction is relatively easy due to the exponential behaviour resulting from the geometric stretching employed. On the other hand, the mesh size decreases only linearly with the number of points in azimuthal and spanwise directions. After extensive two–dimensional tests we used a grid with $166 \times 206 \times 64$ points in radial, azimuthal, and axial direction, respectively. Radial stretching of 4.5% yields $\Delta r_1 = 4.8 \times 10^{-4}$, and the cells were evenly spaced on the cylinder wall in all directions. Experimental results in [42] suggest that the spanwise correlation length is significantly lower in the present regime than for $Re = 3900$ so that we set the spanwise period length to 1. The diameter of the computational domain is 15. The Smagorinsky model was applied with $C_s = 0.1$ and the computation was started from uniform flow. After about 7 shedding periods the statistical quantities were assembled during 16 periods.

Figure 10 collects some of the results. The plot of the instantaneous velocity illustrates the extremely narrow boundary layer and shear layer, the fine scale structures in the wake, as well as spurious oscillations on the sides. The recirculation length is much smaller than for the lower Reynolds number, and no secondary recirculation zone is observed. In this context it is interesting to note that in [37] a secondary vortex with angular extent of 12° was observed for $Re = 10^5$. The velocity gradient from which this was infered can however also result from spanwise flow since the employed device is insensitive to direction. Indeed, the three–dimensionality of the flow is considerable (see Figure 11). The spanwise velocity component locally exceeds 0.9. In Figure 10 we observe that the recirculation length matches the experimental data (the value in Table 2 is obtained from flying–hot–wire data which are less reliable very close to the cylinder [11]). Comparing the velocity on the centerline reveals that $< \bar{u} >$ is overpredicted closely behind the recirculation bubble while being slightly too small around 6 diameters from the cylinder. In the latter region, however, the grid is already quite coarse and the averaging could be improved. Nevertheless, the agreement is quite good. The normal profile of $< \bar{u} >$ in Figure 10d also shows the relatively good agreement with the experiment. Figure 10e finally reports the resolved shear stress $< \bar{u}'\bar{v}' >$, which also agrees satisfactorily. On the other hand, Figure 10a and Table 2 show that the separation angle is too large in the computation. This goes along with a slightly lower drag. The Strouhal number is within the experimental range but relatively large while the back pressure coefficient agrees fairly well. Our interpretation, backed by the wavy streaklines upstream of the separation in Figure 11, is the following: as discussed above the discretization of the boundary layer is relatively coarse (at certain instances the value of y_1^+ locally exceeds 5 in the front part). This induces additional numerical oscillations so that the boundary layer experiences additional "turbulence" as if the Reynolds number were somewhat larger. Indeed, the backward shift of the separation, reduced drag, and increase in Strouhal number all point to a premature tendency towards the critical state.

Finally, we can make a comparison with previous results obtained with statistical models. Figure 10c includes the corresponding curve obtained with a two–layer $k - \epsilon$ turbulence model [16]. It is evident that the model gives unsatisfactory results for the considered flow. This is due to the presence of large scale fluctations which are known to be difficult to model by RANS methods; in particular the $k - \epsilon$ model underpredicts the strength of the periodic shedding motion.

Table 2 Global quantities for cylinder flow computations at $Re = 140000$.

	St	C_D	C_{bp}	L_r/D	Θ
LESOCC	0.217	1.157	−1.33	0.42	93.8
Experiment [11]	0.179	1.237	−1.21	0.5	
[33],[38],[37]	0.2	1.2	−1.34		79

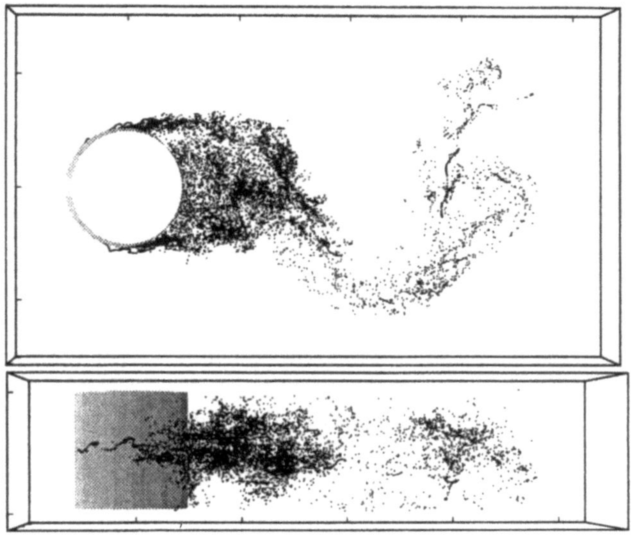

Figure 11 Streaklines in the computation at $Re = 1400000$ starting at two points close to the stagnation line, top and side view at the same instant. The figures show a closeup near the cylinder, tics are placed at unity distance. In the top view the streaklines are partly covered by the representation of the cylinder wall.

4 Conclusions

The paper reports on large-eddy simulations of flow around circular cylinders at two different subcritical Reynolds numbers. The complex physical phenomena occurring in these flows are generally well captured by the calculations and the results agree favourably with experiments. It is the first time that such calculations are reported for the high Reynolds number of Re = 140000; also in this case the results are in most respects satisfactory, but the very thin laminar boundary layer up to separation could not be resolved sufficiently and a better treatment of this boundary layer should be attempted. Altogether, the results show clearly that LES is much more suitable for simulating this type of flow than are RANS models.

A special effort was made to compute the case with Re = 3900 under similar conditions with a structured FV code and an unstructured FE code in order to allow a detailed comparison of the two methods. First of all, the results demonstrate that the unstructured LES code is operational and LES calculations are possible with unstructured FV methods. The comparison has shown that both codes run at similar cost for the case computed: the extra computational effort due to the unstructured nature of the code is roughly compensated by the lower number of grid points that need to be used. However, the comparison of the results also revealed that, in the case of the unstructured grid employed, the reduction of grid points was overdone somewhat as the separated shear layer and the wake were not resolved sufficiently.

Proper construction of a suitable grid for LES calculations of complex flows is a difficult task for both structured and unstructured grids. While the former suffer from the introduction of unnecessary points in certain areas, the latter require a priori knowledge on where fine resolution is necessary and hence on what an ideal discretization should look like. Only with substantially fewer grid points does the higher price per grid point pay off when the unstructured method is used. Hence, great effort has usually to be invested in generating a suitable mesh. In the case of structured grids, better economy with regard to grid points can be achieved by using block-structured grids; this technique is currently implemented into LESOCC.

In any case, with a fixed mesh good a priori knowledge on the properties of the solution is required. However, often such knowledge is lacking and it is increasingly difficult to obtain the more complex the flow situation is. When using an unstructured grid, the natural way out of this dilemma is of course to employ a method that allows to adapt the grid during the calculation, based on certain quality criteria. Certainly, much work is still required in developing such adaptive mesh techniques suitable for large-eddy simulations.

Acknowledgment

The authors thank the Deutsche Forschungsgemeinschaft (DFG) and the Centre National de la Recherche Scientifique (CNRS) for the support of this work within the French–German Research Program 'Numerical Flow Simulation'.

References

[1] E. Balaras, C. Benocci, and U. Piomelli. Two-layer approximate boundary conditions for large–eddy simulations. *AIAA*, 34:1111–1119, 1996.

[2] F. Bastin. Jet noise using LES. In *Annual Research Briefs 1996*, Center for Turbulence Research, 1996.

[3] P. Beaudan and P. Moin. Numerical experiments on the flow past a circular cylinder at sub–critical Reynolds number. Technical Report TF-62, Stanford University, 1994.

[4] J.P. Bertoglio and J. Mathieu. Study of subgrid models for sheared turbulence. *4th Symposium on Turbulent Shear Flows, Karlsruhe*, 1983.

[5] G. Bosch. *Experimentelle und theoretische Untersuchung der instationären Strömung um zylindrische Strukturen.* PhD thesis, Universität Karlsruhe, 1995.

[6] M. Breuer. Numerical and modelling influences on large eddy simulations for the flow past a circular cylinder. In *Proceedings of the 11th Turbulent Shear Flow Conference, Sept. 8-11, 1997, Grenoble, France*, 1997.

[7] M. Breuer, D. Lakehal, and W. Rodi. Flow around a surface mounted cubical obstacle: Comparison of LES and RANS–results. In M. Deville, S. Gavrilakis, and I.L. Ryming, editors, *Computation of 3D Complex flows*, volume 53 of *Notes on Numerical fluid Mechanics*, pages 22–30. Vieweg Verlag, 1996.

[8] M. Breuer and W. Rodi. Large eddy simulation of turbulent flow through a straight square duct and a 180° bend. In P.R. Voke, R. Kleiser, and J.P. Chollet, editors, *Fluid Mech. and its Appl.*, volume 26. Kluwer Acad. publ., 1994.

[9] M. Breuer and W. Rodi. Large eddy simulation of complex turbulent flows of practical interest. In E.H. Hirschel, editor, *Flow simulation with high performance computers II*, volume 52 of *Notes on Numerical Fluid Mechanics*, pages 258–274. Vieweg, Braunschweig, 1996.

[10] W. Cabot. Near–wall models in large eddy simulations of flow behind a backward facing step. In *Annual Research Briefs 1996*, pages 199–210. Center for Turbulence Research, 1996.

[11] B. Cantwell and D. Coles. An experimental study of entrainment and transport in the turbulent near wake of a circular cylinder. *J. Fluid Mech.*, 136:321–374, 1983.

[12] G.S. Cardell. *Flow past a circular cylinder with a permeable splitter plate*. PhD thesis, Graduate Aeronautical Lab., California Inst. of Technology, 1993.

[13] J.P. Chollet and M. Lesieur. Parameterization of small scales of three dimensional isotropic turbulence. *J. Atmos. Sci.*, 38:2747–2757, 1981.

[14] C.K. Chyu. *A study of the Near–Wake Structure from a Circular Cylinder*. PhD thesis, Lehigh University, Bethlehem, Philidelphia, USA, 1995.

[15] C.K. Chyu and D. Rockwell. Evolution of patterns of streamwise vorticity in the turbulent wake of a circular cylinder. *J. Fluid Mech.*, 320:117–137, 1996.

[16] R. Franke. *Numerische Berechnung der instationaeren Wirbelablösung hinter zylindrischen Körpern*. PhD thesis, Universität Karlsruhe, 1991.

[17] K. Jansen. (private communication).

[18] K. Jansen. Unstructured grid large eddy simulation of wall bounded flows. Annual Research Briefs, 1993.

[19] R.H. Kraichnan. Eddy viscosity in two and three dimensions. *J. Atmos. Sci.*, 33:1521–1536, 1976.

[20] L.M. Lourenco and C. Shih. Characteristics of the plane turbulent near wake of a circular cylinder. A particle image velocimetry study. (data taken from Beaudan,Moin(1994)), 1993.

[21] R. Mittal. Large–eddy simulation of flow past a circular cylinder. In *Annual Research Briefs 1995*, pages 107–116. Center for Turbulence Research, 1995.

[22] R. Mittal. Progress of LES of flow past a circular cylinder. In *Annual Research Briefs 1996*, pages 233–241. Center for Turbulence Research, 1996.

[23] C. Norberg. Effects of Reynolds number and low–intensity free–stream turbulence on the flow around a circular cylinder. Technical Report Publ. No. 87/2, Dep. of Appl. Thermosci. and Fluid Mech., Chalmers Univ. of Technology, Sweden, 1987.

[24] L. Ong and J. Wallace. The velocity field of the turbulent very near wake of a circular cylinder. *Experiments in Fluids*, 20:441–453, 1996.

[25] S. Parpais and J.P Bertoglio. A spectral closure for inhomogeneous turbulence applied to turbulent confined flow. 6th European Turbulence Conference, Lausanne, 1996.

[26] D.E. Paxson and R.E. Mayle. Velocity measurements on the forward portion of a cylinder. *J. Fluids Eng.*, 112:243–245, 1990.

[27] G. Pot, L.L. de Sousa, J.P. Gregoire, and Y. Souffez. Improvement of industrial finite element algorithms for CFD code N3S. In Morgan, Onate, Periaux, Peraire, and Zienkiewicz, editors, *Finite Elements in Fluids*, volume 1, pages 527–536. Pineridge Press, 1993.

[28] A. Richter and E. Naudascher. Fluctuating forces on a rigid circular cylinder in confined flow. *J. Fluid Mech.*, 78:561–576, 1976.

[29] W. Rodi. Comparison of LES and RANS calculations of the flow around bluff bodies. *J. Wind Ind. Aerodyn.*, 69-71:55–75, 1997.

[30] W. Rodi. Large-Eddy Simulation and statistical turbulence models: Complementary approaches. In O. Métais and J. Ferziger, editors, *New Tools in Turbulence Modelling*, Les Éditions de Physique, pages 49–72. Springer, 1997.

[31] W. Rodi, J.H. Ferziger, M. Breuer, and M. Pourquié. Status of large eddy simulation: Results of a workshop. *J. Fluid Eng.*, 119:248–262, 1997.

[32] P. Rollet-Miet. Simulation des grandes échelles a partir du code N3S: Ecoulement en canal. Technical Report HE-41/97/036/A, EDF, Chatou, France, 1997.

[33] G. Schewe. On the force fluctuations acting on a circular cylinder in crossflow from subcritical up to transcritical Reynolds numbers. *J. Fluid Mech.*, pages 265–285, 1983.

[34] H. Schlichting. *Grenzschichttheorie*. Braun, 1958.

[35] S. Sebag, V. Maupu, and D. Laurence. Non–orthogonal calculation procedures using second moment closure. In *Eighth Symposium on Turbulent Shear Flows, Munich*, pages 20.3.1–20.3.6. 1991.

[36] O. Simonin and M. Barcouda. Measurements and prediction of turbulent flow entering a staggered tube bundle. Fourth Int. Symp. on App. of Laser Anemometry to Fluid Mechanics, Lisbon, Portugal, 1988.

[37] J. Son and T.J. Hanratty. Velocity gradients at the wall for flow around a cylinder at Reynolds numbers from 5×10^3 to 10^5. *J. Fluid Mech.*, pages 353–368, 1969.

[38] S. Szepessy and B.W. Bearman. Aspect ratio and end plate effects on vortex shedding from a circular cylinder. *J. Fluid Mech.*, pages 191–217, 1992.

[39] T. Wei and C.R. Smith. Secondary vortices in the wake of circular cylinders. *J. Fluid Mech.*, 169:513–533, 1986.

[40] H. Werner and H. Wengle. Large–Eddy Simulation of turbulent flow over and around a cube in a plane channel. In U. Schumann et al., editor, *8th Symp. on Turb. Shear Flows*, 1993.

[41] G.S. West and C.J. Apelt. Fluctuating lift and drag forces on finite lengths of a circular cylinder in the subcritical Reynolds number range. *Journal of Fluids and Structures*, 11:135–158, 1997.

[42] M.M. Zdravkovich. *Flow Around Circular Cylinders*. Oxford University Press, 1997.

Towards an adaptive wavelet-based 3D Navier-Stokes Solver

F. Koster[1], M. Griebel[1],
N. K.-R. Kevlahan[2], M. Farge[2], K. Schneider[3]

[1] Inst. f. Angew. Math. Universität Bonn, Wegelerstr.6, D-53115 Bonn
[2] LMD-CNRS, Ecole Normale Sup., 24 rue Lhomond, F-75231 Paris cedex 05
[3] CPT-CNRS Luminy, Case 907, F-13288 Marseille cedex 09

Summary

In this paper we present results obtained using wavelet methods for the analysis and simulation of two-dimensional turbulence, and a preliminary study of a three-dimensional turbulent channel flow. The two-dimensional results show the efficiency of wavelets for the numerical simulation of turbulence, and suggest new methods for modelling the flow, based on a decomposition into coherent and incoherent parts. The three-dimensional study compares different wavelet type approaches applied to instantaneous velocity, vorticity and pressure fields. At least for the data and the setting considered in our experiments, i.e. flows with dominating boundary layer, the results indicate which of the various forms of the three-dimensional Navier–Stokes Equations (3D-NSE) (primitive variables, vorticity-streamfunction formulation) and which multiscale approach is well suited for the numerical integration of the 3D-NSE. In addition, we present efficient techniques for the implementation and parallelization of an adaptive wavelet-based 3D-NSE solver.

1 Introduction

In both two and three dimensions turbulence is characterized by localized regions of strong variations in the quantities describing the flow. These coherent structures take the form of roughly circular patches of strong vorticity in two dimensions [25] and thin tubes of strong vorticity in three dimensions [23]. The existence of these tubes has been confirmed many times both in numerical [31], [34] and laboratory experiments [3]. The tubes have diameters of the order of the Kolmogorov scale and lengths up to the integral scale. The interest of these patches or tubes is that they are believed to control the dynamics of the flow. This is the foundation of the various vortex-based methods [24] in both two and three dimensions. Furthermore, the localized nature and sometimes complex internal structure of the vortices suggests the use of multiscale methods. Due to their localization in scale and space these methods are good candidates to reduce the complexity of the flow. One example of multiscale methods, wavelets, have been used for the analysis and simulation of two-dimensional turbulence with periodic boundary conditions [11], [13], [5]. The results in two dimensions have been encouraging and suggest new modelling methods (see §3), and it is clear that the next step is to attempt to apply similar methods to three-dimensional flows and flows with boundaries.

This paper has four main sections. The first part (§2) briefly describes multiscale methods in both two and three dimensions. Starting from three particular univariate approaches we will explain two constructions for multivariate methods. One of these approaches is new in the context of wavelets, but gives superior results in three dimensions.

The second part (§3) describes some new results obtained by the French members of our group for the analysis and simulation of two-dimensional turbulence using wavelets. The goal is to demonstrate that wavelet techniques can be efficient and highly accurate for computing two-dimensional turbulent flows. Furthermore, wavelet analysis shows that the flow may be objectively separated into a coherent part (with non-Gaussian vorticity statistics) and an incoherent part (with Gaussian vorticity statistics). This result suggests a new highly efficient modelling method where the coherent part of the flow is calculated using the wavelet solver, while the incoherent part is modelled statistically or merely advected by the coherent part.

In the third part (§4) we applied different multiscale methods to a database of a 3D turbulent channel flow. This work was done in Bonn. By the comparison of compact representations of the velocity and vorticity we aim to answer the following questions:

- Do we need to solve the 3D-NSE in terms of vorticity or velocity in order to exploit the localization of the coherent structures?
- Which of the various multiscale methods yield a good trade-off between numerical efficiency and suitability for turbulence simulations?
- How can we implement an adaptive method efficiently?

The main, but still preliminary, result of this part is that at least for the flow under consideration the formulation of the 3D-NSE in primitive variables seems to be better suited to the integration than the formulation in terms of vorticity. In part, this is due to the fact that in three dimensions the numerical costs for the vorticity formulation are much higher. The second result is that simple multiscale methods (prewavelets) give a very good trade-off between reduction of the complexity of the flow and numerical efficiency. This analysis will form the foundation for the implementation of a wavelet-based 3D-NSE solver by our group. Some algorithmic details for efficient data management and parallelization of the adaptive solver are discussed in §5. Note that the use of such techniques is crucial for adaptive codes, in order to preserve their advantage of reducing the complexity of the physical problem by working with a sparse representation.

2 Multiscale Methods

We will briefly describe some topics from the theory of multiscale methods. These lines are in the spirit of the standard literature, see also the recent introducing paper of Dahmen [7].

2.1 General Concept of Multiscale Analysis

For a multiscale analysis (MSA) of a class H of functions one considers a sequence of nested subspaces $V_j \subset H$, whose union is dense in H, and complementary spaces $W_j = V_{j+1} \ominus V_j$.

$$\{0\} = V_0 \subset ... \subset V_j \subset V_{j+1} \subset ... H \quad ; \quad W_j \oplus V_j = V_{j+1} \quad \text{and} \quad clos \bigcup_{j=0}^{\infty} V_j = H.$$

Let V_j and W_j resp. have bases $\{\phi_l^j\}_{l\in I_j}$ and $\{\psi_l^j\}_{l\in J_j}$, where I_j, J_j are some index sets. For the following explanations it will be convenient, to write the bases as (possibly infinite) column vectors Φ^j and Ψ^j. For the MSA of a given function $u \in H$ one is looking for coefficients $u^j := \{u_l^j\}_{l\in J_j}$ of the expansion

$$u = \sum_{j=0}^{\infty} (u^j)^T \cdot \Psi^j \tag{2.1}$$

With some additional requirements on H and the bases Φ^j and Ψ^j, the coefficients u^j decay with increased *scale* j

$$|u^j|_{l_p} \to 0 \quad , \quad j \to \infty \quad .$$

Further assumptions on H, Φ^j and Ψ^j then will allow for a good approximation of u by an expansion similar to (2.1) using only a very limited number of significant coefficients u_l^j

$$\tilde{u} = \sum_{j,l\ ;\ |u_l^j| > \epsilon(j)} u_l^j \psi_l^j \quad \text{and} \quad \|u - \tilde{u}\|_H \quad \text{is small.} \tag{2.2}$$

In this equation $\epsilon(j)$ is a certain cut-off parameter, which here depends only on the scale j, but may be chosen in a more sophisticated manner.
In practical applications usually one has a representation $u = (c^J)^T \cdot \Phi^J \in V^J$, derived from collocation values of u by, e.g. numerical quadrature methods [33], and one is looking for the coefficients $\{u^j\}_{0\le j\le J}$. This corresponds to the transform $T^J : c^J \mapsto \bigcup_{j=0}^{J-1} u^j$.
Often, besides the sequence V^j of *trial* spaces, one has explicit knowledge of an ascending sequence of *test* spaces V_j^* and their complements W_j^*, such that the bases Φ^{*j}, Ψ^{*j} for a certain dual pairing $< .,. >$ fulfill *biorthogonality* relations

$$< \Phi^j, \Psi^{*j} >= \mathbf{0} \quad ; \quad < \Psi^j, \Phi^{*j} >= \mathbf{0} \quad ; \quad < \Phi^j, \Phi^{*j} >= I_{I_j} \quad ; \quad < \Psi^j, \Psi^{*j} >= I_{J_j}.$$

The nestedness of the spaces $V_j^*, W_j^* \subset V_{j+1}^*$ leads to the existence of *refinement matrices*, such that

$$\Phi^{*j} = \mathcal{H}_j^* \cdot \Phi^{*j+1} \quad ; \quad \Psi^{*j} = \mathcal{G}_j^* \cdot \Phi^{*j+1} \quad .$$

Similar relations hold for the test spaces involving matrices $\mathcal{H}_j$ and $\mathcal{G}_j$. Together with biorthogonality this implies the following recursive scheme (Fast Wavelet Transform)

$$c^J = < u, \Phi_J^* > \quad ; \quad c^j = \mathcal{H}_j^* c^{j+1} \quad \text{and} \quad u^j = \mathcal{G}_j^* c^{j+1} \quad \text{for} \quad 0 \leq j < J \quad .$$

Hence, T^J is formally given by the chain $T^{1,0} \circ .. \circ T^{J,J-1}$ of operators, each $T^{j,j-1}$ mapping the coefficients $\bigcup_{k=j}^{J-1} u^k \bigcup c^j$ to $\bigcup_{k=j-1}^{J-1} u^k \bigcup c^{j-1}$. Whenever the matrices $\mathcal{H}_j^*$ and $\mathcal{G}_j^*$ are sparse, uniformly in j, the transform T^J can be applied very fast. A moments thought yields that the inverse operation $(T^J)^{-1}$ is given by the recursive scheme $c^{j+1} = (\mathcal{H}_j)^T c^j + (\mathcal{G}_j)^T u^j \quad (0 \leq j < J)$.

2.2 Univariate Examples

We will briefly introduce three families of univariate biorthogonal systems. From these we build up the multivariate systems used in our experiments. The three families have in common, that they are generated by dilates and translates of 'mother' functions $\phi^{(*)}$, $\psi^{(*)}$. The families are the hierarchical nodal basis, prewavelets and an example for compactly supported orthogonal wavelets. Figure 1 shows some of the trial functions.

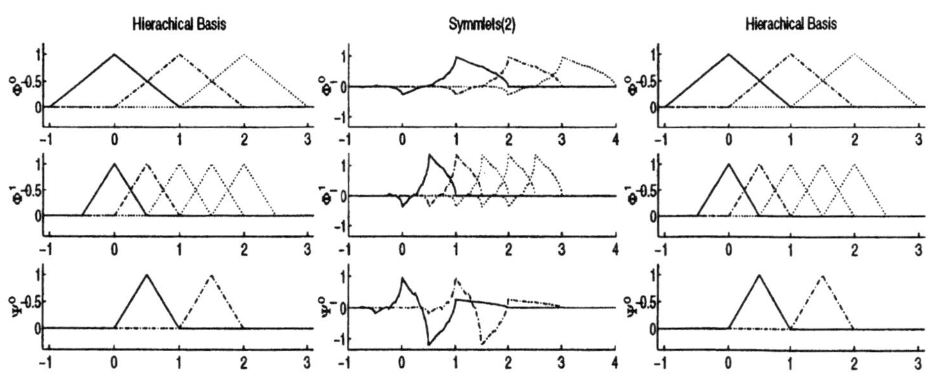

Figure 1: Univariate trial functions

(HB) Hierarchical Nodal Basis for $H = L_2([0,1])$:
Let $\phi(x) = \max\{0, 1 - |x|\}$ denote the well-known 'hat' function. Then, suitable basis functions for V_j and V_j^* are

$$\phi_l^j := \phi(2^j x - l) \quad \text{and} \quad \phi_l^{*j} := \delta(2^j x - l) \quad ; \quad 0 \leq l \leq 2^j).$$

Here δ denotes the Dirac functional. Bases of the complementary spaces are given by

$$\psi_l^{j-1} = \phi(2^j x - 2l + 1) \quad \text{and} \quad \psi_l^{*j-1} = \phi_{2l-1}^{*j} - \frac{1}{2}(\phi_{2l-2}^{*j} + \phi_{2l}^{*j}) \quad ; \quad 1 \leq l \leq 2^j - 1.$$

(OW) Compactly Supported Orthogonal Wavelets for $H = L_2(\mathbf{R})$:
In this case ϕ is the solution of the equation

$$\phi(x) = \sqrt{2} \sum_{l=-L}^{R} h_l \phi(2x - l) ,$$

with some very distinct coefficients h_l. Solutions of such equations and the corresponding coefficients were first given by Daubechies [8], [9]. The L_2-orthogonal basis functions are the dilates and translates $\phi_l^j = \phi_l^{*j} = 2^{j/2}\phi(2^j x - l)$ and $\psi_l^j = \psi_l^{*j} = 2^{j/2}\psi(2^j x - l)$ for $l \in \mathbf{Z}$, where $\psi(x) = \sqrt{2}\sum_{l=-R+1}^{L+1}(-1)^l h_{1-l}\phi(2x - l)$.

In numerical applications one usually deals with bounded domains, intervals in the simplest case, thus the framework for $L_2(\mathbf{R})$ above is not satisfactory and one has to modify at least the basis functions ϕ_l^j, ψ_l^j near the boundaries. Such a construction is given by [15], generalizing and improving previous results of [6], [27] with respect to the numerical stability of the boundary adaptation. In our experiments we used boundary adapted least asymmetric wavelets of the Daubechies family with two vanishing moments (Symmlets(2)).

(PRW) Lifting Prewavelets:
A simple, but very powerful, method for the construction of biorthogonal systems has been given by W. Sweldens [32]. In terms of the refinement matrices biorthogonality reads

$$\begin{pmatrix} \mathcal{H}_j \\ \mathcal{G}_j \end{pmatrix}^T \cdot \begin{pmatrix} \mathcal{H}_j^* \\ \mathcal{G}_j^* \end{pmatrix} = \begin{pmatrix} I & \mathbf{0} \\ \mathbf{0} & I \end{pmatrix} . \tag{2.3}$$

These relations are preserved, if we switch to a biorthogonal system with refinement matrices

$$\begin{pmatrix} \tilde{\mathcal{H}}_j \\ \tilde{\mathcal{G}}_j \end{pmatrix} = \begin{pmatrix} I & \mathbf{0} \\ K & I \end{pmatrix} \cdot \begin{pmatrix} \mathcal{H}_j^* \\ \mathcal{G}_j^* \end{pmatrix} \quad ; \quad \begin{pmatrix} \tilde{\mathcal{H}}_j^* \\ \tilde{\mathcal{G}}_j^* \end{pmatrix} = \begin{pmatrix} I & -K^T \\ \mathbf{0} & I \end{pmatrix} \cdot \begin{pmatrix} \mathcal{H}_j^* \\ \mathcal{G}_j^* \end{pmatrix} .$$

This allows for the modification of existing biorthogonal systems to custom design a new one with special desired features. In our case we choose the primal system corresponding to the hierarchical nodal basis (HB) and K such that the new system exhibits the following properties,

$$< P, \tilde{\Psi}^j > = < P, \tilde{\Psi}^{*j} > = 0 \quad \text{for each polynomial of degree less than 2} .$$

The second condition was also valid for the hierarchical nodal basis, but the first was not. The purpose of this modification comes from the wish for easy preconditioning the Laplace or Helmholtz problems, which have to be solved as part of the integration of the 3D-NSE.

Some properties of these biorthogonal systems are compared in Table 1.

Table 1: Work count/preconditioning for biorth. bases

	relative num. work count	preconditioning in 3D
HB	1	No
PRW	2	Yes
OW	5	Yes

2.3 Multivariate Constructions

In the experiments of the next section, H is a space of functions over the domain $[x_1, x_2] \times [y_1, y_2] \times [z_1, z_2]$. This tensor product structure allows for the following two multivariate approaches. We restrict the presentation to the two-dimensional case, since the generalization to higher dimensions is straightforward.

(ISO) Isotropic approach:
For this construction the bivariate Spaces $\mathbf{V}_j$ are defined by

$$\mathbf{V}_j = V_j \otimes V_j \quad ; \quad \mathbf{W}_j = (V_{j-1} \otimes W_{j-1}) \oplus (W_{j-1} \otimes W_{j-1}) \oplus (W_{j-1} \otimes V_{j-1}).$$

The transform formally is given by $\mathbf{T}^J = \bigcirc_{j=1}^{J-1}(T^{j,j-1} \otimes T^{j,j-1})$.
This means we have to apply at each stage of the transform first the univariate $T^{j,j-1}$ with respect to x for all y, leading to the decomposition $\mathbf{V}_j = (W_{j-1} \otimes V_j) \oplus (V_{j-1} \otimes V_j)$, and then with respect to y for all columns, leading to the final decomposition $\mathbf{V}_j = \mathbf{V}_{j-1} \oplus \mathbf{W}_{j-1}$.
Since the univariate spaces V_j, W_j are usually spanned by equally dilated basis functions, the basis of $\mathbf{W}_j$ consists of functions with equal dilation in both coordinate directions. Simply stated, they are isotropic. This approach is the classical one in the framework of multivariate MSA.

(ANISO) Anisotropic and isotropic basis functions:
In contrast to the ISO approach the ANISO construction works in the same way as the multivariate FFT. First we apply the complete decomposition with respect to x and then with respect to y. Formally this leads to the transform $\mathbf{T}^J = T^J \otimes T^J$. Despite of its easy application, this construction leads to a more complicated hierarchy of subspaces, which now depend on two parameters.

$$\mathbf{V}_{j,J} = \oplus_{k=j+1}^{J}(W_k \otimes V_j) \oplus (V_j \otimes W_k) \quad ; \quad \mathbf{W}_{j,J} = \oplus_{k=j}^{J}(W_k \otimes W_j) \oplus (W_j \otimes W_k).$$

Further information about such multiparametric space splittings are given in [16], [28]. This approach exchanges the 3 types (corresponding $V_{j-1} \otimes W_{j-1}$, $W_{j-1} \otimes W_{j-1}$, $W_{j-1} \otimes V_{j-1}$) of equally dilated basis functions of the complementary spaces with a larger number of different types of basis functions, each having a characteristic dilation in the first and second coordinate direction. Thus there are both, isotropic and anisotropic basis functions. In the experiments of the fourth section it turns out, that this approach is superior compared to ISO, although for fixed J and fixed underlying univariate MSA the numerical cost is larger by a factor of $\frac{7}{4}$ in the three-dimensional case.

2.4 Error Estimation

A simple estimate for the error is

$$\|u - \sum_{j,l;\ |u_l^j|>\epsilon(j)} u_l^j \psi_l^j\|_H \leq \sum_{j,l;\ |u_l^j|\leq\epsilon(j)} |u_l^j| \|\psi_l^j\|_H. \tag{2.4}$$

For our examples, except near the boundaries, the basis functions ψ_l^j are translates of some $\psi_{l_0}^j$. Hence, for inner functions ψ_l^j the norm $\|\psi_l^j\|_H$ does not depend on l. Together with the estimate above this suggests to use the estimators

(L2) $\qquad \epsilon(j) = \epsilon / \|\psi_{l_0}^j\|_{L_2}$

(H1) $\qquad \epsilon(j) = \epsilon / \|\psi_{l_0}^j\|_{H_1}.$

Here ϵ is a predefined cut-off value.

For the ANISO construction there is another simple a priori error estimator. This estimator essentially recovers the idea of so called sparse grids [35], [4]. The three underlying univariate basis functions span the space of polynomials of degree less than 2. This implies that for functions u with bounded mixed derivatives $\|\frac{\partial^4 u}{\partial^2 x \partial^2 y}\|_H$, the following estimate is appropriate

$$|u_l^j| \|\psi_l^j\|_{L_2} \leq a^{j_1+j_2} \quad ; \quad \text{for some } a < 1.$$

This suggests setting

(SPG) $\qquad \epsilon(j) = \begin{cases} 0 & ; \quad j_1 + j_2 \leq L \\ \infty & ; \quad \text{else} \end{cases}$

for an integer cut-off parameter $L \leq 2J$, which controls the compression rate. A very similar technique, which essentially uses the same kind of adaptivity in scale, but in a slightly different manner, has been successfully applied to the simulation of a turbulent pipe flow [17].

At the present point some remarks are in order. The three univariate approaches, introduced in this paper, were built by dilation and translation. This gives rise to some problems when the data is given by collocation values on nonuniform grids, e.g. this is the case for the database used in the fourth section. Of course one could deal with this problem by modifying the quadrature $c^J = < u, \Phi^J >$, but this is rather complicated. Thus we use a much cheaper solution. In a curvilinear NSE solver the data $(\mathbf{u}, p, ...)$ is given on an equidistant, rectangular grid. We treat the data in the same manner and hence, can apply the fast transforms for equidistant grids. However, we must change the error estimator.

Consider data $\{u_i\}$ given on the non-equidistant grid $a = x_1 < ... < x_n = b$. To consider this data as given on an equidistant grid means to identify $\{u_i\}$ with collocation values of $u \circ x(\xi)$, where $x : [0,1] \to [a,b]\ ;\ \xi \mapsto x(\xi)$ is the gradiation function, which satisfies $x' := \frac{\partial x}{\partial \xi} > 0$. Applying the transform yields

$$u \circ x(\xi) = \sum_{j,l} u_l^j \psi_l^j(\xi) = \sum_{j,l} u_l^j \tilde{\psi}_l^j \circ x(\xi) \quad ,$$

with some implicitly defined basis functions $\tilde{\psi}_l^j$ on $[a,b]$. Clearly we should use $||\tilde{\psi}_l^j||_{H([a,b])}$ as weights in (L2) or (H1). Especially for the later we obtain

$$|\tilde{\psi}_l^j|^2_{H_1([a,b])} = \int_a^b \frac{|\psi_l^j \circ x(\xi)|^2}{x'^2(\xi)} dx = \int_0^1 \frac{|\psi_l^j \circ x(\xi)|^2}{x'(\xi)} d\xi = \frac{1}{x'(\xi_l^j)} |\psi_l^j|^2_{H_1([0,1])}$$

for some $\xi_l^j \in supp\ \psi_l^j$. In regions of strong gradiation x' is very small, hence $\epsilon(j,l)$ will be small. This is exactly what one would expect. Note, in contrary to modifying the quadrature, this approach is easy to extend to arbitrary curvilinear transformed domains and thus is well suited for use in *adaptive curvilinear* codes.

The next issue concerns orthogonal wavelets. For these one has $||\psi_l^j||_{L_2} = 1$ and equality in the estimate (2.4), which is optimal for the non linear thresholding. However, application of the orthogonal wavelet transform to $u \circ x$ can be reinterpreted as application of a non orthogonal transform, i.e. Parseval's identity $||u||_0^2 = \sum_{j,l} |u_l^j|^2$ does not hold anymore. Instead of that there are some constants $c(j)$, $C(j)$, such that

$$c(j_0) \sum_{l,\ j \geq j_0} |u_l^j| ||\psi_l^j||^2 \leq || \sum_{l,\ j \geq j_0} u_l^j \psi_l^j ||^2 \leq C(j_0) \sum_{l,\ j \geq j_0} |u_l^j| ||\psi_l^j||^2.$$

The condition number $\max_j \frac{C(j)}{c(j)}$ gives the deviation from L_2-orthogonality. The point is that the constants behave like

$$c(j) = 1 - \mathcal{O}(2^{-j}) \quad ; \quad C(j) = 1 + \mathcal{O}(2^{-j}) \quad \text{, i.e.}$$

for fine scales (large j), where most of the compression happens, the nice features of orthogonal wavelets are almost recovered by using the weighted cut-off values.

3 Two-dimensional isotropic turbulence

3.1 Goals and methods

In this section we present results obtained using wavelets to analyze and calculate two-dimensional isotropic turbulence with periodic boundary conditions. These results make use of the 2D-NSE solver developed by Fröhlich and Schneider [13] and build on the wavelet analysis techniques developed by Farge *et al.* [11]. The goals of the present work were to verify the accuracy and efficiency of the wavelet based 2D-NSE solver against the usual spectral methods, to develop new wavelet-based forcing methods that avoid the problems inherent in the traditional methods, to develop an objective criterion for separating the coherent and incoherent parts of the flow (based on the different statistical properties of the two parts), and to show that the incoherent part of the flow is stable near the coherent vortices.

3.2 Wavelet simulation

The wavelet transform was first introduced as an analysis technique, but numerical methods have been developed recently which use wavelet bases to actually solve partial differential equations [11], [13], [5]. These methods are particularly well-suited to equations, such as the Navier–Stokes equations at high Reynolds number, whose solutions contain isolated multi-scale structures or quasi-singularities. We compared simulations using these wavelet techniques with standard spectral simulations and nonlinearly filtered spectral simulations [30]. The evolution of the vorticity field for each of the four simulation methods is shown in Figure 2. Comparing the four simulations, one notices that the nonlinear Fourier filtering deforms the edge of the vortices and produces spurious oscillations that quickly spread to fill the whole background flow. The error in representing the edges of the vortices is due to the poor ability of Fourier methods to represent localized sharp gradients; this is exacerbated by the filtering. The homogeneous distribution of the error is not surprising since each Fourier mode is completely de-localized in physical space and thus any error is immediately spread over all space.

The results showed that the wavelet methods are very accurate, and require roughly four times fewer active modes than spectral methods. Furthermore, the number of active wavelet modes is approximately constant in time, even during intense nonlinear interactions, whereas the number of active spectral modes peaks when the interactions are most intense (see Figure 3). It is important to note that in this investigation the non-active modes were simply thrown away. If the modes were modelled in some way, the number of wavelet modes actually calculated would be much smaller (of the order of 1%) and thus the wavelet method would become even more efficient. This possibility is discussed below.

3.3 Wavelet forcing

In order to obtain a statistically steady state in a pseudo-spectral simulation the usual method is to force at a few intermediate scale modes. This method is based on the idea (derived from statistical theories of turbulence) that energy is injected at a particular length-scale (or wavenumber) and then (in two dimensions) cascades to larger scales. The problem with this technique is that it applies a statistical idea to the calculation of an individual flow realization. Physically, it is not reasonable to force by injecting energy at a particular length-scale since we know that turbulent flows are actually forced by the production of vorticity by instability at boundaries. This vorticity detaches from the boundary and is advected into the interior of the flow. In practice, forcing at a single wavenumber does not even produce approximately constant total energy and enstrophy: the

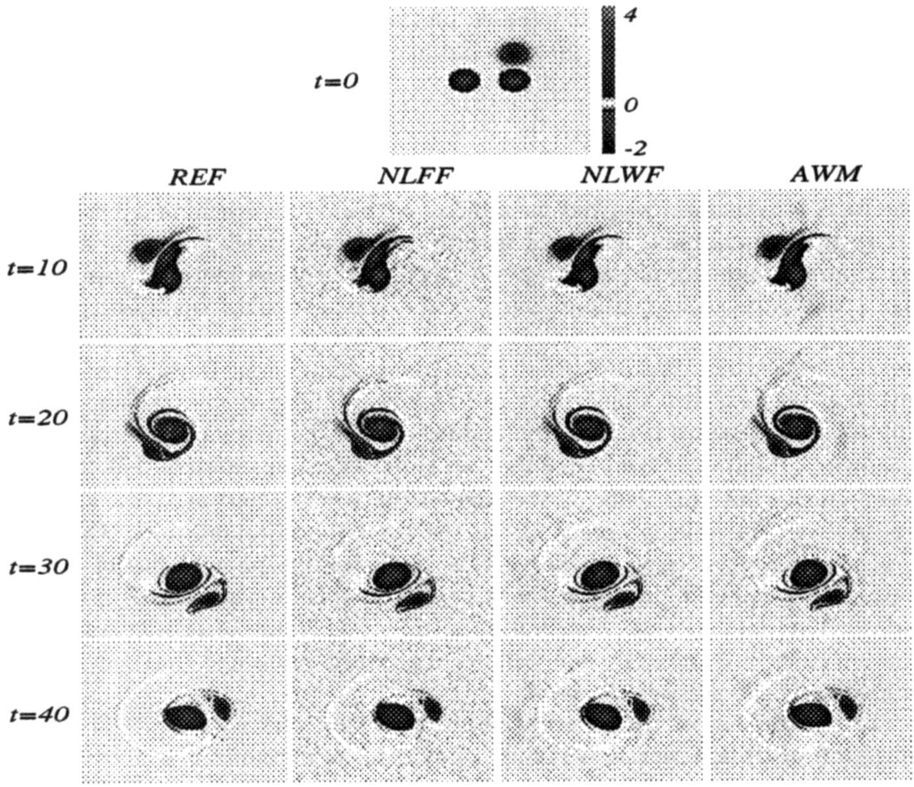

Figure 2: Evolution of the vorticity field as a function of time for each of the four methods. (REF = pseudo-spectral reference simulation, NLFF = nonlinear Fourier filtering, NLWF = nonlinear wavelet filtering, AWM = adaptive wavelet method).

energy and enstrophy fluctuate significantly about a mean. If this mean does not change the flow is said to be statistically stationary. Clearly, this is a rather weak definition of stationarity.

The large wavelet modes correspond to the coherent vortices of the flow. Based on this observation we have introduced a new forcing method that directly amplifies the vortices of the flow by reinforcing the large wavelet modes [29]. This forcing method injects energy and enstrophy into the vortices (rather than into a wavenumber) and thus corresponds more closely to the physical forcing of the turbulent flows than the usual method described above. By forcing in wavelet space we can also control the smoothness of the excited vortices. Wavelet forcing was shown to produce almost constant total enstrophy and energy; a significant improvement on the wavenumber-based method. We also checked that the probability distribution functions (PDF) of vorticity and the energy spectrum do not evolve. This new forcing method should allow the simulation of more realistic

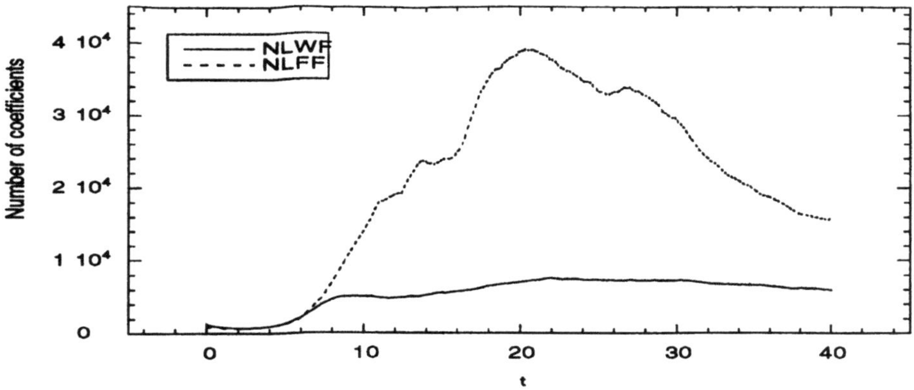

Figure 3: Evolution of the number of active modes for each method (NLFF = nonlinear Fourier filtering, NLWF = nonlinear wavelet filtering and adaptive wavelet simulation).

stationary isotropic two-dimensional turbulent flows, even with periodic boundary conditions.

3.4 Coherent structure eduction

In order to develop effective vortex-based methods for solving or modelling the 2D-NSE it is vital to have an objective and efficient criterion for separating the coherent vortices from the incoherent part of the flow. The separation should ideally ensure that the incoherent part has well-defined simple statistical properties so that it can be easily modelled or calculated.

Recently, Donoho [10] developed a de-noising technique to remove Gaussian white noise from a signal. He showed that if one applies the following threshold $\widetilde{\omega_T}$ to the wavelet modes

$$\widetilde{\omega_T} = (2\log_{10}(N))^{1/2}\sigma/N^{1/2} \tag{3.5}$$

where N is the number of points, and σ is the variance of the noise, then one can extract the signal from the Gaussian background noise. We decided to turn Donoho's theory around and *suppose* that the turbulent vorticity contains a Gaussian component whose variance is simply given by the total variance of the vorticity [12]. Note that the threshold should be constant for a stationary flow.

We have analyzed vorticity fields obtained from direct numerical simulations (DNS) of statistically stationary two-dimensional turbulence where the forcing is done in wavelet space. Using the nonlinear wavelet technique based on an objective universal threshold we separate the vorticity field into coherent structures and background flow. Both components are multi-scale with different scaling laws, and therefore cannot be separated by Fourier filtering. We find that the coherent structures have non-Gaussian statistics (and represent only 1.3%

of the total 128^2 modes) while the background flow is Gaussian (and makes up 98.7% of the modes). This result has important implications for modelling two-dimensional turbulence since it shows that the flow can be divided into a component with Gaussian statistics (which can be easily modelled by an equivalent stochastic process), and a component with non-Gaussian statistics (which will have to be calculated exactly). The fact that the non-Gaussian part can be represented with only a small number of modes means that this decomposition should form the basis for a new highly efficient numerical simulation method for two-dimensional turbulence. The interpretation of the weak wavelet modes as the incoherent part of the flow was reinforced by noting that the coherence function (scatter plot of the vorticity versus the streamfunction) is isotropic for the weak wavelet modes, but has a characteristic *sinh* profile for the strong modes corresponding to the coherent structures.

3.5 Stability of the background flow

The question of the stability of the background flow is important for any vortex method that calculates exactly the evolution of the coherent vortices, but only models the incoherent (or background) part of the flow. If the incoherent part is unstable it generates new vortices, and this process would correspond to a significant forcing of the coherent part by the background flow. Clearly, such an interaction would have to be taken into account in any vortex-based model. Even if the background remains stable it could still affect the evolution of the coherent vortices (e.g. by shielding them from the effect of other vortices), and if this is the case the effect would also have to be included in the model.

We have investigated the stability and effect of the vorticity filaments that are typical of the background flow [21]. To address these questions we used a high-resolution pseudo-spectral DNS to study a simplified model of two-dimensional turbulence: the merging of two vortices accelerated by a third. Vortex merging is the fundamental interaction of two-dimensional turbulence and is also the interaction that generates vorticity filaments. The calculation was carried out on the CRAY C98 of IDRIS (Institut du Développement et des Ressources en Informatique Scientifique) of the CNRS. A wavelet analysis permitted us to cleanly separate the vorticity filaments from the coherent vortices. By comparing the evolution of the filaments with and without the coherent vortices we showed that the filaments are stabilized by the vortices (this comparison is shown in Figure 4). The stabilizing role of the coherent vortices had been suggested on the basis of analytical studies, but had not been verified under realistic conditions. A linear stability analysis and calculation of strain rates then revealed precisely how the stabilization takes place. Finally, by using an asymptotic analysis checked by a DNS we established that the vorticity filaments can protect the vortex they surround from the deforming effects of the weak strain produced by neighbouring vortices.

These results suggest that, provided the density of coherent vortices is sufficient, the background flow should not generate any new coherent vortices. Thus

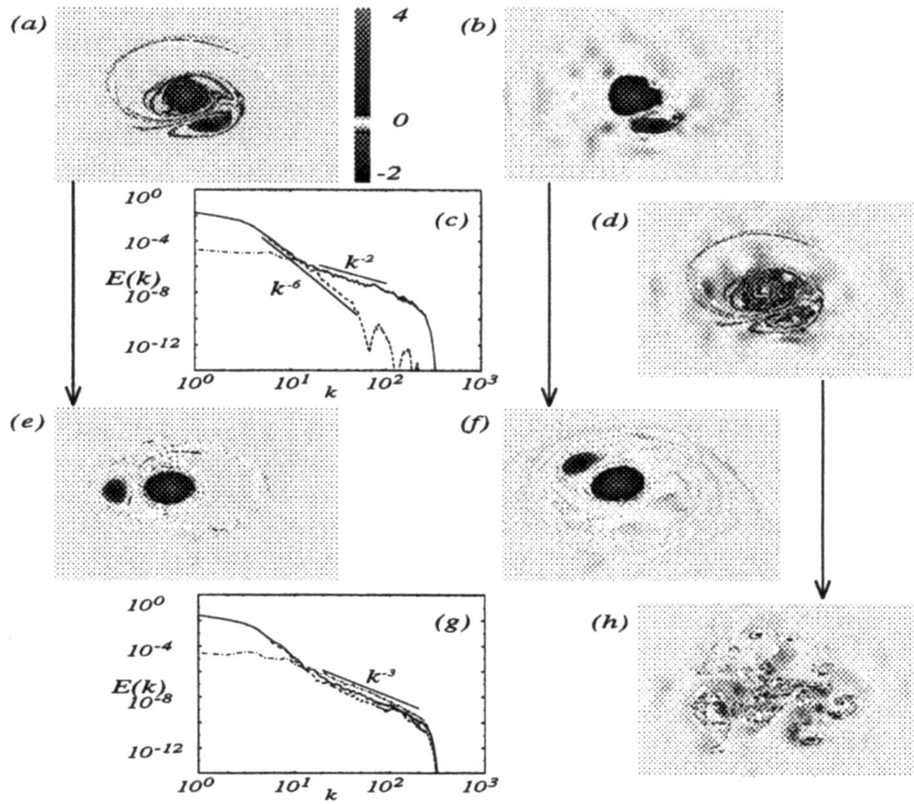

Figure 4: Dynamical analysis of coherent structures and incoherent background flow. (a) Total vorticity at $t = 10$ computed with a resolution $1\,024^2$. (b) Vorticity corresponding to the coherent vortices alone at $t = 10$. They are made up of 31 strong wavelet packet coefficients which contain 83% of the total enstrophy. (c) Energy spectra at $t = 10$: —, the total energy spectrum; ---, the coherent vortices energy spectrum; -·-, the filament energy spectrum. (d) Vorticity corresponding to the filaments alone at $t = 10$. They are made up of 1 048 545 weak wavelet packet coefficients which contain 17% of the total enstrophy. (e) Integration of the total vorticity until $t = 30$. (f) Integration of the coherent vortices alone until $t = 30$. (g) Energy spectra at $t = 30$: —, the total energy spectrum; ---, the coherent vortices energy spectrum; -·-, the filament energy spectrum. (h) Integration of the filaments alone until $t = 30$.

the production of coherent vortices by the background should usually be negligible in any vortex-based model. However, we saw that the background can have an effect on the coherent vortices by reducing their deformation by distant vortices. This shielding effect may reduce the minimum distance for merger and thus change the dynamics of coherent vortex evolution. It remains to be seen whether this effect is important enough that it needs to be included in vortex-based models.

4 Multiscale Analysis of a 3D Turbulent Channel Flow

In order to figure out the potential of multiscale methods for the numerical solution of the 3D Navier Stokes equations, we applied the different methods of the second section to some instantaneous fields of velocity components, pressure and derived quantities of a 3D turbulent channel flow. We choose this flow, because of the simple geometry of the domain and the presence of very different regimes in the flow. It is nearly isotropic in the central flow and strongly anisotropic in the vicinity of the wall. Due to this different behaviour, in general, one uses non equidistant grids for the simulation. For the multiscale methods we used, this leads to a lack of L_2-orthogonality, as described at the end of subsection 2.4. The different flow regimes are representative for most flow configurations with boundaries. Thus, is should be possible to 'extrapolate' the results reported here to other cases of practical interest. However, the case of isotropic turbulence, where equidistant grids are appropriate and allow for L_2-orthogonal wavelets is beyond the scope of the present findings.

4.1 Numerical Experiments

The database we used for our experiments was computed with a second order finite volume code similar to [1] by H.J. Kaltenbach [20], who kindly allowed us to analyze his database. This DNS was run with the usual geometry of upper and

Table 2: Parameter of DNS by Kaltenbach

geometry	grid	Δt $[\frac{H}{U_{bulk}}]$	Re_{bulk}	Re_τ	u_τ
$10h \times 2h \times 4h$	$192 \times 96 \times 192$	0.025	2801	178	0.063546

lower walls separated with a distance of $2h$, and periodic boundary conditions in stream- and spanwise directions. Characteristical quantities of this DNS are given in Table 2. Figure 5 depicts the different flow regimes: long stretched vortices in the vicinity of the wall and a rather isotropic behavior in the central part of the channel. To this database we applied the following procedure:

Figure 5: Isosurfaces of modulus of vorticity

transform of $X \in \{u_1, u_2, u_3, \omega_1, \omega_2, .., p, \Delta p\}$ to obtain coefficients X_l^j with respect to the multiscale basis
threshold the significant coefficients, i.e. $\|X_l^j\| \geq \epsilon(j)$ in (2.2), with respect to one of the criteria L2/H1/SPG and for different rates of compression
inverse transform of the compressed coefficients to obtain $\tilde{X}$
comparison of X and $\tilde{X}$

(Experiment I) In a first series of experiments we run the above procedure for the various quantities. For the comparison in the fourth step we considered mean and r.m.s values of X, $\tilde{X}$ and the error $X - \tilde{X}$.

In addition to these experiments we run a modified type of experiment, where we handle and compress the velocity or vorticity components simultaneously, i.e. we apply the following procedure:

transform of $\mathbf{X} \in \{\mathbf{u} = (u_1, u_2, u_3),\ rot\ \mathbf{u}\}$ to obtain coefficients $\mathbf{X}_{(1)}{}_l^j, .., \mathbf{X}_{(3)}{}_l^j$ with respect to the multiscale basis
threshold in each component $\mathbf{X}_{(1)}{}_l^j, .., \mathbf{X}_{(3)}{}^j$ simultaneously the important coefficients, i.e. $\sqrt{\|\mathbf{X}_{(1)}{}_l^j\|^2 + .. + \|\mathbf{X}_{(3)}{}_l^j\|^2} \geq \epsilon(j)$ in the 3-component analogue of (2.2), with respect to one of the criteria L2/H1/SPG and for different rates of compression
inverse transform of the compressed coefficients to obtain $\tilde{\mathbf{X}}$
comparison of $\mathbf{X}$ and $\tilde{\mathbf{X}}$

(Experiment II) Now, we applied the procedure for the three velocity components. Then, for the comparison we took into account not only mean and r.m.s values of $\mathbf{u}$ and $\tilde{\mathbf{u}}$ but also mean and r.m.s values of the derived quantities $\omega = rot\ \mathbf{u}$ and $rot\ \tilde{\mathbf{u}}$.

(Experiment III) Then, we run the above procedure for the three components of the vorticity. For the comparison in the fourth step we considered mean and r.m.s values of $rot\ \mathbf{u}$ and $\widetilde{rot}\ \mathbf{u}$.

From these experiments we will identify multiscale transforms with a good trade-off between work count and compression properties and good criteria for adaptivity. In addition we will see, whether a direct compression of ω (experiment III) or single components of it (experiment I) allows for much higher compression rates (CR) than the evaluation of the vorticity from a compressed velocity field (experiment II). At least for the flow configuration under consideration this tells, whether a formulation of the NSE in primitive variables $\mathbf{u}, p$ is well suited for numerical algorithms and how much we loose by a simultaneous compression of the three components of $\mathbf{u}$ or ω.

In the following the compression rate (CR) denotes

$$\mathrm{CR} = \frac{\text{number of all coefficients}}{\text{number of retained coefficients}},$$

i.e. CR=50 means that only 2% of the coefficients are significant.

4.2 Results of Multiscale Experiments

Before we go into detail, some general remarks are in order. For all multiscale methods we could observe the following effect. Comparing mean and r.m.s values of the error $e := X - \tilde{X}$ with mean and r.m.s values of the original quantity $mean(X)$, $rms(X)$, it turns out that the usual L_2-norm of the error

$$||e||_0^2 = \int e^2 dxdydz \quad ,$$

is of the same order of magnitude than the total turbulent energy of the flow. This seems to indicate that we can *not* expect a good agreement of $rms(\tilde{X})$ and $rms(X)$. But, in fact the contrary is the case. Even for quite large compression rates CR ≥ 50, we find good agreement of $rms(\tilde{X})$ and $rms(X)$.

$$||X - \tilde{X}||_0 = \mathcal{O}(||X - mean(X)||_0) \quad \text{but} \quad rms(\tilde{X}) \cong rms(X).$$

The analogue observations were also obtained for the 3-component fields, i.e. for $\mathbf{X}$ and $\tilde{\mathbf{X}}$. An example for this behavior is given in Figures 6, where r.m.s. values of the streamwise velocity component are compared with those of the compressed data and the compression error. As usual the r.m.s values are scaled by the shear velocity u_τ and the wall distance y is scaled by ν/u_τ. From this observation we conclude that L_2-norms of the error $X - \tilde{X}$ may be misleading, to judge from the quality of the approximation $\tilde{X}$. For a compression rate of CR=63 we obtain in Figure 8 a very good coincidence of e.g $rms(u_2)$ and $rms(\tilde{u}_2)$, although the relative error $\frac{||u_2 - \tilde{u}_2||_0}{||u_2||_0}$ in this case is 31 %, which normally is absolutely unacceptable.

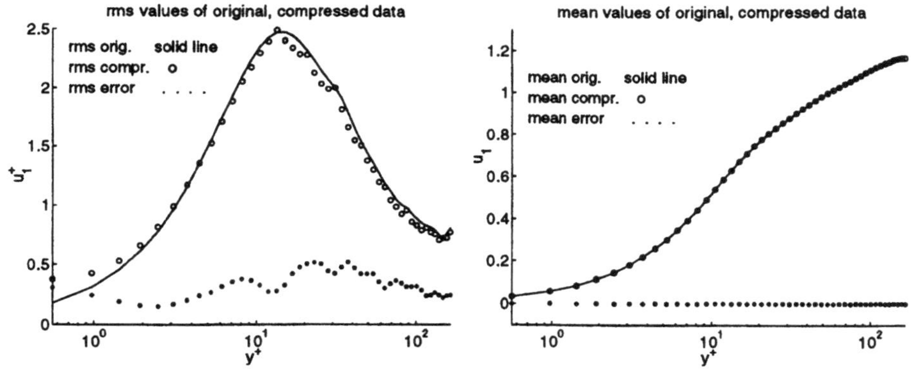

Figure 6: Exp. I /ANISO/PRW/L2 ; CR=48

This phenomenon is due to the property of multiscale basis functions to represent with only a small number of significant coefficients most of the coherent structures in the flow, which seem to govern these statistical quantities (mean, r.m.s). But this is good news, since then the property of compact representation of these important parts should be preserved during the temporal evolution governed by the NSE. However, only experiments with an adaptive NSE solver can give the right answer.

Now we present some more detailed results of our experiments. The most important single parameter for the multiscale transforms was the choice of the multivariate approach ISO or ANISO. As a main result, for all univariate approaches the second construction yielded much better results. One example is given in Figures 7 and 8, which compare the different behavior for prewavelets and the H1 error estimator. For a quite low compression rate of CR=20 the comparison shows, that the isotropic construction ISO is not able to capture the dynamic structures (especially $\overline{u_1' u_2'}$) in the buffer and log-law layer, whereas the second construction ANISO gives very good results even for CR=63.
The analogue result was obtained for the experiment I setting. Of course one might argue, that this is due to the flow regime, with its streaky structures near the wall, which is disadvantageous for the ISO approach. But also in the more central region $y^+ = 40$ to $y^+ = 110$ the r.m.s values of the streamwise velocity component are not recovered. This indicates that the ISO approach performs poorly for physical flows. Hence, in the following comparisons we will only consider results obtained with the ANISO construction.

Another important parameter is the correct error estimation. For modified Symmlets (see the remarks at the end of subsection 2.4) L2 resulted in better approximations, by means of mean and r.m.s. values, than H1. However, for the hierarchical nodal basis and prewavelets in experiments I and III the results for L2 and H1 were similar, while in experiment II the criterion H1 was the best. For both criteria the quality of approximation decreases near the wall. This effect of course is more pronounced for r.m.s values of the vorticity in experiment II (right

Figure 8). Despite of these detractions, mean values are captured satisfactorily (Figure 9). The sparse grid criterion SPG performs not such well. This is the case especially in the buffer and log-law layer. In this region rather coarse and smooth structures generated close to the wall decay to much smaller vortices (Figure 5), which result in significant coefficients in scales, which are simply cut off.

If the multivariate construction is chosen, the main contribution to the numerical costs comes from the underlying univariate approach. Our compression experiments clearly show, that prewavelets perform much better than the hierarchical basis. The additional property of cheap preconditioning Laplace and Helmholtz equations makes them a very promising multiscale technique. Orthogonal wavelets on one hand performed very well in experiments I and III, but in experiment II we could observe significant artifacts near the wall - despite a careful boundary treatment of the transform. If we could solve these problems it might be possible that orthogonal wavelets become an alternative to prewavelets. Experiments, we made with orthogonal wavelets of a higher number of vanishing moments and ignoring the boundary artifacts, seem to indicate this for moderate numbers (2...4) of vanishing moments. At the present state orthogonal wavelets could be applied to the simulation of 3D isotropic turbulence, where equidistant grids allow to completely preserve the advantage of L_2-orthogonality and where boundary artifacts should not be present.

We finish this section with the comparison of the compressed vorticity $\widetilde{rot\ \mathbf{u}}$ and the vorticity of a compressed velocity field $rot\ \tilde{\mathbf{u}}$. As expected the approximation of $rot\ \mathbf{u}$ is somewhat better for $\widetilde{rot\ \mathbf{u}}$ than for $rot\ \tilde{\mathbf{u}}$ (Figures 8, 9 and 12). In particular this holds for r.m.s values immediately near the wall. But the results there are not so much better, to make the vorticity formulation of the NSE competitive to the one in primitive variables, at least from the point of numerical costs. A comparison of the results of experiment I and III, e.g. for ω_3, shows that the simultaneous treatment of all three components is reasonably and does not diminish accuracy (Figures 12, 13).

5 Algorithmical Aspects

From the previous experiments we learned that a relatively small number of coefficients X_l^j (with respect to the multiscale basis and the given quantity $X \in \{\mathbf{u}, p\}$) is sufficient for a reasonable good approximation of X. Thus it is an appealing idea to exploit this for the fast approximative solution of the NSE. Such a spatially adaptive NSE solver computes the approximations $\mathbf{u}_\Lambda$, p_Λ of $\mathbf{u}$, p using only the coefficients $\{(\mathbf{u},p)_l^j \mid (j,l) \in \Lambda\}$. Here, Λ denotes the set indices (j,l) with significant coefficients X_l^j, which fulfill one of the criteria L2/H1. The basic steps of this approach are given in Table 3.

The locality in space or scale of coherent structures, e.g. vortex tubes, induces the locality of *active* coefficients. This is caused by the multiresolution approach, where $X_l^j \psi_l^j$ represents the difference of X from the present level to the next

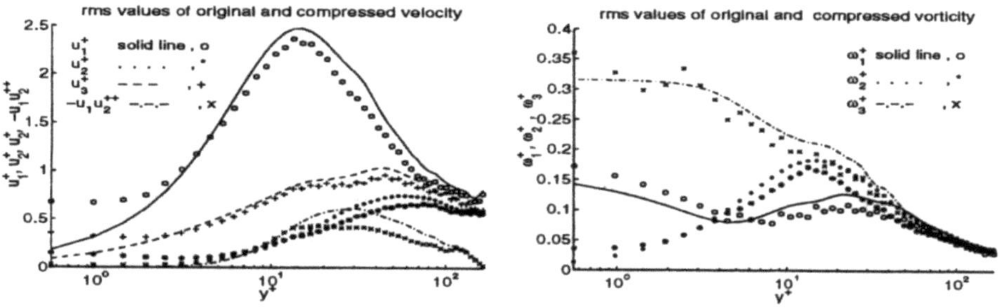

Figure 7: Exp. II/ISO/PRW/H1 ; CR=20

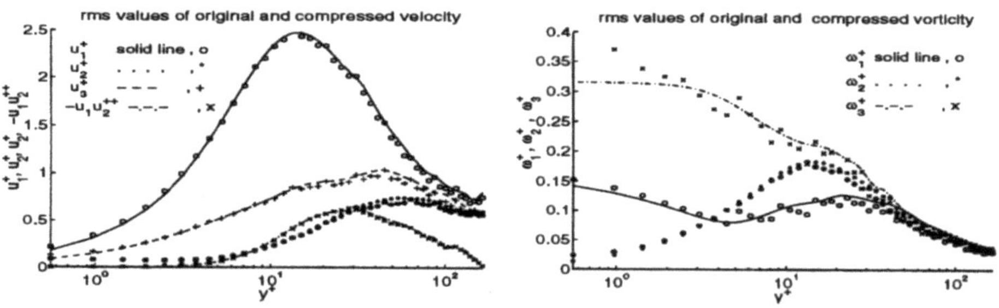

Figure 8: Exp. II/ANISO/PRW/H1 ; CR=63

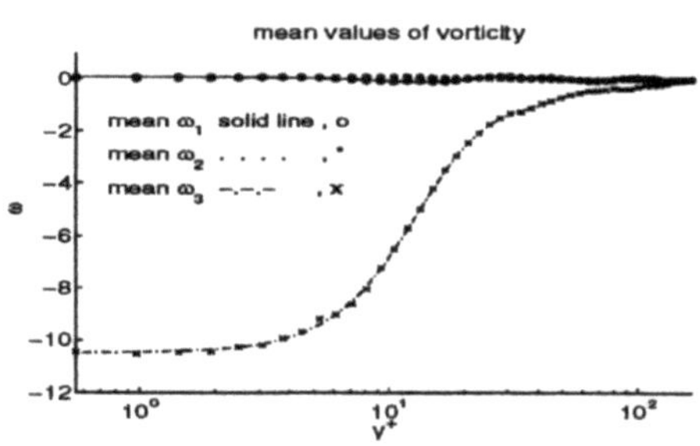

Figure 9: Exp II/ANISO/PRW/H1 ; CR=63

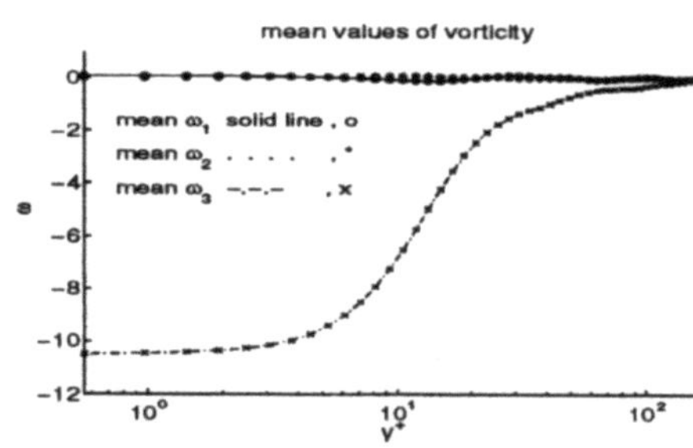

Figure 10: Exp II/ANISO/PRW/SPG ; CR=47

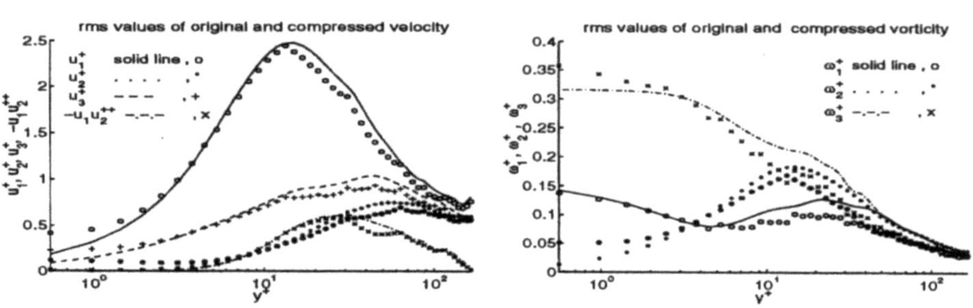

Figure 11: Exp. II/ANISO/PRW/SPG ; CR=47

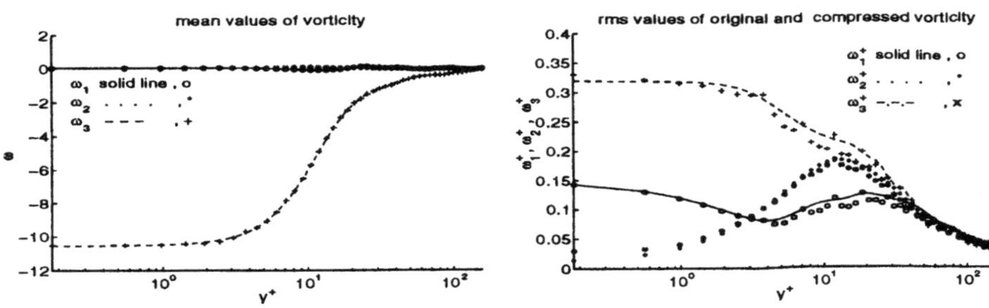

Figure 12: Exp. III /ANISO/PRW/L2 ; CR=63

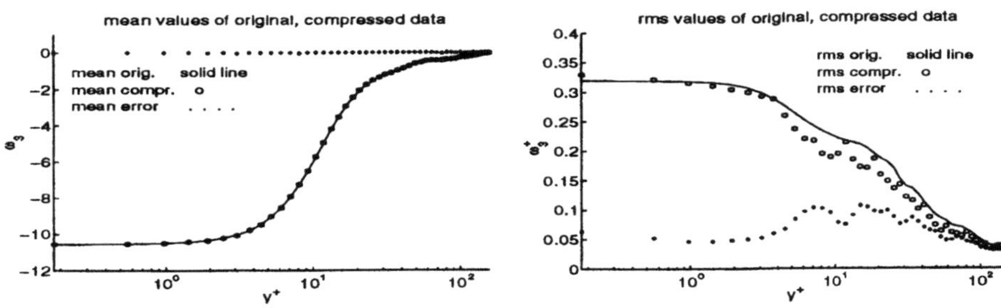

Figure 13: Exp. I /ANISO/PRW/L2 ; CR=63

coarser level $j-1$.

5.1 Hash Table Data Structures

The scattered distribution of active coefficients and their small number compared to all coefficients forbid the use of ordinary arrays for storage. There have to be found other methods. The problem of efficient storage of scattered data is known in computer sciences for decades and there are some well established techniques, which provide more or less efficient storage and fast access/insert/delete of data. The atoms access/insert/delete are exactly what we require in Table 3. Some of these common techniques are lists, tree data structures or hash table data structures [22]. Until recently, state-of-the-art numerical codes used tree data structures only. Instead of that, in [18], [19] the use of hash tables was proposed. Let N be the number of indices in Λ_n, which in applications will be of order $10^4 ... 10^6$, then tree data structures have a $\mathcal{O}(\log(N))$ worst/average case complexity of the atoms access/insert/delete. Especially for often used operations, as the addition of two functions by means of adding their coefficients, the overhead for accessing the data is considerable or even dominating compared to the actual numerical operation.

Hash tables solve this problem. In comparison to the tree approach hash table addressing gives more or less direct access to the data stored, i.e. it is proven to possess a $\mathcal{O}(1)$ complexity with a low constant if a statistical data distribution

Table 3: Basic spatial adaptive algorithm

given the approximation $\mathbf{u}_{\Lambda_0}$, p_{Λ_0} of $(\mathbf{u}, p)(t_0 = 0)$	
	compute $\mathbf{u}^{n+1}_{\Lambda_n}$, $p^{n+1}_{\Lambda_n}$ from $\mathbf{u}^{n}_{\Lambda_n}$, $p^{n}_{\Lambda_n}$
	mark indices (j, l) of important coefficients $(\mathbf{u}^{n+1}, p^{n+1})^j_l$ and their neighbours in space and scale for retaining or insertion and mark small coefficients for deletion to obtain new index set Λ_{n+1}
	insert/delete coefficients to obtain $\mathbf{u}^{n+1}_{\Lambda_{n+1}}$, $p^{n+1}_{\Lambda_{n+1}}$
$n = n + 1$	

is assumed. Furthermore lower additional storage overhead for logical connectivities are required. Let us briefly describe this appealing idea.
Each index (j, l) is mapped by a certain hash function h to a hash key, which is used as an address in the hash table. The hash table is an m-element array of pointers to m different lists, in which all the required data (j, l, X^j_l) is stored. This basic approach is depicted in Figure 14. The lists are necessary, since the

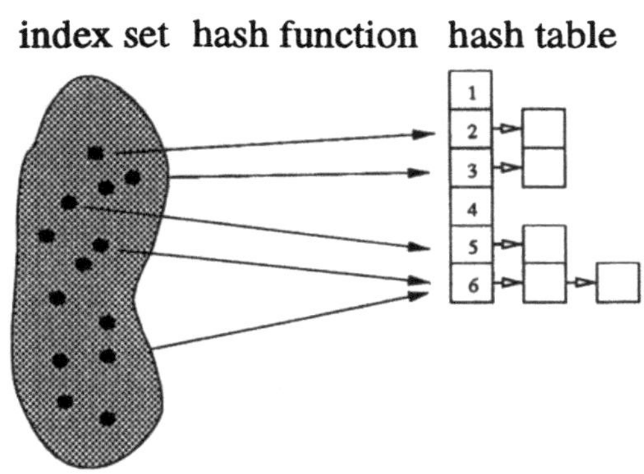

Figure 14: Hashing

number of admissible indices (j, l) is much larger than the number N of active coefficients. Thus h can not be injective. Usually m is of order N to provide short lists and fast average access to the data. As an example we have stored the coefficients with respect to the hierarchical nodal basis of the function shown in Figure 15 with a hash table. Figure 16 depicts the distribution of the lengths of the lists. Apparently the hash function used in this example leads to a very broad distribution of the hash keys and the lists are almost of optimal length.
It seems that hash tables first have been used in adaptive PDE solvers and in the context of adaptive multigrid solvers by [18].

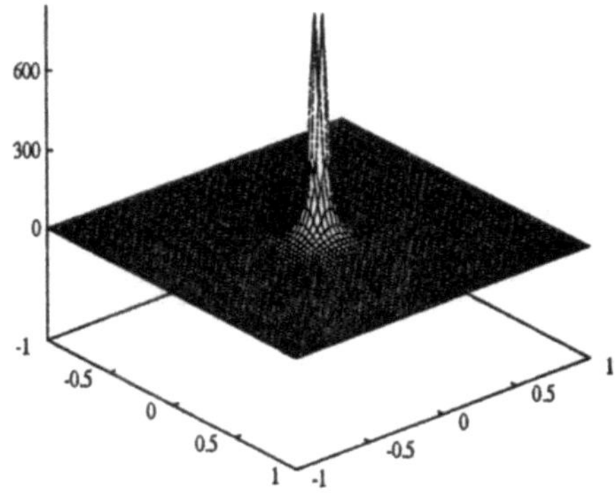

Figure 15: Example

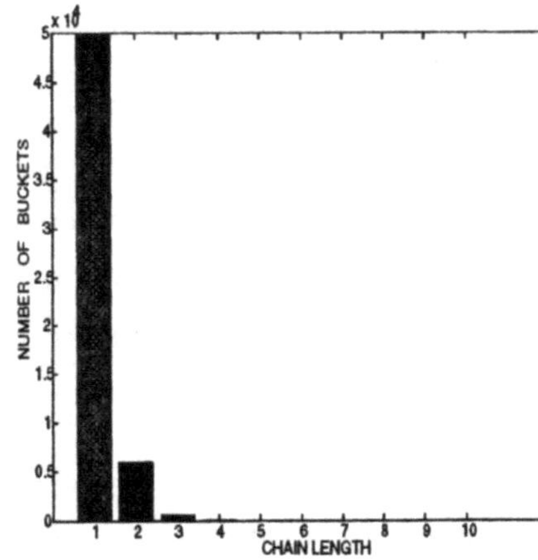

Figure 16: list lengths

5.2 Parallelization with Space-filling Curves

Besides the reduction of the complexity of the physical problem, e.g. by multiscale methods, nowadays one has the opportunity to distribute large problems onto several processors. This speeds up the computations. Another reason might be the huge amount of memory required for the solution, which is not present on a single processor machine. To tackle a new class of problems (think of a great challenge project) it is necessary to combine both approaches.
Since the scatteredness of the data in our case rules out the use of 3D array data structures, the usual domain decomposition technique is no longer advisable. A moving region of strong variations in the flow, for a fixed domain decomposition, would cause a very high load for some processors, while others are waiting and wasting precious computing time. Thus we have to find other load balancing techniques, which hopefully fit to the hash table storage technique. A very promising approach has been introduced in [18], [19]. We shall briefly describe this idea for the bivariate case.

In a first step the indices $(j,l) = ((j_1,j_2),(l_1,l_2))$ are mapped to unique nodes $(x,y) \in [0,1]^2$ of a certain finest grid. Usually this will be something like $(l_1 2^{-j_1}, l_2 2^{-j_2})$. Hence, the node is located within the support of ψ_l^j, which nicely corresponds to the localization of the basis functions. Then, in a second step the values (x,y) are mapped to a value $s \in [0,1]$. This is done by a space-filling curve, which provides us with an injective, continuous mapping $f : [0,1]^2 \bigcap Q \times Q \mapsto [0,1]$. Here, Q is the set of numbers with finite binary expansion.

$$(j,l) \quad \overset{g \text{ (injective)}}{\longleftrightarrow} \quad (x,y) \in [0,1]^2 \quad \overset{f \text{ (injective)}}{\longleftrightarrow} \quad s \in [0,1].$$

The evaluation of both mappings g and f requires only some cheap operations and is fast. After applying these two steps for each $(j,l) \in \Lambda_n$, we have a set

$\{s(j,l) \mid (j,l) \in \Lambda_n\}$. These values are increasingly ordered.

$$s_1 < \ldots < s_N \quad N = \#\Lambda_n .$$

For the load balancing the indices (j,l), and the data associated to them, are distributed among the p processors in the following way.

$$\begin{array}{lllll} \text{P1:} & \{(j,l) \mid s_1 & \leq s(j,l) \leq & s_{N/p-1}\} \\ \vdots & \vdots & & \vdots \\ \text{P}p\text{:} & \{(j,l) \mid s_{(p-1)N/p} & \leq s(j,l) \leq & s_N\}. \end{array}$$

Figure 17 shows an example of a space-filling curve and the distribution of the nodes onto three processors in the non adaptive case.

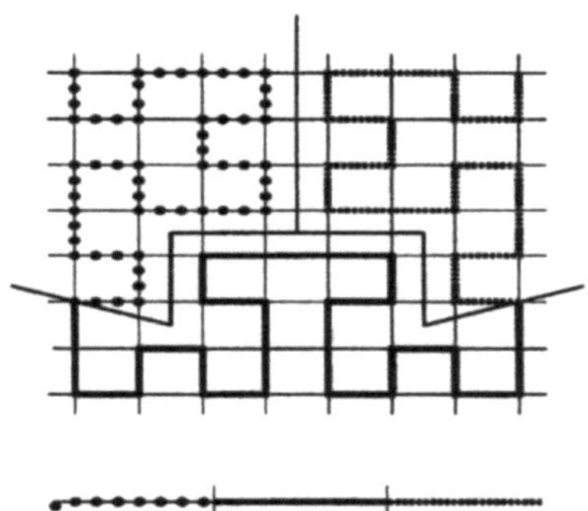

Figure 17: Load balancing using space-filling curves

Clearly this technique leads to an optimal load balancing, while the volume of communication depends on the boundaries of the partitions. These boundaries may sometimes be kinky, depending on the data, and are certainly not optimal, but are of reasonable size. In total, the load balancing is very cheap, parallelizes well and thus can be applied in each cycle of algorithm 3.
In addition there are great similarities between this load balancing technique and hash table storage. Both techniques lead to, in some sense, linear ordered data. E.g. simple modifications of f are good candidates for the local hash function on each processor.

6 Conclusion

The goal of this paper has been two-fold: first to test and extend the wavelet techniques that have been developed for isotropic two-dimensional turbulence, and secondly to evaluate the possibility of using a similar approach for three-dimensional turbulence with boundaries. The results of the first investigation suggest a new way of modelling high Reynolds number two-dimensional turbulence, while the second investigation has helped to clarify the general properties of a multiscale adaptive wavelet-based method for solving the three-dimensional Navier-Stokes equations (3D-NSE) at high Reynolds number.

The adaptive wavelet technique for the turbulent 2D-NSE was found to be highly accurate and to use far fewer active modes than standard or nonlinearly filtered pseudo-spectral techniques. Furthermore, the number of active modes remains constant, even during periods of intensely nonlinear interaction with strong gradients. This suggests that the wavelet representation is well-suited to the dynamics of the 2D-NSE at high Reynolds number.

A new, more physical, forcing technique based on the wavelet representation was proposed and tested. In this method the strongest wavelet coefficients (corresponding to the coherent vortices) are reinforced at each time step. The forcing models the way turbulence is generated in real flows by vorticity production via an instability at the boundaries. This method produces stationary statistics, and can be used to provide a more realistic forcing in simulations of turbulent flows. The method should also work in three dimensions.

Wavelet analysis of two-dimensional turbulence showed that the flow may be divided into a coherent part (the vortices) with non-Gaussian one-point vorticity statistics and an incoherent part (the background flow) with Gaussian one-point statistics. The threshold, dividing the weak and strong wavelet modes, is based on the variance of the vorticity field and the number of points and is thus objective (and constant if the simulation is stationary). The coherent part of the flow represents only 1% of the total number of modes for a resolution of 128^2 and will decrease for higher resolution.

The fact that the background has Gaussian statistics suggests that it could be easily modelled, and in this case the adaptive wavelet method would only used for the coherent vortices. A study of the vorticity filaments which make up the background showed that they should be stabilized by the coherent vortices they are associated with, provided the density of vortices is not too low. However, the filaments do have an effect on the flow since they shield the vortex they surround from the deformation induced by the strain of neighbouring vortices. It remains to be seen whether this effect significantly alters the turbulence dynamics.

In summary, these investigations have established the usefulness and accuracy of wavelet techniques for analyzing and solving the 2D-NSE at high Reynolds numbers. The results also suggest a new way of modelling two-dimensional turbulence. The dynamics of the coherent vortices could be found by directly solving the 2D-NSE using an adaptive wavelet technique, while the effect of the background could be modelled using a simple statistical model (since the background vorticity field has Gaussian one-point statistics).

The second goal of the paper was to evaluate the potential of various multiscale methods for the solution of the three-dimensional 3D-NSE. We showed that, even with some simple error estimators and simple prewavelets, reasonable approximations of the original turbulence data are possible for relatively high compression rates of about 60 at resolution $192 \times 96 \times 192$.

Furthermore, it turned out that for the solution of the 3D-NSE, at least for flows which are governed by boundaries, the formulation in primitives variables seems to be more efficient compared to the formulation in terms of vorticity. In the fifth section we presented efficient methods for the implementation of an fully adaptive solver. These methods produce very efficient data storage and a cheap, but reasonably effective, load balancing technique.

The results presented here have demonstrated the usefulness of adaptive wavelet techniques for the 2D-NSE at high Reynolds numbers, and have laid the foundations of an equivalent technique for the 3D-NSE. The next stages in two and three dimensions are, respectively, the development of a highly efficient turbulence model (where only the coherent vortices are calculated exactly) and the implementation of a solver for the 3D-NSE.

References

[1] K. Akselvoll, P. Moin *An efficient method for temporal integration of the Navier-Stokes equations in confined axisymmetric geometries*; J. Comp. Phys. 125 (1996), pp. 454.

[2] G.K. Batchelor *The theory of homogeneous turbulence*; Cambridge University Press (1953).

[3] D. Bonn, Y. Couder, P. Van Dam *From small scales to larges scales in 3D turbulence*; Phys. Rev E 47 (1993).

[4] H.J. Bungartz *Sparse Grids and their Application for the adaptive Solution of the Poisson Eqn. (in german)*; Thesis TU München (1992).

[5] P. Charton, V. Perrier *A pseudo-wavelet scheme for the two-dimensional Navier-Stokes equation*; Comp. Appl. Math. 15/2 (1996), pp.139–160.

[6] A. Cohen, I. Daubechies, P. Vial *Wavelets on the Interval and Fast Wavelet Transforms*; Appl. Comp. Harm. Anal. 1 (1993).

[7] W. Dahmen *Wavelet and Multiscale Methods for Operator Equations* ; Acta Numerica (1997).

[8] I. Daubechies *Orthonormal bases of compactly supported wavelets* ; Comm. Pure Appl. Math. (1988).

[9] I. Daubechies *Orthonormal bases of compactly supported wavelets II* ; SIAM J. Math. Anal. 24 (1993).

[10] D. L. Donoho *Wavelet Shrinkage and W.V.D. - A Ten-Minute Tour*; Technical Report 416 (1993).

[11] M. Farge et.al. *Wavelets and Turbulence*; Proceedings of the IEEE Vo. 84 No. 4 (1996).

[12] M. Farge, K. Schneider, N. K.-R. Kevlahan *Coherent structure eduction in wavelet-forced two-dimensional turbulent flows*; To appear in *Dynamics of slender vortices* (ed. E. Krause). Cambridge University Press.

[13] J. Fröhlich, K. Schneider *An adaptive wavelet-vaguelette algorithm for the solution of PDEs.*; J. Comput. Phys. 130 (1997) pp.174–190.

[14] J. Fröhlich, K. Schneider *Numerical simulation of decaying turbulence in an adapted wavelet basis*; Appl. Comput. Harm. Anal. 3 pp. 393-397.

[15] M. Griebel, F. Koster *Remarks on orthogonal wavelets on the interval*; Preprint Univ. Bonn , SFB 256 (1998).

[16] M. Griebel, P. Oswald *Tensor product type subspace splittings and multilevel iterative methods for anisotropic problems*; Advances in Comp. Math. (1995).

[17] M. Griebel, W. Huber *Turbulence Simulation on Sparse Grids Using the Combination Method*; TU München SFB Bericht Nr. 342/19/94 (1994).

[18] M. Griebel, G. Zumbusch *Hash-storage techniques for adaptive multilevel solvers and their domain decomposition parallelization*; Proc. of the 10th Int. Conf. on Domain Decomposition Methods, Boulder, USA (1997).

[19] M. Griebel, G. Zumbusch *Parallel multigrid in an adaptive PDE solver based on hashing* ; Parallel Computing (ParCo 97), Bonn, (1997).

[20] H.J. Kaltenbach *Unpublished data*; TU Berlin (1997).

[21] N. K.-R. Kevlahan,M. Farge *Vorticity filaments in two-dimensional turbulence: creation, stability and effect*; J. Fluid Mech. 346 (1997) pp.49–76.

[22] D.E. Knuth *The Art of Computer Programming 3*; Addison Wesley (1973).

[23] R.H. Kraichnan *A theory of turbulence dynamics*; II. Symposium on Naval Hydrodynamics (1958), Office of Naval Research, Washington DC, ACR-38.

[24] A. Leonard *Computation of 3D incompressible flows with vortex elements*; Ann. Rev. Fluid Mech. 17 (1985), pp.523.

[25] J. McWilliams *The emergence of isolated coherent vortices in turbulent flow*; J. Fluid Mech. 146 (1984), pp.2–43.

[26] Y. Meyer *Ondelettes sur l'intervalle*; Rev. Mat. Iberoamericana 7 (1992).

[27] P. Monasse, V. Perrier *Orthonormal Wavelet bases adapted for PDE with boundary conditions*; Preprint LMD Paris (1995).

[28] P. Oswald *Multilevel finite element approximation*; Teubner Verlag (1994).

[29] K. Schneider, M. Farge *Wavelet forcing for numerical simulation of two-dimensional turbulence*; C. R. Acad. Sci. Paris 325 IIb (1997), pp. 263–270.

[30] K. Schneider, N. K.-R. Kevlahan, M. Farge *Comparison of an adaptive wavelet method and nonlinearly filtered pseudo-spectral methods for two-dimensional turbulence*; Theoret. Comput. Fluid Dynamics **9** (1997) pp.191–206.

[31] E.D. Siggia *Numerical study of small-scale intermittency in 3D turbulence*; J. Fluid Mech. 107 (1981), pp. 375-406.

[32] W. Sweldens *The Lifting Scheme: a construction of second generation wavelets*; SIAM J. Math. Anal. (1997).

[33] W. Sweldens, R. Piessens *Quadrature Formulae and Asymptotic Error Expansions*; SIAM J. Num. Anal. 31 (1994).

[34] A. Vincent, M. Meneguzzi *The dynamics of vorticity tubes in homogeneous turbulence*; J. Fluid Mech. 258 (1994), pp.245-254.

[35] C. Zenger *Sparse Grids*; Notes on Numerical Fluid Mechanics Vol. 31, Vieweg (1991).

The Minimal Turbulent Flow Unit as a Test Case for Three Different Computer Codes

M. Manhart[1], G. B. Deng[2], T. J. Hüttl[1], F. Tremblay[1], A. Segal[3], R. Friedrich[1], J. Piquet[2], P. Wesseling[3]

[1] Lehrstuhl für Fluidmechanik, Technische Universität München
Boltzmannstr. 15, 85748 Garching, Germany
[2] Laboratoire de Mécanique des Fluides, Ecole Centrale Nantes
1, Rue de la Noë, 44321 Nantes Cedex 3, France
[3] J. M. Burgers Centre for Fluid Dynamics and Faculty of Technical Mathematics and Informatics, Delft University of Technology
2628 Delft, The Netherlands

Summary

It is the aim of this investigation to compare the accuracy and performance of three codes in simulations of fully developed incompressible turbulent channel flow. Two of the codes use staggered variable arrangements and second order central differencing in space. While one of them is adapted to Cartesian grids, the other uses coordinate invariant formulations of the discrete equations based on contravariant mass fluxes. The third code favours collocated variable arrangements. In order to keep the computational costs moderate, turbulent flow through a minimal channel has been computed and compared. The size of the channel in streamwise, spanwise and wall-normal directions is $\pi h \times 0.3\pi h \times 2h$ and in wall units $416 \times 125 \times 265$ for a Reynolds number based on friction velocity and channel half width h of 132.5. This corresponds to a Reynolds number based on bulk velocity of about 2000. The computations have been conducted on a vector computer. The first- and second-order statistics show that all three codes are comparable in accuracy but not in the computational effort spent to get the results.

1 Introduction

Near-wall turbulence is characterized by peak values in the components of the Reynolds stress tensor as a result of organized fluid motions. In a reference frame moving with the local mean longitudinal velocity these motions manifest themselves in an ordered transport of axial momentum towards and away from the wall, in the so-called streaky structures. These have a spanwise size of about 100 wall units in the mean and a length of about 800. In numerical simulations Jiménez and Moin [4] have identified the minimal set of structures that are necessary to sustain turbulent flow in a channel indefinitely. The resulting so-called minimal flow unit extends from wall to wall and has streamwise and spanwise computational periods of up to 420 and 140 wall units, respectively. It allows to successfully predict certain essential turbulent statistical quantities, such as the mean streamwise velocity and second-order single-point statistics (the latter at least below $y^+ = 40$). Since the streamwise computational period is roughly half that of an orderly structure, streamwise two-point correlation functions cannot be accurately predicted. The same is true for higher than second-order statistical quantities. Varying the box size, Jiménez and Moin [4] found interesting solutions in which the turbulence resides only at a single wall for fairly long periods, while the flow at the other wall is relatively quiescent. These time periods are of the order of $1000h/u_b$, which means that the turbulence structures are quite stable, being convected several thousand wall units.

We have selected this flow case for comparison of the performance of three computer codes of different algorithmic nature, because it involves a relatively small computational effort and is at the same time physically complex. All three computer codes are finite volume codes, designed for flows with different degrees of geometrical complexity. The first code, MGLET, is a Cartesian staggered mesh code with explicit time advancement. It has a long tradition in Large-Eddy-Simulation (LES) and Direct Numerical Simulation (DNS) (see eg. Werner and Wengle [8], Manhart and Wengle [5] and Manhart [6]). The second code tested employs almost the same approach as the code called HORUS developed by the CFD group in Nantes since more than a decade. The HORUS code is a Reynolds Averaged Navier-Stokes equation (RANSE) solver for incompressible flow. Several turbulence models ranging from simple zero equation models to more complex Reynolds-stress transport equation models have been implemented. HORUS has been successfully applied to simulate flow around complex geometries such as ship flow, wing-body junction flow, etc. (see Deng & al. [1], [2]). So far, DNS has been performed mainly on simple geometries. Very efficient algorithms have been especially developed for such kind of simulation. But these algorithms are usually limited to simple geometries. New algorithms need to be developed when we want to perform DNS on complex geometries. It is interesting to see whether an approach used to solve the RANSE can also be used for DNS. To save CPU time, a Cartesian version of a collocated curvilinear code has been developed. It differs from HORUS only in the semi-implicit time advancement and the fully conservative discretization of the momentum equation. The third code, DeFT, is a fully implicit code on a staggered, non-orthogonal grid. It is designed for solving the Navier-Stokes equations in geometrically complicated flow domains, using domain decomposition and boundary-fitted coordinates.

2 Description of numerical approaches

2.1 The code MGLET

MGLET, is based on a finite volume formulation of the Navier-Stokes equations on a staggered Cartesian non-equidistant grid. The spatial discretization is of second order (central) for the convective and diffusive terms. For the time advancement of the momentum equations an explicit second-order time step (leapfrog with time-lagged diffusion term) is used, i.e.:

$$u^{n+1} = u^{n-1} + 2\Delta t \left[C\left(u^{n}\right) + D\left(u^{n-1}\right) - G\left(p^{n+1}\right)\right] \tag{2.1}$$

where C, D and G represent the discrete convection, diffusion and gradient operators, respectively.
The pressure at the new time level p^{n+1} is evaluated by solving the Poisson equation

$$Div\left[G\left(p^{n+1}\right)\right] = \frac{1}{2\Delta t} Div\left(u^{*}\right) \tag{2.2}$$

where u^* is an intermediate velocity field, calculated by omitting the pressure term in equation 2.1. By applying the velocity correction

$$u^{n+1} = u^{*} - 2\Delta t G\left(p^{n+1}\right) \tag{2.3}$$

we arrive at the divergence-free velocity field u^{n+1} at the new time level.
The solution of the Poisson equation is done either in a direct way or iteratively. The direct solver, which has been used for the calculations presented in this paper, uses FFTs in homogeneous streamwise and spanwise directions resulting in independent tridiagonal systems in the inhomogeneous wall normal direction for each wavenumber combination (k_x, k_y), (see Schmitt [7]). The iterative solver is a point-wise velocity-pressure iteration like that described in Hirt et al. [3]. It is used as a single-grid iteration or, alternatively, as a smoother in a multigrid cycle. The direct solver reduces the divergence of the velocity field u^{n+1} down to machine accuracy. When using the iterative solver the remaining divergence depends on the computational time spent, (see below).

2.2 The HORUS code: a non-staggered approach

HORUS uses a non-staggered approach where the pressure and the Cartesian velocity components are defined at the center of the grid cell. The flux interpolation technique proposed by Rhie & Chow is employed to evaluate the fluxes at the control volume faces. The momentum equations are solved by a semi-implicit time-advancement scheme with the Adams-Bashforth method for the convection terms and the Crank-Nicolson method for the diffusion terms. The discretized equations are:

$$Div\left(u_{face}^{n+1}\right) = 0 \tag{2.4}$$

$$\frac{u^{n+1} - u^n}{\Delta t} + C\left(\frac{3}{2}u^n - \frac{1}{2}u^{n-1}\right) = -G\left(p^{n+1}\right) + D\left[\frac{1}{2}\left(u^{n+1} + u^n\right)\right] \tag{2.5}$$

where Div, C, G and D are discrete operators representing the divergence, the convection, the gradient and the diffusion terms, respectively. They are all evaluated by using second-order central differencing in conservative form. As mentioned above, the velocity component u_{face} is evaluated by the Rhie & Chow flux interpolation approach that is described below. From the discrete momentum equation (2.5), we can obtain an expression for the velocity:

$$u = \widetilde{u} - c_p G(p), \text{ with } c_p = \frac{1}{\frac{1}{\Delta t} - \frac{d_c}{2}} \tag{2.6}$$

where d_c is the diagonal coefficient of the diffusion operator. At the center of the control volume face, the velocity component is calculated as:

$$u_{face} = \overline{\widetilde{u}} - \overline{c_p} G_{face}(p) \tag{2.7}$$

where an over-line represents a spatial average operator, and G_{face} is the discrete gradient operator formulated at the center of the control volume face.

The coupling between the velocity and the pressure is solved by using the PISO algorithm which consists of a prediction and a correction step. In the prediction step, we solve the momentum equation by using the pressure field at the previous time step.

$$\frac{u^* - u^n}{\Delta t} + C\left(\frac{3}{2}u^n - \frac{1}{2}u^{n-1}\right) = -G\left(p^n\right) + D\left[\frac{1}{2}\left(u^* + u^n\right)\right]. \tag{2.8}$$

In the correction step, we update the pressure field by solving the following pressure Poisson equation:

$$Div\left[\overline{c_p}G_{face}(p^{n+1})\right] = Div\left(\overline{\widetilde{u}^*}\right) \tag{2.9}$$

where $\widetilde{u}^*$ is determined by

$$\widetilde{u}^* = u^* + c_p G(p^n). \tag{2.10}$$

Finally, we correct the velocity field by

$$u^{n+1} = \widetilde{u}^* - c_p G(p^{n+1}). \tag{2.11}$$

The momentum equation (2.8) is solved by using the CGSTAB conjugate gradient method, while the pressure Poisson equation (2.9) is solved either by an algebraic multi-grid solver, or by the CGSTAB solver using the Stone Implicit Procedure as a preconditioner. When using the iterative solver, we reduce the residual of the pressure Poisson equation by a factor of 1000.

2.3 The code DeFT

DeFT (Delft Flow and Transport) uses an invariant finite volume discretization to solve the Navier-Stokes equations in 2D and 3D with an arbitrary number of transport equations. The mesh is block-structured and boundary-fitted, allowing the computation of flows in complex geometries. The contravariant components of the mass fluxes and the pressure are the unknowns and have a staggered arrangement in order to avoid pressure oscillations. The spatial discretization is second-order accurate and central. The spatially discrete momentum equations are integrated in time using a second-order accurate pressure correction method in which the nonlinear convection terms are Newton-linearized without loosing order of accuracy. DeFT can be viewed as a generalization of MGLET from Cartesian to general coordinates. If the time stepping method would be the same, DeFT would reduce to MGLET on Cartesian grids. But to avoid severe stability restrictions on strongly non-uniform grids, time marching is achieved by an unconditionally stable implicit method, namely the Θ-method:

$$\begin{aligned} M\tfrac{u^{n+1}-u^n}{\Delta t} + \Theta S(u^{n+1}) + (1-\Theta)S(u^n) + \Theta G p^{n+1} + (1-\Theta)Gp^n \\ = \Theta F^{n+1} + (1-\Theta)F^n. \end{aligned} \tag{2.12}$$

Here M is a diagonal matrix, S represents the discretization of the stress and convective terms, G is the gradient matrix and F is a source term. u and p are the velocity and pressure, respectively.

For $\Theta = 0.5$ the Crank-Nicolson method results which is second-order accurate, but is known to produce oscillating solutions. For this reason, in the current calculation Θ is set to 0.55, a value that retains an accuracy very close to second order without oscillations. As in MGLET, equation (2.12) is not solved as it stands, but the pressure correction method is used.

The momentum equations are solved with the GMRES method and an ILUD postconditioner. The pressure equation is solved with the preconditioned GMRES method using one multigrid V-cycle as a preconditioner. The iterative solver is stopped when the residual is reduced by four orders of magnitude for the momentum equations and six for the pressure. For a detailed survey of the characteristics of DeFT, the reader should consult [14, 12, 9]. Two-equation turbulence modeling and compressible flows can be accounted for ([10, 11, 13]).

2.4 Statistical evaluation

Since the three codes have different traditions there are different strategies for the statistical evaluations. In the codes HORUS and DeFT instantaneous velocity and pressure fields are saved on hard disk and the statistics is built later by a separate code. This has the advantage that a) selection of the time steps used for statistics is possible a posteriori, b) one can redo the statistical evaluation without running the simulation again and c) the data are available for more detailed analysis. One severe drawback of this strategy is that for bigger simulations a huge amount of disk space is necessary for storing the instantaneous velocity fields.

This drawback can be overcome by the strategy used by the code MGLET. Here the statistical evaluation is done during the running simulation. (This is the only possible way to evaluate big simulations in the range of $50 \cdot 10^6$ grid points.) But this strategy has two main drawbacks: a) one has to select the statistical variables before running the simulation and b) a selection of time steps used for statistics is not possible. During the course of the minimal channel simulations an additional complication showed up with this strategy. Since in the code MGLET the second-order statistics are built using the fluctuating part of the variables, they cannot be evaluated before the first-order statistics is converged. Normally, in a fully developed turbulent flow, the second-order statistics need about four times the number of (statistically independent) samples. But in the very low Reynolds number flow presented in the present paper periods of relaminarisation occurred (see below) and very long computational times for evaluating the first order statistics have been necessary.

3 Computational details of the turbulent flow unit

3.1 Configuration

We are dealing with a turbulent plane channel flow with small dimensions in streamwise and spanwise directions. The Reynolds number has been chosen to $Re_b = 2000$ based on the bulk velocity and the channel half width h or $Re_{cl} = 3000$ when the centerline velocity of a laminar profile with the same flux is used as reference quantity. This corresponds to a Reynolds number $Re_\tau = 132.5$ based on friction velocity. In the streamwise, spanwise and wall-normal directions (x, y, z) the flow unit has the dimensions $l_x = \pi h$, $l_y = 0.3\pi h$ and $l_z = 2h$. We have chosen two different grids in order to check the dependence of the solutions on the resolution (see table 1).

Table 1 Parameters of the numerical grids used:

case	Nx	Ny	Nz	Δx^+	Δy^+	Δz^+_{min}	Δz^+_{max}
coarse	32	16	128	13.0	7.8	0.9	3.8
fine	64	32	128	6.5	3.9	0.9	3.8

The grid is equidistant in x- and y-directions. In wall-normal direction, close to the wall a clustering of grid points is achieved according to

$$z_i = \alpha h \frac{tanh\,(\gamma\,(N-1))) - tanh\,(\gamma\,(N-i))}{tanh\,(\gamma\,(N-1))} \tag{3.13}$$

$$N = 0.5\,(n_z + 1) \quad , \qquad i = 1, \ldots, \left(\frac{n_z + 1}{2}\right) \tag{3.14}$$

where n_z is the total number of points in the z-direction. The coefficients α, γ have the values 1, 0.0175, respectively. They ensure that 9 points are below $z^+ = 10$ and that the first point is located below $z^+ = 1$.

From a physical point of view, the flow is driven by a pressure gradient in streamwise direction which can be treated as a source term in the momentum equation. Two different approaches have been tested to drive the flow in the computation that led to two qualitatively different behaviours of the flow.

In the first approach, the mean streamwise pressure gradient is kept constant which is equivalent to a specification of the mean wall shear stress, leaving the bulk velocity open. The Reynolds number Re_τ based on channel half width h and friction velocity u_τ is chosen as 132.5.

The second approach is keeping the mass flux constant by specifying a time dependent pressure gradient using the control scheme

$$\frac{\partial p}{\partial x} = \frac{1}{2h}\left[\tau_w^u - \tau_w^l\right] - \alpha\frac{u_b - u_m}{\Delta t} \qquad (3.15)$$

where u_m and u_b are the measured and desired bulk velocity, respectively, and τ_w^l and τ_w^u are the plane averaged wall shear stresses at the lower and upper wall, respectively. We used a relaxation parameter α to prevent strong temporal variations of $\partial p/\partial x$. A value $\alpha = \Delta t$ worked fine and kept the variation of the mass flux below 10^{-5}. In the constant mass flux approach, that has also been used by Jiménez and Moin [4], Re_b has been specified and Re_τ has been obtained as a simulation result. These two procedures should be equivalent if the flow is fully developed turbulent. It is not a priori clear whether this is also true if the flow is intermittent, as in the present case.

3.2 Computational costs

When comparing the CPU-times and the memory requirements of the three codes one has to bear in mind that all three codes have been designed for the calculation of flows with different characteristics and complexities. MGLET uses a staggered Cartesian grid and an explicit time step, HORUS is the Cartesian version of a curvilinear collocated grid code with a semi-implicit time step and DeFT uses a non-orthogonal staggered grid and a fully implicit time advancement. In the implicit treatment the solution of the linear system requires storing of the matrices which is reflected in the memory requirement of the codes, which is compared in table 2.

Table 2 Memory requirement of the three codes (double precision).

	MGLET	HORUS	DeFT
Words/node	12	90	243
MByte for coarse grid	9	47	127
MByte for fine grid	31	190	510

In the code DeFT a big part of the memory (150 Words/node) is mainly used for the implicit time step. The curvilinear version of HORUS with Reynolds stress model implemented uses about 200 Words/node. Since in a curvilinear code one has the choice to keep geometrical coefficients in core memory or to recalculate them every time step, there is always a compromise between memory and CPU-time requirements.

The advantage of the implicit time advancement is that it allows for a bigger time step compared to the explicit formulation due to *stability* reasons. We have made some tests with the DeFT code (coarse grid) and found that a time step of $\Delta t = 0.03 U_{cl}/h$ is the maximum due to *accuracy requirements*. Using larger time steps resulted in a considerable smoothing of the flow field compared to the solution derived with a very small time step. The time step restriction of the explicit code MGLET is about $0.03U_{cl}/h$ and $0.015U_{cl}/h$ for the coarse and fine grids, respectively. Nevertheless, we used a time step of only 0.005 for all the simulations with MGLET and HORUS, to avoid stability problems during the initial phase of the simulations.

Table 3 CPU-time requirements of the three codes on a Fujitsu VPP700.

	MGLET		HORUS		DeFT
	coarse	fine	coarse	fine	coarse
Δt	0.005	0.005	0.005	0.005	0.03
$sec/\Delta t$	0.13	0.44	0.27	1.16	7.6
$sec/(1.0U_{cl}/h)$	26	88	54	232	253

The CPU-time requirements for the codes are listed in table 3. All three codes have been vectorized and run at about 300-400 Mflop/s on a Fujitsu VPP700. The computational efficiency is not yet optimal due to the small problem size, but comparable between the three codes. The fraction of CPU-time spent in the different stages of one time step are listed in table 4. One can see that the solution of the Poisson equation for the pressure is the most time consuming part of each of the algorithms, unless the direct solver is used. Unfortunately, the direct solver requires two periodic directions and cannot easily be parallelized due to the FFT's in two directions. Computer time can be saved predominantly by the choice of the solver for the linear systems or by the choice of the convergence criterion that should be 'as accurate as needed and as inaccurate as possible'. For example the code MGLET has been run with single precision accuracy and the resulting divergence after solving the Poisson equation by the direct solver to machine accuracy was in the range of $10^{-5}h/U_{cl}$.

Table 4 Approximate fractions of CPU-time spent during the different stages of the solution algorithm.

	MGLET	HORUS	DeFT
Momentum equation	38	29	40
Poisson equation	62	71	60

4 Comparison of results

4.1 General behavior of the solution

Constant pressure gradient. In the constant pressure gradient case, the Reynolds number of $Re_\tau = 132.5$ based on the channel half width h and friction velocity u_τ is adjusted by the mean pressure gradient. The bulk-velocity varies in time according to the following formula:

$$\frac{\partial u_b}{\partial t} = -\frac{\partial p}{\partial x} + \frac{1}{2h}\left[\tau_w^u - \tau_w^l\right] \tag{4.16}$$

where τ_w^u and τ_w^l are the instantaneous plane averaged wall shear stresses at the upper and lower walls, respectively.

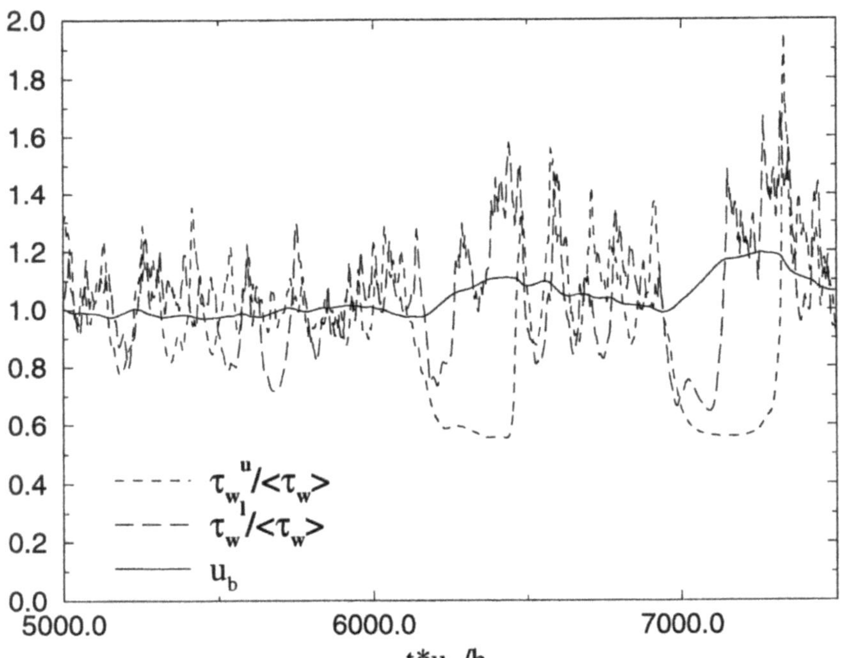

Figure 1 Time history of wall shear stress and bulk velocity in case $dp/dx =$ *const.*, coarse grid (Code MGLET).

Fig. 1 shows a typical history of the bulk velocity and the wall shear stress for the coarse grid (MGLET). There are periods of very low wall shear stress (as low as 60% of the time mean) during which a considerable rise of the bulk velocity can be observed. An inspection of instantaneous velocity fields shows that during these periods the flow becomes nearly laminar on one side of the channel. There are periods of about $2000u_{cl}/h$ without any laminar event. Those events occurred in the coarse grid as well as in the fine grid runs.

The random occurrence of these laminar flow events makes it very hard to ensure correct statistics of the flow field. The rise of the bulk velocity during these events results in a steeper gradient of the velocity profile in the logarithmic region and a higher centerline velocity. But, even worse, there appears a high peak in the streamwise rms-values of the fluctuations, because the switching between laminar and turbulent states is interpreted as fluctuation. This peak can only be avoided if laminar periods are excluded from the statistics.

Constant bulk velocity When the bulk velocity is kept constant during the simulation, the acceleration of the flow during the laminar periods, that has been observed in the constant pressure gradient case, cannot occur. That means that this case is much easier in terms of statistical evaluation. The wall shear stress history is similar to the other case, exhibiting valleys of below 60% of the long term average (see Fig. 2).

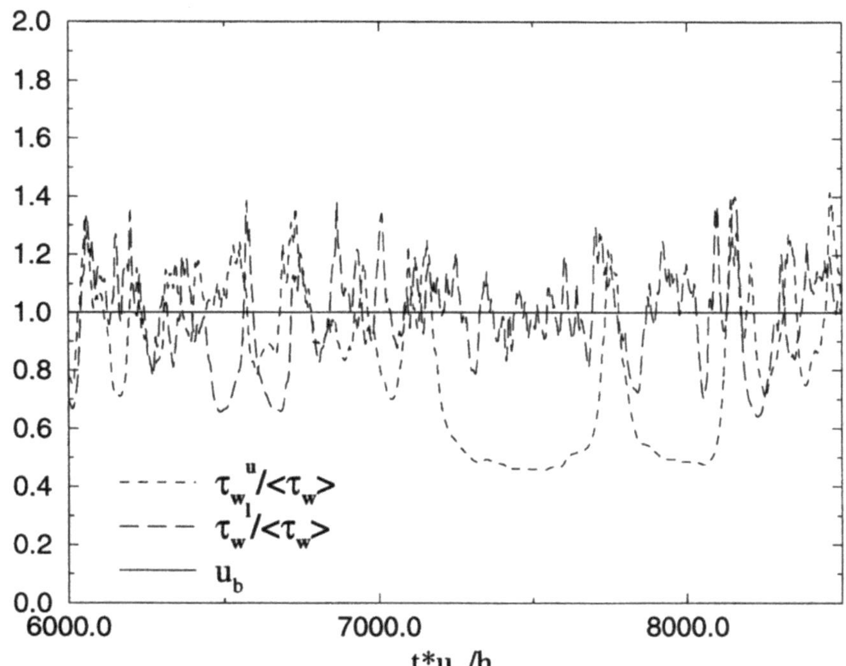

Figure 2 Time history of wall shear stress and bulk velocity in case $u_b =$ *const.*, coarse grid (Code MGLET).

Keeping the flux constant during these periods results in a different time averaged velocity profile compared to the other case (see Fig. 1). Since the bulk velocity remains constant even during the laminar periods, these periods cannot affect the time average of the streamwise velocity profile as much as if it could vary. For that reason this case is better suited for statistical evaluation, and all of the results presented in the following have been computed using a constant bulk velocity.

4.2 Global quantities

In the following, we check the ability of the three codes to predict the time averaged shear stress when a constant bulk velocity is given. In table 5 we compare long term averages of the wall shear stress calculated by the three different codes. The empirical correlation for turbulent plane channel flow would give a value of about $\tau_w \approx 0.00207$. We see that there is a trend to values below the empirical correlation. This can be due to the laminar events during which a value of less then 60% of the longterm average is achieved (compare figure 2). But all codes predict a value that lies within 5% variation.

Table 5 Long term average of the wall shear stress $< \tau_w^l - \tau_w^u > \cdot h/U_{cl}$.

	MGLET	HORUS	DeFT
Coarse grid	0.00192	0.00184	0.00193
Fine grid	0.00190	0.00194	

4.3 Time averaged velocity profiles

The next check concerns the time averaged streamwise velocity profile. We are able to compare our results against those obtained by a spectral code (Jiménez and Moin [4]). In figure 3 we have plotted the coarse grid runs and in figure 4 the fine grid runs. A remarkable accordance between all three codes can be observed close to the wall. On the other hand, all the profiles lie somewhat higher than that of the spectral code. We notice that the fine grid results move towards the spectral results in a region below $z^+ \approx 50$, see figure 4. But still a higher centerline velocity has been obtained compared to the spectral result. One reason might be that in the spectral results there were no laminar events during the sampling time of $750 u_{cl}/h$. In our case we didn't remove the laminar periods from the statistics that occurred several times during the whole computation of about $10000 u_{cl}/h$.

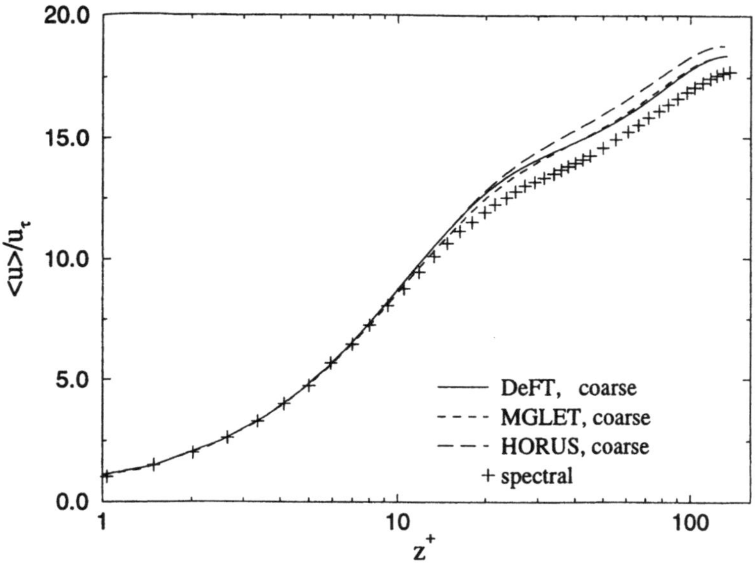

Figure 3 Mean streamwise velocity profiles in inner coordinates for coarse grid runs. Spectral results from Jiménez and Moin [4].

4.4 Velocity fluctuations

As mentioned before, the occurrence of laminar events contaminates the rms-values of the velocity fluctuations in a strong way. In figures 5 and 6, the streamwise velocity fluctuations are compared with the spectral results of Jiménez [4]. All finite volume codes give considerably higher values when compared to the spectral results, but are in good accordance when compared to each other. Also, there seems to be no significant grid dependence in the streamwise velocity fluctuations.

In the wall normal velocity fluctuations, however, considerable grid dependences can still be observed (figures 7 and 8). The coarse grid results lie about 5% lower and the fine grid results about 5% higher than the spectral results. Again, a remarkable accordance between the coarse grid results of the finite volume codes may be noted.

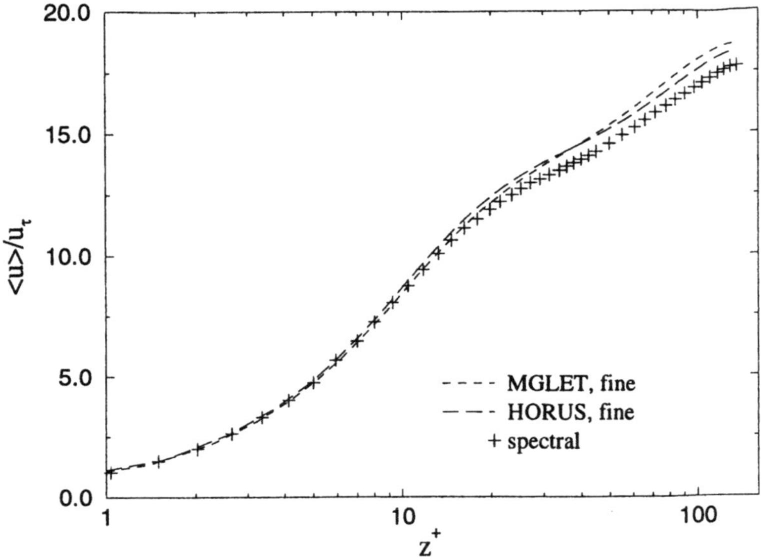

Figure 4 Mean streamwise velocity profiles in inner coordinates for fine grid runs. Spectral results from Jiménez and Moin [4].

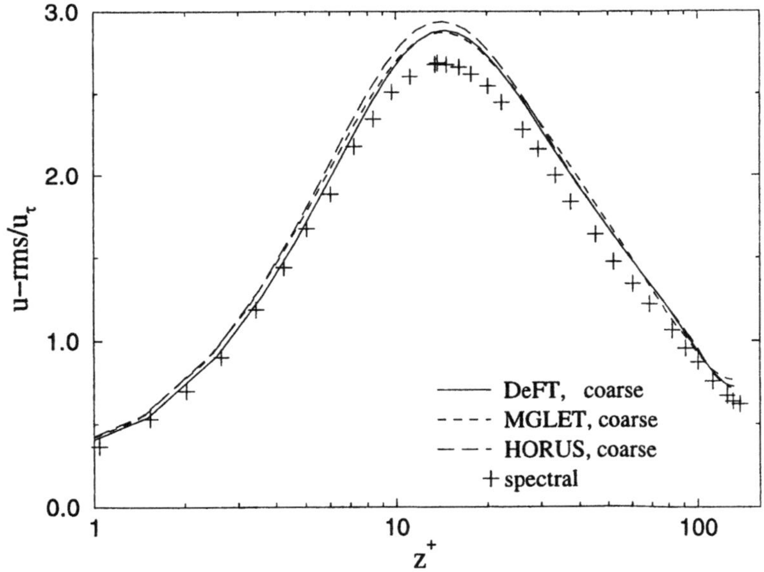

Figure 5 Streamwise rms velocity fluctuations in inner coordinates for coarse grid runs. Spectral results from Jiménez and Moin [4].

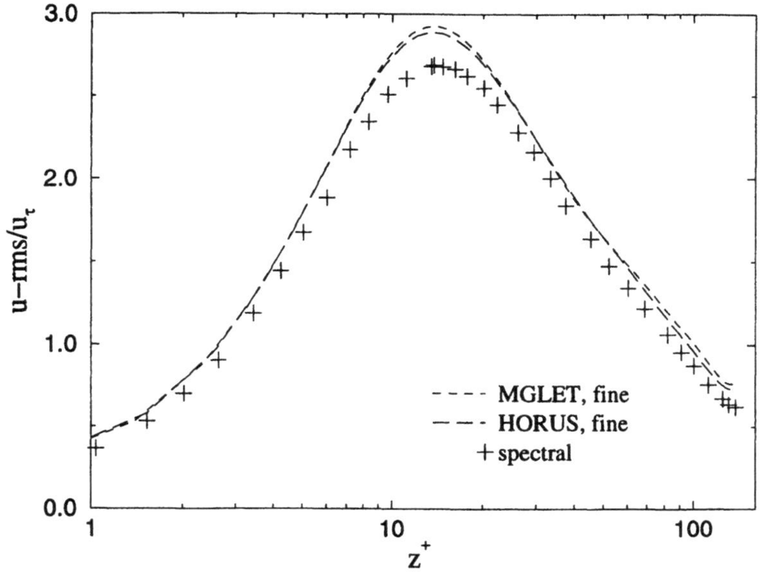

Figure 6 Streamwise rms velocity fluctuations in inner coordinates for fine grid runs. Spectral results from Jiménez and Moin [4].

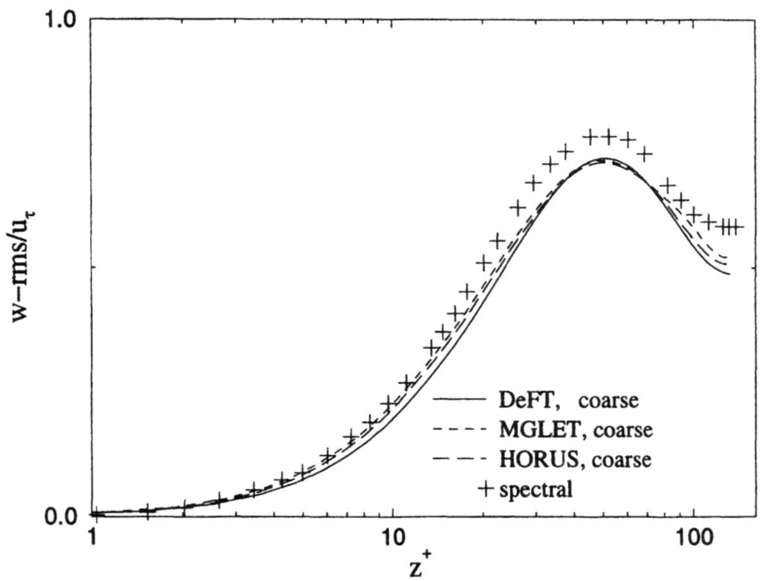

Figure 7 Wall normal rms velocity fluctuations in inner coordinates for coarse grid runs. Spectral results from Jiménez and Moin [4].

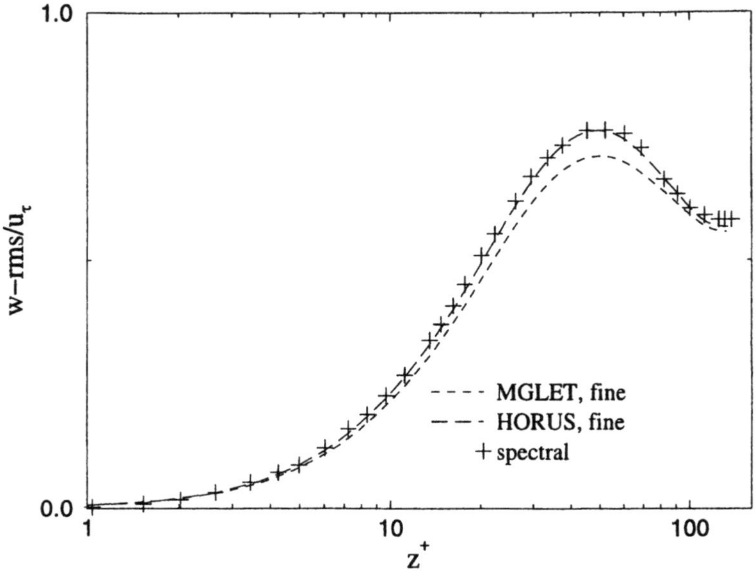

Figure 8 Wall normal rms velocity fluctuations in inner coordinates for fine grid runs. Spectral results from Jiménez and Moin [4].

5 Conclusions

Comparing the first- and second-order statistics we made the following observations. All time-stepping algorithms seem to be suited for the DNS of turbulent flows from an accuracy point of view. The accuracy of collocated and staggered grids seems to be equivalent. There are still some uncertainties in the statistics due to the laminar events, so definite conclusions are not possible in the present stage of the investigation. The best test case would be a fully developed channel flow without the inconvenience of partial relaminarisation.

The comparison of the computational requirements leads us to the following conclusions. The solution of the Poisson equation for the pressure is always the most expensive task. The curvilinear formulation is always time and memory consuming but is required for complex geometries. The implicit time step provides no advantage over an explicit formulation due to time step limitations by accuracy for this kind of flow. An implicit time step also leads to very large memory requirements. But implicit time stepping may be efficient on strongly nonuniform grids, if locally small cells impose severe stability restrictions on the time step.

References

[1] G.B. Deng, J. Piquet, P. Queutey and M. Visonneau: Three Dimensional Full Navier-Stokes Solvers for Incompressible Flows Past Arbitrary Geometries. Int. J. for Num. Method in Eng., Vol.31, pages 1427-1451, 1991.

[2] G.B. Deng, and M. Visonneau: Computation of a Wing-body Junction Flow with a Reynolds Stress Turbulence Model. Twenty-Second Symposium on Naval Hydrodynamics, Washington DC, 1998.

[3] C.W. Hirt, B.D. Nichols, N.C. Romero: SOLA - A numerical solution algorithm for transient fluid flows. Los Alamos: Los Alamos Sci. Lab., (LA-5852), 1975.

[4] J. Jiménez and P. Moin: The minimal flow unit in near-wall turbulence. *J. Fluid Mech.*, 255, pp. 213-240, 1991.

[5] M. Manhart and H. Wengle. Large-eddy simulation of turbulent boundary layer flow over a hemisphere. In Voke P.R., L. Kleiser, and J-P. Chollet, editors, *Direct and Large-Eddy Simulation I*, pages 299–310, Dordrecht, March 27-30 1994. ERCOFTAC, Kluwer Academic Publishers.

[6] M. Manhart. Zonal direct numerical simulation of turbulent plane channel flow. In *5th French-German Workshop on Computation and visualization of three-dimensional vortical and turbulent flows.* Technische Universität München, Dezember 1996.

[7] L. Schmitt: Grobstruktursimulation turbulenter Grenzschicht-, Kanal- und Stufenströmungen. Dissertation (phd-thesis), Lehrstuhl für Fluidmechanik, TU München, 1982.

[8] H. Werner and H. Wengle. Large-eddy simulation of turbulent flow over a square rib in a channel. In H.H. Fernholz and H.E. Fiedler, editors, *Advances in Turbulence*, volume 2, pages 418–423. Springer-Verlag, Berlin, 1989.

[9] P. Wesseling, A. Segal, C.G.M. Kassels and H. Bijl. Computing flows on general two-dimensional nonsmooth staggered grids. *J. Eng. Math.*, 1998, *To appear.*

[10] M. Zijlema, A. Segal and P. Wesseling. Invariant discretization of the k-ε model in general co-ordinates for prediction of turbulent flow in complicated geometries. *Computers and Fluids*, vol. 24, pp. 209–225, 1995.

[11] M. Zijlema, A. Segal and P. Wesseling. Finite volume computation of incompressible turbulent flows in general coordinates on staggered grids. *Int. J. Num. Meth. Fluids*, vol. 20, pp. 621–640, 1995.

[12] P. Wesseling, A. Segal, J.J.I.M. van Kan, C.W. Oosterlee and C.G.M. Kassels. Finite volume discretization of the incompressible Navier-Stokes equations in general coordinates on staggered grids. *Comp. Fluid Dynamics Journal*, vol. 1, pp. 27–33, 1992.

[13] H. Bijl and P. Wesseling. A numerical method for the computation of compressible flows with low Mach number regions In J.-A. Désidéri, C. Hirsch, P. Le Tallec, M. Pandolfi and J. Périaux, editors, *Computational Fluid Dynamics '96*, pp. 206–212, 1996, Wiley, Chinchester.

[14] A. Segal, P. Wesseling, J. van Kan, C.W. Oosterlee and K. Kassels. Invariant Discretization of the Incompressible Navier-Stokes equations in Boundary Fitted Co-ordinates. *Int. J. Num. Meth. Fluids*, vol. 15, pp. 411–426, 1992.

APPLICATIONS OF A 4th-ORDER HERMITIAN SCHEME FOR NON-EQUIDISTANT GRIDS TO LES AND DNS OF INCOMPRESSIBLE FLUID FLOW

Adnan Meri and Hans Wengle
Institut für Strömungsmechanik und Aerodynamik
Universität der Bundeswehr München
D-85577 Neubiberg, Germany

Anne Dejoan, Eric Védy and Roland Schiestel
Institut de Recherche sur les Phénomènes Hors d'Equilibre, IRPHE
UMR 6594 CNRS/Universités d'Aix-Marseille I & II
La Jetée - Technopôle de Châteaux-Gombert
38 rue Frédéric Joliot Curie
F-13451 Marseille Cedex 20

Summary

A fourth-order Hermitian, or compact scheme has been implemented into the production programs on both sides of the french-german cooperation. On the french side the numerical scheme is applied in its equidistant form on a transformed equidistant grid which gives the desired non-equidistant grid in the physical space. On the german side the fourth-order scheme has been formulated directly for a non-equidistant and staggered grid. The significant improvements in phase error and amplitude error behaviour of the higher-order scheme in comparison to the second-order central scheme is demonstrated in solving simpler test problems and in addition, very complex flow problems such as the non-equilibrium turbulent channel flow and the transitional or fully turbulent backward-facing step flow. The selected complex flow problems are solved by DNS for Re = 3000 and by LES for Re = 13800. The results from LES of periodically forced turbulent channel flow (with hysteresis effects) indicate the necessity of using improved subgrid scale models.

1 Introduction

To calculate the three-dimensional and time-dependent structure of a turbulent flow there are ***two basic simulation concepts*** available. On the one hand, the so-called *Direct Numerical Simulation* (DNS) requires the resolution of *all* the relevant scales in a turbulent flow, but its range of application is limited to relatively small Reynolds numbers (often too small from a practical engineering point of view). On the other hand, the *Large-Eddy Simulation* (LES) is capable to deliver directly the spatial and temporal behaviour of at least the large-scale structures of high Reynolds number flows, and only the effects of the small-scale motions which cannot be resolved on a given computational mesh need to be modelled with a so-called subgrid-scale model.

The application of higher-order (spatial) discretization schemes is motivated, with respect to DNS, by the desire to provide a better representation of the smaller length scales on a given computational mesh and, with regard to LES, to provide a more accurate representation of the convective transport of the resolved scales, and to provide a better separation between the effects of the subgrid-scale modelling scheme and the discretization errors of the numerical solution scheme. From an engineering point of view the higher-order schemes should be applicable without too many complications to more complex flow cases, with non-periodic boundary conditions and non-trivial geometry.

In this paper, results from a french-german cooperation are presented. A numerical scheme which is formally of fourth-order accuracy on a non-equidistant grid has been derived, implemented and tested for incompressible flow cases. The fourth-order scheme described in this paper is mentioned in the literature under different names, such as Hermitian, compact, or Padé scheme. The method is characterized by considering a linear combination of the values of the derivatives and of the function itself at three neighbouring grid points. The name *compact* scheme refers to the fact that, in comparison to a second-order finite-difference scheme, fourth-order accuracy can be achieved on the same compact stencil using the Hermitian approximation. However, the implicit formulation inherent in the Hermitian approximation requires the numerical solution of tridiagonal systems of unknowns.

The main objectives of this study are

- to shortly characterize the derivation of the fourth-order scheme,
- to verify the accuracy of the fourth-order scheme by solving test problems, and
- to apply the fourth-order method to DNS and LES, respectively, of non-equilibrium flow cases such as the perturbed plate channel flow and the transitional and turbulent flow over a backward-facing step for a low Reynolds number (DNS) and a high Reynolds number (LES).

2 Formulation of a fourth-order Hermitian scheme

In more complex flow cases the use of non-equidistant grids is unavoidable. In general, there are two different strategies of achieving a non-equidistant distribution of the discrete values of the numerical solution: (a) a proper transformation of a non-uniform grid in the physical space provides an *equidistant* grid in the transformed space and the Hermitian relations for an equidistant grid are applied (see 2.1), or (b) the Hermitian relations are derived directly for the given non-equidistant grid in physical space (see 2.2). Examples of formulations of Hermitian approximations can be found, for example, in Lele [8] for an equidistant grid and in Adam [1] for a non-equidistant grid.

2.1 Formulation on equidistant grids and using a coordinate transformation

In the numerical scheme developed by Schiestel and Viazzo [11] the spatial derivatives in inhomogeneous directions are approximated by a fourth-order Hermitian finite-difference scheme and in homogeneous directions by a Fourier pseudo-spectral scheme. If necessary, strong mesh refinement is accomplished by applying a proper coordinate transformation in the inhomogeneous directions, i.e. the Hermitian scheme is used in the transformed space in a form appropriate for an equidistant grid. The governing equations are solved on a staggered grid and the non-linear terms are discretized from a skew-symmetric form. Time advancement is carried out with a second-order Adams-Bashforth scheme for the non-linear terms and with a Crank-Nicolson scheme for the viscous term. Pressure-velocity coupling is solved via an approximate Poisson-equation that guarantees fourth-order accuracy with additional internal iterations [11] .

One of the test problems studied by Viazzo and Schiestel [12] to verify the fourth-order Hermitian scheme is the temporal evolution of small perturbations in channel flow. The energy of the perturbed flow, E(t), has been obtained by integration, and compared with the solution of the corresponding Orr-Sommerfeld equation for the disturbed flow: $E(t)/E(0) = exp(2\omega_i t)$. In **figure 1** the results for differently chosen spatial discretizations are compared with this solution of the linear stability theory, and with corresponding solutions from Malik et al. [9] using second-order finite-differences and a spectral method, respectively. From figure 1 it can be concluded that the fourth-order Hermitian method needs significantly less grid points in comparison to a second-order finite-difference scheme. It is capable of providing results as accurate as a spectral method using about the same number of grid points.

As a conclusion from the work and experience collected on the french side it was decided to implement a 4th-order Hermitian scheme into the production program on the german side as an additional option.

2.2 Direct formulation on non-equidistant grids

In contrast to the approach on the french side, on the german side the derivation of the fourth-order Hermitian scheme has been accomplished *directly* for a non-equidistant grid, i.e. no transformation of the original cartesian coordinate system has been used. The reason for this approach was to avoid a serious restriction in the applicability of our general program for flow cases with flow obstacles of general geometry (defined on a cartesian coordinate system). The governing equations are solved on a staggered grid using a classical finite-volume discretization. Time advancement is carried out either by an explicit Adams-Bashforth scheme or alternatively, by an explicit leapfrog scheme (second order). Pressure-velocity coupling is solved to second-order accuracy via an iterative point-by-point relaxation or alternatively, by a multigrid solver for the Poisson equation for the pressure correction. The implementation of the fourth-order (compact) scheme had to consider the given structure of the program, i.e. the basic finite-volume discretization on a staggered grid. Discretizing the integral form of the balance equations, fourth-order interpolations are required for the convection terms and fourth-order approximations for the first-derivatives are required for the diffusion terms. For example, to discretize the x-momentum equation with a control volume centered around the u-velocity component defined at index $i + \frac{1}{2}$, i.e at the side surface of the basic grid cell (with index i), interpolations of the velocity components and their derivatives are required to locations where they are needed but not defined. The following sketch illustrates the problem (x-momentum equation):

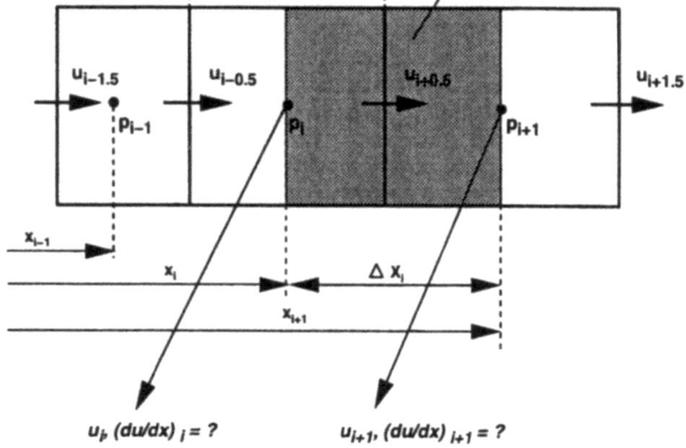

Approximation of first derivatives at interior nodes

Assuming the discrete values of a function f are given at the boundaries (index $i+\frac{1}{2}, i-\frac{1}{2}$) of the basic grid cell (with index i at the center of the cell) of size DDx_i. Then we start from an approximation for the discrete values of the derivatives $F=\frac{\partial f}{\partial x}$ at the locations x_i, x_{i-1}, x_{i+1} of the following form:

$$\alpha_i F_{i-1} + F_i + \beta_i F_{i+1} = \frac{\gamma_i f_{i+\frac{1}{2}} + \lambda_i \hat{f}_i + \phi_i f_{i-\frac{1}{2}}}{DDx_i}. \tag{2.1}$$

Note, the triangle hat of the function value $\hat{f}_i$ signalizes that this value is not defined at the location i, and therefore, it must be provided by a corresponding fourth-order interpolation. Relations between the coefficients α_i, β_i, γ_i, λ_i und ϕ_i are derived by matching the coefficients of the Taylor series for the function values and the values of their derivatives, respectively. If the matching is done with increasing order, the following constraints for these coefficients can be derived:

$$\begin{aligned}
\gamma_i + \lambda_i + \phi_i &= 0.0 \\
\alpha_i + \beta_i - \frac{1}{2}K_i(\gamma_i S_i - \phi_i) &= -1.0 \\
-\alpha_i + \beta_i S_i - \frac{1}{8}K_i(\gamma_i S_i^2 + \phi_i) &= 0.0 \\
\alpha_i + \beta_i S_i^2 - \frac{1}{24}K_i(\gamma_i S_i^3 - \phi_i) &= 0.0 \\
-\alpha_i + \beta_i S_i^3 - \frac{1}{64}K_i(\gamma_i S_i^4 + \phi_i) &= 0.0
\end{aligned} \tag{2.2}$$

with $K_i = \frac{\Delta x_{i-1}}{DDx_i}$ and the *stretching factor of the grid* $S_i = \frac{\Delta x_i}{\Delta x_{i-1}}$ The solution of this system of equations leads to:

$$\begin{aligned}
\phi_i &= \frac{(e-a)g - (c+a)p}{qg - ph} \\
\gamma_i &= \frac{c + a - h\phi_i}{(ad - bc)} \\
\lambda_i &= -\gamma_i - \phi_i \\
\beta_i &= \frac{(-1.0 - \frac{3}{8}K_i\phi_i - b\gamma_i)}{a} \\
\alpha_i &= -1.0 - \beta_i - 0.5K_i\phi_i + 0.5K_i S_i \gamma_i
\end{aligned} \tag{2.3}$$

with

$$a = S_i + 1.0, \quad c = S_i^2 - 1.0, \quad e = S_i^3 + 1.0$$

$$b = K_i S_i(-\frac{1}{8}S_i - \frac{1}{2}), \quad d = K_i S_i(-\frac{1}{24}S_i^2 + \frac{1}{2}), \quad f = K_i S_i(-\frac{1}{64}S_i^3 - \frac{1}{2})$$

$$g = ad - bc, \quad p = af - be, \quad h = -\frac{3}{8}K_i c - \frac{11}{24}K_i a, \quad q = -\frac{3}{8}K_i e + \frac{31}{64}K_i a\,.$$

The first unmatched coefficient in the Taylor series determines the formal truncation error (T.E.). Here, we are constructing a fourth-order scheme on a non-equidistant grid with an arbitrary stretching factor S_i with the following truncation error:

$$T.E._{neq} = \frac{1}{24}[(\frac{\alpha_i}{S_i^4} + \beta_i) - \frac{1}{160}K_i(\gamma_i S_i - \frac{\phi_i}{S_i^4})](\Delta x_i)^4 \frac{\partial^5 f}{\partial x^5}\,. \tag{2.4}$$

For a grid with constant stretching factor ($S_i = S = const$). it follows:

$$T.E._{neq} = C_i(\Delta x_i)^4 \frac{\partial^5 f}{\partial x^5} \tag{2.5}$$

$$C_i = \frac{1}{24}[(\frac{\alpha_i}{S^4} + \beta_i) - \frac{1}{160}K_i(\gamma_i S - \frac{\phi_i}{S^4})] = const. \tag{2.6}$$

Finally, for an equidistant grid ($S_i = 1$) the coefficients in the approximation (2.1) are (see Lele [8]):

$$\alpha = \beta = \frac{1}{22}, \quad \gamma = -\phi = \frac{24}{22} \quad \text{and} \quad \lambda = 0.0\,. \tag{2.7}$$

With these coefficients the truncation error for an equidistant grid merges to the fourth-order value derived by Lele [8] (Appendix B, table VI):

$$T.E._{eq} = \frac{17}{5280}(\Delta x)^4 \frac{\partial^5 f}{\partial x^5}\,. \tag{2.8}$$

Non-periodic boundary conditions: formulation for the first and the last grid point

For the first derivative *at the first grid point (l)* within the computational domain third-order accuracy is provided by applying the following non-centered formulation:

$$F_l + \beta_l^* F_{l+1} = \frac{1}{DDx_l}[a_l f_{l-\frac{1}{2}} + b_l f_{l+\frac{1}{2}} + c_l f_{l+\frac{3}{2}}]\,. \tag{2.9}$$

For the first derivative *at the last grid point (m)* a corresponding non-centered formulation is used:

$$\alpha_m^* F_{m-1} + F_m = \frac{1}{DDx_m}[a_m^* f_{m+\frac{1}{2}} + b_m^* f_{m-\frac{1}{2}} + c_m^* f_{m-\frac{3}{2}}]. \qquad (2.10)$$

Tridiagonal system of linear equations

The approximations (2.1), (2.9) and (2.10) lead to the following tridiagonal system of equations (for the first derivatives) which can be solved , e.g. by a so-called Thomas algorithm:

$$\begin{pmatrix} 1.0 & \beta_l^* & & & & & \\ \alpha_{l+1} & 1.0 & \beta_{l+1} & & & & \\ & \ddots & \ddots & \ddots & & & \\ & & \alpha_i & 1.0 & \beta_i & & \\ & & & \ddots & \ddots & \ddots & \\ & & & & \alpha_{m-1} & 1.0 & \beta_{m-1} \\ & & & & & \alpha_m^* & 1.0 \end{pmatrix} \cdot \begin{pmatrix} F_l \\ F_{l+1} \\ \vdots \\ F_i \\ \vdots \\ F_{m-1} \\ F_m \end{pmatrix} =$$

$$\begin{pmatrix} \frac{a_l f_{l-\frac{1}{2}} + b_l f_{l+\frac{1}{2}} + c_l f_{l+\frac{3}{2}}}{DDx_l} \\ \frac{\gamma_{l+1} f_{l+\frac{1}{2}} + \lambda_{l+1} \hat{f}_{l+1} + \phi_{l+1} f_{l+\frac{3}{2}}}{DDx_{l+1}} \\ \vdots \\ \frac{\gamma_i f_{i+\frac{1}{2}} + \lambda_i \hat{f}_i + \phi_i f_{i-\frac{1}{2}}}{DDx_i} \\ \vdots \\ \frac{\gamma_{m-1} f_{m-\frac{1}{2}} + \lambda_{m-1} \hat{f}_{m-1} + \phi_{m-1} f_{m-\frac{3}{2}}}{DDx_{m-1}} \\ \frac{a_m^* f_{m+\frac{1}{2}} + b_m^* f_{m-\frac{1}{2}} + c_m^* f_{m-\frac{3}{2}}}{DDx_m} \end{pmatrix}. \qquad (2.11)$$

On the right-hand side of (2.11) and of (2.1) the function values characterized by a triangle hat, i.e. $\hat{f}_i$, are not defined at the location with index i and therefore, before solving the system (2.11) they must be interpolated to fourth-order accuracy using approximations in analogy to (2.1). Further details can be found in [10].

2.3 Solution of test problems on non-equidistant grids

Results for an identical 2D laminar backward-facing step problem evaluated with the two different approaches on the french and the german side, respectively, did not exhibit any significant difference. Then, the new option for the fourth-order Hermitian scheme formulated directly for a non-equidistant grid has been tested on the german side solving selected test problems with increasing complexity.

(a) 1D and time dependent convective transport of sinus waves

First we solved the time-dependent convective transport of sinus waves on a 1D computational domain with periodic and non-periodic boundary conditions, respectively. For example, **figure 2** shows a comparison of numerical solutions (of 2nd-order and 4th-order, respectively) on an equidistant grid with the corresponding analytical solution. This simple example already demonstrates clearly the significantly smaller phase errors of the fourth-order Hermitian method. In a second example two non-equidistant segments (first a compressed grid then a expanded grid with identical stretching factor S) in the middle of an equidistant grid (grid spacing Δx_o) with periodic boundary conditions. **Figure 3** demonstrates that the local error integrated over the whole computational domain decays with 4th-order and 2nd-order, respectively, for the two schemes compared. In a third example, **figure 4**, we used different non-equidistant grids with different constant stretching factors $S_i = \Delta x_{i+1}/\Delta x_i = const.$ All grids start with the same initial grid spacing Δx_o. For example, on a grid with stretching factor $S = 1.007$ the local error in the amplitude for the 4th-order method is about a factor of 4 lower than for the 2nd-order scheme. With increasing stretching factor this error increases for both methods, and for $S = 1.05$ (5 % stretching) the error for the 4th-order method is about 75 % of the error of the 2nd-order scheme. Note, these numbers will be different if we choose a different initial grid spacing Δx_o.

In summary: during the first phase of testing pure advective transport it has been verified that the method is of fourth-order accuracy on a non-equidistant grid and it has been demonstrated *how* the error of the method depends on the expansion ratio $S_i = \Delta X_{i+1}/\Delta X_i$ of the grid. The stretching factor S of the grid should stay below about 5 % to gain a significant benefit in comparison to a 2nd-order method. In contrast to the second-order central differencing scheme, the fourth-order Hermitian scheme exhibits significantly smaller phase errors. In addition, an accurate outflow boundary condition can be realized by using fourth-order extrapolation instead of applying the (normal-gradient = 0)-condition which corresponds to first-order extrapolation only for the first grid point outside the physical domain. The fourth-order extrapolation did not produce any "wiggles" at the outflow cross-section.

(b) 2D manipulated flat-plate boundary layer

In the next stage of comparing the performance of the fourth-order Hermitian scheme with the second-order central differencing scheme the classical laminar-turbulent transition in a 2D boundary layer has been calculated. For perturbations with very small amplitudes, the first stage of the transition process is governed by the amplification of 2D disturbances. The growth or decay of these so-called Tollmien-Schlichting waves is well described by the linear stability theory. Direct numerical solutions of the Navier-Stokes equation have been carried out for this 2D stage by numerically simulating small-amplitude blowing/suction disturbances on a small strip (between X1 and X2 in **figure 5**) at the beginning of the developing laminar boundary layer. The amplification and damping of the disturbances have been evaluated for two different grids with (NX,NZ)=(694,68) and (1386,136) grid points, respectively, and using the second-order differencing scheme, as well as the fourth-order Hermitian scheme. **Figure 6** shows the spatial (downstream) development of the perturbation amplitude A (normalized by its respective minimum A_0, see figure 5). Here, the downstream location is characterized by the local Reynolds number $Re^* = U_\infty \delta_1/\nu$, based on the local displacement thickness $\delta_1(x)$.

In summary: the results for the laminar-turbulent transition of a 2D boundary layer showed that the second-order central scheme requires about four times more grid points in comparison to the fourth-order Hermitian scheme to produce results in accordance with the linear stability theory. Again, no disturbances are created at the outflow section in the case of the fourth-order scheme.

3 Applications to DNS and LES of channel flow and of backward-facing step flow

3.1 DNS of *equilibrium* turbulent plate channel flow at $Re_h = 3300 (Re_\tau = 180)$

From the experience on the german side with DNS of turbulent plate channel flow ($Re_\tau = 180$) , (using second-order central differencing and 0.5 million grid points) it is known that it is possible to achieve good agreement with the reference data from DNS of Kim, Moin and Moser [6] (using spectral methods and 4 million grid points) if the first grid point next to rigid walls is well within the linear-law-of-the-wall. The only variable not showing satisfying agreement is the skewness of the wall-normal velocity close to the wall. In **figure 7** the RMS-values of the three velocity components are compared with the DNS data of Kim, Moin and Moser. There are small differences between the results from the two discretization methods, and the fourth-order solutions are a little bit closer to the reference solutions.

Much more critical is the comparison with the skewness of the wall-normal velocity: the second-order central differencing scheme solutions with 0.5 million grid points is far off the reference solution, and about 4 million grid points are required to reach the same quality as the fourth-order Hermitian solution using 0.5 million grid points (figure 7). None of the two solutions is capable to reproduce the second zero-crossing of the skewness close to the wall. However, if we also use 4.0 million grid points together with the 4th-order Hermitian scheme, the result comes pretty close to the results in [6] using a Chebyshev-Fourier spectral scheme.

3.2 LES of *non-equilibrium* turbulent plate channel flow at $Re_h = 13800$

Large-eddy simulations at high Reynolds number have been carried out on the french side to investigate the response of turbulence to periodic pressure gradient oscillations and to study the effects of imposed wall perturbations on the large-scale flow structure. Both non-equilibrium flow states are considered in order to test the ability of classical statistical models and more particularly the subgrid-scale models to represent unsteady turbulent flows. For the results presented here we used a fine resolution with (NX,NY,NZ)=(256,128,62) grid points, and the dimensions of the computational domain were $(x/h, y/h, z/h) = (4\pi, \pi, 2)$. The boundary conditions were periodic in the longitudinal and transversal directions.

(a) Turbulent channel flow forced by pressure gradient oscillations:

In this case the unsteady periodic plate channel flow is disturbed by forced pressure gradient oscillations. Binder and Kueny [3] have shown that the so-called Stokes thickness $l_s = (\nu/\pi f)^{1/2}$ (made dimensionless with the viscous length $l_\nu = \nu/u_\tau$) is the determining parameter in such unsteady flows. Here, f is the frequency of the imposed oscillations and u_τ is the wall friction velocity. A value of $l_s{}^+ = l_s/l_\nu = 12.9$ has been selected to create a situation for which the oscillating flow interacts with the turbulence. The relative amplitude of the superimposed oscillation velocity a_{uc} is equal to 0.05 (at channel center line). The statistical treatment has been carried out on seven periods to obtain the phase-averaged quantities.

The results confirm that the time-mean velocity field and the time-mean turbulent intensity are not significantly affected by the imposed oscillation. The periodic velocity oscillations are analysed in **figure 8** . Figure 8a and figure 8b, respectively, compare the amplitude A_1 and the phase shift Φ_1 of the first mode of the velocity, respectively, versus the non-dimensional distance from the wall $z_s = z/l_s$. The calculations give a steep gradient of the amplitude near the wall which is characteristic of low frequency cases while further away from the wall a local maximum is found like in Stokes flow. The phase shift presents a rapid decrease with distance from the wall that also recalls a Stokes flow behaviour while measured quantities are more scattered. In figure 8c it can be observed that the turbulent intensity variations referred to the local amplitude of the velocity is approximately twice as large as that of the ratio $\sqrt{\overline{u'^2}}/\bar{U} = 0.24$. It is also seen that the comparison with the experiment is satisfactory. Figure 8d represents the variation of the phase of the turbulent intensity modulation with distance from the wall. The large phase shifts observed suggest that the use of subgrid-scale models assuming the instantaneous alignment of the implicit scales with the explicit strain-rate tensor may fail in unsteady situations.

(b) Turbulent channel flow with imposed wall perturbations:

In this case the flow is disturbed by periodic blowing and suction in time and space applied on the lower wall. The perturbation has been implemented via an imposed wall-normal velocity:

$$w_{wall}(x,y) = A_o sin(2\pi f t) sin(\kappa_1 x)$$

where f is the frequency of the perturbation and $2\pi/\kappa_1$ is the longitudinal wavelength. The values chosen are: $A_o = 0.075 u_\tau, f = 2u_\tau/h$ and $\kappa_1 = 1/2$. Thirteen periods of time have been used to collect the phase-averaged quantities. In **Figure 9a** hysteresis curves of the turbulent shear stress are plotted for several distances from the wall. It appears that the effect of the periodic disturbance is confined to a region close to the wall. **Figure 9b** presents hysteresis curves of the pressure gradient, the wall shear stress, the turbulent shear stress and the turbulent energy at the distance $z^+ = 0.88$: all the quantities are strongly influenced by the perturbation and lag effects are visible in the turbulent shear stress and the energy quantities at the x-position considered (corresponding to the maximum of the amplitude of the imposed perturbation).

In summary: non-equilibrium flow states in plane channel flow are created by forced pressure gradient oscillations and by forced temporal and spatial perturbations of the wall-normal velocity. The detailed instantaneous turbulence characteristics are affected as soon as the frequency of the oscillation is increased past a critical level and the interaction between the oscillating motion of the disturbance and the turbulent flow structure is enabled. The classical time-averaged quantities seem to be unaffected by the perturbation. However, phase-averaged quantities may exhibit strong hysteresis effects which may be important in engineering applications. A fine mesh resolution is required to treat the problem. However, the resulting lag effects will not be captured properly in a large-eddy simulation using a Smagorinsky subgrid-scale model. A transport equation subgrid-scale model may be more adequate under such circumstances.

3.3 DNS of a transitional backward-facing step flow at $Re_h = 3000$

From investigations of transitional backward-facing step flow ($Re_h = 3000$) on the german side a statistical-steady-state solution is available, using second-order central differencing and about 13 million grid points [2]. Starting from an identical initial flow field, two runs have been carried out to answer the question how different the *instantaneous* flow fields will be using a second-order finite-differencing scheme and a fourth-order Hermitian scheme, respectively. In **figure 10** instantaneous fields of the lateral vorticity component are shown after T=6.0, 8.0, 10.0, 12.0, 14.0 reference times. Here, a reference time is the time a marker particle needs to travel with reference velocity 1.0 over a reference distance 1.0 (which is equal to the step height).

A first obvious conclusion from figure 10 is that the typical transport velocity of the large-scale lateral rollers is slower in the case of using second-order differencing. This is a consequence of the larger phase errors of that method in comparison to the fourth-order Hermitian scheme (see also figure 2).

Finally, the complete statistics up to second order has been collected for the 4th-order Hermitian case to compare it with the available statistics using a 2nd-order central differencing scheme. The time-averaged results for the mean flow field do not show significant differences, and in both cases the mean recirculation length is $x_r/h = 6.3$ which corresponds to the experimental value of $x_r/h = 6.4$ [4]. From comparing the 2nd-order statistics it can be concluded that the two results are pretty close to each other downstream of about five obstacle heights ($x_r/h = 5.0$). However, in the region from $x/h = 1.0$ to $x/h = 4.0$, i.e. in the region of laminar-turbulent transition in the free shear layer bounding the recirculation zone, the more accurate numerical scheme produces vertical profiles of $< uu >$ and $< ww >$ closer to the experimental data, in particular close to the bottom wall (see **figure 11**).

3.4 LES of turbulent backward-facing step flow at $Re_h = 13800$

As a first test for a program version on the french side with a fourth-order scheme in *two* inhomogeneous flow directions an LES for a turbulent backward-facing step flow has been carried out. The Reynolds number was $Re_h = 13800$. Instantaneous velocity profiles from a previous LES of plate channel flow were used as inflow condition, and convective type conditions were applied at the outflow section. The dimensions of the computational domain are $(x/h, y/h, z/h) = (4\pi, \pi, 2)$. The grid contains (NX,NY,NZ)=(96,64,62) grid points and the mesh is refined near rigid walls using an analytical transformation. The Smagorinsky model (with a constant $C_s = 0.2$) and near-wall modifications in the viscous sublayer is used as a subgrid-scale model.

In **figure 12** some results are compared with corresponding data from a DNS by Le, Moin and Kim [7] ($Re_h = 5100$) and experimental data from Jovic and Driver [5] (Re_h = 10400 and 25500). The mean reattachment length x_r/h determined from the distribution of the friction coefficient C_f is slightly underpredicted. The change of the reattachment length with Reynolds number is not very strong above $Re_h = 5000$. However, there are stronger effects of the boundary layer thickness (at the edge of the step) and of the expansion ratio on x_r/h. In the relaxation region the mean velocity profiles progressively recover the standard logarithmic and linear laws of the wall. In the recirculation zone no universal law prevails and a comparison is made with the empirical Simpson's law $U/U_N = 1 - A[z/N - log(z/N) - 1]$, where N is the distance from the wall of the maximum reverse flow velocity U_N and A is a constant. A value of A between 0.2 and 1.0 is suggested to fit the present results. The evolution of the second-order velocity correlations exhibit the correct behaviour in comparison to the DNS data of Le et al. [7], with an asymptotic limit approached at the end of the computational domain.

On the overall, the results from the first test run show satisfying agreement with available data from the literature. The new program version using a fourth-order Hermitian scheme in two (inhomogeneous) directions is considered to be validated and is now open for further extensions, e.g. the implementation of the fourth-order scheme in all three directions, of options for different subgrid-scale models, for different boundary and inflow conditions and for different coordinate transformation to create different non-equidistant distributions of variables in physical space.

4 Conclusions

Using a Hermitian approximation on a compact computational stencil a fourth-order accurate numerical scheme can be formulated directly for a non-equidistant and staggered grid. An alternative is a formulation on an equidistant grid in a transformed space which gives a proper non-equidistant grid in the physical space. From solving test problems it can be concluded that the fourth-order scheme exhibits significantly lower phase errors and total amplitude errors for a grid stretching less than about 5% in comparison to a second-order central differencing scheme. In applying the higher-order method to the numerical solution of complex fluid mechanical flow problems such as the laminar-turbulent transition in a 2D boundary layer, the classical equilibrium and the periodically forced plate channel flow, the transitional and the fully turbulent backward-facing step flow, the improved capability in representing the small scales is visible close to rigid walls, in particular for the higher-order statistics. The solutions from LES of the flow cases with hysteresis effects indicate that improved subgrid-scale models, e.g. a transport equation model, may be more adequate under such circumstances. For future direct numerical simulations the fourth-order option formulated directly for a non-equidistant grid will be a valuable alternative in a production program for solving more complex flow cases. In comparison to the second-order option the additional cost was about 40% for a flow case with 13 million grid points on a single scalar processor.

References

[1] Y. Adam. A hermitian finite difference method for the solution of parabolic equations. *Comp. and Maths. with Appls.*, 1:393–406, 1975.

[2] G. Bärwolff, H. Wengle, and H. Jeggle. Direct numerical simulation of transitional backward-facing step flow manipulated by oscillating blowing/suction. In W. Rodi and G. Bergeles, editors, *Engineering Turbulence Modelling and Experiment 3*, pages 219–228, Amsterdam, 1996. Elsevier Science Publishers.

[3] G. Binder and J.L. Kueny. Measurements of the periodic velocity oscillations near the wall in unsteady turbulent channel flow. *Michel, R. Cousteix, J. and Houdeville, R. (edts.): IUTAM Symposium 'Unsteady Turbulent Shear Flows', Toulouse, France, Springer-Verlag*, pages 100–108, 1981.

[4] A. Huppertz and G. Janke. Preliminary experiments on the control of three-dimensional modes in the flow over a backward-facing step. In L. Machiels S. Gavrilakis and P.A. Monkewitz, editors, *Advances in Turbulence VI*, pages 461–464, Dordrecht, 1996. Kluwer Academic Publishers.

[5] S. Jovic and D. Driver. Reynolds number effect on the skin friction in separated flows behind a backward-facing step. *Experiments in Fluids*, 18(6):464–467, 1995.

[6] J. Kim, P. Moin, and R. Moser. Turbulence statistics in fully developed channel flow at low reynolds number. *J. Fluid Mech.*, 177:133–166, 1987.

[7] H. Le, P. Moin, and J. Kim. Direct numerical simulation of turbulent flow over a backward-facing step. *J. Fluid Mech.*, 330:349–374, 1997.

[8] S.K. Lele. Compact difference schemes with spectral-like resolution. *Journal of Computational Physics*, 103:16–42, 1992.

[9] M. Malik, T. Zang, and M. Hussaini. A spectral collocation method for the navier-stokes equation. *J.Comp. Physics*, 61:64–88, 1985.

[10] A. Meri. *Anwendung eines Kompaktverfahrens vierter Ordnung für nicht-äquidistante Gitter zur direkten numerischen Simulation turbulenter Strömungen.* Bericht Nr. 98/4, Institut für Strömungsmechanik und Aerodynamik, Universität der Bundeswehr München, 85577 Neubiberg, 1998.

[11] R. Schiestel and S. Viazzo. A hermitian-fourier numerical methods for solving the incompressible navier-stokes equations. *Int.J. Computers and Fluids*, 24:739–752, 1995.

[12] S. Viazzo and R. Schiestel. Simulation des grandes échelles turbulentes en canal à l'aide d'un schéma hermitien. *C.R.Acad.Sci.*, 321(IIb):225–232, 1995.

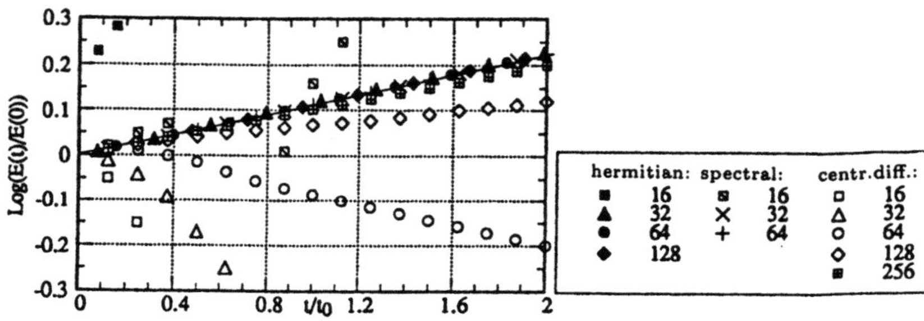

Figure 1: Energy growth rate of small perturbations in turbulent channel flow

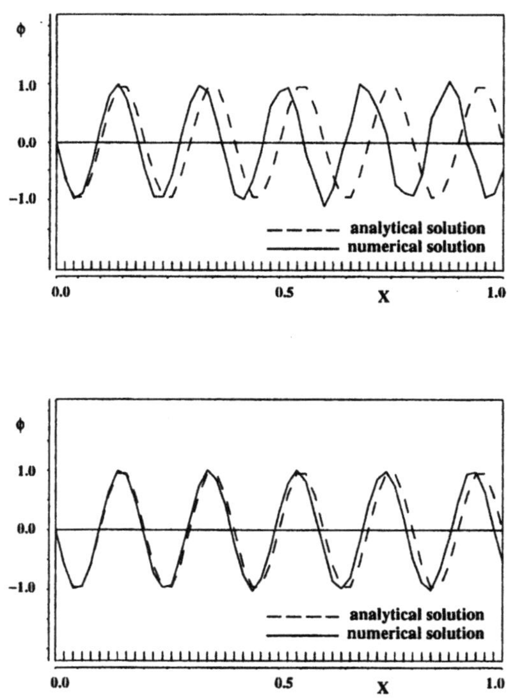

Figure 2: Numerical and analytical solution of the transport equation: $(\frac{\partial\phi}{\partial t} = -u_c\frac{\partial\phi}{\partial x})$
above: 2nd-order central differencing scheme,
below: 4th-order Hermitian scheme

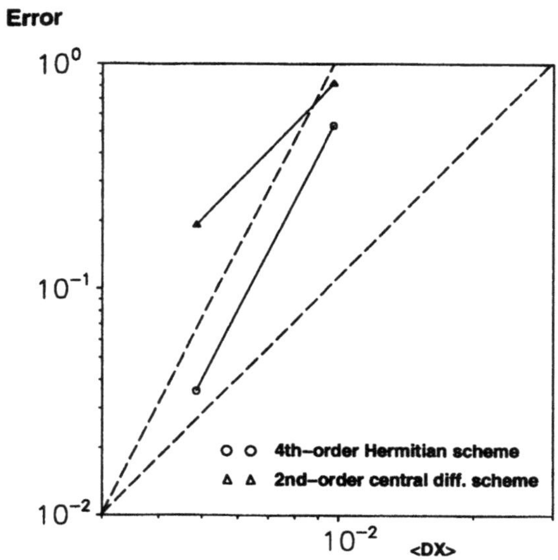

Figure 3: Comparison of the total amplitude error F_{tot} for a non-equidistant grid (periodic boundary condition) as function of the grid size ($< DX >= \bar{\Delta x} = \frac{LX}{NX}$)

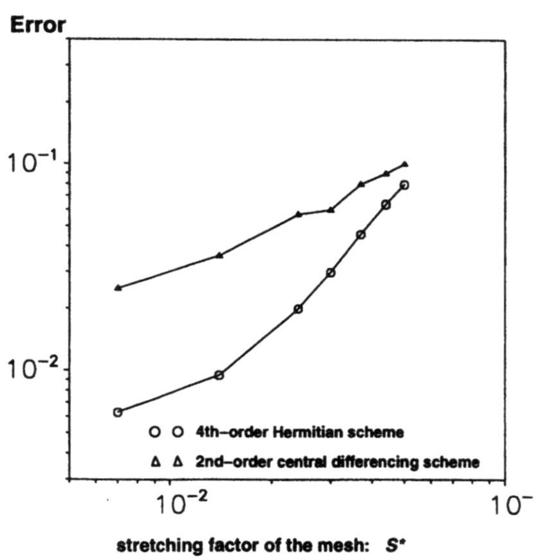

Figure 4: Comparison of the total amplitude error F_{tot} for a non-equidistant grid with stretching factor $S = \frac{\Delta x_i}{\Delta x_{i-1}}$ as a function of $S^* = S - 1.0$

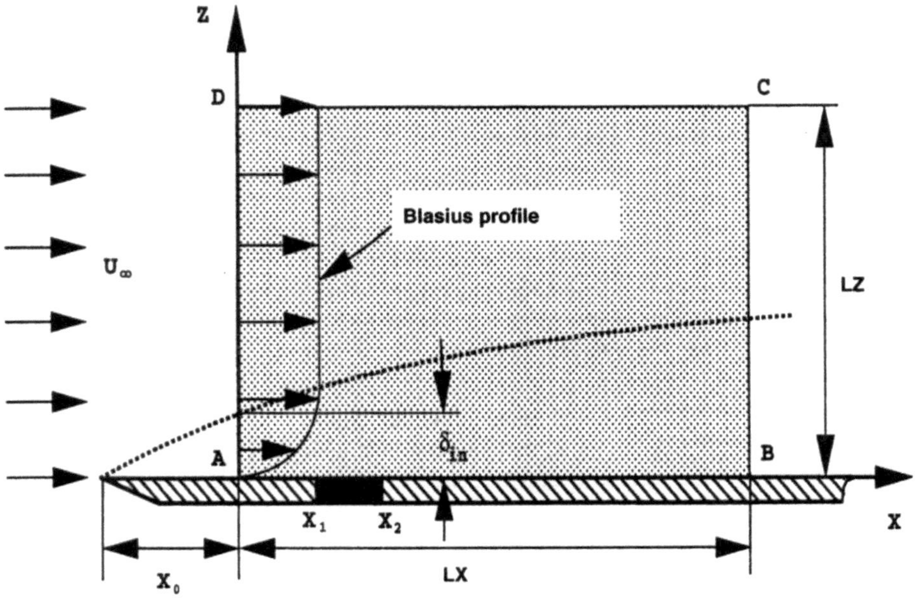

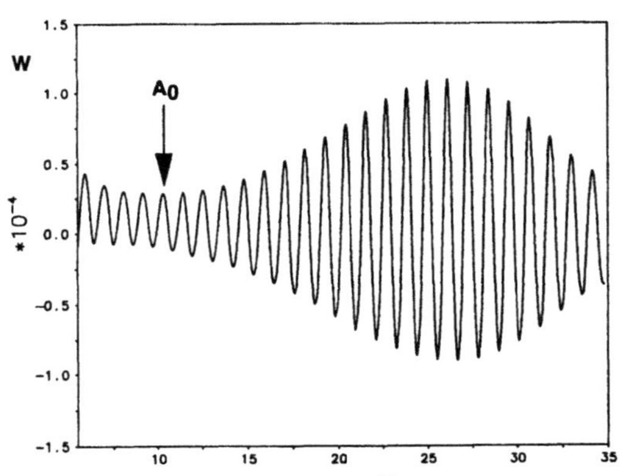

Figure 5: Manipulated 2D boundary layer:
above: sketch of the flow problem
below: vertical velocity component W downstream of the manipulation strip close to the bottom wall

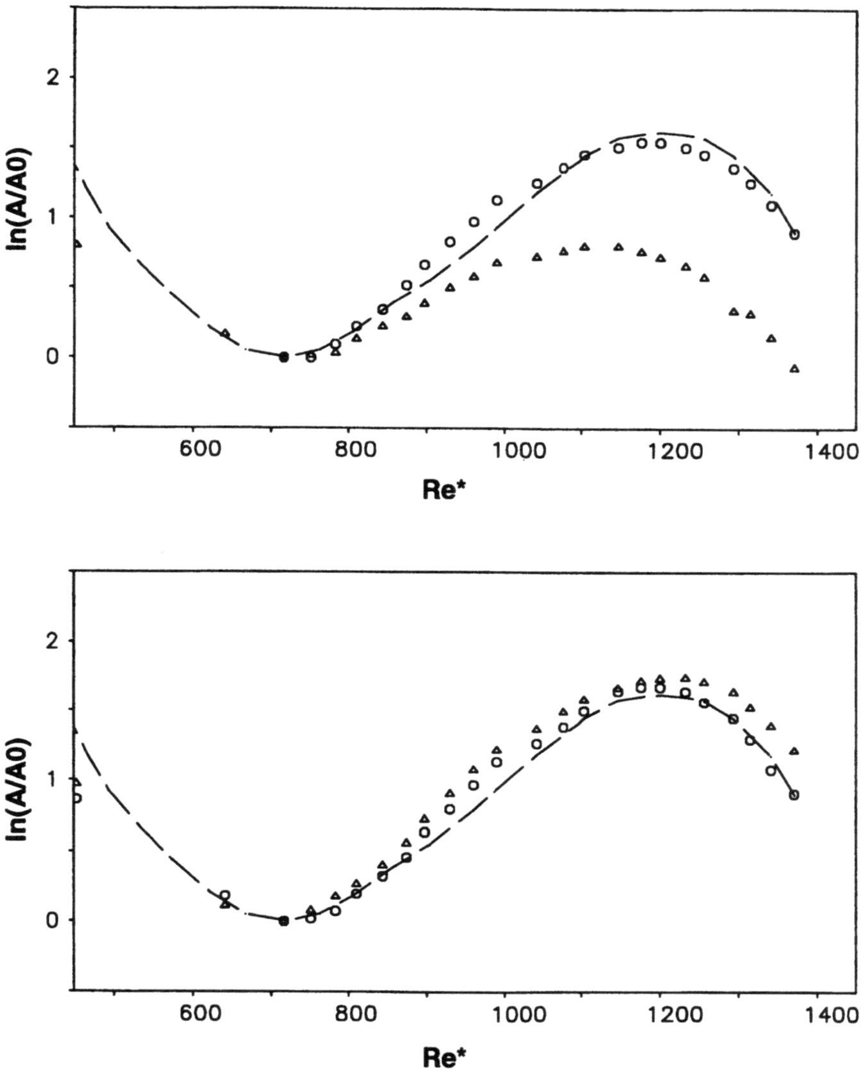

Figure 6: Comparison of the amplification rate $ln\frac{A}{A_0}$:
above: (694, 68) grid points, below: (1386, 136) grid points
circles: 4th-order Hermitian scheme
triangles: 2nd-order central differencing scheme
dashed line: result from linear stability theory

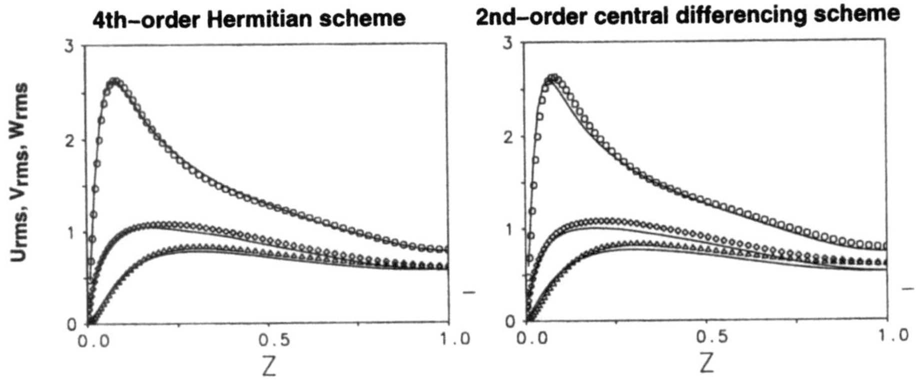

Kim, Moin & Moser (with 4 million grid points and spectral scheme, 1987):
o o o U_{rms}, ◇ ◇ ◇ V_{rms}, △ △ △ W_{rms}

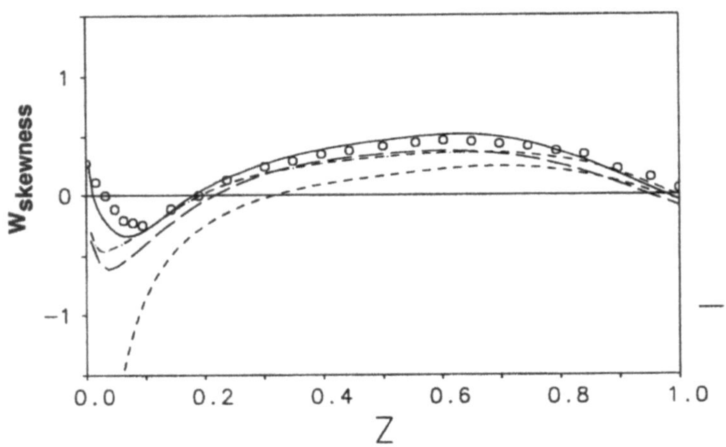

o o o	**Kim, Moin & Moser (1987)**
—·—·—·—	**2nd-order central differencing scheme (4.0 mill. grid points)**
————	**4th-order Hermitian scheme (4.0 mill. grid points)**
- - - - - - - - -	**2nd-order central differencing scheme (0.5 mill. grid points)**
— — — —	**4th-order Hermitian scheme (0.5 mill. grid points)**

Figure 7: Comparison between 2nd-order central differencing and 4th-order Hermitian scheme:
above: U_{rms}, V_{rms}, W_{rms} of the coarse grid (0.5 mill. grid points)
below: skewness of the W-fluctuations (fine and coarse grid)

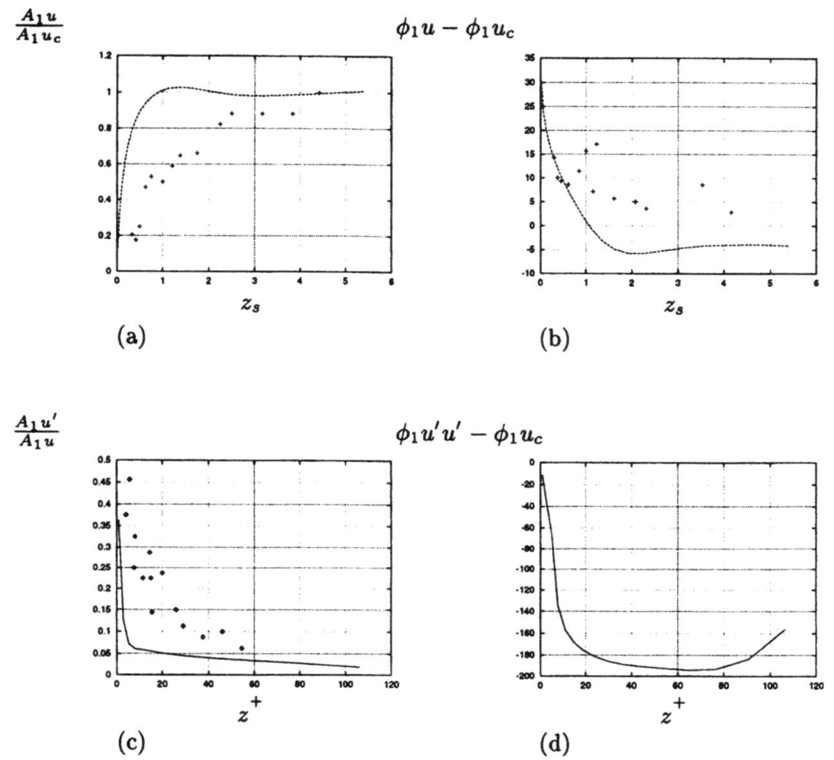

Figure 8: Turbulent channel flow forced by pressure gradient oscillations:
(a) relative amplitude of fundamental mode of the periodic velocity oscillations vs non-dimensional distance z_s from the wall.
(b) phase of fundamental mode of the periodic velocity oscillations vs non-dimensional distance z_s from the wall.
(c) relative amplitude of the turbulent intensity vs z^+.
(d) phase of fundamental mode of the turbulent intensity vs z^+.
symbols: Binder and Kueny [3]

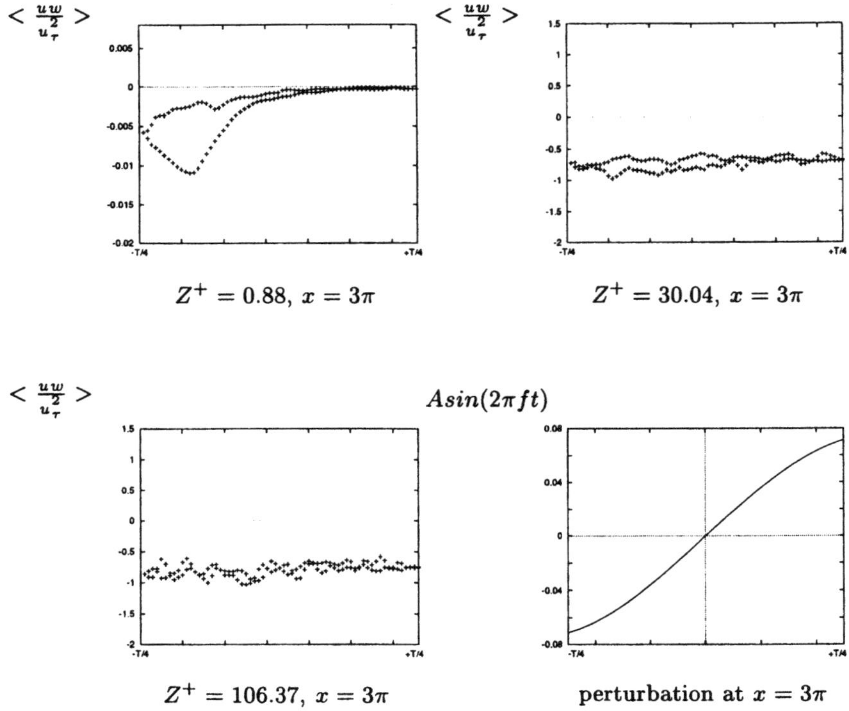

Figure 9a: Turbulent channel flow with imposed wall perturbation: hysteresis plots of the turbulent shear stress at $x = 3\pi$ at several distances z^+ from the lower wall.

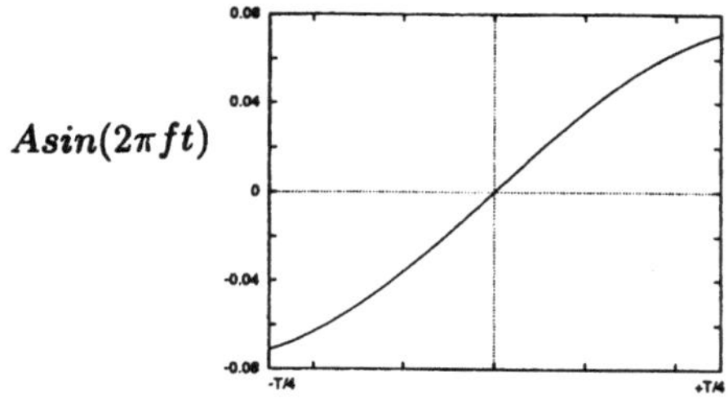

perturbation: $x = 3\pi$

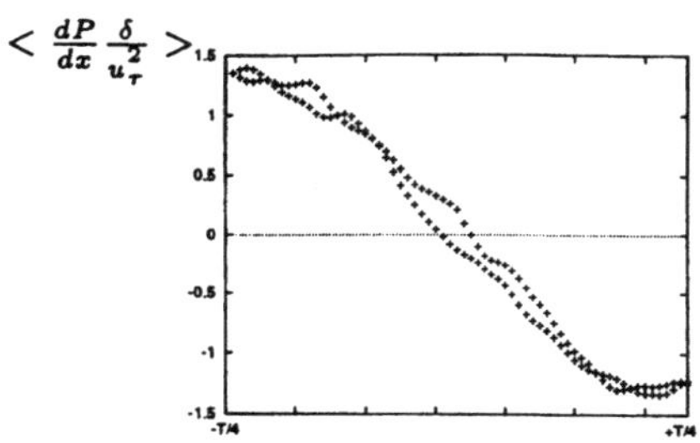

$Z^+ = 0.88,\ x = 3\pi$

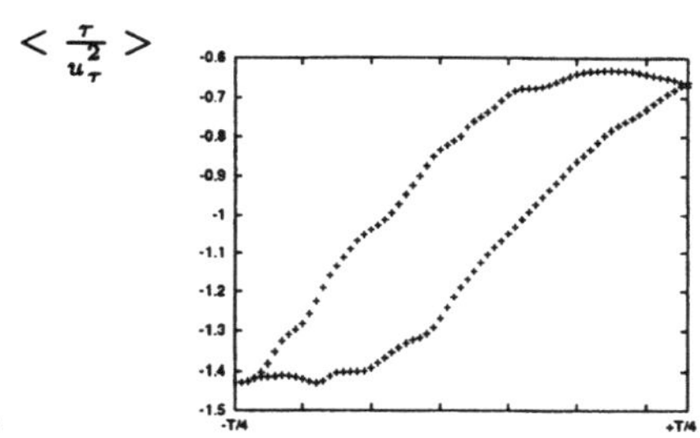

$Z^+ = 0.88,\ x = 3\pi$

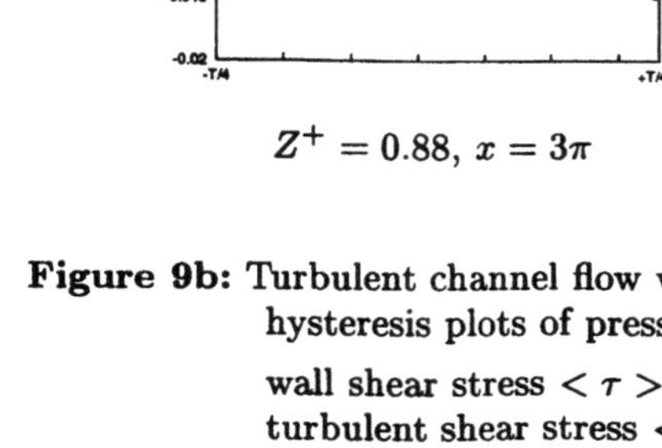

$Z^+ = 0.88,\ x = 3\pi$

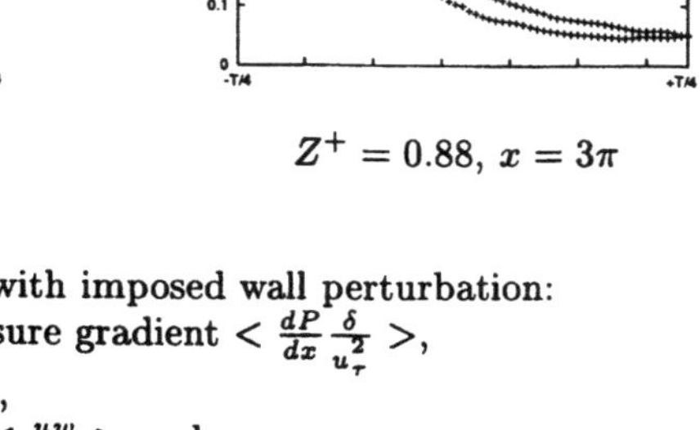

$Z^+ = 0.88,\ x = 3\pi$

Figure 9b: Turbulent channel flow with imposed wall perturbation:
hysteresis plots of pressure gradient $< \frac{dP}{dx} \frac{\delta}{u_\tau^2} >$,
wall shear stress $< \tau >$,
turbulent shear stress $< \frac{uw}{u_\tau^2} >$ and
turbulence kinetic energy $< \frac{k}{{u_\tau}^2} >$.

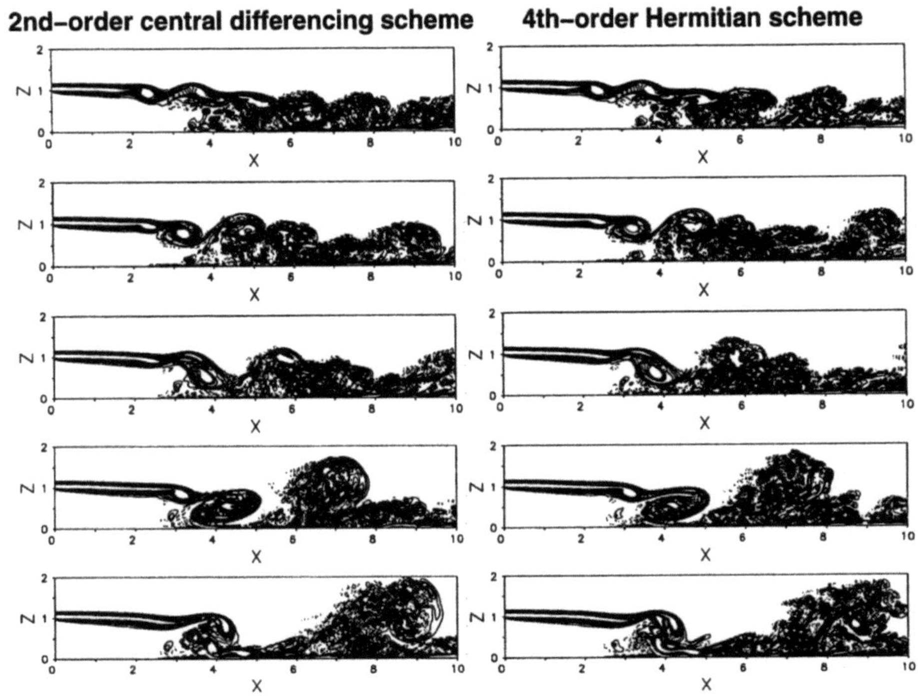

Figure 10: Lateral vorticity ω_y in flow over a backward-facing step:
left: 2nd-order central differencing scheme
right: 4th-order Hermitian scheme
from above to below: dimensionless time $T = 6.0, 8.0, 10.0, 12.0, 14.0$

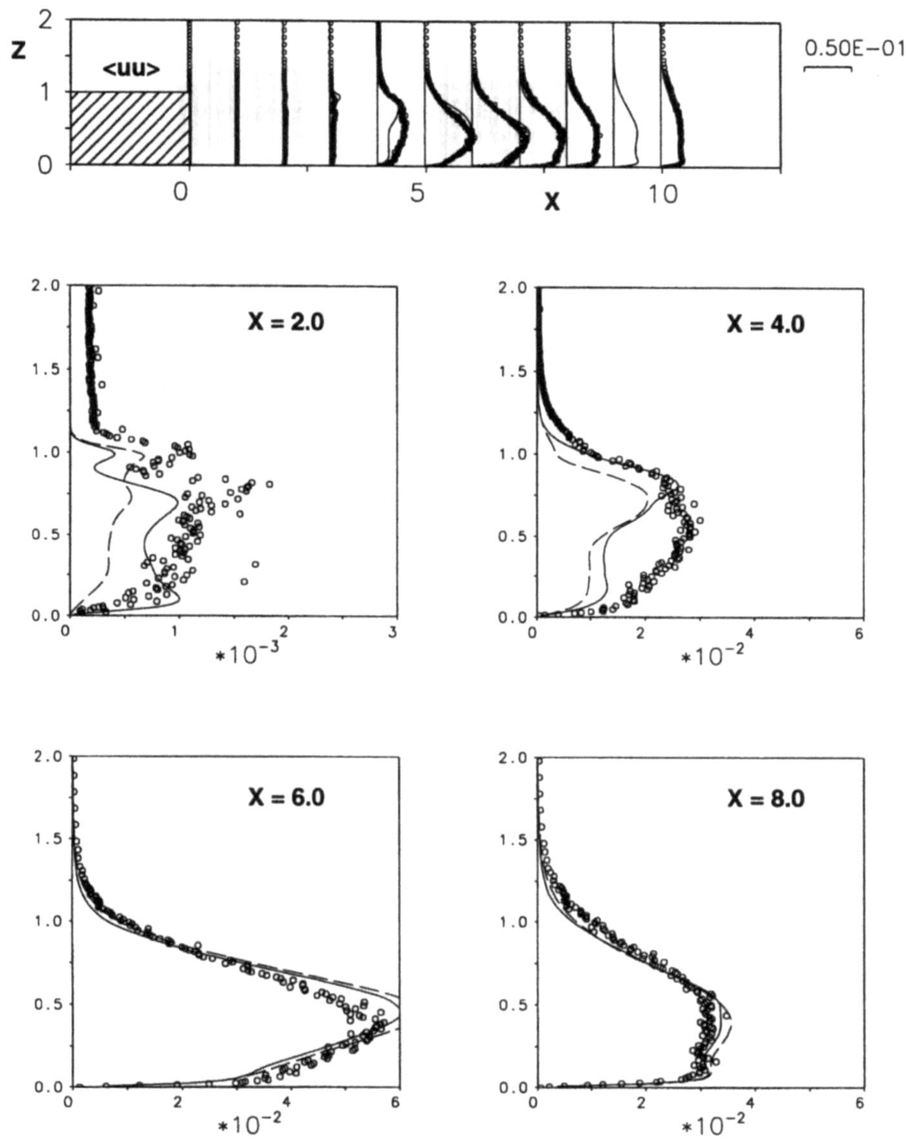

Figure 11: Vertical profiles of longitudinal normal stress $< uu >$
at different downstream positions $X = x/h$
broken line: 2nd-order central differencing scheme
full line: 4th-order Hermitian scheme
symbols: experimental data from Huppertz and Janke [4]

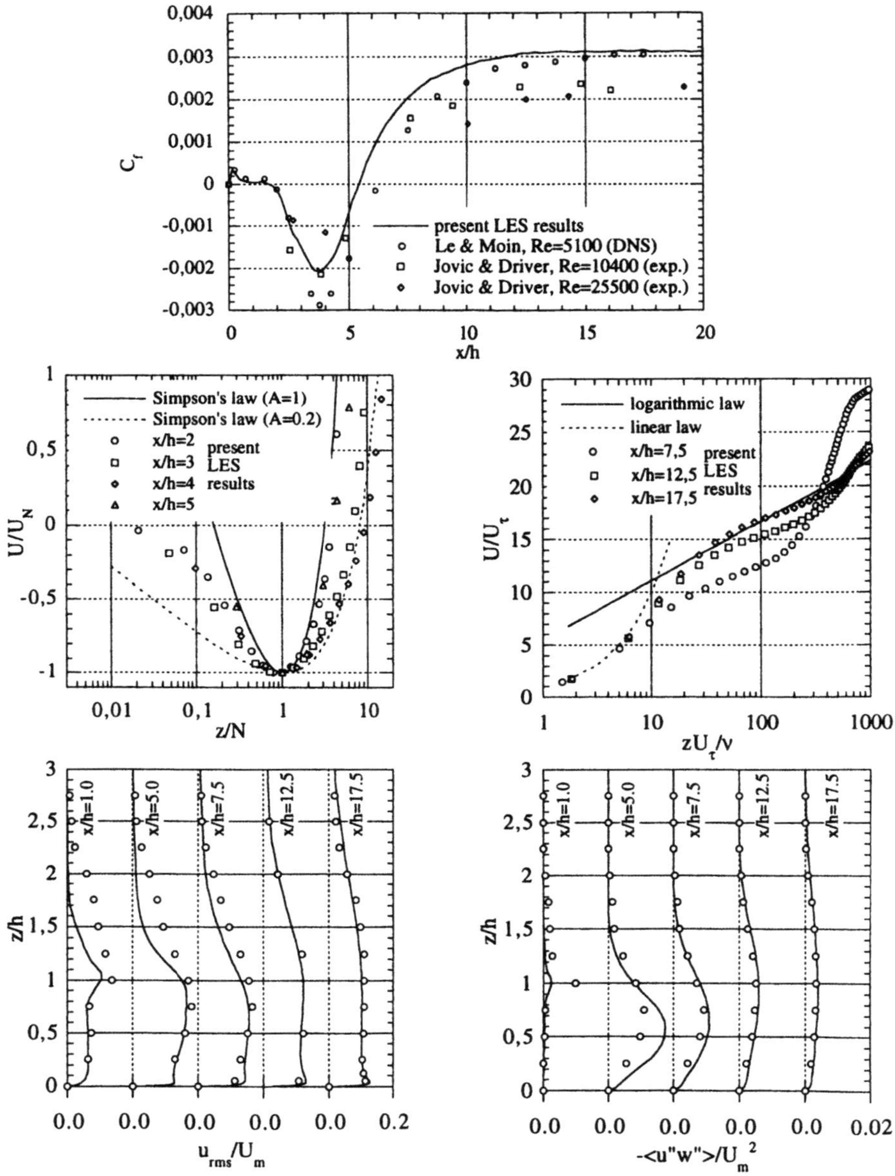

Figure 12: LES of turbulent backward-facing step flow at $Re_h = 13800$.

Computation of Complex Turbulent Flows

Schmid M.[1], Deng G. [2], Seidl V. [1], Visonneau M. [2], Perić M.[1]
[1] Technische Universität Hamburg-Harburg
Arbeitsbereich Fluiddynamik und Schiffstheorie
Lämmersieth 90, D-22305 Hamburg
[2] Ecole Centrale de Nantes
Laboratoire de Mécanique des Fluides UMR 6598 CNRS
1, Rue de la Noë, B.P. 92101 44321 Nantes Cedex 3 France

1 Summary

This paper presents the results of a joint effort by two research groups to improve the understanding and the predictability of complex turbulent flows. On one hand, the effort was made to reliably evaluate modelling errors of several popular low-Reynolds-number, two-equation turbulence models by making sure that all other errors in the computations were negligibly small. On the other hand, methods suitable for direct numerical simulation of turbulent flows in complex geometries have been thoroughly evaluated by performing a systematic grid and time step refinement, evaluating discretization errors, and comparing the results with established solutions of other groups for flows in simple geometries. Finally, some complex flows have been analysed using both direct numerical simulation and turbulence modelling approaches.

2 Introduction

The joint effort of the two research groups was concentrated on three major topics:

- evaluation of modelling errors for several turbulence models by applying them to a recirculating flow over a two-dimensional (2D) hill, making sure that all other errors are negligibly small, and comparing final solutions with a well-documented experimental data;
- evaluation of second-order implicit methods based on pressure-correction approach and colocated grids for their suitability of performing direct numerical simulation of turbulent flows in complex geometries;
- computation of some flows in complex geometries.

Since the two groups use different discretization methods embedded in different computer codes, the emphasis was on ensuring that in both cases the results do converge towards the same grid-independent solution when the same turbulence model is used. Although this may sound as a superfluous task, several recent international workshops showed that the difference between solutions obtained by different groups using nominally the same turbulence model were as large as the difference between solutions obtained by one group using different turbulence models. Indeed, each group found several unwanted reasons for differences in solutions obtained initially, so the final solutions – which were the same within narrow tolerance – can be regarded as numerically accurate.

For the discretization of the incompressible Navier-Stokes equations for Newtonian fluids both groups used a Finite-Volume-Method (FV). The integral form of the conservation equations for mass and momentum,

$$\frac{\partial}{\partial t}\int_V \rho \, \mathrm{d}V + \int_S \rho \boldsymbol{v} \cdot \boldsymbol{n} \, \mathrm{d}S = 0 \,, \tag{2.1}$$

$$\frac{\partial}{\partial t}\int_V \rho u_i \, \mathrm{d}V + \int_S \rho u_i \boldsymbol{v} \cdot \boldsymbol{n} \, \mathrm{d}S = \int_S \tau_{ij} \boldsymbol{i}_j \cdot \boldsymbol{n} \, \mathrm{d}S - \int_S p \boldsymbol{i}_i \cdot \boldsymbol{n} \, \mathrm{d}S \,, \tag{2.2}$$

serve as the starting point.

The method used in Hamburg can use unstructured grids with control volumes (CVs) of an arbitrary shape and is based on simple second-order integral approximations using midpoint rule, linear interpolation, and central differences. No coordinate transformation is used and the derivatives are always computed with respect to the Cartesian coordinates, which simplifies the implementation of turbulence models. The grid can be locally refined by subdividing one CV into several smaller ones, in which case the non-refined neighbour CVs become equivalent to polyhedra with more than six faces. Also, the grid may be made of blocks which do not match at interfaces. All variables are stored at CV-centers. Detailed description of the method can be found in Demirdžić *et al.* (1997).

In Nantes, block-structured grids and a partial coordinate transformation is used, with a more sophisticated discretization of momentum equations (consistent physical interpolation, CPI; see Deng *et al.*, 1994).

Although both methods are of FV-type, the differences in discretization approach – especially in computing convective fluxes through cell faces – are substantial. In both cases the momentum equations are linearized and solved sequentially using preconditioned conjugate-gradient type solvers. The pressure is obtained from a pressure-correction equation, which is derived from the discretized forms of the mass and momentum conservation equations. The pressure-correction equations and the solutions algorithms are also slightly different in the two methods, due to the differences in mass-flux approximations; see Demirdžić *et al.* (1997) and Deng *et al.* (1994) for a detailed comparison.

3 Evaluation of Turbulence Models

When evaluating turbulence models, one has to make sure that all other numerical errors (programming, iteration convergence and discretization errors) are substantially smaller than the modelling errors. We have chosen the 2D flow over a hill, which was studied experimentally by Almeida *et al.* (1993) and served as a benchmark test case for several workshops, to evaluate three low-Re, two-equation turbulence models. In order to systematically validate computer codes and exclude all errors other than modelling ones, the laminar flow was computed first using the same sequence of four systematically refined grids at both sites. The finest grid had 133 632 CVs, and the solutions obtained on it by both codes were almost indistinguishable. The separation point, the reattachment point, the velocity and pressure profiles – all quantities were found to converge towards the same values within the iteration convergence tolerance (0.1%). For some quantities, the two methods were found to be approaching the grid-independent solution from different sides, so on coarse grids larger differences can be obtained.

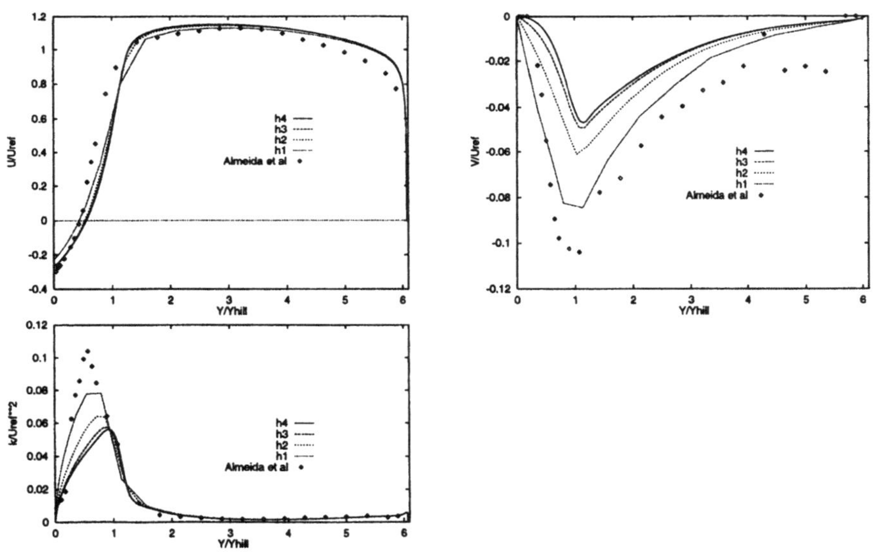

Fig. 1: Normalized profiles of the streamwise velocity (U), normal velocity (V), and turbulent kinetic energy (k) at x=0.90 m, as predicted by the Nantes group on four grids using the k-ω model of Wilcox (1988).

The models used belong to the family of two-equation models, based on the Boussinesq approximation and assuming isotropy. Two additional equations need to be solved: one for the kinetic energy k and one for the dissipation rate ε or ω. The equations differ slightly from the standard k-ε model and contain some additional damping functions which are necessary in the vicinity of solid walls to allow integration through the viscous sublayer. The eddy viscosity μ_t is added to the molecular viscosity and the sum is treated as an effective diffusion coefficient. The models tested are the low-Re k-ε model versions of Launder & Sharma (1974) and of Chien (1982), and the Wilcox notation of the k-ω model (1988).

The same sequence of four systematically refined grids was used as in the case of laminar flow; however, here are the differences between solutions on various grids substantially larger. As Fig. 1 shows, the numerical solutions are converging towards a grid-independent solution, the difference between solutions on the two finest grids being much smaller than the difference between the most accurate solution and experimental data. However, the solutions on the coarsest grid agree best with experimental data! This is an example of a situation in which the modelling and discretization errors can partially cancel out, leading to an impression that already on a coarse grid good agreement between computation and experiment is achieved.

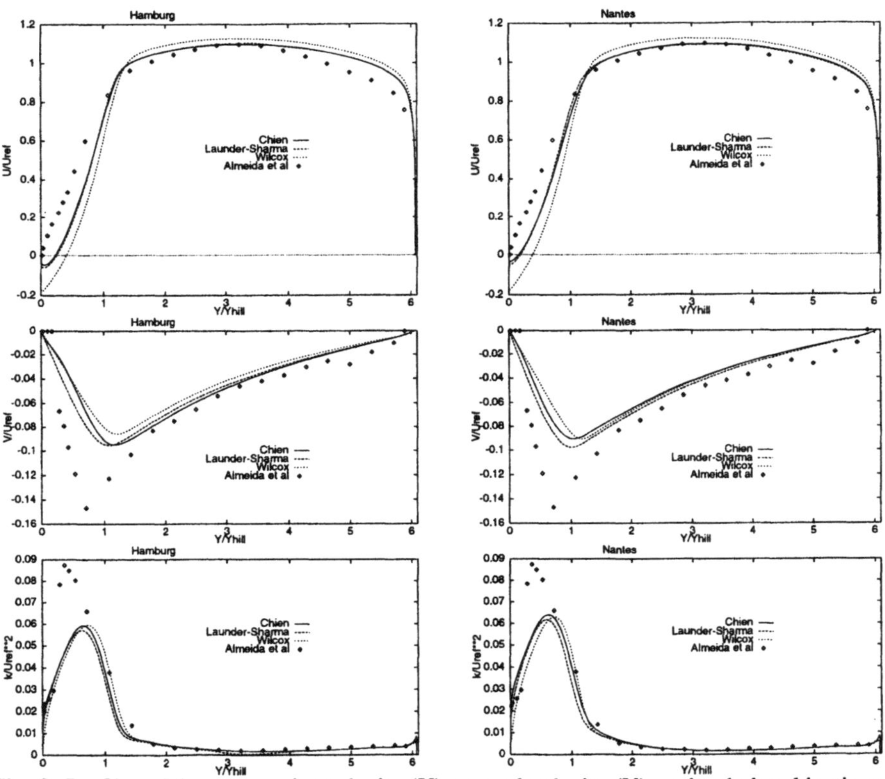

Fig. 2: Profiles of the streamwise velocity (U), normal velocity (V), and turbulent kinetic energy (K) at x=0.134 m, as predicted by the two groups on the finest grid using three different turbulence models.

A proper evaluation of turbulence models requires that the discretization and convergence errors (assuming that programming errors are absent) are demonstrated to be small enough. In this case the iterations were carried out until residual norms in all equations were reduced by four to five orders of magnitude, ensuring four-digit accuracy of iterative solutions on each grid. Comparison of solutions on various grids shows that the difference between the grid $h4$ (finest, 133 632 CV) and grid $h3$ is roughly quarter of the difference between solutions on grids $h3$ and $h2$, as expected of a second-order method. The Richardson extrapolation estimates the discretization error on the finest grid to be of the order of one third of the difference between solutions on the two finest grids, which is of the order of 0.1% of the normalized variable.

Only numerically accurate solutions should be compared with experimental data to evaluate the modelling errors. Figure 2 shows such comparison at one cross-section. None of the models describes the profiles of velocity components or the kinetic turbulent energy quantitatively correctly; in particular, the length of the recirculation zone is predicted too large, and the level of turbulent kinetic energy is too low.

The extent of modelling errors for each quantity can now be evaluated by assuming that the measurement errors are also substantially smaller than the modelling errors (which is not true for each quantity). The turbulence models involve empirical and/or computer-optimized constants which need to be tuned by comparing predictions with experimental data. They can always be tuned to fit data for one particular case, but the idea is to obtain a set of parameters which will be applicable at least to a class of turbulent flows (there is little hope for finding a universal turbulence model for all flow types). Most of the model constants were obtained several years ago and are based on both numerical solutions on coarser grids and experimental data from simpler flows. As the above example shows, tuning of model parameters by using numerical solutions from coarse grids can lead to modelling not only the turbulence, but also the discretization errors (cf. Fig. 1)! It is perhaps appropriate to suggest that a major effort needs to be invested to re-tune the model parameters by using the powers of parallel computers to obtain numerically accurate solutions for many turbulent flows, involving some more complex ones like the flow from this example, and many plausible parameter settings, in order to choose the optimum ones.

4 Direct Numerical Simulation of Turbulence

Direct numerical simulation (DNS) of turbulent flows in simple geometries is usually performed using specialized high-order methods (spectral methods, compact finite differences). However, when flows in complex geometries are considered, such methods can not be used. Choi *et al.* (1994) used a second-order fractional-step method based on a staggered variable arrangement to perform a DNS of flow over riblets and showed that this approach was meaningful. However, recently several research groups expressed concern that the non-staggered (colocated) arrangement of variables – which is preferred when flows in complex geometries are computed – is not suitable for DNS, due to the fact that additional damping terms are introduced in the discretized continuity equation to prevent pressure-velocity decoupling.

It was therefore the aim of this study to perform – in parallel to a DNS of flow around sphere as a generic example of complex turbulent flows – also a detailed analysis of performance of a second-order pressure-correction scheme based on colocated variable arrangement in a DNS of a turbulent channel flow. For this flow there are benchmark DNS data provided by Kim *et al.* (1987), which is believed to be very accurate. In addition to a systematic grid and time step refinement, this data is used to assess the quality of prediction of the colocated pressure-correction scheme.

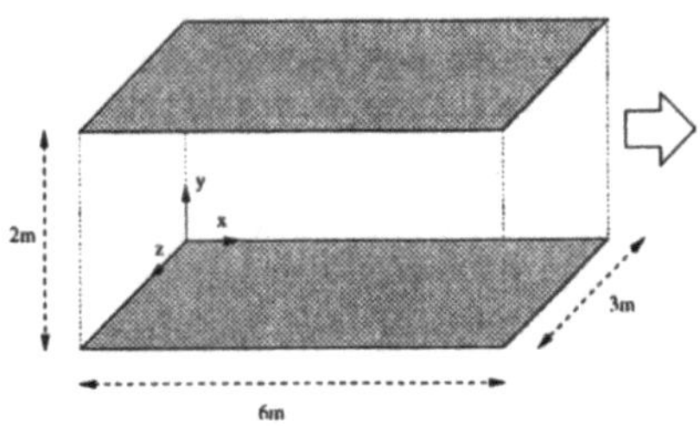

Fig. 3: Channel configuration.

The configuration of the plane channel flow is shown in Fig. 3. Solid walls (with a no-slip condition) bound the flow in one direction, whereas the other two are homogeneous with periodic boundary conditions. In the streamwise direction a pressure drop was imposed according to $\mathrm{Re}_\tau = 180$. The Reynolds number based on the mean velocity and channel height is Re= 5 750.

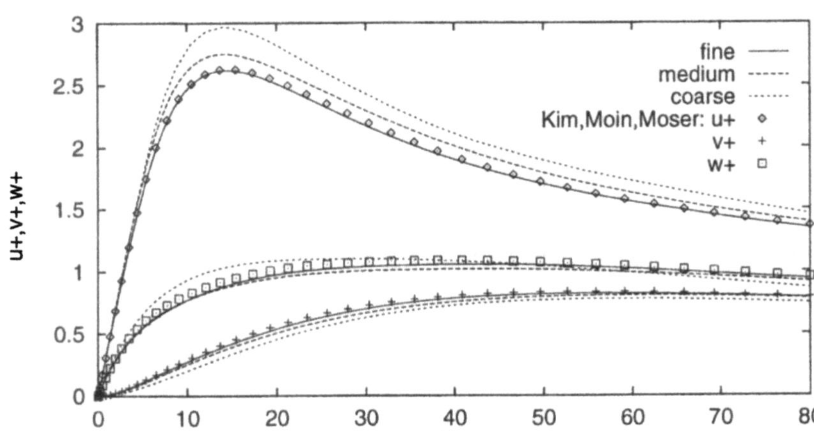

Fig. 4: Comparison of normalized Reynolds stress $\sqrt{\overline{u_i u_i}/u_\tau^2}$ with reference data of Kim *et al.* (1987).

The simulation was first performed on a coarse grid with 42×68×28 CV in x,y, and z-direction, respectively, using different variable arrangements: colocated and staggered. Although the same grid and time step size was used for both computations, there are two major differences which can lead to differently large numerical errors:

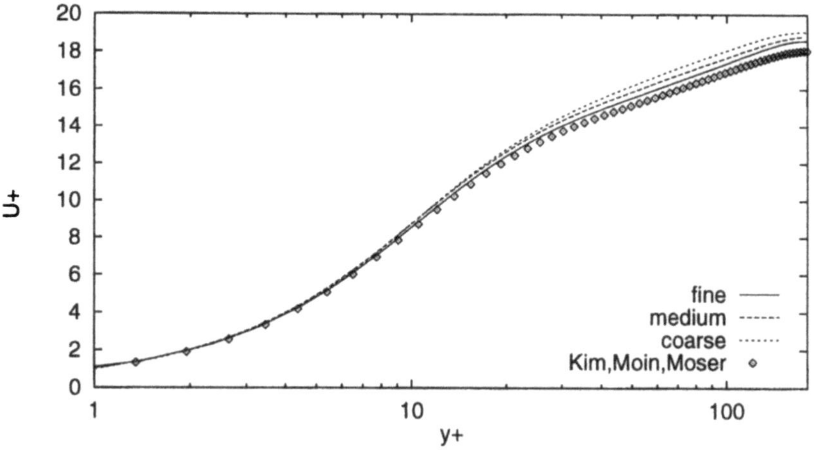

Fig. 5: Comparison of mean streamwise velocity profile with reference data of Kim *et al.* (1987).

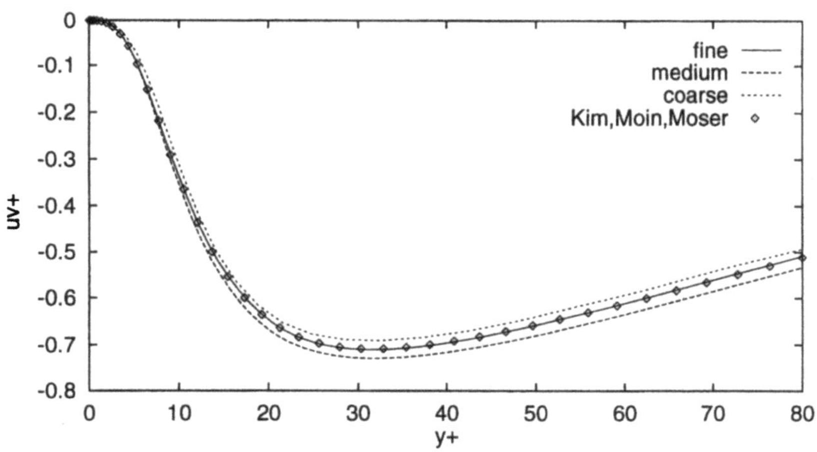

Fig. 6: Comparison of normalized Reynolds shear stress $\overline{uv}/u_\tau$ with reference data of Kim *et al.* (1987).

one is the variable arrangement, and the other is the time integration scheme. In the case of a colocated arrangement, cell-face velocities used to compute mass fluxes are obtained by linearly interpolating cell-center velocities and adding a correction term which is proportional to the product of a discretized third derivative of pressure (using central differences) and the square of mesh spacing (see Ferziger and Perić, 1996, for more details). In the staggered version, no interpolation is involved, as all variables are stored where needed for the second-order approximations of surface integrals used here. Thus, the colocated version involves additional interpolations of both velocities and pressure, plus the damping term mentioned above which should smooth out pos-

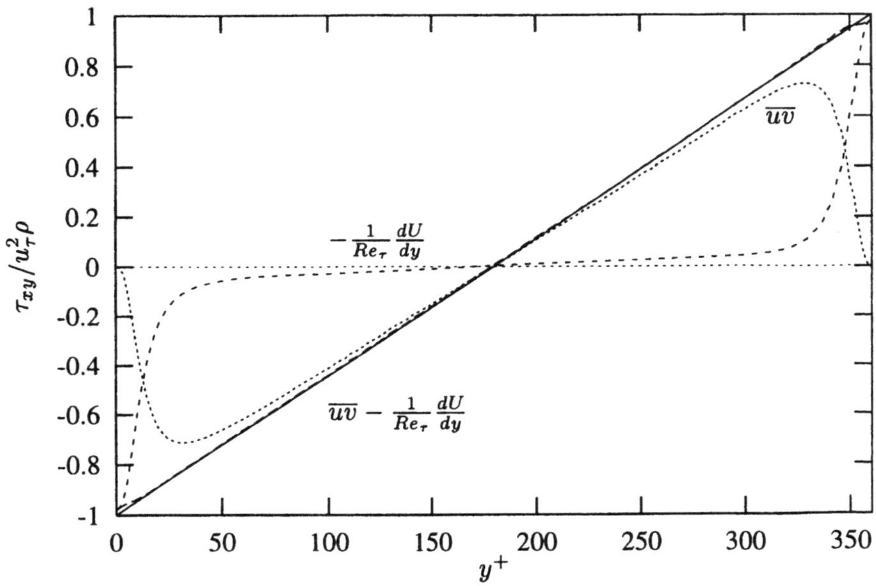

Fig. 7: The normalized shear stress over the channel width.

sible oscillations due to decoupling of pressure and velocities. For the staggered version, Crank-Nicolson scheme was used for time integration, while in the colocated version, three-time-levels implicit scheme was used. Both schemes are of second order, but the latter has a larger truncation error (by a factor of four). Also, the colocated version needs pressure at wall boundaries (which is obtained by linear extrapolation from near-wall nodes), while the staggered version does not.

The results obtained in Nantes on staggered grid compared more favourably with the reference DNS data than those obtained on a colocated grid in Hamburg. The energy spectra for the results obtained on a colocated grid (cf. Fig. 12) reveals poor spatial resolution especially in the spanwise direction of the flow. This is probably due to the fact that the additional discretization errors in the colocated version do not cancel out; on the contrary, the damping term used to compute the cell-face velocities seems to be smoothing out also the turbulent pressure fluctuations.

Since the first grid is very coarse, the results can not be expected to be accurate anyway (except if the discretization errors coincidentally act in the staggered version as a kind of a subgrid-scale turbulence model, thus leading to an unexpectedly good agreement of results with reference DNS data). Computations on finer grids should show whether the errors are diminishing with grid refinement and how fine the grid needs to be in order to achieve results of sufficient accuracy using colocated arrangement.

A second grid is created by a systematic refinement of the first one and contains thus $6.4 \cdot 10^5$ CVs ($84 \times 136 \times 56$). The results are improved, as can be seen in Figs. 4-6. Further grid refinement is done systematically only in the spanwise direction, since it was felt that the flow was already well resolved in the other two directions (there were already 16 CVs in the viscous sublayer, $y^+ \leq 5$, which is more than sufficient (Unger, 1994). The finest grid had $2.1 \cdot 10^6$ CVs ($120 \times 160 \times 112$). The computations were done using 32 nodes on a Cray T3E computer in Stuttgart (HLRS) for the second grid level and 32 nodes on a Cray T3E computer in Berlin (ZIB). For the finest grid the computation took 450 hours with 67 200 CV per node, and the averaging time was approximately 5600 ν/u_τ^2 or 30 non-dimensional time units (tu_τ/δ). The grid is equidistant in the homogeneous directions and slightly non-uniform normal to walls.

As Figs. 4 to 6 show, the solution obtained on the finest grid (the time step was the same for all grids) agrees quite well with the reference DNS data of Kim *et al.* (1987), which was obtained on a still finer grid and using a spectral method of high accuracy. The total shear stress (which is the sum of the Reynolds stress $\overline{uv}$ and the viscous stress $\mu \partial U/\partial y$), shown across the channel in Fig. 7, deviates only near the wall from the expected linear profile. The reason may lye in the fact that more samples are needed to obtain good mean values in the near-wall region, where low-frequency oscillations may not have been filtered out by the averaging.

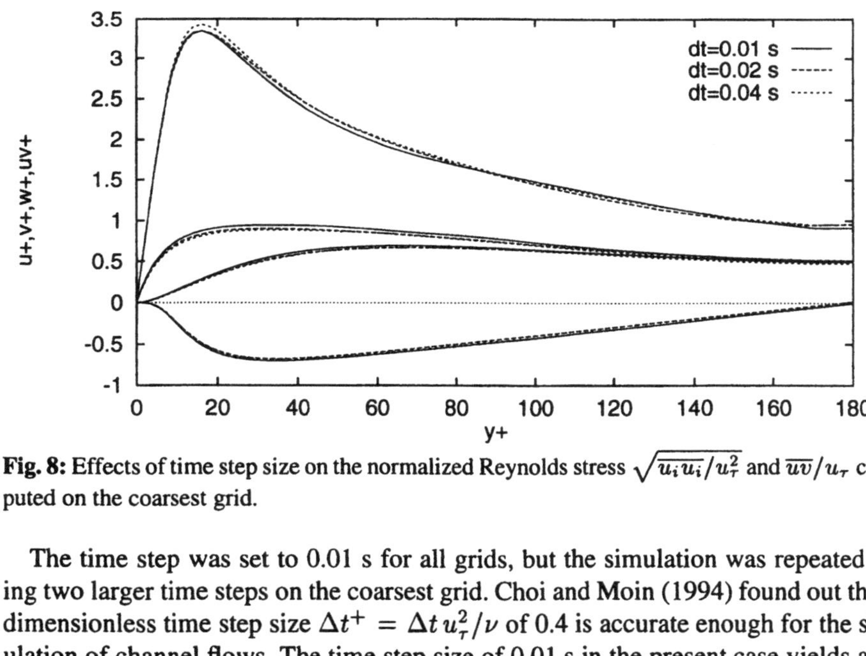

Fig. 8: Effects of time step size on the normalized Reynolds stress $\sqrt{\overline{u_i u_i}/u_\tau^2}$ and $\overline{uv}/u_\tau$ computed on the coarsest grid.

The time step was set to 0.01 s for all grids, but the simulation was repeated using two larger time steps on the coarsest grid. Choi and Moin (1994) found out that a dimensionless time step size $\Delta t^+ = \Delta t\, u_\tau^2/\nu$ of 0.4 is accurate enough for the simulation of channel flows. The time step size of 0.01 s in the present case yields a dimensionless time step of 0.108. The Kolmogorov time scale $\tau^+ = (u_\tau^4/\varepsilon\nu)^{1/2}$ equals 2.4 in the sublayer of a channel flow. A dimensionless time step of 0.4 is thus about six times smaller than the time scale which is to be resolved, while the time step used here is even smaller. Therefore, the simulation was repeated with time steps of 0.02 and 0.04 s, leading to dimensionless values of 0.216 and 0.432.

For each simulation the computation was performed over 20 000 time steps, and

4 000 time samples were taken. Additionally, the averaging was performed in the two homogeneous directions as well. The Reynolds stresses in Fig. 8 show that the time step size influences the results much less than the spatial resolution (see Fig. 4). Also, if more samples were taken, the differences would have become even smaller. Thus, present findings correspond to those of Choi and Moin (1994), who used a different solution method (staggered grid, fractional step method with second-order discretization, and the Crank-Nicolson time integration scheme).

The convergence of outer iterations for large time steps is worse than for the smaller ones, as expected. In order to reduce all residual norms within each time step by at least three orders of magnitude, eight, ten, and sixteen outer iterations per time step were necessary for the three time step sizes, respectively. The calculation time per time step increases thus with time step size, but the integration interval is larger. It is expected that the iteration convergence errors were much smaller than the discretization errors; optimization of computational efficiency would require adjustments of the convergence criterion to reflect the larger temporal discretization error when the time step is enlarged.

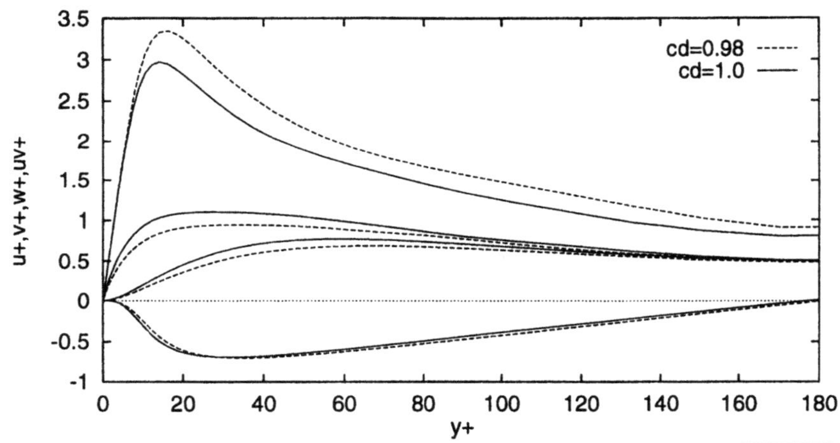

Fig. 9: Effects of the numerical diffusion on the normalized Reynolds stress $\sqrt{\overline{u_i u_i}/u_\tau^2}$ and $\overline{uv}/u_\tau$ computed on coarse grid.

For the evaluation of viscous terms, central differences are widely used (also in this study); they lead to a satisfactory accuracy, second-order convergence, and create no numerical problems. For the convective terms, linear interpolation to cell-face centers corresponds to central differences applied to first derivatives in finite difference methods; it is the simplest approach with second-order accuracy, but when the grid is coarse and Reynolds number is high, it may lead to artificial oscillations in the solution. Many authors use therefore upwind-biased interpolation of higher order, but such an approach is not so simple on unstructured grids with cell-wise local refinement. One possibility to get rid of oscillations – which is often practiced when solving the Reynolds-averaged Navier-Stokes equations – is to blend the linearly interpolated

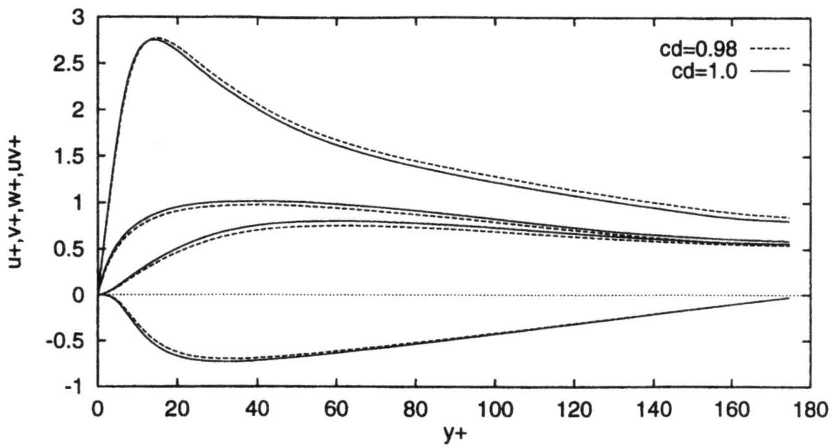

Fig. 10: Effects of the numerical diffusion on the normalized Reynolds stress $\sqrt{\overline{u_i u_i}/u_\tau^2}$ and $\overline{uv}/u_\tau$ computed on medium grid.

cell-face value with the upwind nodal value; usually, only a small fraction of the upwind value is used. The upwind-value brings in a first-order error term and with it the numerical diffusion, which smoothes out the oscillations.

Although it is clear that on coarse grids oscillations do occur and the solution is inaccurate anyway (so why bother about oscillations?), we wanted to investigate what effect the numerical diffusion arising from first-order error terms has on DNS. Such terms may arise from approximations of viscous fluxes as well when arbitrary, non-smooth grids are used in complex geometries (ensuring second or higher order of all approximations under all grid conditions may be difficult for arbitrary geometries). We therefore repeated simulations on the coarse and medium grid by replacing only 2% of the linearly-interpolated cell-face value in convective fluxes by the upwind value. As Figs. 9 and 10 show, there is a huge effect on the coarsest grid, but a rather moderate effect on the medium grid. From this tendency one can conclude that on the finest grid there would be no appreciable effect. Of course, if more than 2% of upwind value were used, the effect would be larger even on fine grids; it is thus essential to use approximations of consistently second or higher order for all terms when DNS is performed, but a small amount of first-order approximations may be tolerable if the grid is fine enough.

All the terms in the equations for non-vanishing Reynolds stresses were also calculated. The transport equations for Reynolds stresses can be written in differential form as:

$$\underbrace{\frac{\partial}{\partial t}\overline{u_i u_j}}_{L_{ij}} + \underbrace{\frac{\partial}{\partial x_k}(U_k \overline{u_i u_j})}_{C_{ij}} = \underbrace{\frac{\partial}{\partial x_k}\overline{u_k u_i u_j}}_{D^t_{ij}} \underbrace{-\overline{u_i u_k}\frac{\partial U_j}{\partial x_k} - \overline{u_j u_k}\frac{\partial U_i}{\partial x_k}}_{P_{ij}}$$

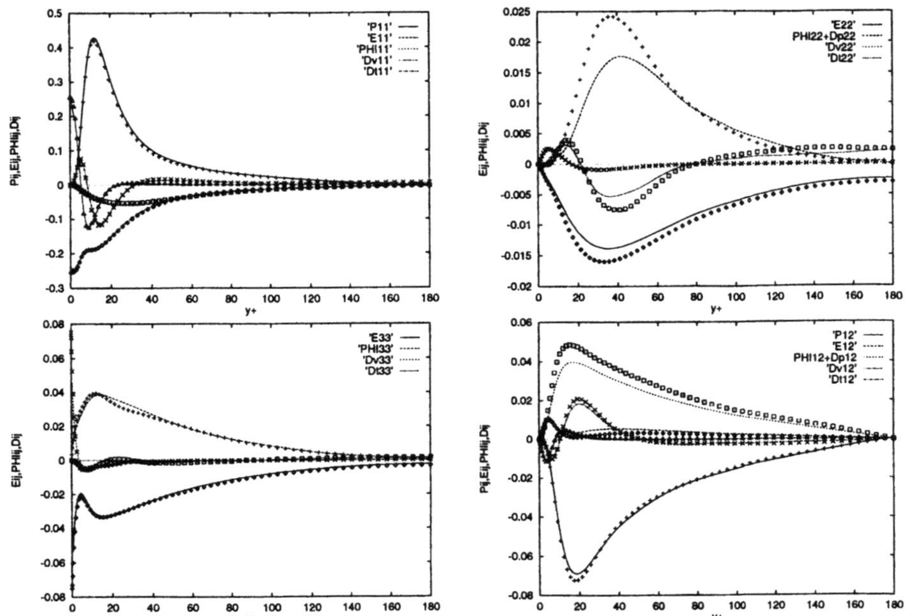

Fig. 11: Balance of the terms in the transport equations for $\overline{uu}$ (top left), $\overline{vv}$ (top right), $\overline{ww}$ (bottom left) and $\overline{uv}$ (bottom right) normalized with u_τ^4/ν, for the finest grid compared with reference data of Mansour *et al.* (1988).

$$+\underbrace{\frac{1}{\rho}\frac{\partial}{\partial x_k}(\overline{pu_j}\delta_{ik}+\overline{pu_i}\delta_{jk})}_{D^p_{ij}}+\underbrace{\frac{\partial}{\partial x_k}\left(\nu\frac{\partial}{\partial x_k}\overline{u_iu_j}\right)}_{D^\nu_{ij}}$$

$$+\underbrace{\overline{\frac{p}{\rho}\left(\frac{\partial u_i}{\partial x_j}+\frac{\partial u_j}{\partial x_i}\right)}}_{\phi_{ij}}\underbrace{-2\nu\overline{\frac{\partial u_i}{\partial x_k}\frac{\partial u_j}{\partial x_k}}}_{\varepsilon_{ij}}. \tag{4.3}$$

For terms involving only velocity correlations, the two homogeneous directions (x,z) can also be used for averaging. The terms involving pressure-correlations can be averaged only in one homogeneous direction (z), but this was not done. The streamwise pressure gradient was prescribed, and the sum of pressure and $\rho\overline{vv}$ must be constant in wall-normal direction, which was satisfied in the present simulation. Also, computation of the higher-order correlations in the above equations introduces additional numerical errors.

As Fig. 11 shows, the profiles agree qualitatively well with the reference data, cf. Mansour *et al.* (1988). Only the terms in the equations for $\overline{vv}$ and $\overline{uv}$ – especially those involving pressure correlations – show appreciable differences. However, all the terms in the equation for $\overline{vv}$ are about 20 times smaller than those in the equation for $\overline{uu}$; smaller terms are more affected by the interpolation and numerical differentiation errors introduced during post-processing than the larger ones. In the light of the coarser grid and a less accurate numerical scheme used for the present simulations compared to those used for obtaining the reference DNS, the agreement can be viewed as quite satisfactory. More samples and a finer grid would have certainly lead to a perfect agreement with the reference data.

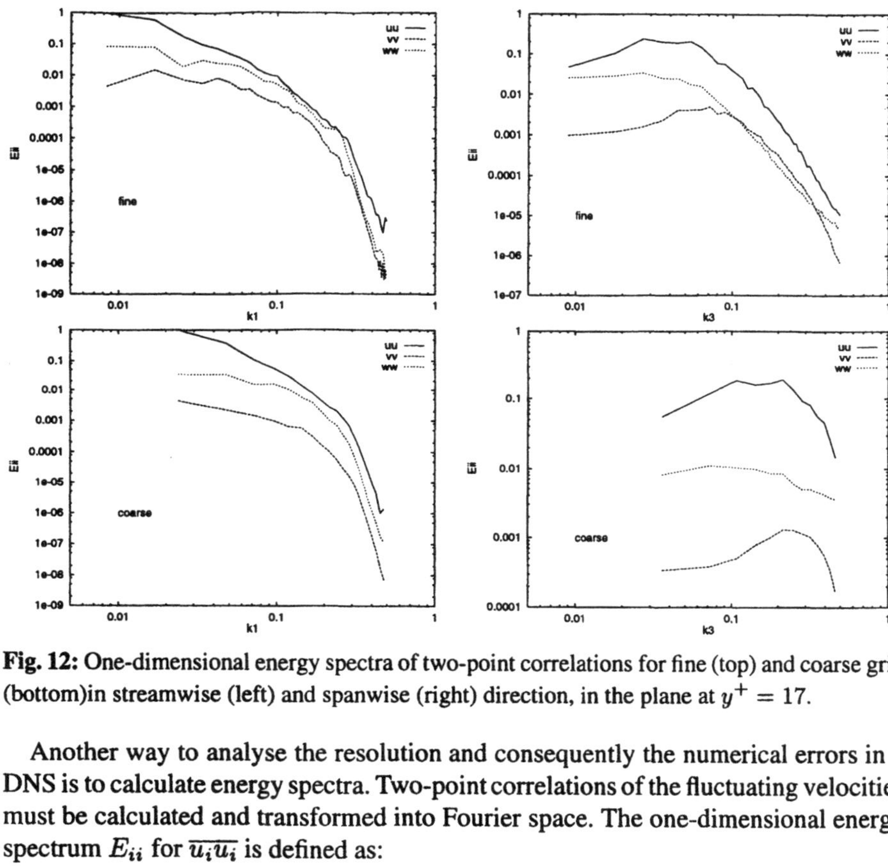

Fig. 12: One-dimensional energy spectra of two-point correlations for fine (top) and coarse grid (bottom)in streamwise (left) and spanwise (right) direction, in the plane at $y^+ = 17$.

Another way to analyse the resolution and consequently the numerical errors in a DNS is to calculate energy spectra. Two-point correlations of the fluctuating velocities must be calculated and transformed into Fourier space. The one-dimensional energy spectrum E_{ii} for $\overline{u_i u_i}$ is defined as:

$$E_{ii}(\boldsymbol{x},k,t) = \frac{1}{2\pi} \int_{-\infty}^{+\infty} \overline{u_i(\boldsymbol{x},t) u_i(\boldsymbol{x}+r,t)}\; \mathrm{e}^{-\mathrm{i}kr} \mathrm{d}r \;, \qquad (4.4)$$

where r is the distance between the two points and k are the frequencies of the spectrum for one direction. During the simulation, the fluctuating velocities were written to files, and in a post-processing session E_{ii} was computed as a function of k for midlines in streamwise and spanwise direction in several planes (in viscous sublayer, log-layer, and transition layer). Figure 12 shows such spectra for $\overline{uu}$, $\overline{vv}$, and $\overline{ww}$ in the plane at $y^+ = 17$ for the fine and the coarse grid. When the grid is fine enough, the energy is properly dissipated down to the smallest scales as k increases; insufficient grid resolution is reflected by an increase in E_{ii} at high values of k. In the present case, only the energy spectrum in streamwise direction does show a small increase at the curve end, which is most probably due to the fact that the averaging period for the two-point velocity correlation was too small (less than ten percent of the whole simulation time). On a coarse grid where more samples have been taken, as it is not as expensive as on the fine grid, almost no increase of energy is visible at higher frequencies. Because of the poor grid solution the spanwise energy spectrum drops only one decimal power for the coarse grid.

This exercise shows that the pressure-correction method based on a colocated variable arrangement and a second-order discretization in space and time is suitable for DNS. The same method was applied in Hamburg to perform a DNS of flow around a sphere attached to a stick; computations of this and some other complex flows are presented in the next section.

5 Computation of Complex Flows

Bodies encountered in aero- and hydrodynamics are usually streamlined, but nevertheless they usually involve flow separation from smooth surfaces, strong secondary flows, and a complex wake structure. All three features are difficult to predict correctly using Reynolds-averaged Navier-Stokes equations and turbulence models; this is why such flows were selected as another topic of the collaborative research of the Hamburg and Nantes group. Several typical flow configurations have or will be studied; some results of simulations performed so far are presented below.

The sphere represents a geometrically simple bluff body, but the flow is not as simple as one might expect. Laminar flows at low Reynolds numbers (less than 1 000) have been extensively studied in the past, but there are not many detailed studies of flows at high Reynolds numbers. It was therefore decided to analyse in more detail turbulent flow around sphere at various Reynolds numbers. The selected geometrical configuration is shown in Fig. 12.

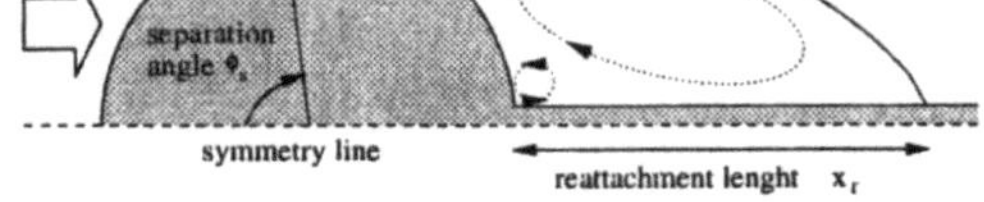

Fig. 13: Sphere flow configuration.

The sphere is held by a backward stick of diameter $d = 0.1\,D$, where D is the sphere diameter. The stick is necessary for *future* experimental investigations in order to enable more reliable estimation of modelling errors at high Reynolds numbers.

The solution domain extends 10 diameters ahead and aside of sphere, and about 15 diameters behind it. Undisturbed flow is assumed at the inlet boundary. The unstructured grid made of hexahedra is coarse far from the sphere and is cell-wise locally refined in regions of large variable changes (near the wall, in the free shear layer, and in the wake). This approach saves many grid nodes and computation time, without sacrificing the accuracy.

Direct numerical simulation of the flow at Re = 5 000 has been performed by Seidl (1997) using the same numerical scheme as described above for the channel flow. The finest grid had $6.3 \cdot 10^6$ CV; 128 nodes of Cray T3D parallel computer in Berlin (ZIB) were used in the simulation.

At the stagnation point on the sphere front, the static pressure equals the stagnation pressure $p_{st} = \frac{1}{2}\rho U^2$, where U is the undisturbed flow velocity far from sphere. As the flow is accelerated, the pressure drops to a minimum somewhat ahead of equator and then starts rising again. The adverse pressure gradient on the sphere surface forces the laminar boundary layer to separate close to equator. In the free shear layer, Kelvin-Helmholtz instability leads to the formation of large vortex rings. The breakdown of these rings feeds the turbulent wake at the back of the sphere and finally the flow reattaches the wall at the stick.

The most important integral values of engineering interest are the angle of separation, θ_s (as measured from the front stagnation point), the reattachment of the flow at the stick, x_r, and the drag coefficient, C_D. The latter is defined as the sum of pressure and viscous forces on the sphere per unit cross-sectional area, normalized by the pressure at the stagnation point.

No detailed experimental data is available for this flow so far, but based on the experience from channel flow simulations and from comparison of solutions obtained on two different grids, it is expected that the DNS data obtained is accurate within few percents (depending on the quantity looked at). The dependence of sphere drag on Reynolds number known from literature shows a local minimum at Re $= 5\,000$; the simulation predicted the mean drag coefficient to be $C_D = 0.38$, which agrees well with literature data (Roos *et al*, 1971).

A flow visualization study performed in the towing tank of the Institute of Shipbuilding in Hamburg shows the same flow patterns as observed in the simulation, cf. Fig. 14. Dye was introduced through the stick and released through two holes at the sphere surface, one about 25° ahead and one 25° behind the equator, in order to visualize the flow separation, the shear layer, and the wake. The large amount of dye accumulated around the second hole is due to the almost stagnant fluid there, as predicted by the simulation. Both the DNS and the experiment show the separation close to the equator and the typical instability in the free shear layer. The vortex sheet rolls-up into vortex rings, which later break down into the wake. The energy containing scales are well captured as the energy spectra in Fig. 15 show; only the smallest scales in the azimuthal direction have not been resolved. More details can be found in the thesis of Seidl (1997).

Reynolds-averaged Navier-Stokes equations were also solved for the axi-symmetric mean flow around sphere, using low-Re turbulence models investigated earlier for the

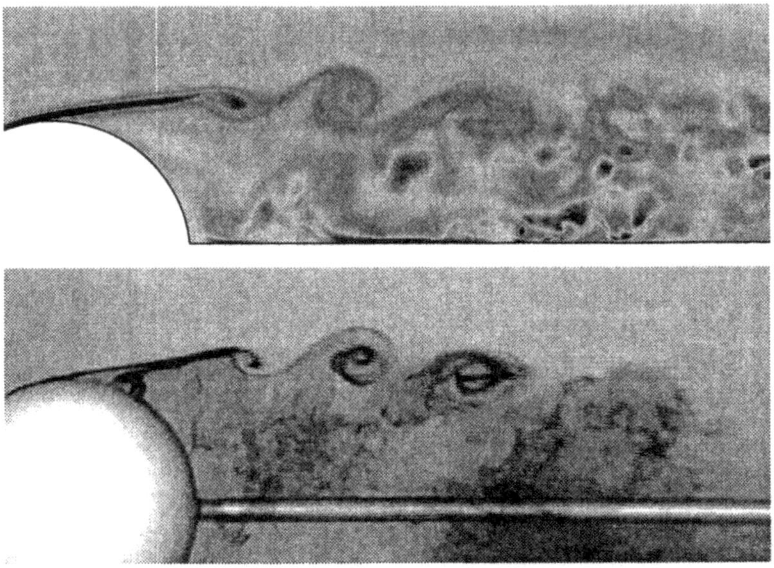

Fig. 14: Flow around sphere at Re = 5 000: predicted vorticity in one cross-section (above) and a snapshot from an experimental flow visualization by towing the sphere in a water tank and releasing dye from two holes (below).

flow over a hill. At Re = 5 000, all three models predict the separation point and the drag coefficient reasonably well, only the reattachment length is underpredicted. However, none of the models predicts accurately the changes of flow features with increasing Reynolds number (to 50 000 and over); in particular, the drag is predicted too low. A strong back flow behind the sphere in all axi-symmetric (two-dimensional) computations leads to a higher pressure at the back and lower drag. After the reattachment of the flow at the stick the development of the velocity profile is slower than predicted by simulation. Figure 16 shows results of the ongoing computations. The inlet values for the turbulent kinetic energy and the dissipation rate that define the turbulent length scale need to be small so that the flow in front of the sphere is not influenced by the model and predictions match the laminar velocity profiles there. As both depend on each other, the results are influenced by their inlet values to some extent. The impact is small for small inlet values; for the chosen set of parameters a length scale of 0.0632 D results. Detailed analysis is in progress in Nantes and Hamburg for various Reynolds numbers; the results will be presented in future reports.

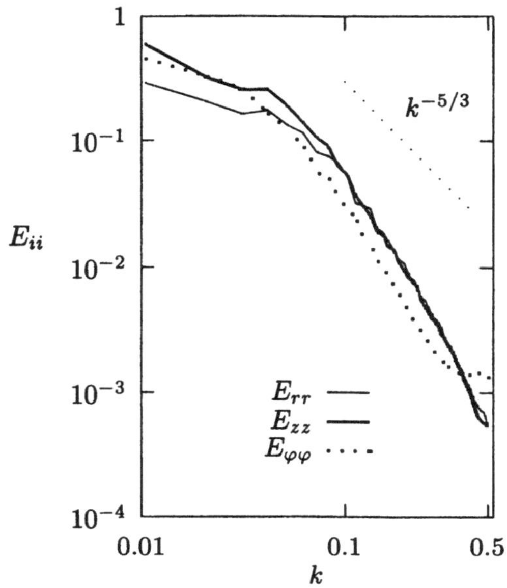

Fig. 15: One dimensional power spectra of two-point correlations for the flow around the sphere at Re = 5 000.

6 Conclusions

It has been demonstrated that it is essential to reliably estimate the numerical errors (iteration convergence, discretization) before one can judge the performance of turbulence models. For the two-dimensional flow over a hill, computations on coarse grids showed good agreement with measurement data but, when really grid-independent solutions are obtained, large discrepancies and thus large modelling errors become evident. The computer-optimized model parameters may need to be re-tuned by making sure that grid-independence is achieved.

The relatively large discrepancy between DNS-data obtained using staggered and colocated arrangements of variables in conjunction with a second-order spatial and temporal discretization on coarse grids proved to be no reason for concern, since with grid refinement the colocated version reproduced well the reference data of Kim *et al.* (1987). The pressure-correction based methods using colocated variable arrangement have therefore no disadvantages compared to fractional-step type of methods using staggered arrangement and approximations of the same order.

Finite-volume methods based on pressure-correction approach and a colocated variable arrangement are suitable for the computation of flows in complex geometries, since they allow a cell-wise local grid refinement and a simple implementation of turbulence models. Future effort will be directed towards analysis of flows past three-dimensional bodies using DNS, LES and RANSE computations, with the emphasis

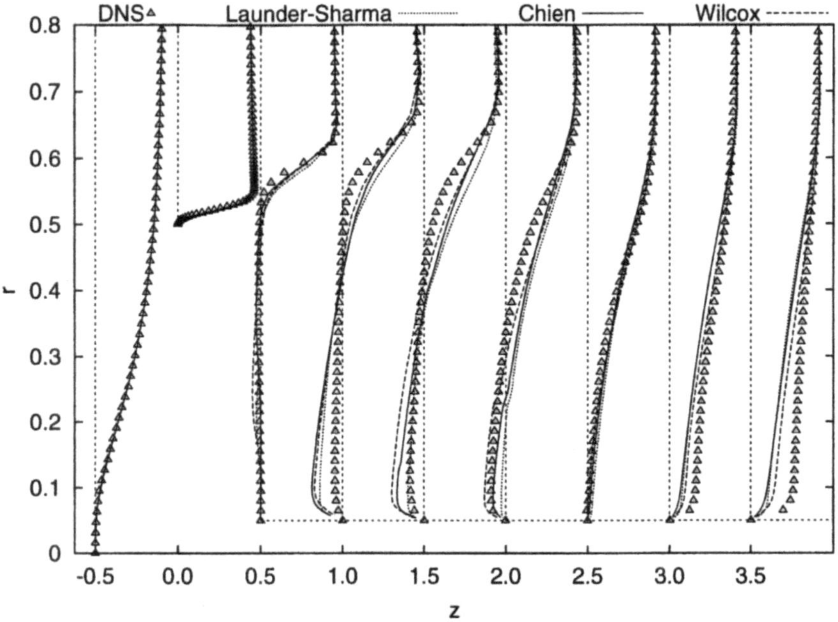

Fig. 16: Mean streamwise velocity profiles of flow around sphere at cross sections using three different low-Re turbulence models compared with the reference data of Seidl (1997).

of ensuring numerical accuracy and reliably evaluating turbulence models.

7 Acknowledgements

The computations were done using up to 128 nodes on Cray T3D/T3E parallel supercomputers at Konrad-Zuse-Zentrum für Informationstechnik (ZIB) in Berlin and on Cray T3E at the high performance computing center in Stuttgart (HLRS). We appreciate gratefully the provided CPU time and the support of both centers.

References

E. Achenbach, Vortex Shedding from Spheres, *J. Fluid. Mech.*, **62**, 209–221 (1974).
G. P. Almeida, D. F. G. Durao, and M. V. Heitor, Wake flows behind two-dimensional model hills. *Exp. Thermal and Fluid Science*, **7**, 87, (1993).
K. -Y. Chien, Predictions of channel and boundary-layer flows with a low-Reynolds-number turbulence model, *AIAA J.*, **20**, 33–38, (1982).
H. Choi, P. Moin, J. Kim, Turbulent drag reduction: studies of feedback control over

riblets, *J. Fluid. Mech.*, **255**, 503–539, (1994).
H. Choi, P. Moin, Effects of the Computational Time Step on Numerical Solutions of Turbulent Flow. *Journal of Computational Physics*, **113**, 1–4, (1994).
I. Demirdžić, S. Muzaferija, M. Perić, Advances in computation of heat transfer, fluid flow, and solid body deformation using finite volume approach, chap. 2 in W.J. Minkowycz, E.M. Sparrow (eds.), *Advances in Numerical Heat Transfer*, pp. 59–96, Taylor and Francis, New York, 1997.
G. B. Deng, J. Piquet, P. Queutey, M. Visonneau, Incompressible Flow Calculations with a Consistent Physical Interpolation using the CPI Method, *Computers & Fluids*, Vol. 23,**8**, 1020–1047, (1994).
J. H. Ferziger and M. Perić, *Computational Methods for Fluid Dynamics*, Springer, Berlin, 1996.
J. Kim, P. Moin, R. Moser, Turbulence statistics in fully developed channel flow at low Reynolds number, *J. Fluid Mech.*, **177**, 133–166, (1987).
B. E. Launder, B. I. Sharma, Application of the Energy Dissipation Model of Turbulence to the Calculation of Flow Near a Spinning Disc, *Letters in Heat and mass transfer*, Vol. 1, **2**, 131–138, (1974).
N. N. Mansour, J. Kim, P. Moin, Reynolds-stress and dissipation-rate budgets in a turbulent channel flow. *J. Fluid Mech.*, **194**, 15–44, (1988).
F. Roos and W. Willmarth, Some Experimental Results on Sphere and Disk Drag, *AIAA Journal*, **9**, 285–291 (1971).
M. Schäfer, S. Turek, R. Rannacher, M. Breuer, F. Durst, W. Rodi, Defintion of Benchmark Problem (Incompressible turbulent flow), 1996.
V. Seidl, Entwicklung und Anwendung eines parallelen Finite-Volumen-Verfahrens zur Strömungssimulation auf unstrukturierten Gittern mit lokaler Verfeinerung, Dissertation, 1997.
D. C. Wilcox, Reassesment of the Scale-Determing Equation for Advanced Turbulence Models, *AIAA Journal*, **26**, 1299–1310, (1988).

Addresses of the Editors of the Series "Notes on Numerical Fluid Mechanics"

Brief Instruction for Authors

Manuscripts should have well over 100 pages. As they will be reproduced photomechanically they should be produced with utmost care according to the guidelines, which will be supplied on request. In print, the size will be reduced linearly to approximately 75 per cent. Figures and diagrams should be lettered accordingly so as to produce letters not smaller than 2 mm in print. The same is valid for handwritten formulae. Manuscripts (in English) or proposals should be sent to the general editor, Prof. Dr. E. H. Hirschel, Herzog-Heinrich-Weg 6, D-85604 Zorneding.

MIX
Papier aus verantwortungsvollen Quellen
Paper from responsible sources
FSC® C105338

If you have any concerns about our products,
you can contact us on
ProductSafety@springernature.com

In case Publisher is established outside the EU,
the EU authorized representative is:
Springer Nature Customer Service Center GmbH
Europaplatz 3, 69115 Heidelberg, Germany

Printed by Libri Plureos GmbH
in Hamburg, Germany